Analysis of Electrical Circuits and Networks

ANALYSIS OF ELECTRICAL CIRCUITS AND NETWORKS

JAYDEEP CHAKRAVORTY
(Reader)
Deptt. of Electrical & Electronics Engineering
Sikkim Manipal Institute of Technology
Majitar, Sikkim

UNIVERSITY SCIENCE PRESS
(An Imprint of Laxmi Publications Pvt. Ltd.)
BANGALORE • CHENNAI • COCHIN • GUWAHATI • HYDERABAD
JALANDHAR • KOLKATA • LUCKNOW • MUMBAI • PATNA
RANCHI • NEW DELHI

Printed and bound in India

First Edition : 2009; *Reprint:* 2011 *Edition* : 2025

ISBN : 978-81-31807-40-8

Branches

✆	Bengaluru	080-26 75 69 30
✆	Chennai	044-24 34 47 26
✆	Guwahati	0361-254 36 69
✆	Hyderabad	040-27 55 53 83
✆	Kochi	0484-405 13 03
✆	Kolkata	033-40 04 77 79
✆	Lucknow	0522-430 36 13
✆	Ranchi	0651-224 24 64

PUBLISHED IN INDIA BY

UNIVERSITY SCIENCE PRESS

(An Imprint of Laxmi Publications Pvt. Ltd.)

An ISO 9001:2015 Company

113, GOLDEN HOUSE, GURUDWARA ROAD, DARYAGANJ,

NEW DELHI - 110002, INDIA

Corporate Office : B-33, Sector-5, Noida,

Gautambuddha Nagar (U.P.) - 201 301, INDIA

Telephone : +91-120-44 46 900

www.laxmipublications.com

info@laxmipublications.com

The book is dedicated to my parents, my brother and his wife, my in-laws, and to my sweet, loving wife

CONTENTS

Preface

This book has been written with the intention to make electrical circuit easy to understand for the students. Electrical circuit is one of the most important part of electrical engineering, and thus a clear and thorough knowledge is required in this field. This book explains all the chapters in simple and easy language and more emphasis has been given to solved problems. In this book each and every topic has been explained with sufficient amount of solved examples, so that the students can easily grasp the topics.

This book comprises of 11 chapters. *Chapter-1* provides the basic introduction to Electrical circuits. *Chapter-2* deals with the different methods of Circuit transformations. *Chapter-3* deals with the Circuit analysis methods. In *chapter-4*, Network theorems (D.C. Circuits) excitation has been covered exclusively. In *chapter-5*, Network theorems (A.C. Circuits) excitation has been covered exclusively. In *chapter-6*, concept of Coupled Circuit, Transformer Circuit and Tuned Circuit has been discussed. *Chapter-7* focuses on Theory and concepts of Series and Parallel Resonance in detail. *Chapter-8* highlights the Laplace Transform which is one of the most important tools for the electrical engineers. *Chapter-9* covers the concept of Transient analysis and its solution by the application of Laplace transformations exclusively. *Chapter-10* focuses on Theory and problems of Fourier series and Fourier transforms and solution of electrical circuits by applying Fourier analysis and in *chapter-11*, the concept of two Port Networks, and inter-relationship between the different networks has been discussed in detail.

While preparing this book more stress has been given in worked out examples. Each chapter contains sufficient number of solved examples for the student to understand the subject thoroughly. The book contains over 400 solved problems for the students.

I hope this book will be of great help to the students in understanding the concept of circuit and network analysis.

Any suggestions and criticisms are welcome for the better development of this book.

Author

ACKNOWLEDGEMENT

I am thankful to the Director, Sikkim Manipal Institute of Technology, Sikkim, Dean Sikkim Manipal Institute of Technology, Sikkim and HOD Electrical and Electronics Engineering Department, HOD Electronics and Communication Engineering Department, HOD Mechanical Engineering Department, HOD Computer Science and Engineering Department, HOD Information Technology Department, all the teaching and non-teaching staffs of Electrical and Electronics Engineering Department, and of Sikkim Manipal Institute of Technology for their constant encouragement and support provided in this endeavour.

I am also very grateful to my parents Mr. P. Chakravorty and Mrs. J. Chakravorty for their constant encouragement while I was writing this book. I am also thankful to my brother Mr. S. Chakravorty and his wife Mrs. S. Chakravorty for their constant encouragement. I am very thankful to my wife Mrs. S. Chakravorty for helping me in editing and also in typing this book.

Author

Chapter 1

BASICS OF ELECTRICAL ENGINEERING

1.1 INTRODUCTION

This chapter deals with the basics of electric circuit and networks. Preliminary concepts has been discussed here for the beginners.

1.2 CIRCUIT DEFINITION

(a) *Electrical circuit (or network)* : An electrical circuit or network is interconnection of electric circuit elements or branches, containing active or passive elements. Figure 1.1(*a*) and Figure 1.1(*b*) indicates an electrical circuit.

(b) *Circuit elements* : Any individual circuit component (such as inductor, capacitor, resistor etc) with two terminals by which it can be connected to other circuit elements. In Figure 1.1(*a*) and Figure 1.1(*b*) R_1, R_2, R_3 and R_4 are the circuit elements.

(c) *Branch* : A group of elements connected in series or in parallel and having two terminals is called a branch.

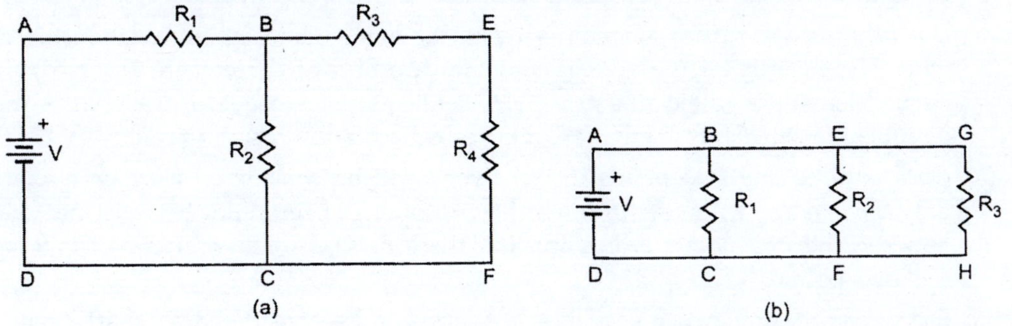

Fig. 1.1

(*d*) *Loop* : It is a closed path in a circuit in which no elements or node is encountered more than once. The circuit shown in Figure 1.1(*a*) has three loops—loop *ABCD*, loop *BEFC*, and loop *AEFD*. Similarly in the circuit shown in Figure 1.1(*b*) there are six loops—loop *ABCD*, loop *BEFC*, loop *EGHF*, loop *AEFD*, loop *BGHC* and loop *AGHD*.

(*e*) *Mesh* : It is a loop that contains no other loop in it. The circuit shown in Figure 1.1(*a*) has only two mesh—mesh *ABEF* and mesh *BCDE*. Similarly for the circuit shown in Figure 1.1(*b*) has only three mesh—mesh *ABCD*, mesh *BEFC* and mesh *EGHF*.

(*f*) *Independent and Non-independent loop* : If a loop contains at least one new element which is not in the other loop of the network then it is called independent loop. Otherwise it is called a non-independent loop.

(*g*) *Node* : It is a part of the network at which at least two or more than two branches meet.

(*h*) *Linear element* : A linear element is one in which voltage and current relation is a linear differential equation with constant coefficient. *E.g.*, resistor. If a network is made up of linear elements then that network is called a linear network.

(*i*) *Nonlinear element* : If any circuit element doesn't follow the linear relationship between voltage and current, then such an element becomes a nonlinear element. Any circuit made up of nonlinear elements is called a nonlinear circuit.

(*j*) *Lumped and Distributed network* : Lumped network is one in which all the circuit elements (resistor, inductor, capacitor) can be physically separated. Where as in a distributed network, elements cannot be physically separated or isolated. Overhead transmission line is an example of distributed network. Where as the network as shown in Figure 1.1(*a*) and Figure 1.1(*b*), is an example of lumped network.

(*k*) *Active and Passive network* : If a network contains one or more than one energy producing source along with other circuit elements then it is called an active network. Where as a network with out any energy producing source is called a passive network.

(*l*) *Bilateral element* : If for a circuit element, the voltage and current relationship remain unaltered when the polarity of the voltage and the direction of current are reversed, then the circuit element is called bilateral, *e.g.*, resistors. If a circuit is made up of bilateral elements then the circuit is called a bilateral circuit.

(*m*) *Unilateral element* : If for a circuit element, the voltage and current relationship changes when the polarity of the voltage and the direction of current are reversed, then the circuit element is called unilateral, *e.g.*, semiconductor diode. If a circuit is made up of unilateral elements then the circuit is called a unilateral circuit.

(*n*) *Open circuit* : Any two points in a network will be considered as open circuit if and only if the current flowing through the two points is zero and the resistance measured between the two points is infinity. But there can be some value of voltage between the two points.

(*o*) *Short circuit* : Any two points in a network will be considered as short circuit if and only if the voltage measured between the two points is zero and also the resistance measured between the two points is zero. But there can be some value of current flowing through the two points.

1.3 ENERGY SOURCES

They actually supply energy to the network or circuit to which they are connected. If a circuit or network contains at least one or more than one energy source or sources then it is called an active circuit or network. And if a circuit or network contains no energy source then it is called a passive network. An energy source can be broadly classified into two categories :

(*i*) Independent energy source.

(*ii*) Dependent energy source.

1.3.1 Independent Energy Source

It supplies constant voltage or current that is completely independent of other circuit elements. An independent energy source can be classified into :

(*i*) Independent voltage source.

(*ii*) Independent current source.

1.3.2 Ideal Voltage Source

It give constant voltage across the terminals irrespective of the current drawn from the terminals. The voltage source is represented in the circuit by the symbol as shown in Figure 1.2(*a*) and Figure 1.2(*b*). The *V-I* characteristics for an ideal voltage source is shown in Figure 1.2(*c*).

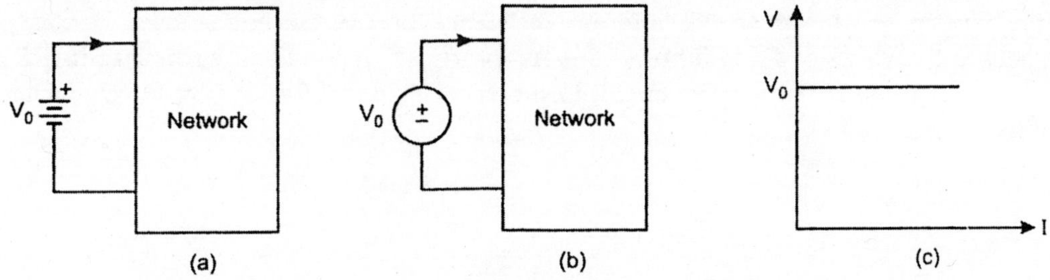

Fig. 1.2

1.3.3 Ideal Current Source

It gives constant current across its terminals irrespective of the voltage appearing across its terminals. The current source is represented in the circuit by the symbol as shown in Figure 1.3(*a*). The arrow indicates the direction of current in the circuit. The *V-I* characteristics for an ideal current source is shown in Figure 1.3(*b*).

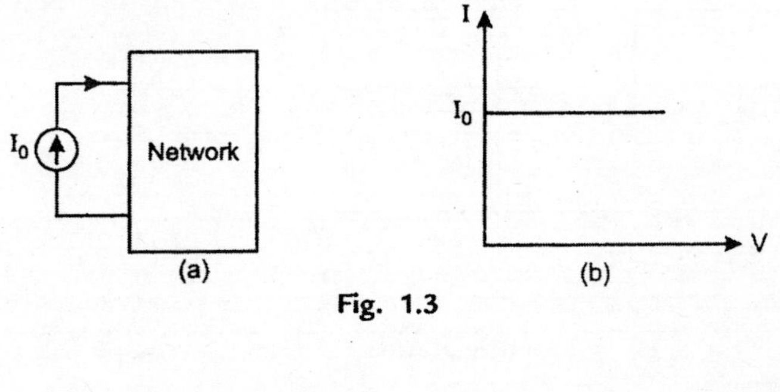

Fig. 1.3

1.3.4 Practical or Non-ideal Voltage Source

Every voltage source offers some resistance to the flow of current. This effect is represented in the circuit by a resistance connected in series across its terminals. This resistance R_V is called the source resistance or the internal resistance of the voltage source. For an ideal voltage source the value of R_V is zero. But for practical or non-ideal voltage source R_V has some value, which is very small. The circuit is shown in Figure 1.4(*a*) and its *V-I* characteristics is shown in Figure 1.4(*b*).

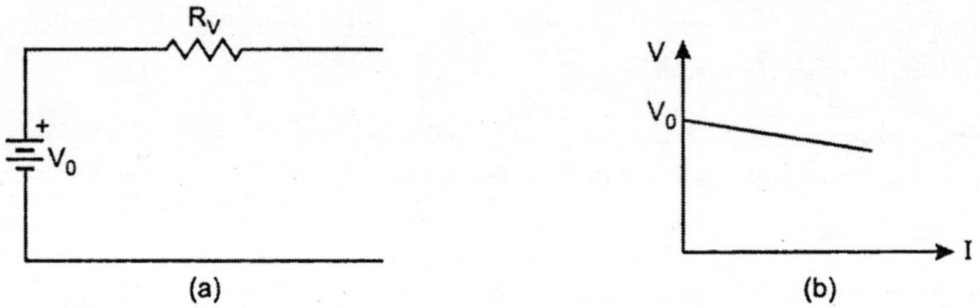

Fig. 1.4

1.3.5 Practical or Non-ideal Current Source

Every current source offers some resistance to the flow of current. This effect is represented in the circuit by a resistance connected in parallel across its terminals. This resistance R_C is called the source resistance or the internal resistance of the current source. For an ideal current source the value of R_C is infinity. But for practical or non-ideal current source R_C has some value, which is very large. The circuit is shown in Figure 1.5(*a*) and its *V-I* characteristics is shown in Figure 1.5(*b*).

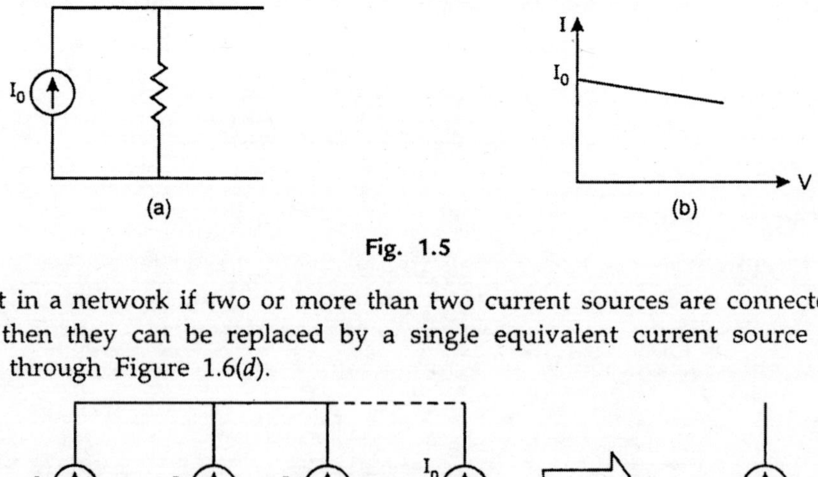

Fig. 1.5

Note that in a network if two or more than two current sources are connected in parallel or in series then they can be replaced by a single equivalent current source as shown in Figure 1.6(*a*) through Figure 1.6(*d*).

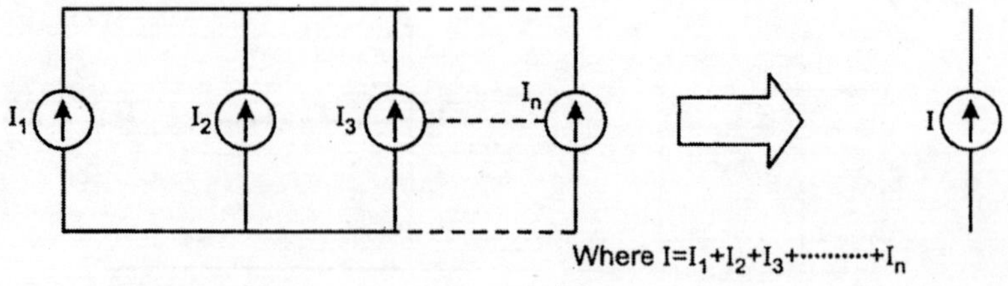

(a)

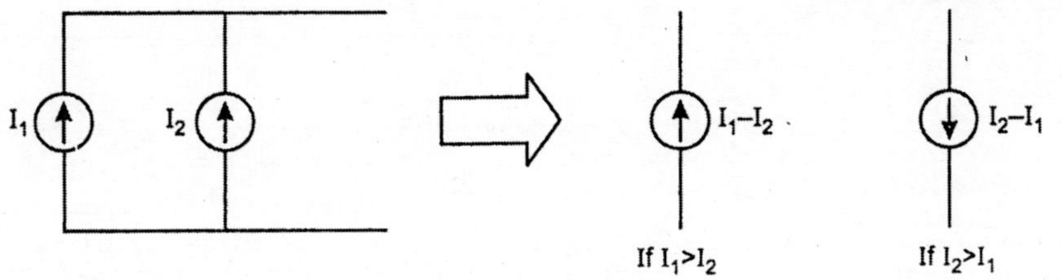

(b)

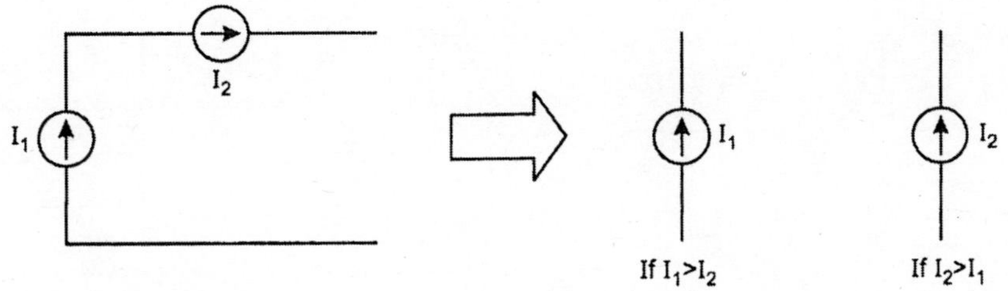

(c)

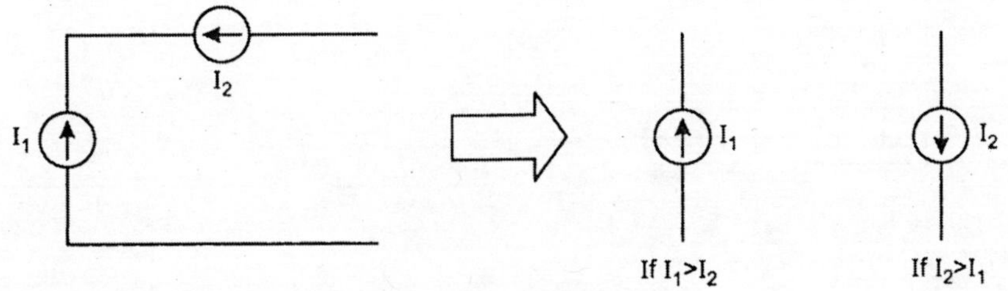

(d)

Fig. 1.6

Similarly, if two or more than two voltage sources are connected in series or in parallel then they can be replaced by a single equivalent voltage source as shown in Figure 1.7(*a*) through Figure 1.7(*d*).

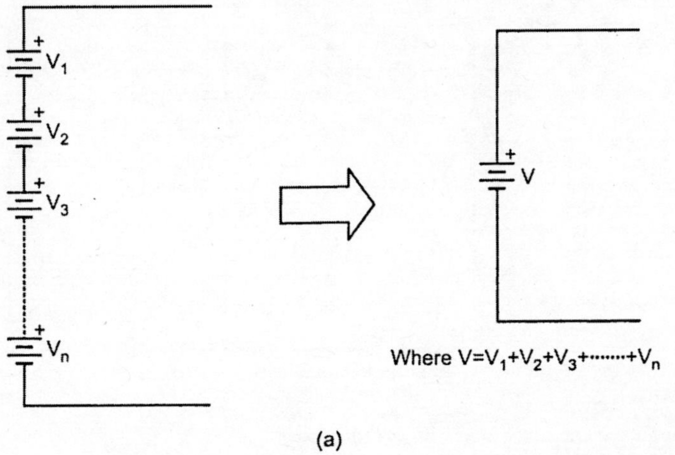

(a)

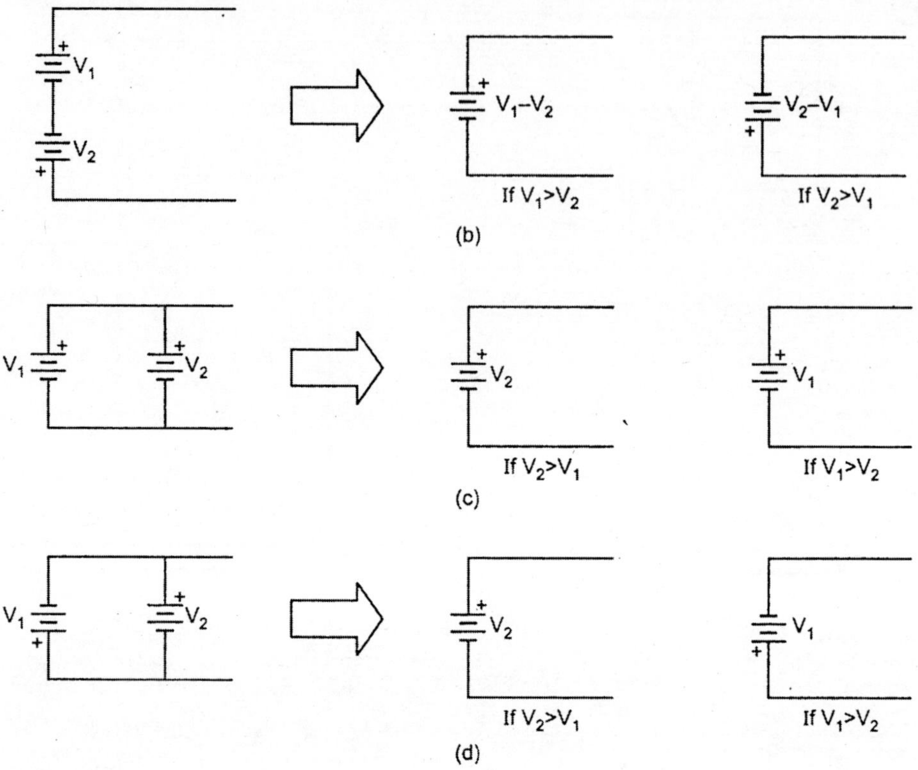

Fig. 1.7

1.3.6 Dependent Energy Sources

It is an active element in which the source quantity is controlled by another voltage or current. Dependent sources are usually designated by a diamond shaped symbol as shown in the Figures 1.8(*a*) to 1.8(*d*).

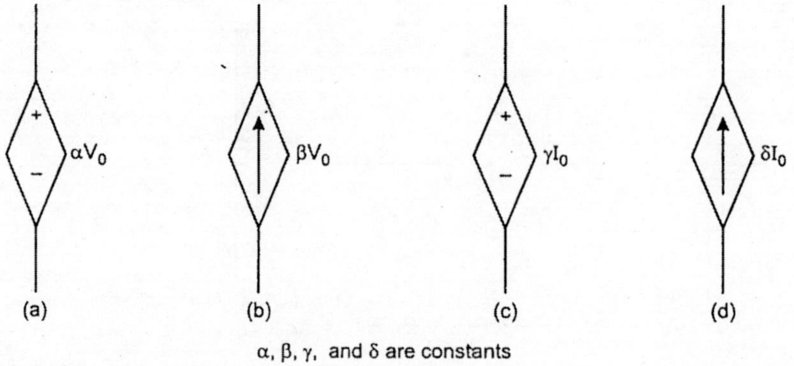

Fig. 1.8

There are possible types of dependent source :

(*i*) Voltage Dependent Voltage Source (VDVS) Figure 1.8(*a*)

(*ii*) Voltage Dependent Current Source (VDCS) Figure 1.8(*b*)
(*iii*) Current Dependent Voltage Source (CDVS) Figure 1.8(*c*)
(*iv*) Current Dependent Current Source (CDCS) Figure 1.8(*d*).

1.4 BASIC CIRCUIT LAWS

1.4.1 Ohm's Law

It states that :

The potential difference or the voltage drop (V) between any two points on a conductor is directly proportional to the current (I) flowing through the conductor, provided the temperature and the other physical conditions is unaltered.

That is $V \propto I$

or $V = R \times I$

where R is a constant called resistance.

The above relation is valid for only metallic conductors or conductors which are a good conductor of electricity. This law is also not valid for unilateral elements.

1.4.2 Kirchhoff's Laws

1. *Kirchhoff's Current Law* (KCL) : It is also called Kirchhoff's Point Law. It states that :

Across any node in an active network, the algebraic sum of all the currents (current entering the node + the current leaving the node) is zero. Consider a node N as shown in Figure 1.9. Into this node N, currents I_2, I_6, and I_7 are entering the node and currents I_1, I_3, I_4, I_5, and I_8 are leaving the node.

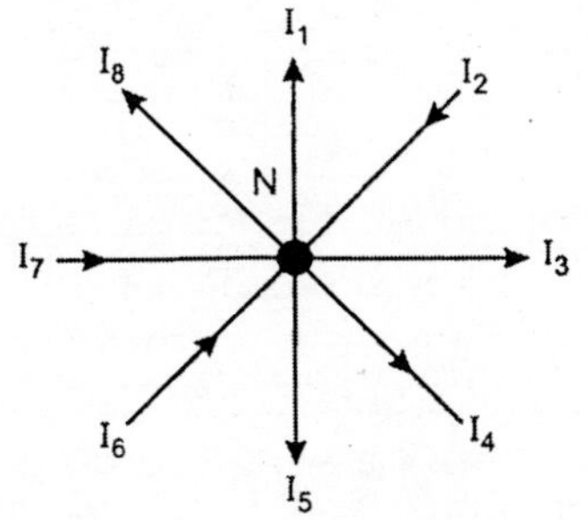

Fig. 1.9

Then if we apply KCL here then we can write

$$I_2 + I_6 + I_7 - I_1 - I_3 - I_4 - I_5 - I_8 = 0 \quad ...(1.1)$$

In the equation (1.1) the assumptions made are

(*i*) The currents entering the node are taken as positive.

(*ii*) The currents leaving the node are taken as negative.

From equation (1.1) we can conclude, KCL states that in a node.

Sum of all incoming currents = Sum of all outgoing currents.

2. *Kirchhoff's Voltage Law* (*KVL*) : It is also called Kirchhoff's Mesh Law. It states that: In an active network, the algebraic sum of all the voltage drops (drop across each circuit elements + the e.m.fs) in a closed path (a mesh or a loop)is zero.

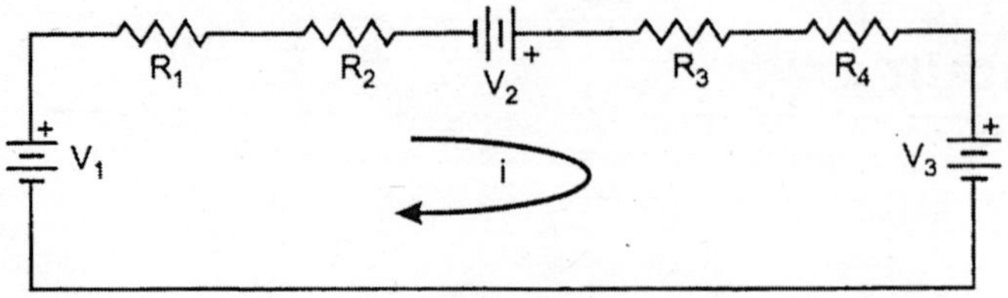

Fig. 1.10

Consider a single loop circuit shown in Figure 1.10. The direction of the flow of current is also indicated in the circuit. The following assumptions are made before the application of KVL.

(*i*) Travelling in the direction of the current, if we enter the negative terminal of the battery, then the potential is taken as negative.

(*ii*) Travelling in the direction of the current, if we enter the positive terminal of the battery, then the potential is taken as positive.

(*iii*) Travelling towards the direction of the currents, all the resistive drops is assumed to be positive.

Let the drop across R_1, R_2, R_3 and R_4 be e_1, e_2, e_3 and e_4 respectively.

Then applying KVL in the circuit of Figure 1.9, we get,

$$-V_1 + e_1 + e_2 - V_2 + e_3 + e_4 + V_3 = 0. \qquad ...(1.2)$$

1.5 CIRCUIT ELEMENTS

1.5.1 Resistance (R)

It is the property of any material to oppose the flow of current and also consumes energy. They are poor conductors. There are many different ways to make a resistor. Some are just a coil of wire made of a material that is a poor conductor. The most common and inexpensive type is made from powdered carbon and a glue-like binder. Such carbon composition resistors usually have a brown cylindrical body with a wire lead on each end, and colored bands that indicate the value of the resistor. It is denoted by capital letter R and its unit is "ohm", denoted by the symbol Ω. Resistances in the circuit can be connected either in series or in parallel or in a series–parallel combination.

1.5.1.1 Series Connection of Resistances

Resistances when connected in a series as shown in Figure 1.10, same amount of current flows through all the resistances. In the circuit shown in Figure 1.11, four resistances are connected in series across a voltage source of V volts. Due to this source a current of I amp flows through the circuit in the direction as indicated by the arrow.

Let the drop across R_1, R_2, R_3 and R_4 be e_1, e_2, e_3 and e_4 respectively.

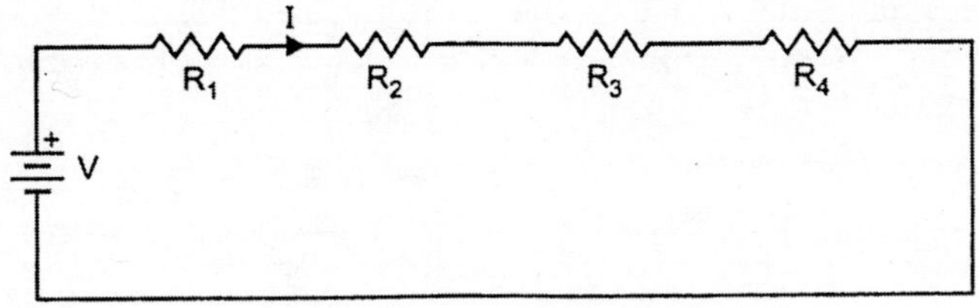

Fig. 1.11

Then from KVL we can say the drops across the resistors must add up to the total voltage supplied by the battery. That is,

$$V = e_1 + e_2 + e_3 + e_4 \quad ...(1.3)$$

Again from ohm's law we know $V = IR$, then equation (1.3) can be wr

$$V = IR_1 + IR_2 + IR_3 + IR_4$$

or,
$$V = I \times (R_1 + R_2 + R_3 + R_4)$$

or
$$V = I \times R_{eq}$$

where,
$$R_{eq} = R_1 + R_2 + R_3 + R_4$$

Thus in general, the equivalent resistance of resistors connected in series is the sum of their resistances. That is,

$$R_{eq} = \sum R_n, \quad ...(1.4)$$

where n is the number of resistances connected in series.

Equation (1.4) can also be written in terms of conductance (G), which is the reciprocal of resistance.

$$\frac{1}{G_{eq}} = \sum \frac{1}{g_n}$$

where n is the number of resistances connected in series.

1.5.1.2 Parallel Connection of Resistances

If the resistances in a circuit are connected in such a way that current have several alternative path to flow, then such type of connection is called parallel connection. The circuit shown in Figure 1.12 is a circuit where all the resistances are connected in parallel.

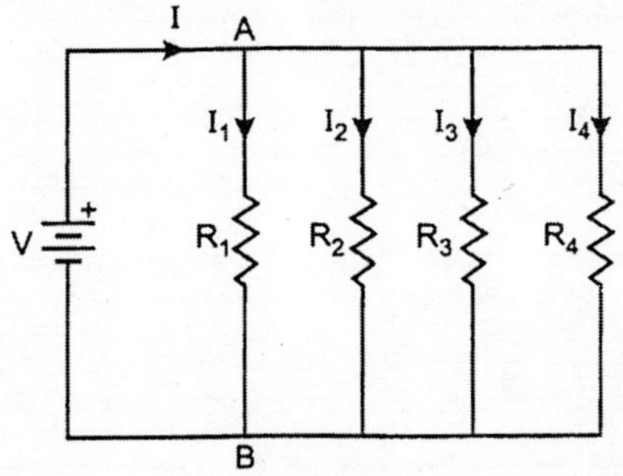

Fig. 1.12

Here the source current I split up into I_1, I_2, I_3 and I_4 as it travels from node A to node B. Again at node A the potential must be the same for each resistor. Similarly, at node B the potential again must be the same for each resistor. That is, each of the four resistors in the parallel circuit must have the same voltage. That is,

$$V = e_1 = e_2 = e_3 = e_4 \qquad ...(1.5)$$

where, e_1, e_2, e_3 and e_4 is the drop across R_1, R_2, R_3 and R_4 respectively. Again from KCL we know,

$$I = I_1 + I_2 + I_3 + I_4 \qquad ...(1.6)$$

Applying ohm's law we get,

$$I = \frac{e_1}{R_1} + \frac{e_2}{R_2} + \frac{e_3}{R_3} + \frac{e_4}{R_4} \qquad ...(1.7)$$

Applying equation (1.5) we get,

$$I = V \times \left(\frac{1}{R_1} + \frac{1}{R_2} + \frac{1}{R_3} + \frac{1}{R_4} \right)$$

or,

$$I = V \times \frac{1}{R_{eq}}$$

where,

$$\frac{1}{R_{eq}} = \left(\frac{1}{R_1} + \frac{1}{R_2} + \frac{1}{R_3} + \frac{1}{R_4} \right)$$

Thus in general we can write, if n number of resistances are connected in parallel, then the equivalent resistance is given by,

$$\frac{1}{R_{eq}} = \sum \frac{1}{R_n} \qquad ...(1.8)$$

In terms of conductance we can represent the equation (1.8) as,

$$G_{eq} = \sum G_n \text{, where } G = \frac{1}{R}.$$

1.5.1.3 *Series-parallel Combination*

Generally in a network resistances are connected in series–parallel combinations, as shown in Figure 1.13.

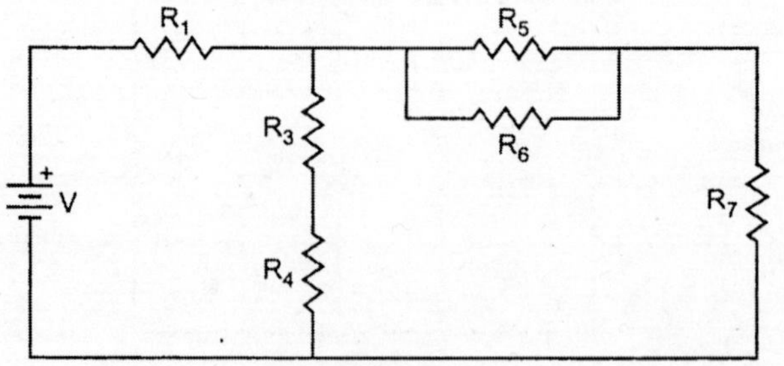

Fig. 1.13

Here, the circuit of Figure 1.13, is a combination of series and parallel connected resistances. Let us find out the equivalent resistance of the circuit as seen from the source terminals. Resistors R_3 and R_4 are connected in series, thus it can be replaced by a single equivalent resistance R_{eq1}, given by,

$$R_{eq1} = R_3 + R_4. \quad ...(1.9)$$

Similarly, resistors R_5 and R_6 are connected in parallel, thus it can be replaced by a single equivalent resistance R_{eq2}, given by,

$$R_{eq2} = \frac{R_5 \times R_6}{R_5 + R_6}. \quad ...(1.10)$$

Thus the circuit of Figure 1.12 can be redrawn as shown in Figure 1.13(*a*).

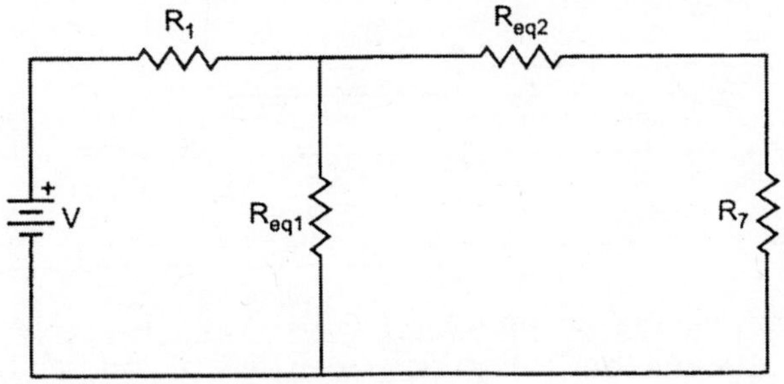

Fig. 1.13(a)

From Figure 1.13(*a*), we can see that R_{eq2} and R_7 are connected in series and thus it can be replaced by an equivalent resistance R_{eq3}, given by,

$$R_{eq3} = R_{eq2} + R_7. \quad ...(1.11)$$

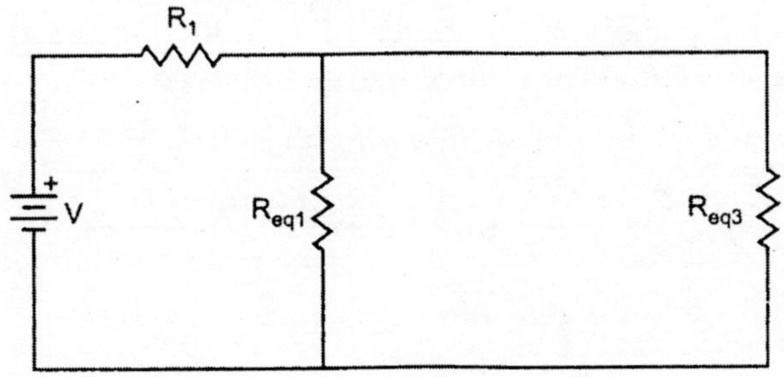

Fig. 1.13(b)

The circuit of Figure 1.13(*a*) can be redrawn as shown in Figure 1.13(*b*).

Again from Figure 1.13(*b*), we see that resistors R_{eq1} and R_{eq3} are connected in parallel, thus it can be replaced by a single equivalent resistance R_{eq4}, given by,

$$R_{eq4} = \frac{R_{eq1} \times R_{eq3}}{R_{eq1} + R_{eq3}}. \quad ...(1.12)$$

Thus the circuit of Figure 1.13(*b*) can be redrawn as shown in Figure 1.12(*c*)

Thus from Figure 1.13(c), the equivalent resistance as seen from the source terminal is

$$R_{eq} = R_1 + R_{eq4} \qquad ...(1.13)$$

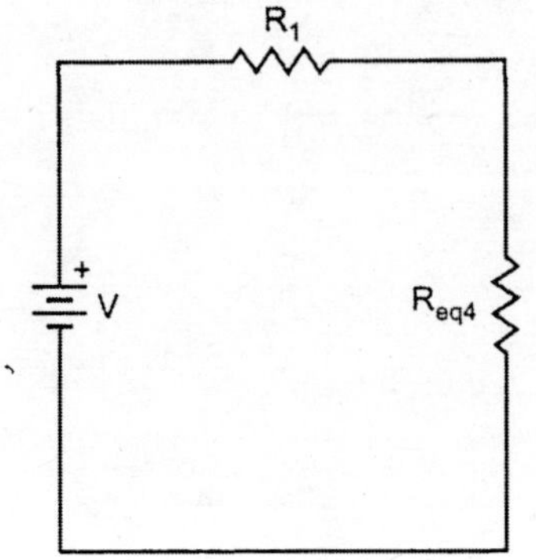

Fig. 1.13(c)

.ıductance (L)

It is the property of any material to oppose any chance in magnitude of current or chance in direction of current through the circuit. It is denoted by capital letter L and its unit is "henry", denoted by the symbol H. Inductances in the circuit can be connected either in series or in parallel or in a series–parallel combination.

1.5.2.1 Series Connection of Inductances

Inductances when connected in a series as shown in Figure 1.14, same amount of current flows through all the inductances. In the circuit shown in Figure 1.14, four inductances are connected in series across an a.c voltage source of V volts. Due to this source a current of i amp flows through the circuit in the direction as indicated by the arrow.

Let the drop across L_1, L_2, L_3 and L_4 be e_1, e_2, e_3 and e_4 respectively,

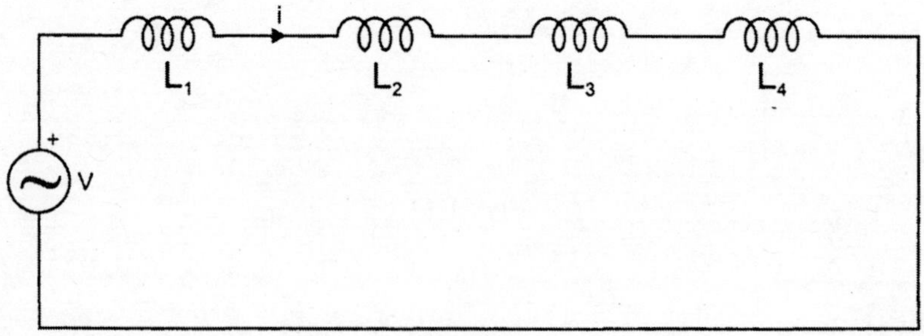

Fig. 1.14

where,

$$e_1 = L_1\frac{di}{dt},\; e_2 = L_2\frac{di}{dt},\; e_3 = L_3\frac{di}{dt},\; \text{and } e_4 = L_4\frac{di}{dt} \qquad ...(1.14)$$

Then from KVL we can say the drops across the inductors must add up to the total voltage supplied by the source. That is,

$$V = e_1 + e_2 + e_3 + e_4. \quad \text{...(1.15)}$$

Substituting the values of e_1, e_2, e_3, and e_4 from equations (1.14), we get,

$$V = L_1\frac{di}{dt} + L_2\frac{di}{dt} + L_3\frac{di}{dt} + L_4\frac{di}{dt}$$

or

$$V = (L_1 + L_2 + L_3 + L_4)\times\frac{di}{dt}$$

or

$$V = L_{eq}\times\frac{di}{dt}$$

where,

$$L_{eq} = L_1 + L_2 + L_3 + L_4$$

Thus in general, we can write,

$$L_{eq} = \sum L_n, \quad \text{...(1.16)}$$

where n is the number of inductances connected in series.

1.5.2.2 Parallel Connection of Inductances

If the inductances in a circuit are connected in such a way that current have several alternative path to flow, then such type of connection is called parallel connection. The circuit shown in Figure 1.15 is a circuit where all the inductances are connected in parallel.

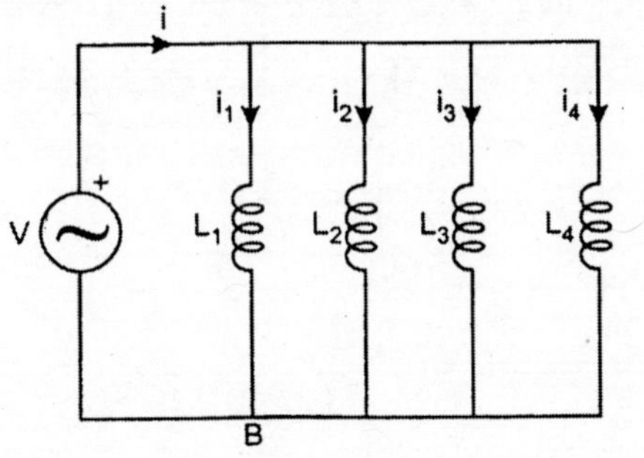

Fig. 1.15

Here the source current i split up into i_1, i_2, i_3 and i_4 as it travels from node A to node B. Again at node A the potential must be the same for each inductance. Similarly, at node B the potential again must be the same for each inductance. That is, each of the four inductances in the parallel circuit must have the same voltage. That is,

$$V = e_1 = e_2 = e_3 = e_4 \quad \text{...(1.17)}$$

where, e_1, e_2, e_3 and e_4 is the drop across L_1, L_2, L_3 and L_4 respectively.

or

$$V = L_1\frac{di_1}{dt} = L_2\frac{di_2}{dt} = L_3\frac{di_3}{dt} = L_4\frac{di_4}{dt}$$

Again from KCL we know,

$$i = i_1 + i_2 + i_3 + i_4 \quad \text{...(1.18)}$$

Applying ohm's law we get,

$$i = \frac{1}{L_1}\int Vdt + \frac{1}{L_2}\int Vdt + \frac{1}{L_3}\int Vdt + \frac{1}{L_4}\int Vdt \quad ...(1.19)$$

$$i = \left(\frac{1}{L_1} + \frac{1}{L_2} + \frac{1}{L_3} + \frac{1}{L_4}\right) \times \int Vdt$$

or

$$i = \frac{1}{L_{eq}}\int Vdt$$

where,

$$\frac{1}{L_{eq}} = \left(\frac{1}{L_1} + \frac{1}{L_2} + \frac{1}{L_3} + \frac{1}{L_4}\right)$$

Thus in general we can write, if n number of inductances are connected in parallel, then the equivalent inductance is given by,

$$\frac{1}{L_{eq}} = \sum \frac{1}{L_n} \quad ...(1.20)$$

1.5.2.3 Series-parallel Combination

Generally in a network inductances are connected in series-parallel combinations, as shown in Figure 1.16.

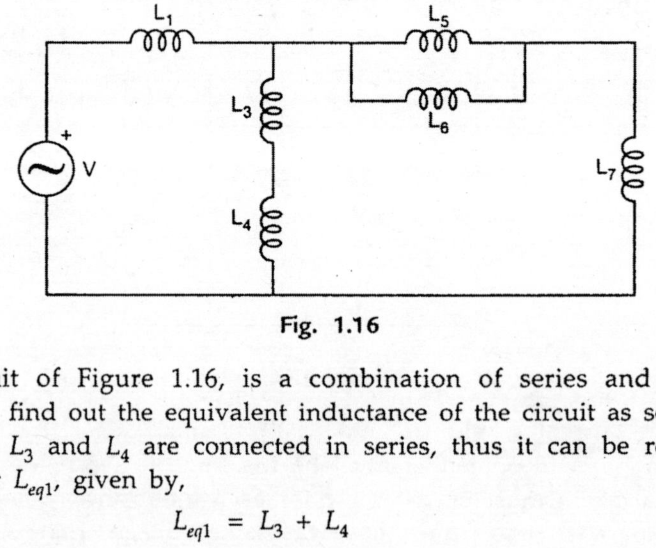

Fig. 1.16

Here the circuit of Figure 1.16, is a combination of series and parallel connected inductances. Let us find out the equivalent inductance of the circuit as seen from the source terminals. inductor L_3 and L_4 are connected in series, thus it can be replaced by a single equivalent inductor L_{eq1}, given by,

$$L_{eq1} = L_3 + L_4 \quad ...(1.21)$$

Similarly, inductors L_5 and L_6 are connected in parallel, thus it can be replaced by a single equivalent inductor L_{eq2}, given by,

$$L_{eq2} = \frac{L_5 \times L_6}{L_5 + L_6} \quad ...(1.22)$$

Thus the circuit of Figure 1.16 can be redrawn as shown in Figure 1.16(*a*).

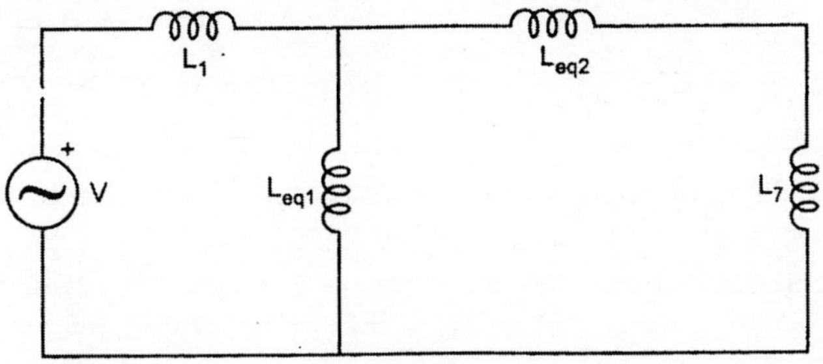

Fig. 1.16(a)

From Figure 1.16(*a*), we can see that L_{eq2} and L_7 are connected in series and thus it can be replaced by an equivalent inductor L_{eq3}, given by,

$$L_{eq3} = L_{eq2} + L_7 \quad ...(1.23)$$

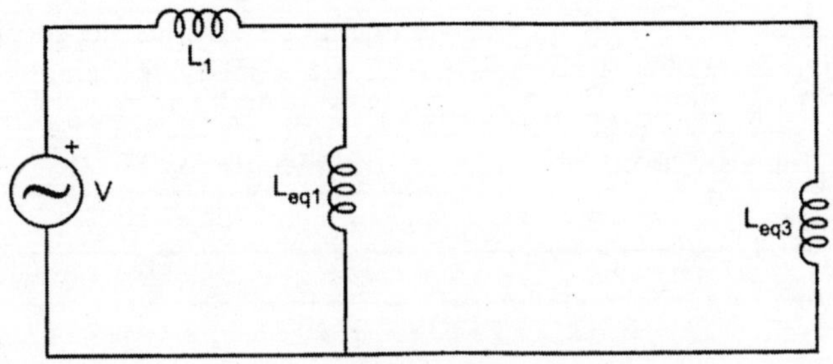

Fig. 1.16(b)

The circuit of Figure 1.16(*a*) can be redrawn as shown in Figure 1.16(*b*).

Again from Figure 1.16(*b*), we see that inductors L_{eq1} and L_{eq3} are connected in parallel, thus it can be replaced by a single equivalent inductor L_{eq4}, given by,

$$L_{eq4} = \frac{L_{eq1} \times L_{eq3}}{L_{eq1} + L_{eq3}} \quad ...(1.24)$$

The circuit of Figure 1.16(*b*) can be redrawn as shown in Figure 1.16(*c*).

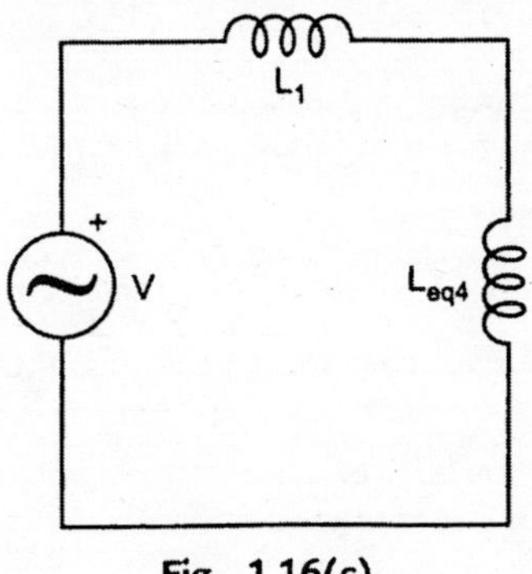

Fig. 1.16(c)

Thus from Figure 1.16(c), the equivalent inductance as seen from the source terminal is,

$$L_{eq} = L_1 + L_{eq4} \quad ...(1.25)$$

1.5.3 Capacitors (C)

Capacitors are another element used to control the flow of charge in a circuit. The name derives from their capacity to store charge, rather like a small battery. Capacitors consist of two conducting surfaces separated by an insulator; a wire lead is connected to each surface. You can imagine a capacitor as two large metal plates separated by air, although in reality they usually consist of thin metal foils or films separated by plastic film or another solid insulator, and rolled up in a compact package. It is denoted by capital letter C and its unit is "farad" or "coulomb/volt", denoted by the symbol *F*. Capacitors in the circuit can be connected either in series or in parallel or in a series-parallel combination.

1.5.3.1 Series Connection of Capacitors

Capacitors when connected in a series as shown in Figure 1.17, same amount of current flows through all the capacitance. In the circuit shown in Figure 1.17, four capacitors are connected in series across an a.c. voltage source of *V* volts. Due to this source a current of *i* amp flows through the circuit in the direction as indicated by the arrow.

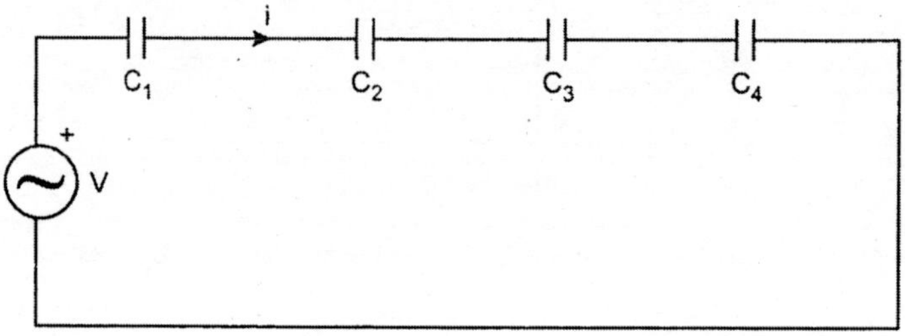

Fig. 1.17

Let the drop across C_1, C_2, C_3 and C_4 be e_1, e_2, e_3 and e_4 respectively.

Then from KVL we can say the drops across the capacitor must add up to the total voltage supplied by the source. That is,

$$V = e_1 + e_2 + e_3 + e_4 \quad ...(1.26)$$

or

$$V = \frac{Q}{C_1} + \frac{Q}{C_2} + \frac{Q}{C_3} + \frac{Q}{C_4} \quad ...(1.27)$$

where Q is the charge on the capacitors.

or

$$V = Q \times \left(\frac{1}{C_1} + \frac{1}{C_2} + \frac{1}{C_3} + \frac{1}{C_4} \right) \quad ...(1.28)$$

or

$$V = Q \times \frac{1}{C_{eq}} \quad ...(1.29)$$

where,

$$\frac{1}{C_{eq}} = \left(\frac{1}{C_1} + \frac{1}{C_2} + \frac{1}{C_3} + \frac{1}{C_4} \right) \quad ...(1.30)$$

Thus in general we can write, if n number of capacitors are connected in series, then the equivalent capacitance is given by,

$$\frac{1}{C_{eq}} = \sum \frac{1}{C_n}. \qquad ...(1.31)$$

1.5.3.2 Parallel Connection of Capacitors

If the capacitors in a circuit are connected in such a way that current have several alternative path to flow, then such type of connection is called parallel connection. The circuit shown in Figure 1.18 is a circuit where all the capacitors are connected in parallel.

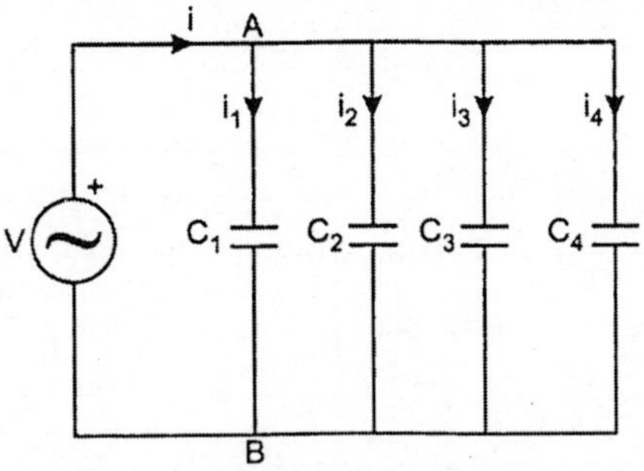

Fig. 1.18

Here the source current i split up into i_1, i_2, i_3 and i_4 as it travels from node A to node B. Again at node A the potential must be the same for each capacitor. Similarly, at node B the potential again must be the same for each capacitor. That is, each of the four capacitors in the parallel circuit must have the same voltage. That is,

$$V = e_1 = e_2 = e_3 = e_4 \qquad ...(1.32)$$

where, e_1, e_2, e_3 and e_4 is the drop across C_1, C_2, C_3 and C_4 respectively. Again from KCL we know,

$$i = i_1 + i_2 + i_3 + i_4 \qquad ...(1.33)$$

or

$$i = C_1\frac{dV}{dt} + C_2\frac{dV}{dt} + C_3\frac{dV}{dt} + C_4\frac{dV}{dt} \qquad ...(1.34)$$

or

$$i = (C_1 + C_2 + C_3 + C_4) \times \frac{dV}{dt} \qquad ...(1.35)$$

or

$$i = C_{eq} \times \frac{dV}{dt} \qquad ...(1.36)$$

where,

$$C_{eq} = C_1 + C_2 + C_3 + C_4 \qquad ...(1.37)$$

Thus in general, we can write,

$$C_{eq} = \sum C_n, \qquad ...(1.38)$$

where n is the number of capacitors connected in parallel.

1.5.3.3 Series-parallel Combination

Generally in a network capacitors are connected in series-parallel combinations, as shown in Figure 1.19.

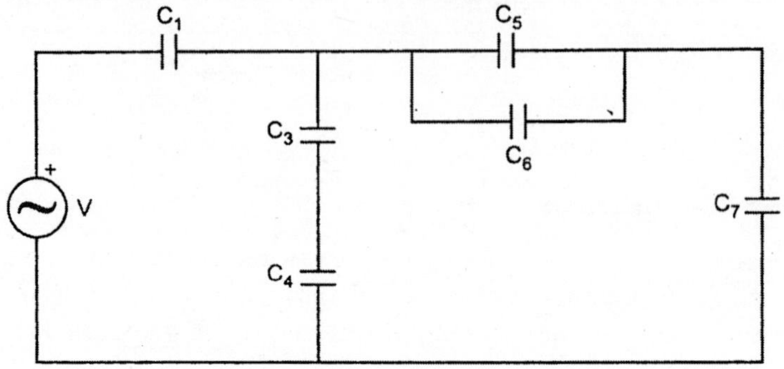

Fig. 1.19

Here the circuit of Figure 1.19 is a combination of series and parallel connected capacitors. Let us find out the equivalent capacitance of the circuit as seen from the source terminals. Capacitors C_3 and C_4 are connected in series, thus it can be replaced by a single equivalent capacitance C_{eq1}, given by,

$$C_{eq1} = \frac{C_3 \times C_4}{C_3 + C_4} \quad ...(1.39)$$

Similarly, capacitors C_5 and C_6 are connected in parallel, thus it can be replaced by a single equivalent resistance C_{eq2}, given by,

$$C_{eq2} = C_5 + C_6 \quad ...(1.40)$$

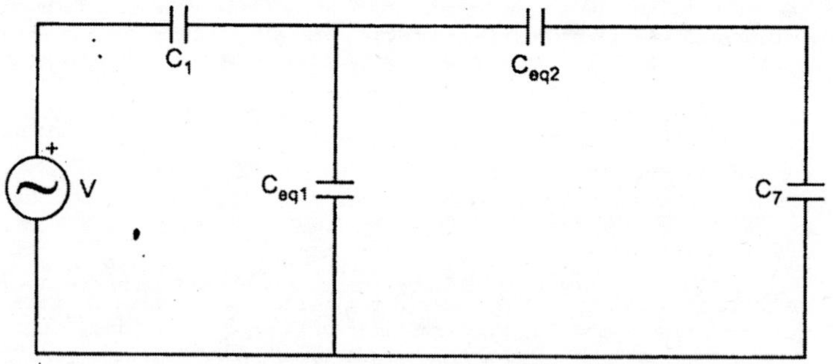

Fig. 1.19(a)

Thus the circuit of Figure 1.19 can be redrawn as shown in Figure 1.19(*a*).

From Figure 1.19(*a*), we can see that C_{eq2} and C_7 are connected in series and thus it can be replaced by an equivalent capacitance C_{eq3}, given by,

$$C_{eq3} = \frac{C_{eq2} \times C_7}{C_{eq2} + C_7} \quad ...(1.41)$$

The circuit of Figure 1.19(*a*) can be redrawn as shown in Figure 1.19(*b*)

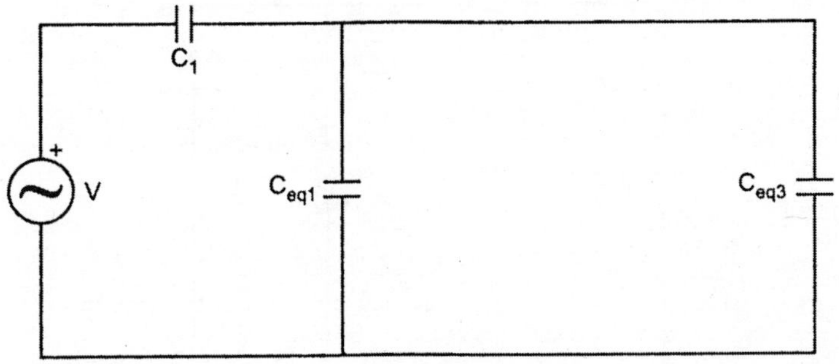

Fig. 1.19(b)

Again from Figure 1.19(*b*), we see that capacitor C_{eq1} and C_{eq3} are connected in parallel, thus it can be replaced by a single equivalent capacitor C_{eq4}, given by,

$$C_{eq4} = C_{eq1} + C_{eq3} \qquad ...(1.42)$$

Thus the circuit of Figure 1.19(*b*) can be redrawn as shown in Figure 1.19(*c*).

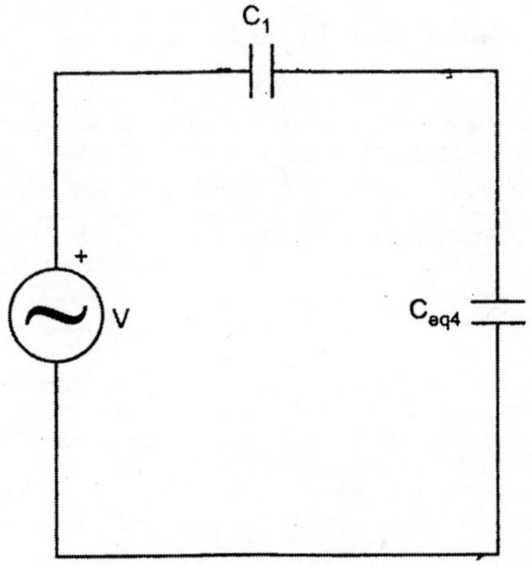

Fig.1.19(c)

Thus from Figure 1.19(*c*), the equivalent capacitance as seen from the source terminal is

$$C_{eq} = \frac{C_{eq4} \times C_1}{C_{eq4} + C_1}. \qquad ...(1.43)$$

1.6 CURRENT DIVISION RULE

The current division rule is applied to parallel circuits, where the currents gets divided in parallel branches. In the circuit shown in Figure 1.20, the source current I gets divided into two currents I_1 and I_2 in the two parallel branches.

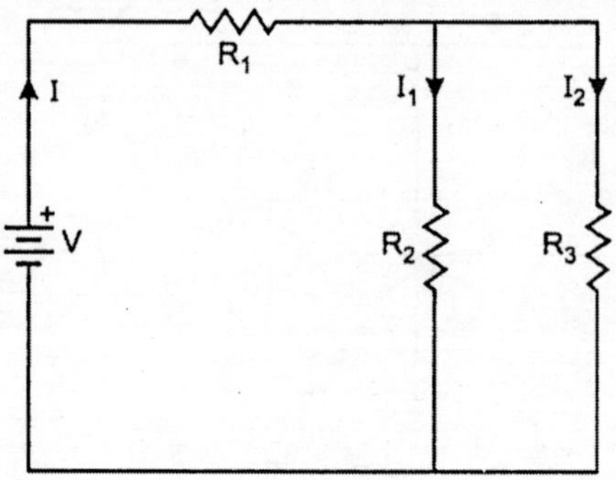

Fig. 1.20

According to current division rule,

$$I_1 = I \times \frac{R_3}{R_2 + R_3} \quad \text{and} \quad I_2 = I \times \frac{R_2}{R_2 + R_3}$$

Thus in general,

$$\text{Current in any parallel branch} = \text{Total current} \times \frac{\text{Opposite branch resistance}}{\text{Sum of parallel resistances}}$$

Proof. The equivalent resistance as seen from the source terminal is

$$R_{eq} = R_1 + \frac{R_2 \times R_3}{R_2 + R_3} \qquad \text{...(1.44)}$$

Then the source current $$I = \frac{V}{R_{eq}} \qquad \text{...(1.45)}$$

Drop at the node A is $$V_A = V - IR_1 = V - \frac{V}{R_{eq}} R_1 = V\left(\frac{R_{eq} - R_1}{R_{eq}}\right)$$

Then Current through R_2 is $$I_2 = \frac{V_A}{R_2} = V\left(\frac{R_{eq} - R_1}{R_{eq} R_2}\right) = \frac{V}{R_{eq}}\left(\frac{R_{eq} - R_1}{R_2}\right) \qquad \text{...(1.46)}$$

Substituting equations (1.44) and (1.45) in equation (1.46) we get

$$I_2 = I\left(\frac{R_3}{R_2 + R_3}\right)$$

Similarly, current through R_3 is

$$I_3 = \frac{V_A}{R_3} = V\left(\frac{R_{eq} - R_1}{R_{eq} R_3}\right) = \frac{V}{R_{eq}}\left(\frac{R_{eq} - R_1}{R_3}\right) \qquad \text{...(1.47)}$$

Substitutir equations (1.44) and (1.45) in equation (1.47) we get

$$I_3 = I\left(\frac{R_2}{R_2 + R_3}\right). \qquad \text{...(1.48)}$$

SOLVED EXAMPLES

EXAMPLE 1.1. *For the circuit shown in Figure E1.1, find the equivalent resistance as seen from the terminal A-B.*

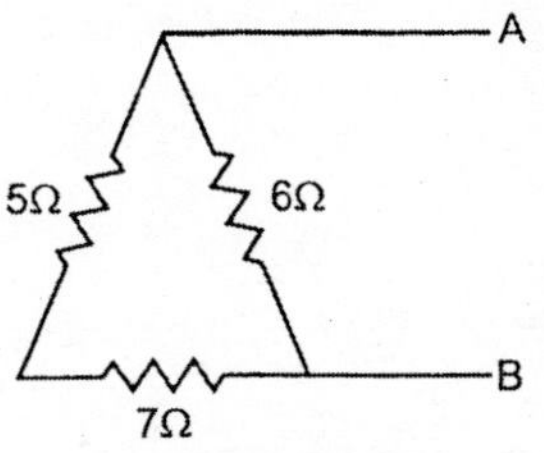

Fig. E1.1

SOLUTION. The circuit of Figure E1.1 can be redrawn as shown in Figure E1.1(*a*). 5 Ω and 7 Ω resistors are in series and can be replaced by an equivalent resistance, as shown in Figure E1.1(*b*), where $R_{eq1} = 5\ \Omega + 7\ \Omega = 12\ \Omega$.

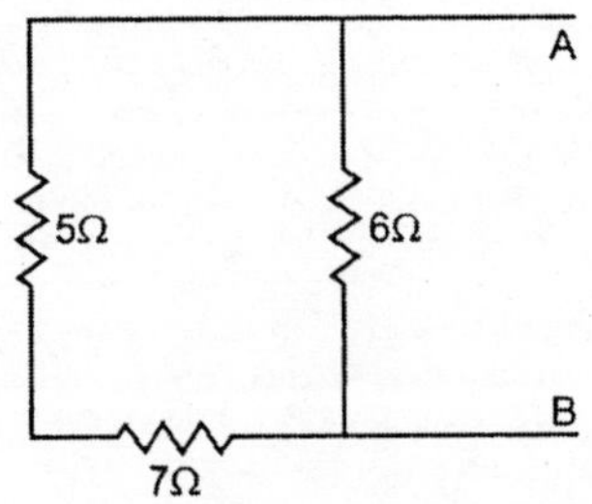

Fig. E1.1(a)

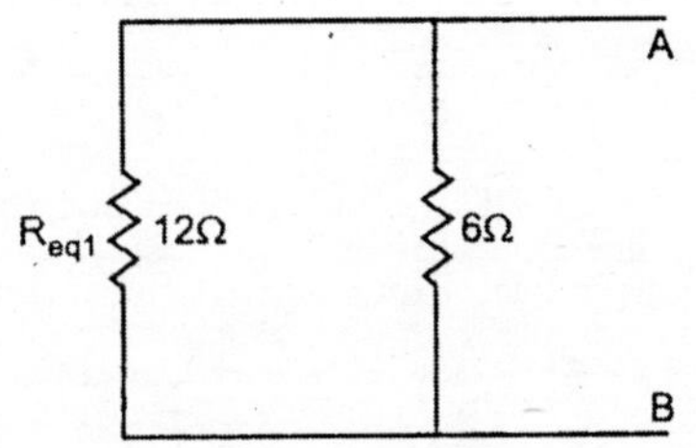

Fig. E1.1(b)

Again resistors 12 Ω and 6 Ω are in parallel. So the equivalent resistance as seen from the terminal *A-B* is

$$R_{eq} = \frac{12 \times 6}{12 + 6} = 4\ \Omega.$$

EXAMPLE 1.2. *For the circuit shown in Figure E1.2, find the equivalent resistance seen from the terminal A-B.*

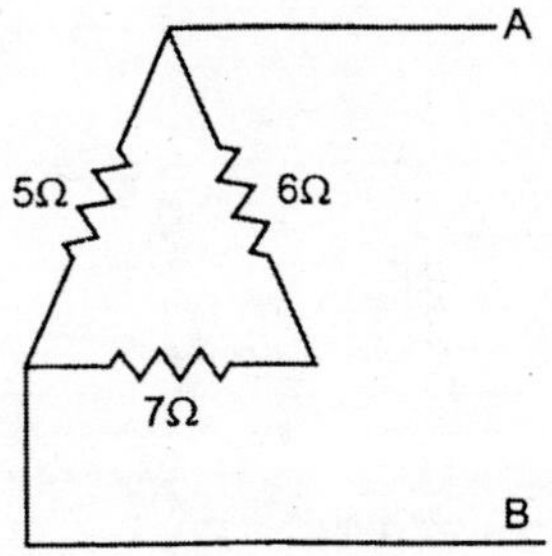

Fig. E1.2

SOLUTION. The circuit of Figure E1.2 can be redrawn as shown in Figure E1.2(*a*), 6 Ω and 7 Ω resistors are in series and can be replaced by an equivalent resistance, as shown in Figure E1.2(*b*), where $R_{eq1} = 6\ \Omega + 7\ \Omega = 13\ \Omega$.

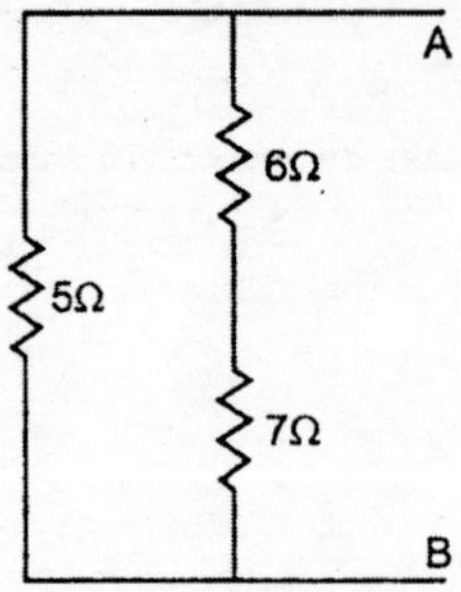

Fig. E1.2(a)

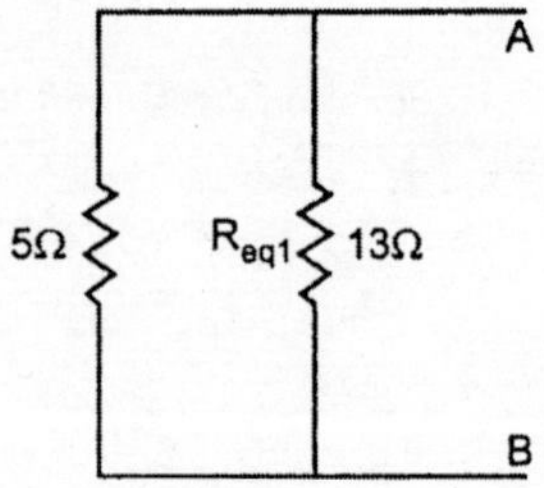

Fig. E1.2(b)

Again resistors 13 Ω and 5 Ω are in parallel. So the equivalent resistance as seen from the terminal *A-B* is

$$R_{eq} = \frac{13 \times 5}{13 + 5} = 3.61\ \Omega.$$

EXAMPLE 1.3. *For the circuit shown in Figure E1.3, find the equivalent resistance as seen from the terminal A-B.*

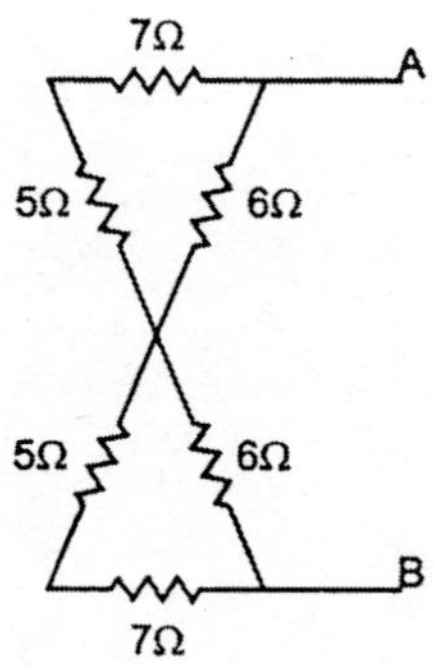

Fig. E1.3

SOLUTION. The circuit of Figure E1.3 can be redrawn as shown in Figure E1.3(*a*). Here 5 Ω and 7 Ω resistors in both the loops are in series and can be replaced by an equivalent resistance, as shown in Figure E1.3(*b*),where R_{eq1} = 5 Ω + 7 Ω = 12 Ω.

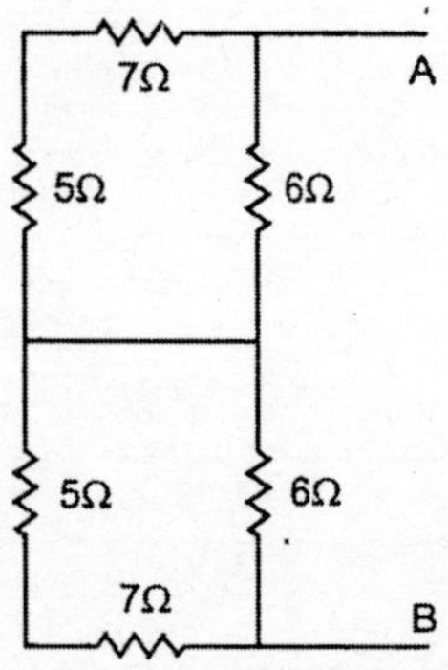

Fig. E1.3(a)

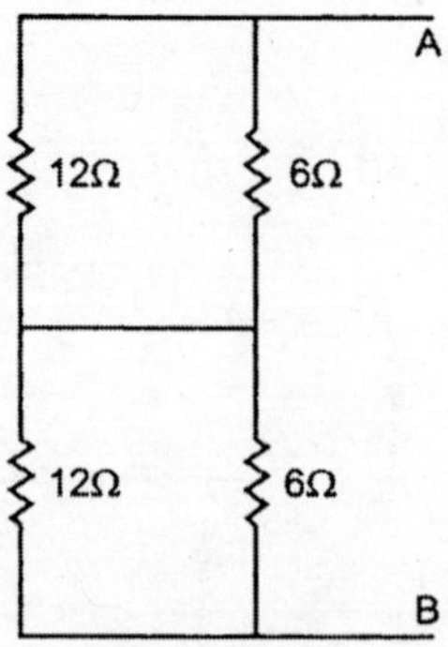

Fig. E1.3(b)

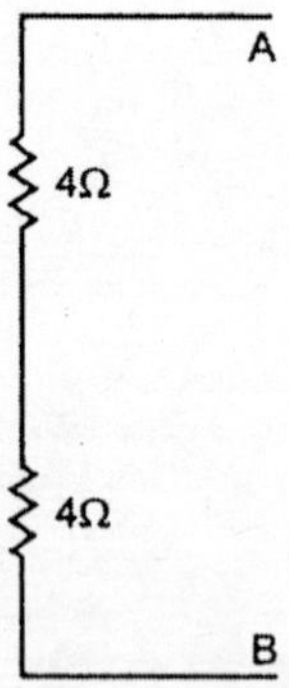

Fig. E1.3(c)

Again resistors 12 Ω and 6 Ω resistor in both the loops are in parallel, and thus can be replaced by an equivalent resistance, as shown in Figure E.1.3(*c*)

$$R_{eq} = \frac{12 \times 6}{12 + 6} = 4\ \Omega$$

So the equivalent resistance as seen from the terminal *A-B* is

$$R_{eq} = 4 + 4 = 8\ \Omega.$$

EXAMPLE 1.4. *For the circuit shown in Figure E1.4, find the equivalent resistance as seen from the terminal A-B.*

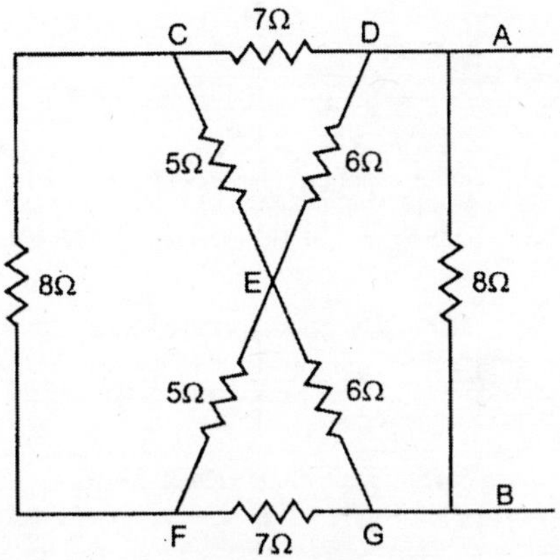

Fig. E1.4

SOLUTION. The circuit of Figure E1.4 can be redrawn as shown in Figure E1.4(*a*). Again if we consider only the triangles *CDE* and *EFG*, then the circuit is same as of the example 3, and thus can be replaced by its equivalent resistance, *i.e.*, 8 Ω, as calculated in the example 3, thus the circuit becomes, as shown in Figure E1.4(*b*)

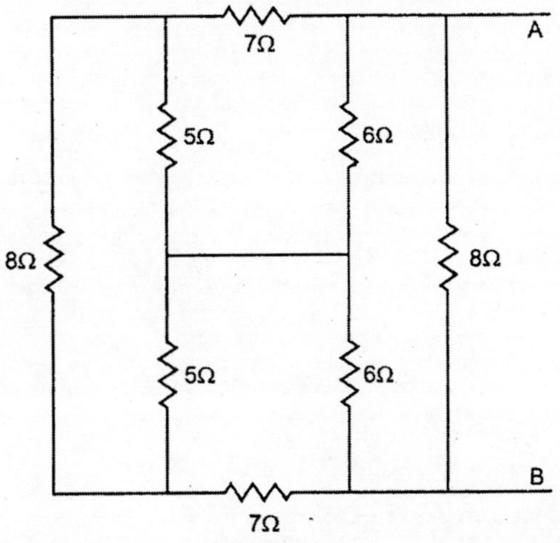

Fig. E1.4(a)

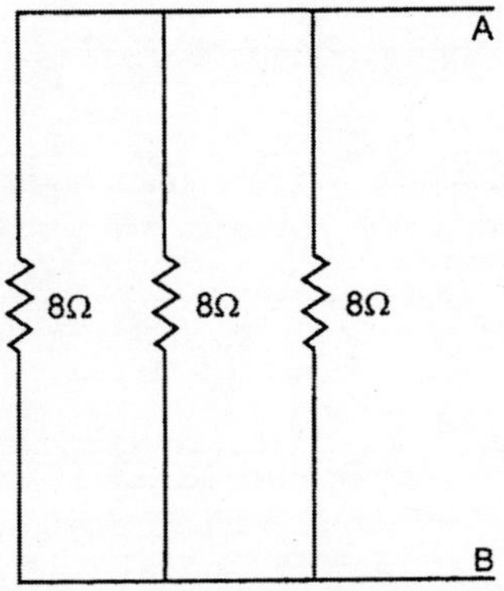

Fig. E1.4(b)

Thus from Figure E1.4(*b*), the equivalent resistance is

$$\frac{1}{R_{eq}} = \frac{1}{8}+\frac{1}{8}+\frac{1}{8}$$

thus $$R_{eq} = 2.67\ \Omega.$$

EXAMPLE 1.5. *For the circuit shown in Figure E1.5, find the current through each resistor.*

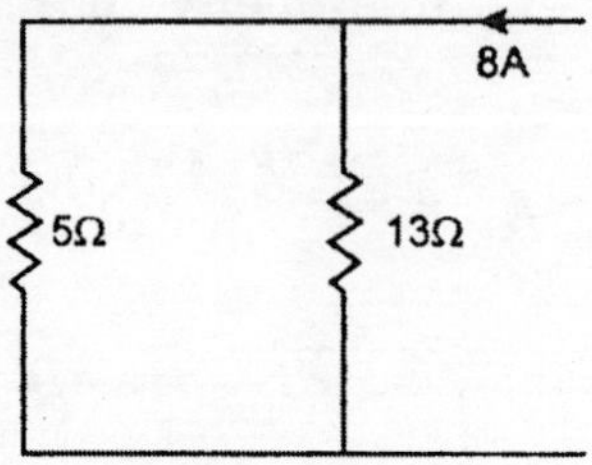

Fig. E1.5

SOLUTION. Let the current through 5 Ω and 13 Ω be i_1 and i_2 respectively as shown in Figure E1.5(*a*). Again since the resistors are connected in parallel, the drop across each resistor is same, that is,

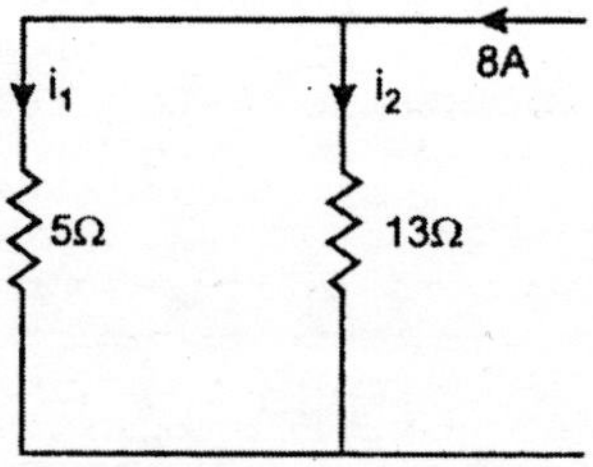

Fig. E1.5(a)

$$5i_1 = 13i_2$$

or $$5i_1 - 13i_2 = 0 \qquad ...(i)$$

again from KCL we know,

$$i_1 + i_2 = 8 \qquad ...(ii)$$

Solving equations (*i*) and (*ii*) we get,

$$i_1 = 5.78\text{ A} \qquad i_2 = 2.22\text{ A}$$

EXAMPLE 1.6. *For the circuit shown in Figure E1.6, find the value of the equivalent resistance as seen from the terminal A-B.*

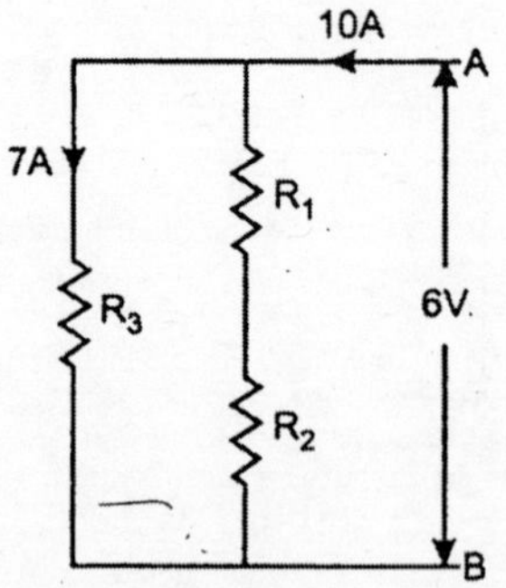

Fig. E1.6

SOLUTION. From the Figure E1.6, the current through the series resistors R_1 and R_2 is $10 - 7 = 3$ A.

The drop across R_3 is 6 V. Thus

$$R_3 = \frac{V}{I} = \frac{6}{7} = 0.857\ \Omega$$

Again $\quad 3 \times (R_1 + R_2) = 6$

or $\quad R_1 + R_2 = 2$

Thus

$$R_{eq} = \frac{(R_1 + R_2) \times R_3}{R_1 + R_2 + R_3} = \frac{2 \times 0.875}{2 + 0.875}$$

$$= \frac{2 \times 0.875}{2 + 0.875}$$

$$= 0.6\ \Omega.$$

EXAMPLE 1.7. *Consider the circuit shown in Figure E1.7, the values of R_1, R_2 and R_3 are 5 Ω, 4 Ω and 10 Ω respectively. The supply voltage is 30 V and the circuit current is 1 A. Find the equivalent resistance of the circuit.*

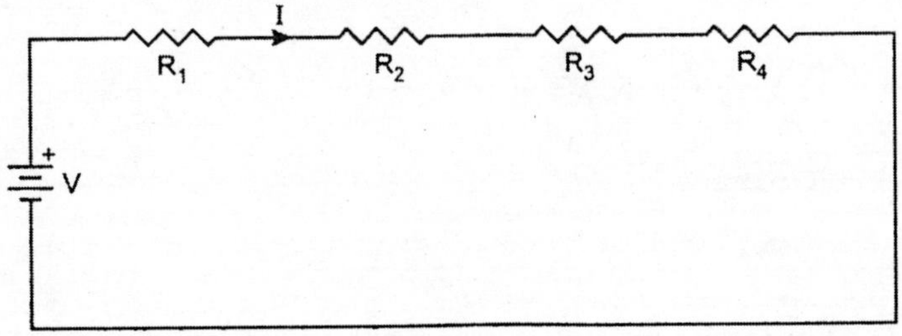

Fig. E1.7

SOLUTION. Let the drop across R_1, R_2, R_3 and R_4 be e_1, e_2, e_3 and e_4 respectively.

Then,

$$e_1 = I \times R_1 = 1 \times 5 = 5\ \text{V}$$

$$e_2 = I \times R_2 = 1 \times 4 = 4\ \text{V}$$

$$e_3 = I \times R_3 = 1 \times 10 = 10\ \text{V}$$

then,

$$e_4 = V - (e_1 - e_2 - e_3)$$

$$= 30 - 5 - 4 - 10 = 11\ \text{V}$$

thus,

$$R_4 = \frac{e_4}{I} = \frac{11}{1} = 11\ \Omega$$

thus the equivalent resistance is

$$R_{eq} = R_1 + R_2 + R_3 + R_4$$

$$R_{eq} = 5 + 4 + 10 + 11 = 30\ \Omega.$$

EXAMPLE 1.8. *Two resistances $R_1 = 3$ Ω and $R_2 = 7$ Ω are connected in series across a 10 V battery as shown in Figure E1.8. Find the current through the circuit when,*

(i) the switch S is open

(ii) the switch S is closed.

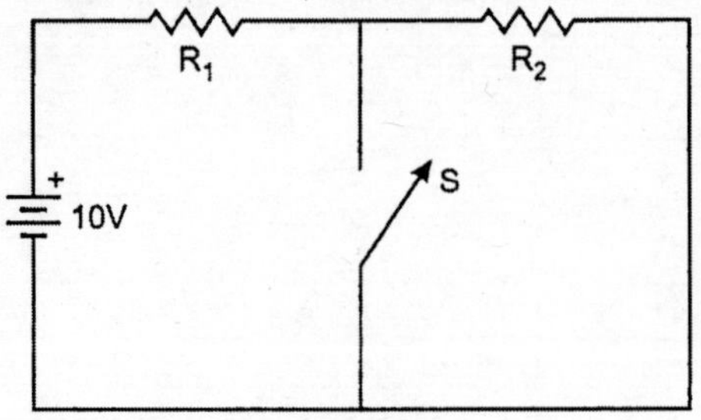

Fig. E1.8

SOLUTION. (*i*) When switch S is open.

$$I = \frac{10}{7+3} = 1 \text{ A}$$

(*ii*) When switch S is closed.

No current will flow through R_2, thus,

$$I = \frac{10}{3} = 3.33 \text{ A.}$$

EXAMPLE 1.9. *Six 2 V batteries having internal resistance of 1 Ω each are connected all in series across a 2 Ω resistance. Find the circuit current.*

SOLUTION. The circuit current is

$$I = \frac{6\times2}{6\times1+2} = \frac{12}{8} = 1.5 \text{ A.}$$

EXAMPLE 1.10. *For the circuit shown in Figure E1.10, find the equivalent capacitance across A-B.*

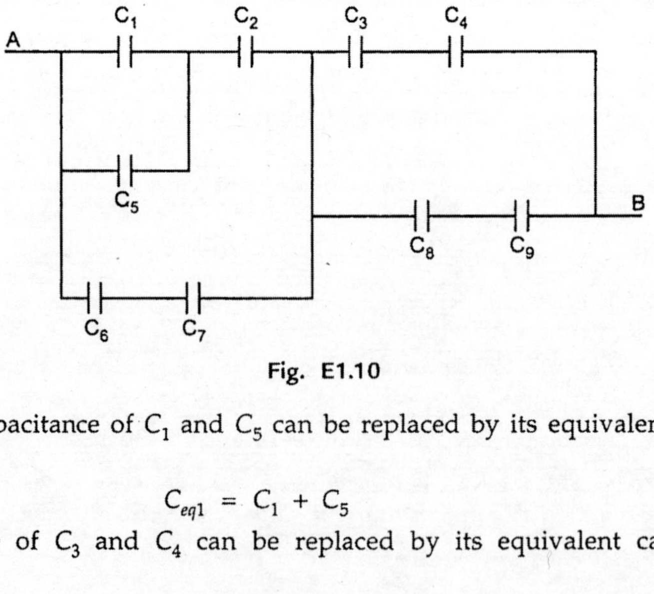

Fig. E1.10

SOLUTION. The capacitance of C_1 and C_5 can be replaced by its equivalent capacitance given by,

$$C_{eq1} = C_1 + C_5 \qquad ...(i)$$

The capacitance of C_3 and C_4 can be replaced by its equivalent capacitance as given below,

$$C_{eq2} = \frac{C_3 \times C_4}{C_3 + C_4} \qquad ...(ii)$$

The capacitance of C_8 and C_9 can be replaced by its equivalent capacitance as given below,

$$C_{eq3} = \frac{C_8 \times C_9}{C_8 + C_9} \qquad ...(iii)$$

The capacitance of C_6 and C_7 can be replaced by its equivalent capacitance as given below,

$$C_{eq4} = \frac{C_6 \times C_7}{C_6 + C_7} \qquad ...(iv)$$

Then the circuit of Figure E1.10 can be redrawn as shown in Figure E1.10(a).

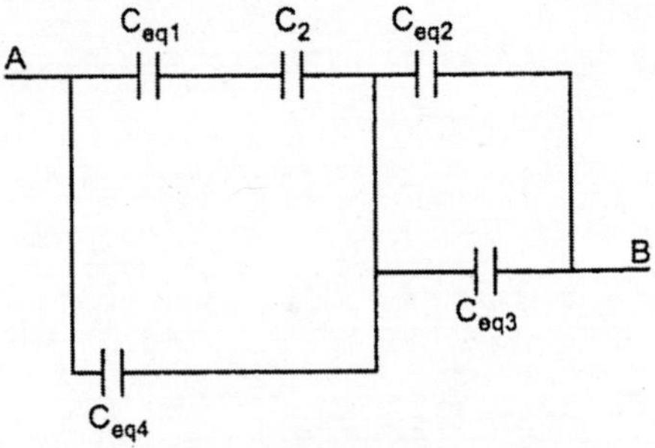

Fig. E1.10(a)

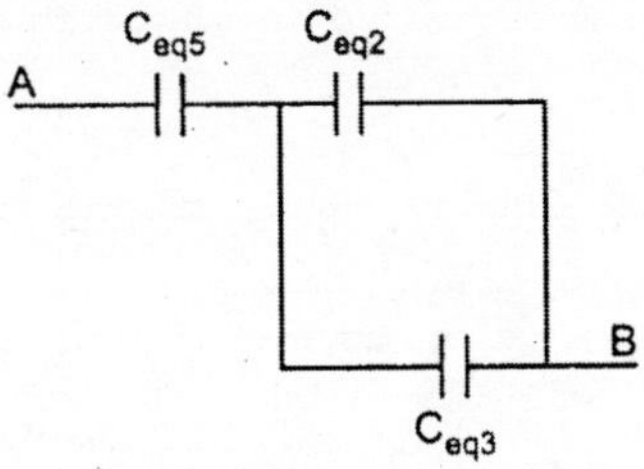

Fig. E1.10(b)

The capacitance of C_{eq1}, C_2 and C_{eq4} can be replaced by its equivalent capacitance as given below,

$$C_{eq5} = \frac{C_{eq1} \times C_2}{C_{eq1} + C_2} + C_{eq4} \qquad ...(v)$$

Then the circuit reduces to as shown in Figure E1.10(*b*). Then the equivalent capacitance across *A-B* is

$$C_{eq} = \frac{(C_{eq2} + C_{eq3}) \times C_{eq5}}{C_{eq2} + C_{eq3} + C_{eq5}}. \qquad ...(vi)$$

EXAMPLE 1.11. *For the circuit shown in Figure E1.11, find the equivalent capacitance across A-B. All the capacitances have the same value 2 μF.*

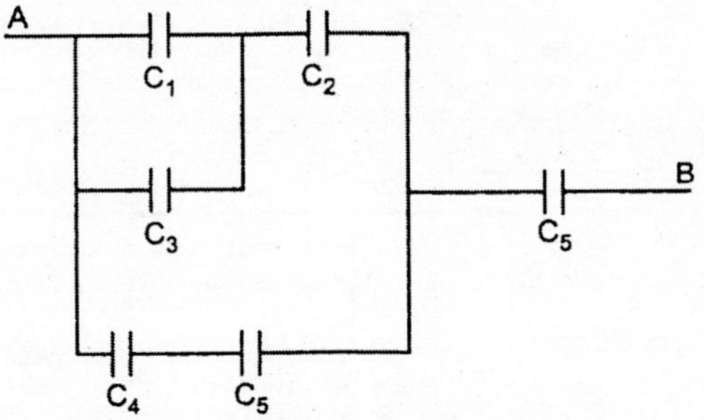

Fig. E1.11

SOLUTION. The capacitance of C_1, C_3 and C_2 can be replaced by its equivalent capacitance as given below,

$$C_{eq1} = \frac{(C_1 + C_3) \times C_2}{C_1 + C_3 + C_2} = \frac{(2+2) \times 2}{2+2+2} = 1.33\ \mu F$$

The capacitance of C_4 and C_5 can be replaced by its equivalent capacitance as given below,

$$C_{eq2} = \frac{C_4 \times C_5}{C_4 + C_5} = \frac{2 \times 2}{2+2} = 1\ \mu F$$

Then the circuit of Figure E1.11 can be redrawn as shown in Figure E1.11(*a*).

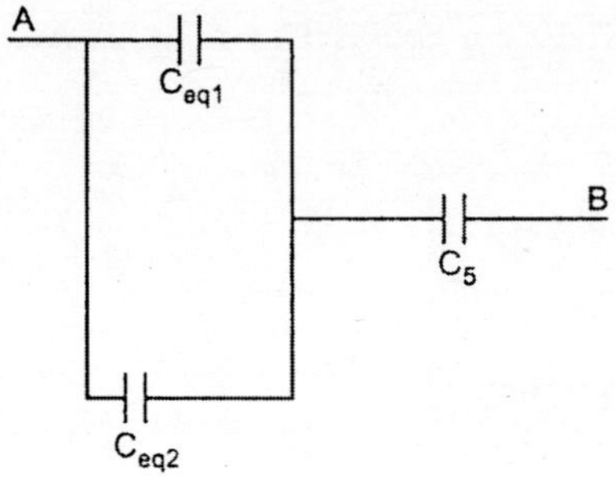

Fig. E1.11(a)

The equivalent capacitance is given by,

$$C_{eq} = \frac{(C_{eq1} + C_{eq2}) \times C_5}{C_{eq1} + C_{eq2} + C_5} = \frac{(1.33 + 1) \times 2}{1.33 + 1 + 2} = 1.08\ \mu F$$

EXAMPLE 1.12. *For the circuit shown in Figure E1.12, find the current through all the branches.*

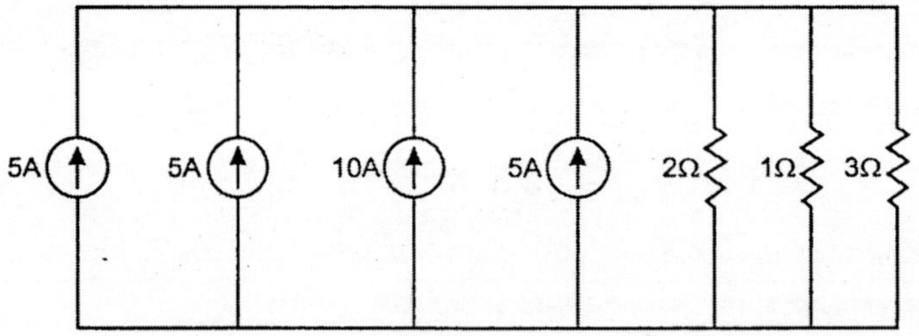

Fig. E1.12

SOLUTION. Adding all the current sources and replacing by a single current source we get the circuit as shown in Figure E1.12(*a*).

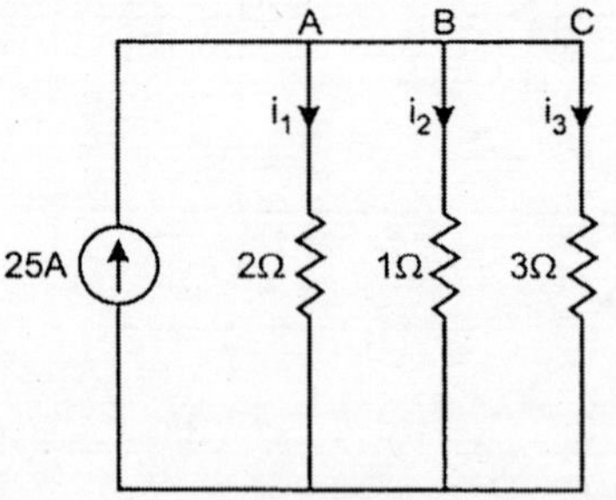

Fig. E1.12(a)

Using current division method current through all the branches is determined by

$$i_1 = \frac{25\times0.75}{2+0.75} = 6.81 \text{ A}$$

$$i_2 = \frac{25\times1.2}{1.2+1} = 13.63 \text{ A}$$

$$i_3 = \frac{25\times0.67}{0.67+3} = 4.56 \text{ A}$$

NOTE If you solve the above circuit in the similar fashion as explained in article 1.6. We will get the above three formulae. The derivation of the above formulae is left for the students as an assignment.

EXAMPLE 1.13. *For the circuit shown in Figure E1.13, find the current through all the branches using current division rule.*

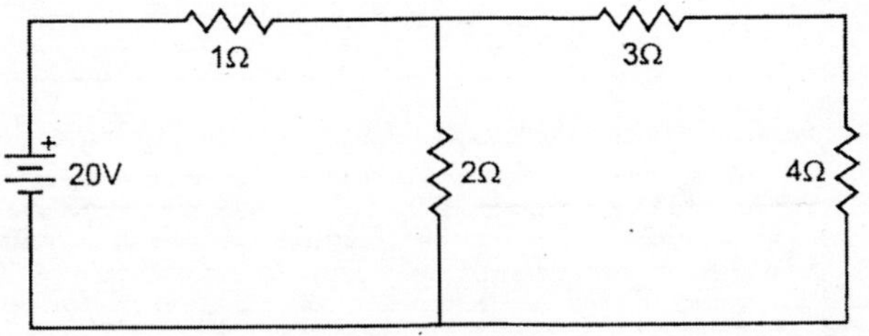

Fig. E1.13

SOLUTION. The equivalent resistance as seen from the source terminal is

$$R_{eq} = 2+\frac{(3+4)\times2}{3+4+2} = 3.56 \ \Omega$$

then the source current is $$I = \frac{V}{R_{eq}} = \frac{20}{3.56} = 5.62 \text{ A}$$

So the current through 1 Ω is 5.62 A

The current through $$2 \ \Omega = 5.62\times\frac{7}{9} = 4.37 \text{ A}$$

Therefore the current through the other parallel branch is = 5.62 – 4.37 = 1.25 A.

EXAMPLE 1.14. *For the circuit shown in Figure E1.14, find the drop across A-B.*

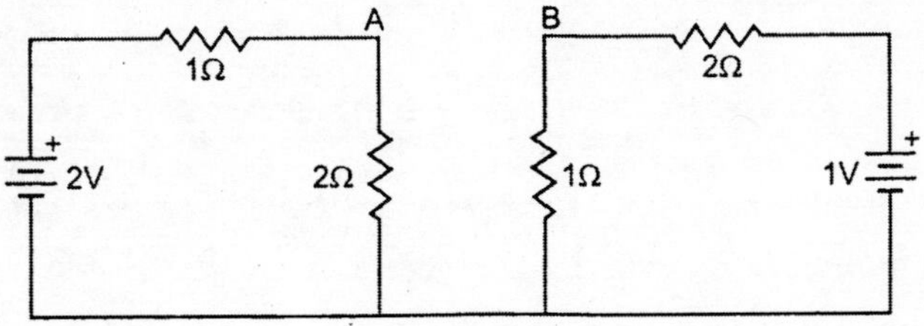

Fig. E1.14

SOLUTION. The loop currents are shown in Figure E1.14(*a*)

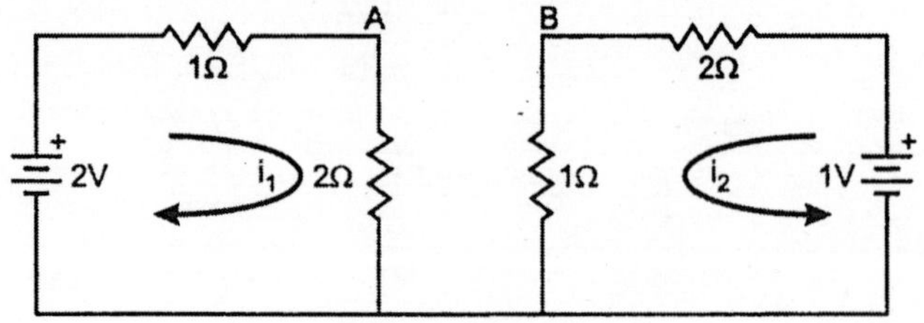

Fig. E1.14(a)

Then $\qquad i_1 = \frac{2}{3} = 0.667 \text{ A}$ and $i_2 = \frac{1}{3} = 0.333 \text{ A}$

The drop at point *A*, $\qquad V_A = 2i_1 = 2 \times 0.667 = 1.334 \text{ V}$

The drop at point *B*, $\qquad V_B = 1i_2 = 1 \times 0.333 = 0.333 \text{ V}$

The drop across *A-B* is $\qquad = V_A - V_B = 1.334 - 0.333 = 1 \text{ V}.$

EXAMPLE 1.15. *For the circuit shown in Figure E1.15, find the drop across A-B.*

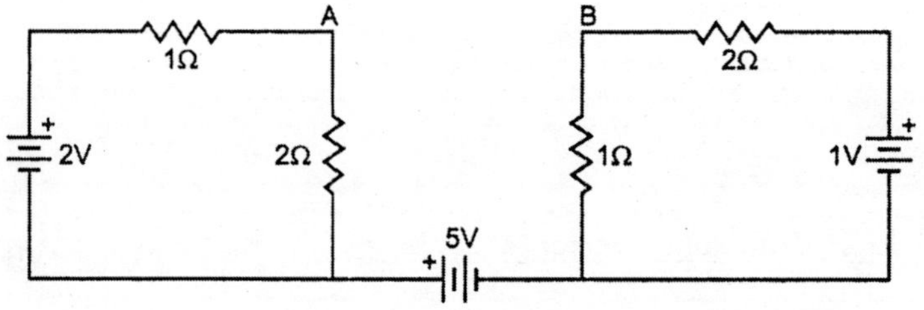

Fig. E1.15

SOLUTION. The loop currents are shown in Figure E1.15(*a*)

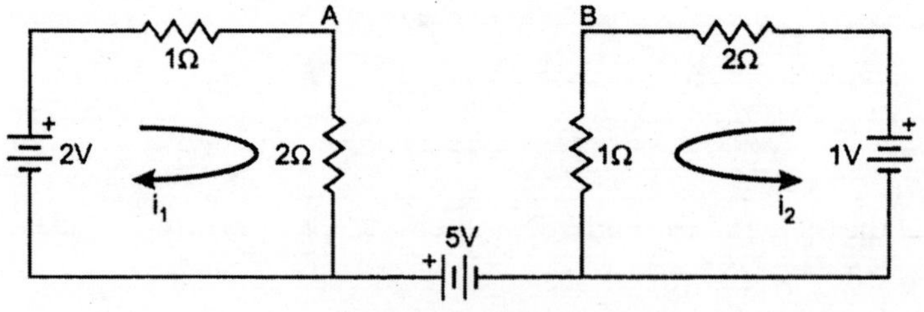

Fig. E1.15(a)

Then $\qquad i_1 = \frac{2}{3} = 0.667 \text{ A}$ and $i_2 = \frac{1}{3} = 0.333 \text{ A}$

The drop at point *A*, $\qquad V_A = 2i_1 = 2 \times 0.667 = 1.334 \text{ V}$

The drop at point *B*, $\qquad V_B = 1i_2 = 1 \times 0.333 = 0.333 \text{ V}$

The drop across *A-B* is $\qquad = V_A + 5 - V_B = 1.334 + 5 - 0.333 = 6 \text{ V}.$

EXAMPLE 1.16. *For the circuit shown in Figure E1.16, find the drop across A-B.*

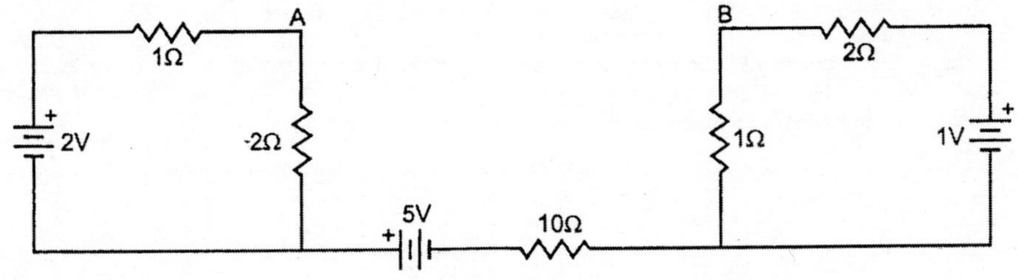

Fig. E1.16

SOLUTION. The loop currents are shown in Figure E1.16(*a*)

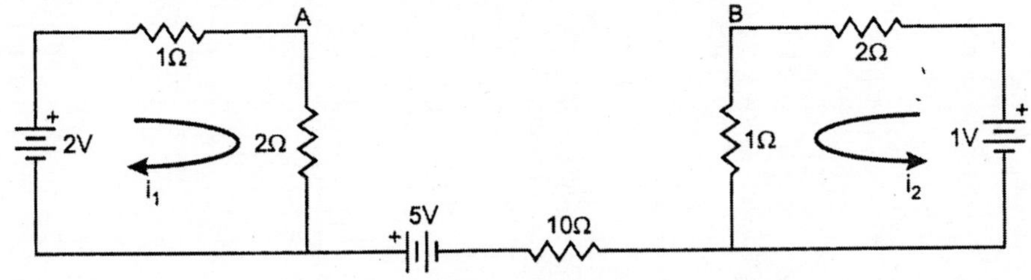

Fig. E1.16(a)

Then $i_1 = \frac{2}{3} = 0.667$ A and $i_2 = \frac{1}{3} = 0.333$ A

The drop at point *A*, $V_A = 2i_1 = 2 \times 0.667 = 1.334$ V

The drop at point *B*, $V_B = 1i_2 = 1 \times 0.333 = 0.333$ V

The drop across *A-B* is $= V_A + 5 - V_B = 1.334 + 5 - 0.333 = 6$ V

Note that there will be no drop across 10 Ω resistance since no current is flowing through it.

EXERCISES

1. For the circuit shown in Figure Q1.1, find the equivalent resistance as seen from the terminal *A-B*.

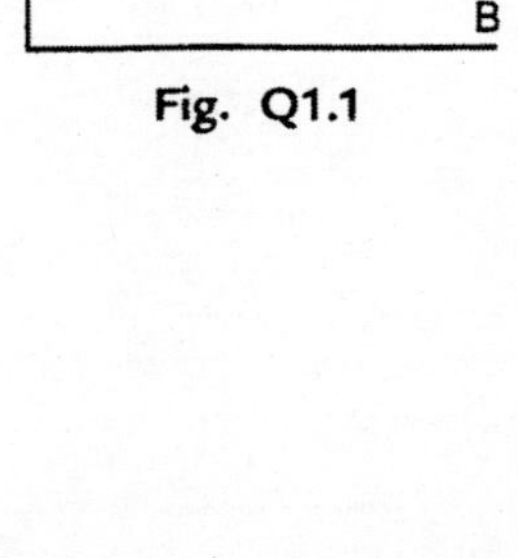

Fig. Q1.1

2. For the circuit shown in Figure Q1.2, find the equivalent capacitance across *A-B*.

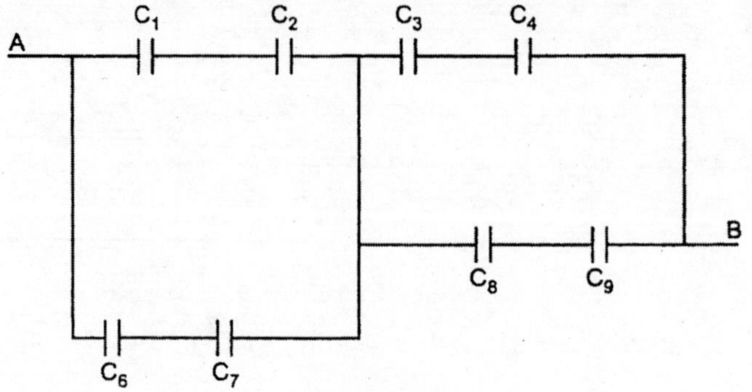

Fig. Q1.2

3. For the circuit shown in Figure Q1.3, find the drop across *A-B*.

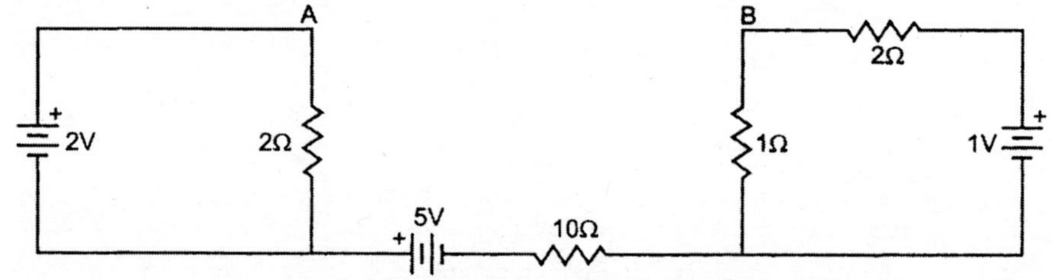

Fig. Q1.3

4. Differentiate between the following
   (*i*) Loop and Mesh
   (*ii*) Independent and non independent loop
   (*iii*) Lumped and distributed network
   (*iv*) open and short circuit.
5. State and explain Ohm's law and Kirchhoff's law.
6. What do you mean by internal resistance? What is the internal resistance of an ideal current source and an ideal voltage source?

Chapter 2

CIRCUIT TRANSFORMATION

2.1 INTRODUCTION

Source transformation and star delta or delta star transformation are used to analyze a complicated network or circuit. In these two methods the actual complicated circuit is reduced into a simple circuit to analyze them.

2.2 SOURCE TRANSFORMATION

Let us consider a practical voltage source and a practical current source. Let to both the sources a load resistance Z be connected across the terminals A-B, respectively, as shown if Figure 2.1(a) and Figure 2.1(b).

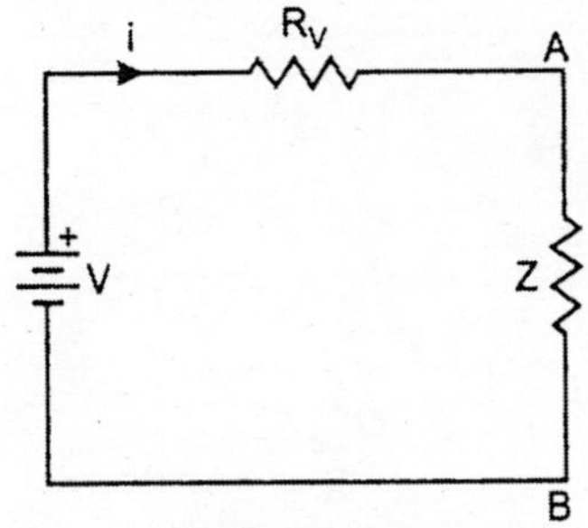

Fig. 2.1(a)

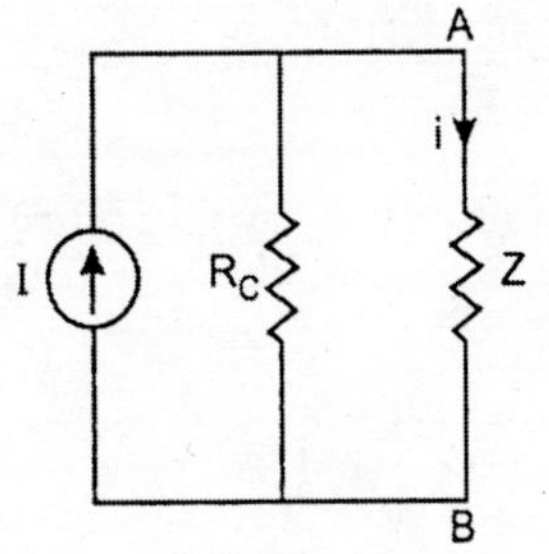

Fig. 2.1(b)

From Figure 2.1(a), the current through load resistance Z is given by

$$i = \frac{V}{R_V + Z} \qquad \text{...(2.1)}$$

similarly from Figure 2.1(*b*), the current through load resistance Z is given by

$$i = I\frac{R_C}{R_C + Z} \quad ...(2.2)$$

Since the load resistance Z is same in both the circuits, the current flowing through it should also be same for the two circuits to be identical. Thus we can write

$$\frac{V}{R_V + Z} = I\frac{R_C}{R_C + Z} \quad ...(2.3)$$

Now let us remove the load resistance Z from both the circuits. Then the circuit will look like as shown in Figure 2.1(*c*) and Figure 2.1(*d*) respectively.

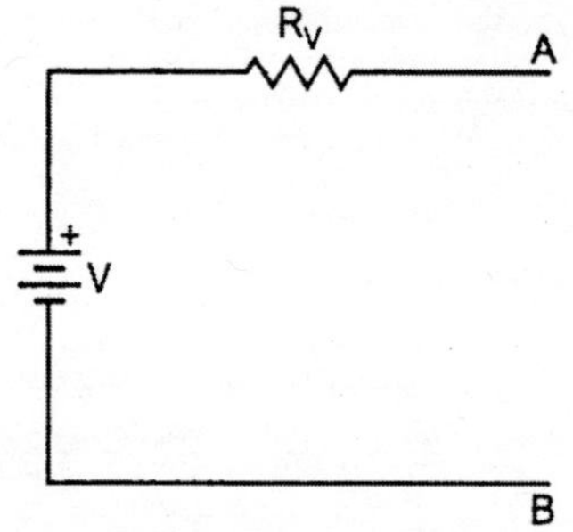

Fig. 2.1(c)

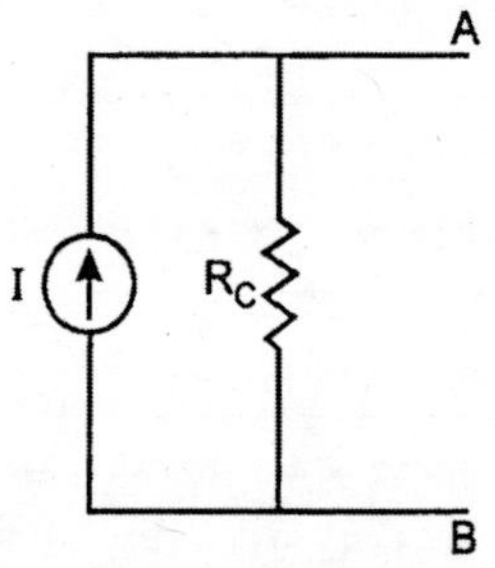

Fig. 2.1(d)

For both the circuit to be identical, the voltage drop across the terminal *A-B*, should be same for Figure 2.1(*c*) and Figure 2.1(*d*). Then from Figure 2.1(*c*)

$$V_{AB} = V \quad ...(2.4)$$

From Figure 2.1(*d*)

$$V_{AB} = IR_C \quad ...(2.5)$$

From equations (1.4) and (1.5) we get

$$V = IR_C \quad ...(2.6)$$

then from equations (1.3) and (1.4) we get

$$R_C = R_V \quad ...(2.7)$$

Thus a practical voltage source with voltage V and series resistance R_V can be replaced by a practical current source of current I with parallel resistance R_C, if and only if $R_V = R_C$.

SOLVED EXAMPLES

EXAMPLE 2.1. *Determine the equivalent voltage source for the circuit shown in Figure E2.1.*

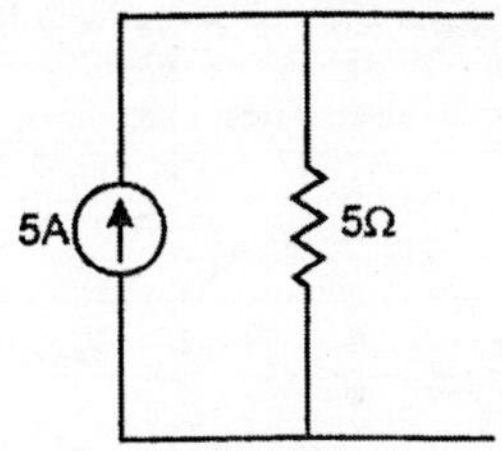

Fig. E2.1

SOLUTION. The equivalent voltage source is shown in Figure E2.1(*a*), where $V = 5 \times 5 = 25$ V and $R_V = 5\Omega$

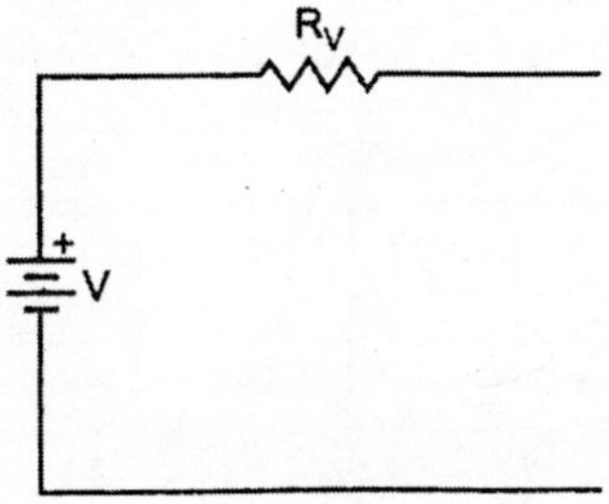

Fig. E2.1(a)

EXAMPLE 2.2. *Determine the equivalent current source for the circuit shown in Figure E2.2.*

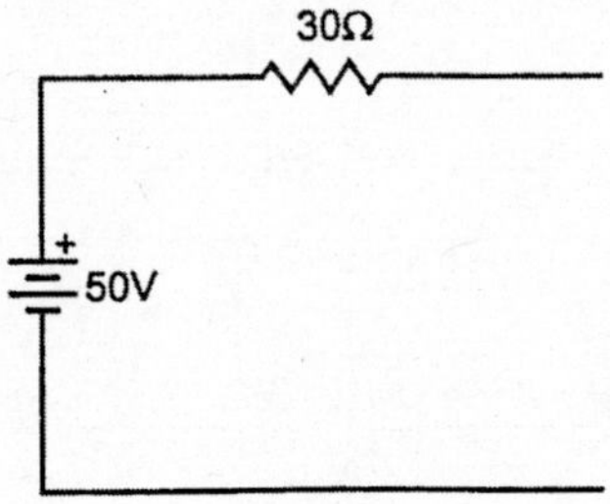

Fig. E2.2

SOLUTION. The equivalent voltage source is shown in Figure E2.1(*a*), where $I = \frac{50}{30} = 1.66$ Amp and $R_C = 30\ \Omega$.

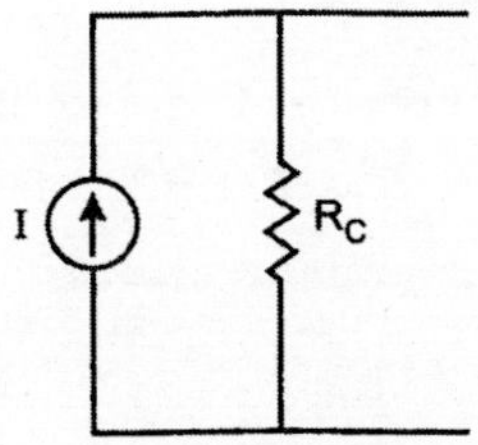

Fig. E2.2(a)

EXAMPLE 2.3. *Using source transformations convert the circuit shown in Figure E2.3 into a single source network.*

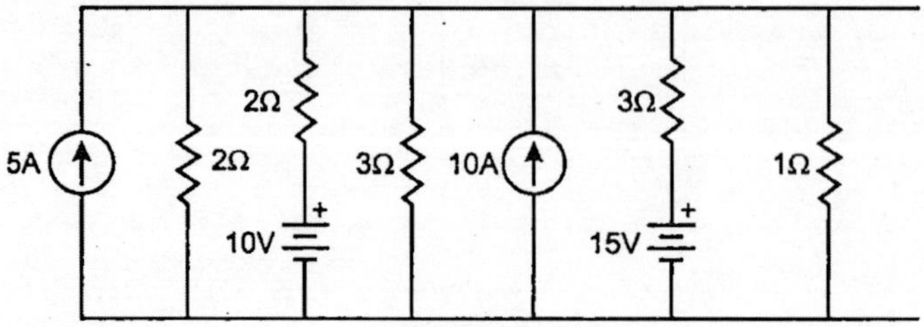

Fig. E2.3

SOLUTION. Converting the voltage source into current source we get the circuit as shown in Figure E2.3(*a*). Here the 10 V source with 2 Ω resistance in series has been replaced by 5 A current source with 2 Ω resistance in parallel. Similarly the 15 V source with 3 Ω resistance in series has been replaced by 5 A current source with 3 Ω resistance in parallel.

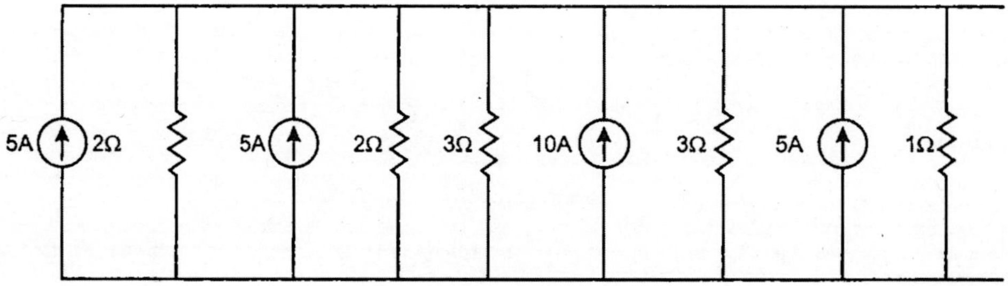

Fig. E2.3(a)

Now rearranging the circuit elements we get the circuit shown in Figure E2.3(*b*).

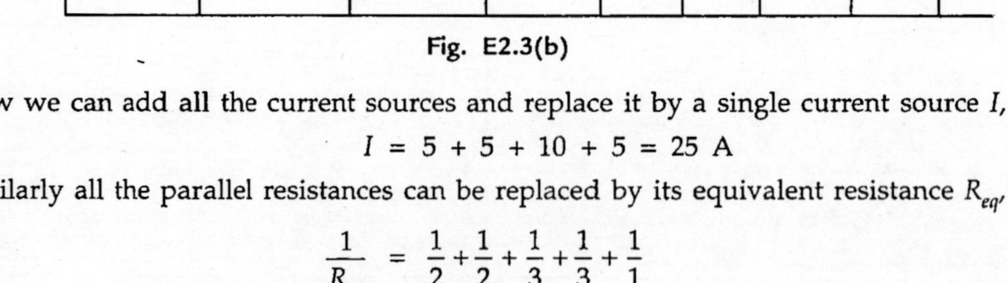

Fig. E2.3(b)

Now we can add all the current sources and replace it by a single current source I, where,

$$I = 5 + 5 + 10 + 5 = 25 \text{ A}$$

Similarly all the parallel resistances can be replaced by its equivalent resistance R_{eq}, where,

$$\frac{1}{R_{eq}} = \frac{1}{2} + \frac{1}{2} + \frac{1}{3} + \frac{1}{3} + \frac{1}{1}$$

$$R_{eq} = 0.375 \ \Omega$$

Then the above circuit reduces to as shown in Figure E2.3(*c*), which is a single source network. If required we can also convert the circuit shown in Figure E2.3(*c*) into an equivalent voltage circuit as shown in Figure E2.3(*d*)

where $$V = 25 \times 0.375 = 9.375 \text{ V}$$

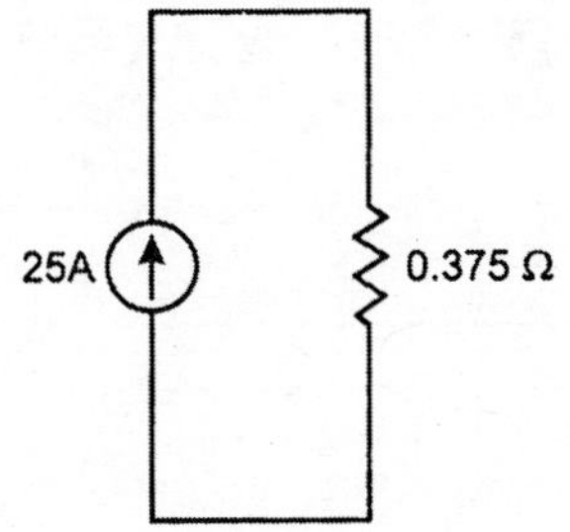

Fig. E2.3(c)

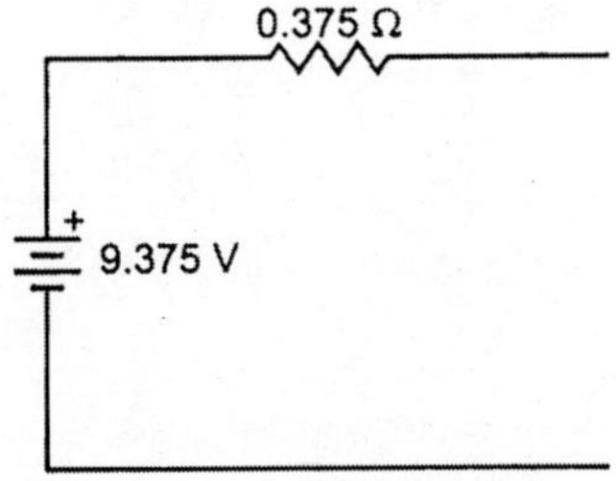

Fig. E2.3(d)

EXAMPLE 2.4. *Using source transformations find the current through 5 Ω resistor for the circuit shown in Figure E2.4.*

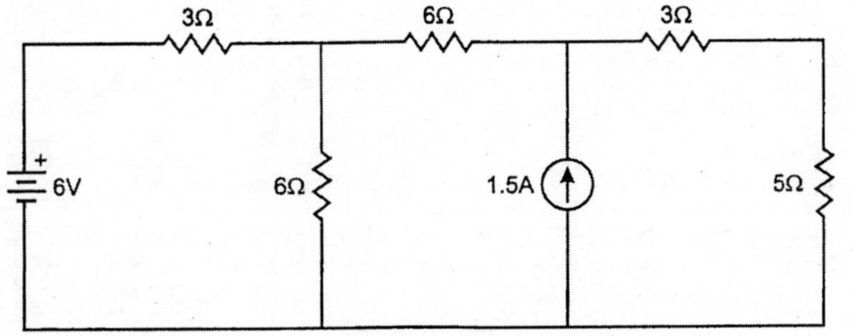

Fig. E2.4

SOLUTION. Converting the voltage source into current source we get Figure E2.4(*a*). Here 6 V source with 3 Ω series resistance has been replaced by a 2 A current source with 3 Ω resistance in parallel. Where $I = \frac{6}{3} = 2$ A

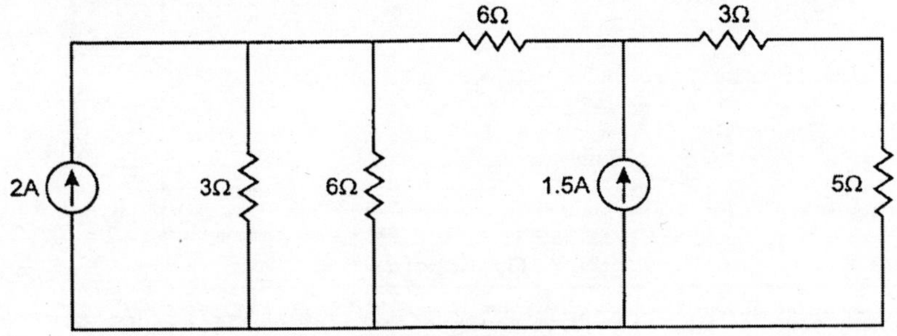

Fig. E2.4(a)

Replacing the two parallel resistances 3 Ω and 6 Ω by its equivalent resistance we get a circuit as shown in Figure E2.4(*b*). Here

$$R_{eq} = \frac{3 \times 6}{3 + 6} = 2 \ \Omega$$

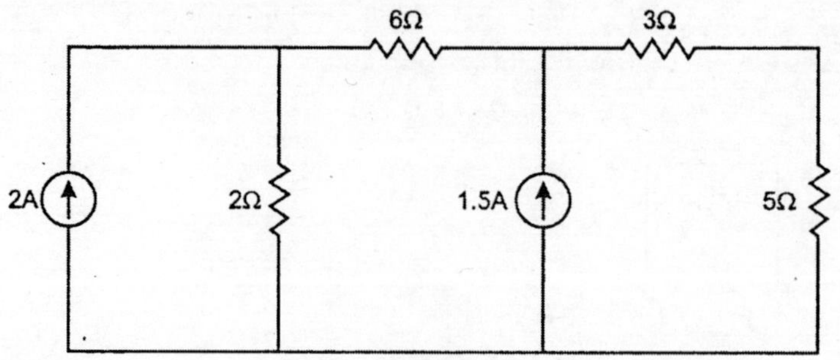

Fig. E2.4(b)

Now converting the 2 A current source into its equivalent voltage source we get the circuit as shown in Figure E2.4(*c*). Here 2 A current source with 2 Ω parallel resistance has been replaced by 4 V source with 2 Ω series resistance.

Where $$V = I \times R = 2 \times 2 = 4 \text{ V}$$

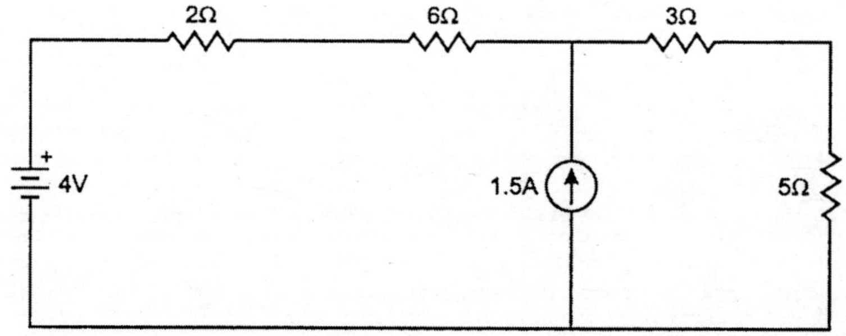

Fig. E2.4(c)

Now adding the two series resistances 2 Ω and 6 Ω we get the circuit as shown in Figure E2.4(*d*).

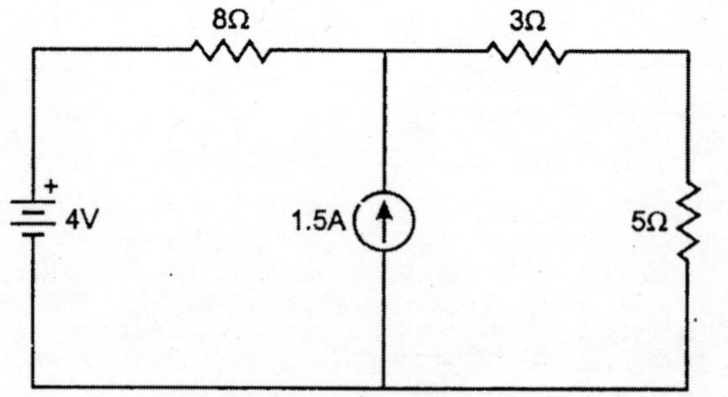

Fig. E2.4(d)

Converting the voltage source into current source we get Figure E2.4(*e*). Here 4 V source with 8 Ω series resistance has been replaced by a 0.5 A current source with 8 Ω resistance in parallel.

Where $$I = \frac{4}{8} = 0.5A$$

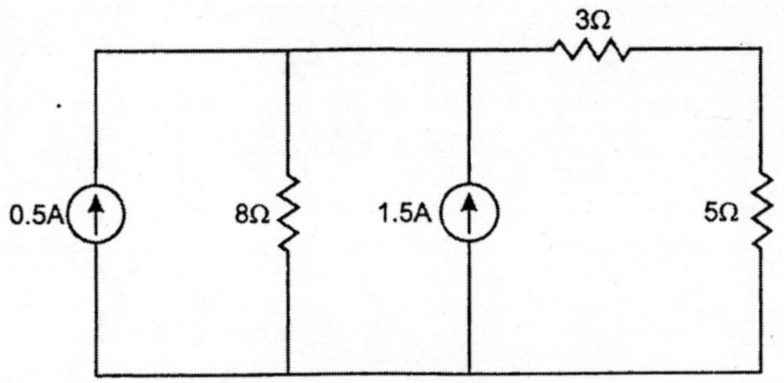

Fig. E2.4(e)

Adding the two current sources of Figure E2.4(*e*) and replacing it by an equivalent current source we get the circuit as shown in Figure E2.4(*f*). Where, $I = 1.5 + 0.5 = 2$ A

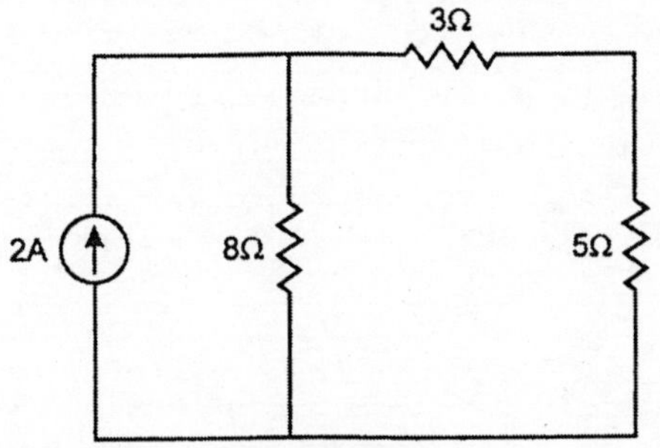

Fig. E2.4(f)

Converting the 2 A current source into its equivalent voltage source we get the circuit as shown in Figure E2.4(*g*). Here 2 A current source with 8 Ω parallel resistance has been replaced by 16 V source with 8 Ω series resistance.

Where $$V = I \times R = 2 \times 8 = 16 \text{ V}$$

Then from Figure E2.4(*g*), the current through 5 Ω = $\dfrac{16}{8+3+5} = 1$ A

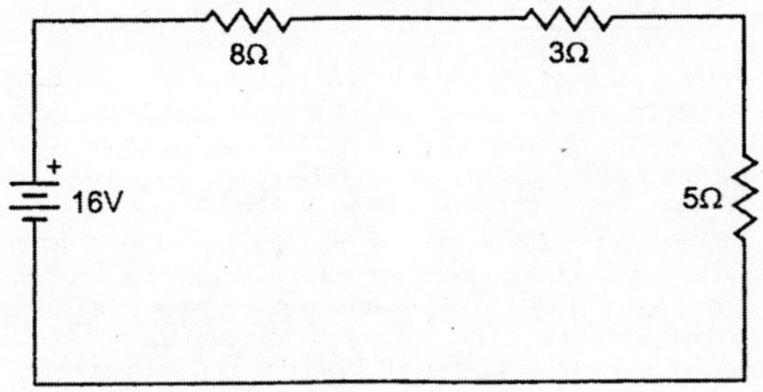

Fig. E2.4(g)

EXAMPLE 2.5. *Using source transformations find the current through 3 Ω resistor for the circuit shown in Figure E2.5.*

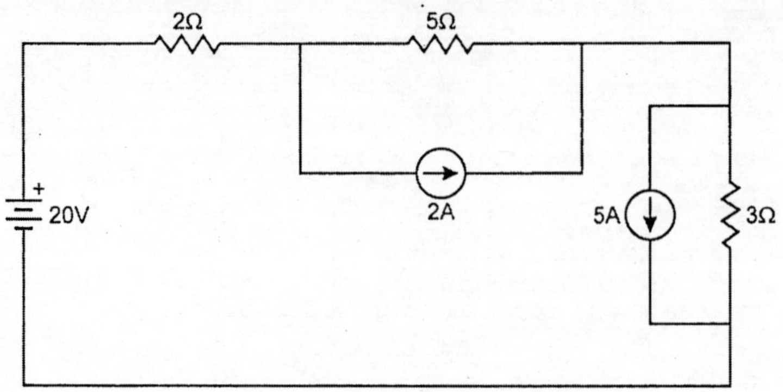

Fig. E2.5

SOLUTION. Converting the two current sources into voltage source we get the circuit as shown in Figure E2.5(*a*). Here 2 A source with parallel 5 Ω resistance has been replaced 10 V source with 5 Ω resistance in series. Similarly 5 A source with 3 Ω resistance has been replaced by 15 V source with 3 Ω resistance in series.

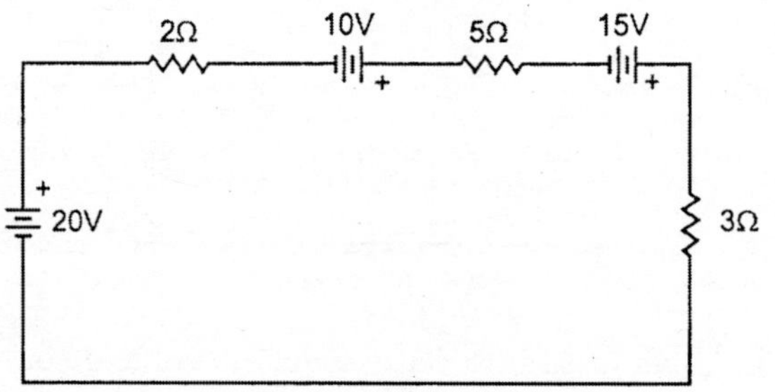

Fig. E2.5(a)

Then from Figure E2.5(*a*) the current through 3 Ω resistor is $I = \dfrac{20 + 10 + 15}{2 + 5 + 3} = 4.5$ A

EXAMPLE 2.6. *Convert the circuit shown in Figure E2.6, into a single source network.*

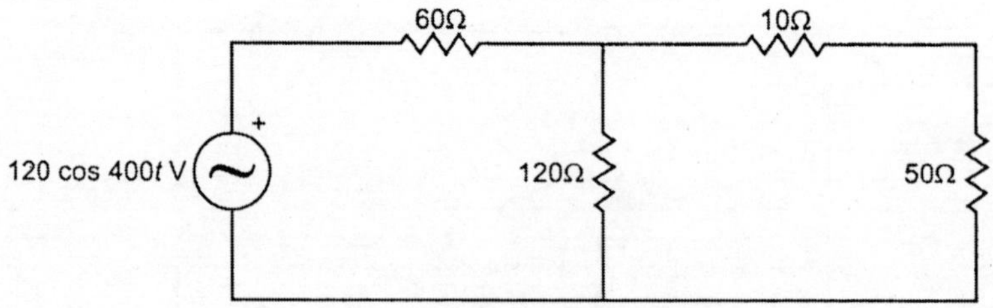

Fig. E.2.6

SOLUTION. Converting the voltage source into current source we get Figure E2.6(*a*). Here 120 cos 400*t* V source with 60 Ω series resistance has been replaced by 2 cos 400*t* A current source with 60 Ω resistance in parallel.

Where $$I = \frac{120 \cos 400t}{60} = 2 \cos 400t \text{ A}$$

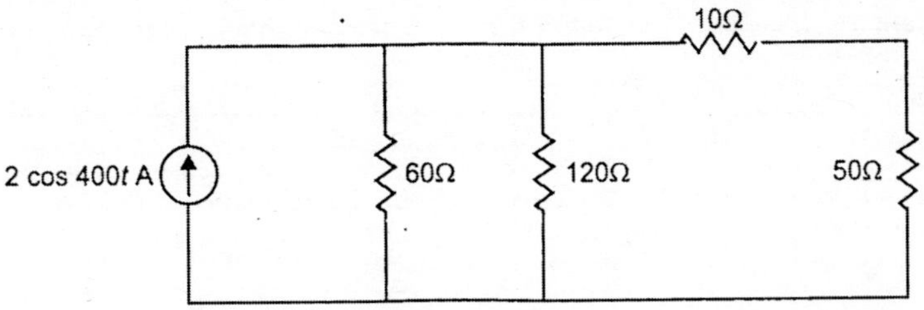

Fig. E2.6(a)

Replacing the two parallel resistances 60 Ω and 120 Ω by its equivalent resistance we get a circuit as shown in Figure E2.6(*b*). Here

$$R_{eq} = \frac{60 \times 120}{60 + 120} = 40\ \Omega$$

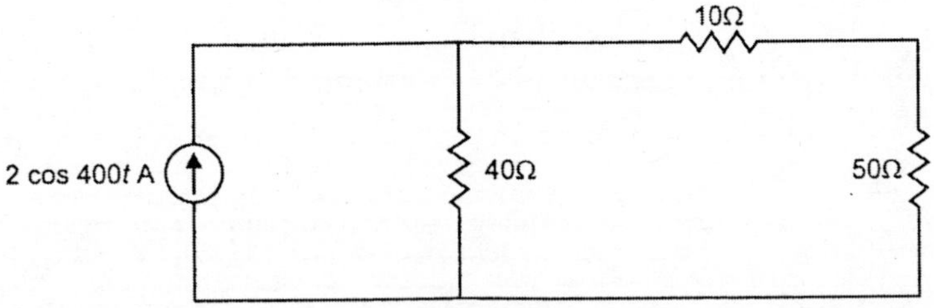

Fig. E2.6(b)

Converting the 2 cos 400*t* A current source into its equivalent voltage source we get the circuit as shown in Figure E2.6(*c*). Here 2 cos 400*t* A current source with 40 Ω parallel resistance has been replaced by 80 cos 400*t* V source with 40 Ω series resistance.

Where $$V = I \times R = 2 \cos 400t \times 40 = 80 \cos 400t \text{ V}$$

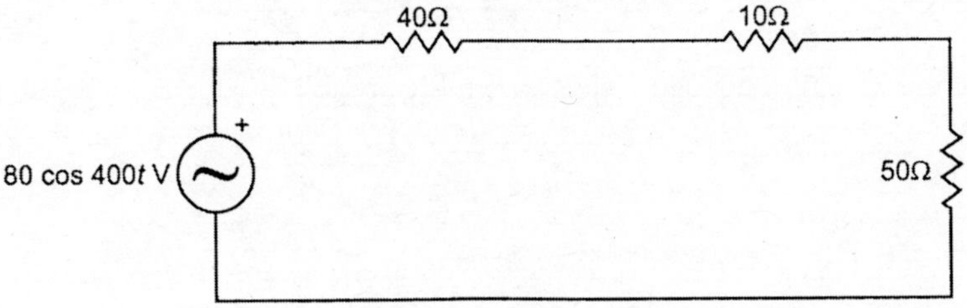

Fig. E2.6(c)

Adding the three series resistances and replacing it by the equivalent resistance we the circuit as shown in Figure E2.6(*d*), which is a single source circuit.

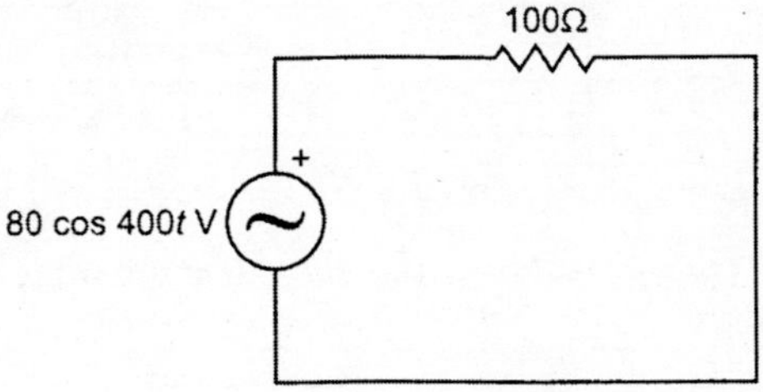

Fig. E2.6(d)

EXAMPLE 2.7. *Convert the circuit shown in Figure E2.7, into a single source network.*

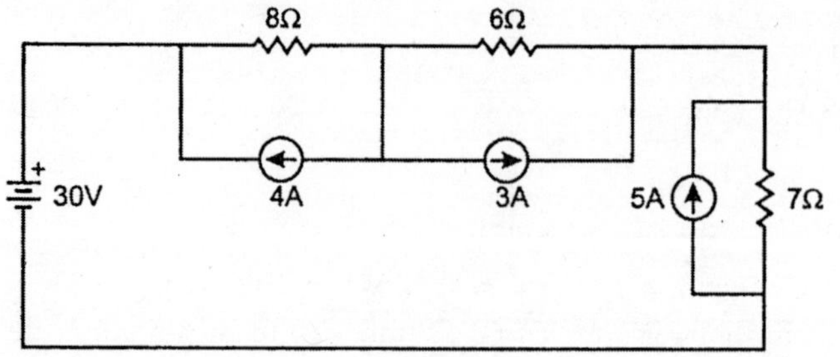

Fig. E2.7

SOLUTION. Converting the three current sources into voltage source we get the circuit as shown in Figure E2.7(*a*). Here 4 A source with parallel 8 Ω resistance has been replaced 32 V source with 8 Ω resistance in series. Similarly 3 A source with 6 Ω resistance has been replaced by 18 V source with 6 Ω resistance in series and 5 A source with 7 Ω resistance has been replaced by 35 V source with 7 Ω resistance in series.

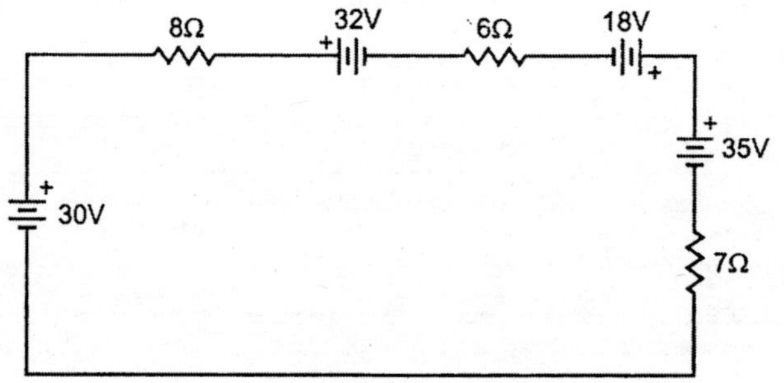

Fig. E2.7(a)

The single source circuit has been shown in Figure E2.7(*b*).

Where $V = -30 + 32 - 18 + 35 = 19$ V

and $R = 8 + 6 + 7 = 21\ \Omega$

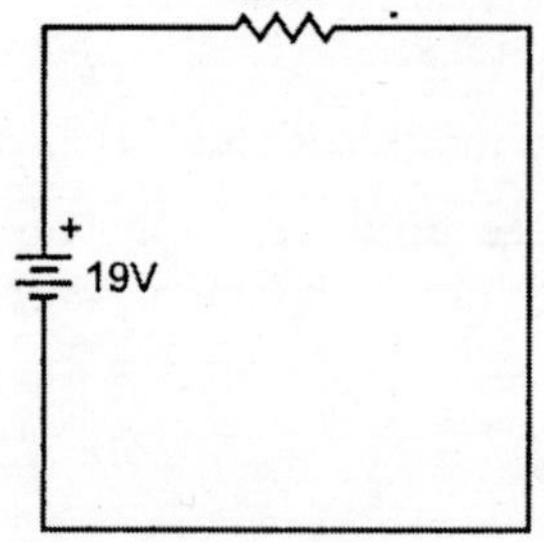

Fig. E2.7(b)

EXAMPLE 2.8. *Using source transformations find the current through R_1, for the circuit shown in Figure E2.8.*

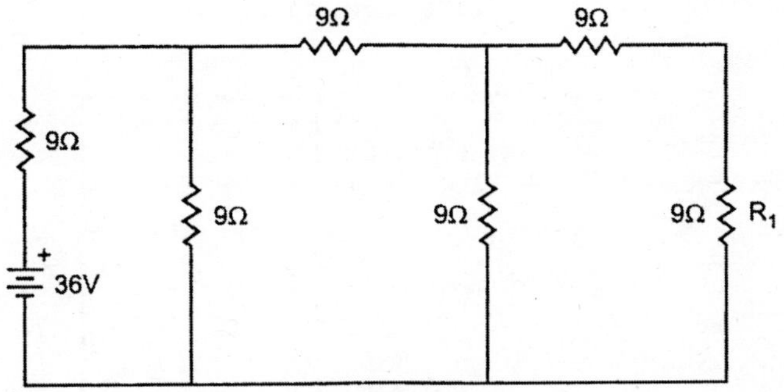

Fig. E2.8

SOLUTION. Converting the voltage source into current source we get Figure E2.8(*a*). Here 36 V source with 9 Ω series resistance has been replaced by a 4 A current source with 9 Ω resistance in parallel.

Where $$I = \frac{36}{9} = 4 \text{ A}$$

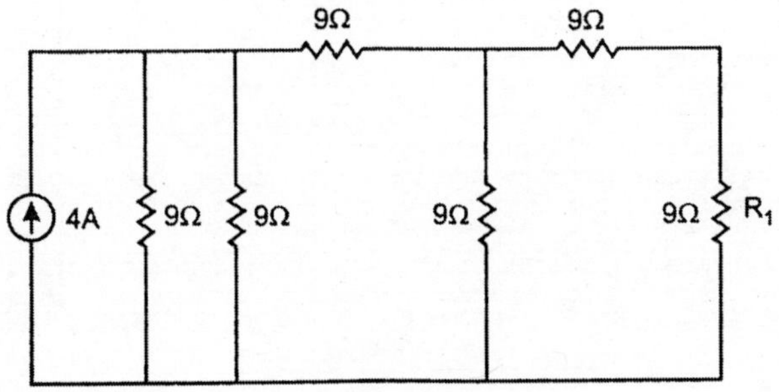

Fig. E2.8(a)

Replacing the two 9 Ω parallel resistances by its equivalent resistance we get a circuit as shown in Figure E2.8(*b*). Here

$$R_{eq} = \frac{9 \times 9}{9 + 9} = 4.5 \ \Omega$$

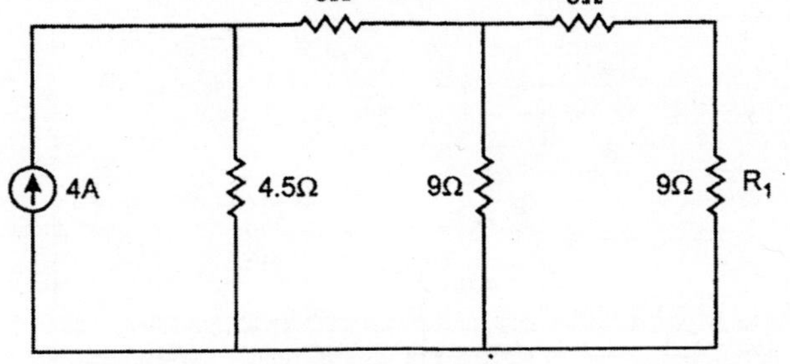

Fig. E2.8(b)

Converting the 4 A current source into its equivalent voltage source we get the circuit as shown in Figure E2.8(*c*). Here 4 A current source with 4.5 Ω parallel resistance has been replaced by 18 V source with 4.5 Ω series resistance.

Where $$V = I \times R = 4 \times 4.5 = 18 \text{ V}$$

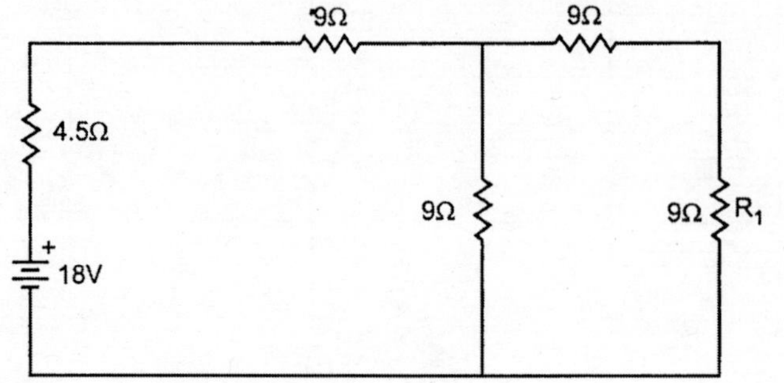

Fig. E2.8(c)

Adding the two series resistances and replacing it by the equivalent resistance we the circuit as shown in Figure E2.8(*d*).

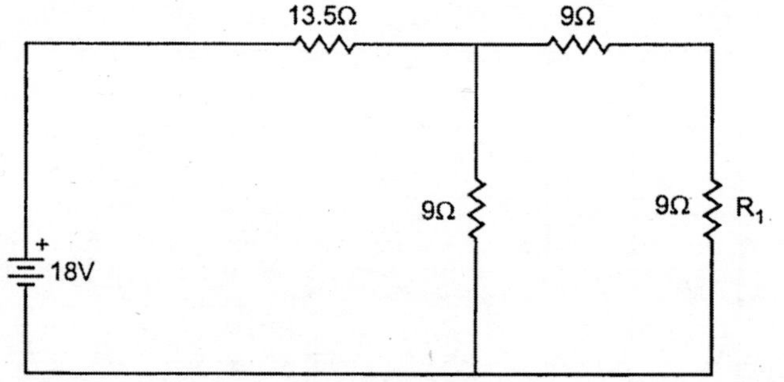

Fig. E2.8(d)

Converting the voltage source into current source we get Figure E2.8(*e*). Here 18 V source with 13.5 Ω series resistance has been replaced by a 1.33 A current source with 13.5 Ω resistance in parallel.

Where $$I = \frac{18}{13.5} = 1.33 \text{ A}$$

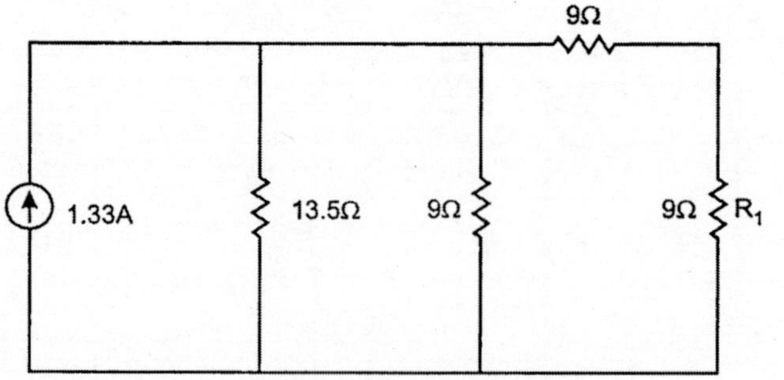

Fig. E2.8(e)

Replacing the two parallel resistances 13.5 Ω and 9 Ω by its equivalent resistance we get a circuit as shown in Figure E2.8(*f*). Here

$$R_{eq} = \frac{13.5 \times 9}{13.5 + 9} = 5.4 \ \Omega$$

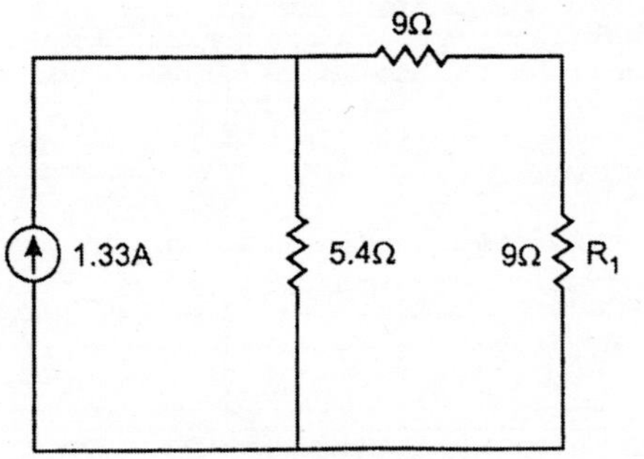

Fig. E2.8(f)

Converting the 1.33 A current source into its equivalent voltage source we get the circuit as shown in Figure E2.8(*g*). Here 1.33 A current source with 5.4 Ω parallel resistance has been replaced by 7.18 V source with 5.4 Ω series resistance.

Where $$V = I \times R = 1.33 \times 5.4 = 7.18 \text{ V}$$

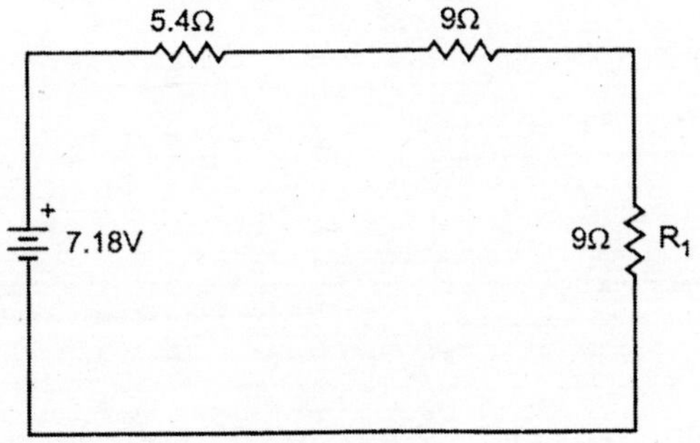

Fig. E2.8(g)

Then the current through R_1 is

$$I = \frac{7.18}{5.4 + 9 + 9} = 0.31 \text{ A}$$

EXAMPLE 2.9. *Convert the circuit shown in Figure E2.9 into a single source circuit.*

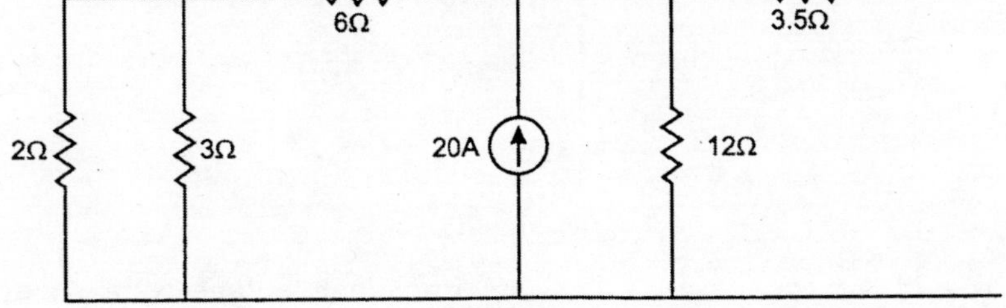

Fig. E2.9

SOLUTION. Replacing the two parallel resistances 2 Ω and 3 Ω by its equivalent resistance we get a circuit as shown in Figure E2.9(*a*). Here

$$R_{eq} = \frac{2 \times 3}{2 + 3} = 1.2 \text{ } \Omega$$

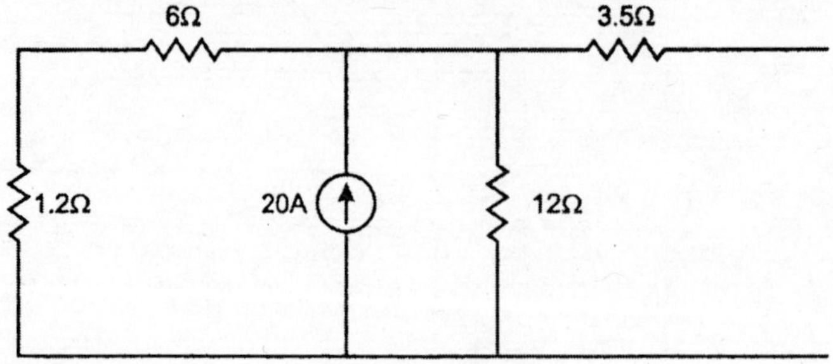

Fig. E2.9(a)

Replacing the two series resistances 1.2 Ω and 6 Ω by its equivalent resistance we get a circuit as shown in Figure E2.9(*b*). Here

$$R_{eq} = 1.2 + 6 = 7.2\ \Omega$$

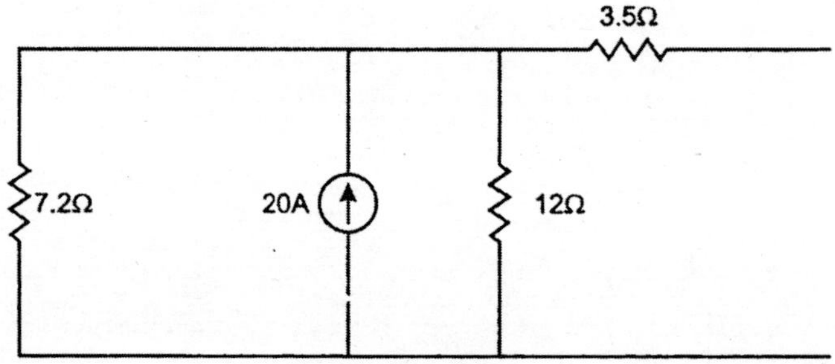

Fig. E2.9(b)

Replacing the two parallel resistances 7.2 Ω and 12 Ω by its equivalent resistance we get a circuit as shown in Figure E2.9(*c*). Here

$$R_{eq} = \frac{7.2 \times 12}{7.2 \times 12} = 4.5\ \Omega$$

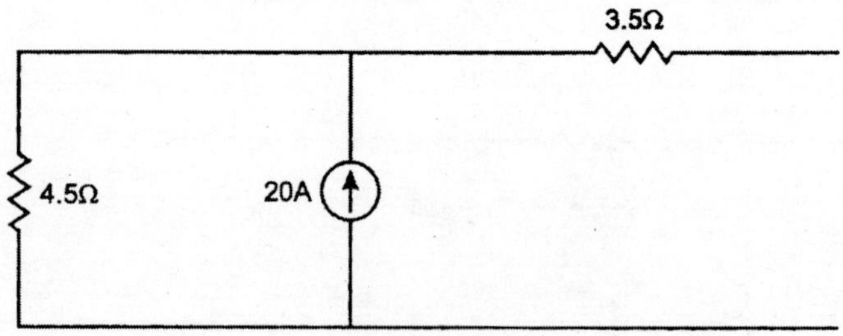

Fig. E2.9(c)

Converting the 20 A current source into its equivalent voltage source we get the circuit as shown in Figure E2.9(*d*). Here 20 A current source with 4.5 Ω parallel resistance has been replaced by 90 V source with 4.5 Ω series resistance.

Where $V = I \times R = 20 \times 4.5 = 90\ \text{V}$

Fig. E2.9(d)

Adding the two series resistances 4.5 Ω and 3.5 Ω, we get a circuit as shown in Figure E2.9(*e*). Here,

$$R_{eq} = 4.5 + 3.5 = 8\ \Omega$$

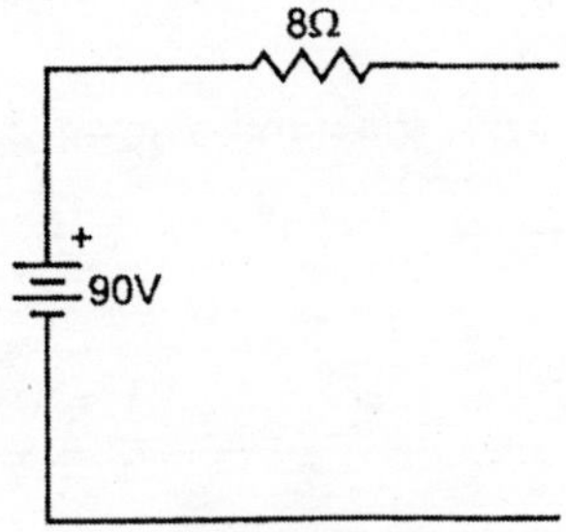

Fig. E2.9(e)

EXAMPLE 2.10. *Find the power across R_1 for the circuit shown in Figure E2.10.*

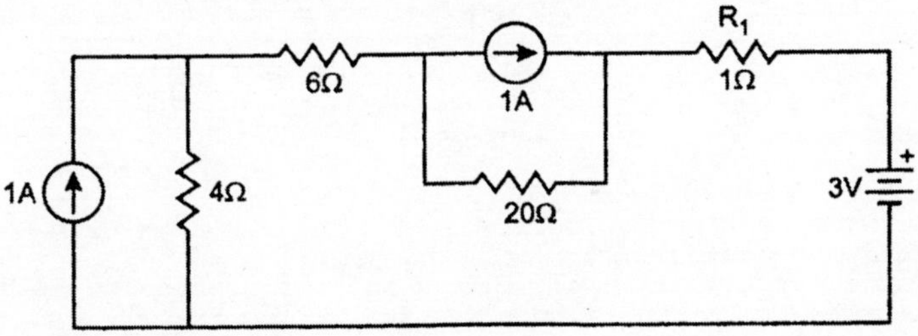

Fig. E2.10

SOLUTION. Converting the two current sources into its equivalent voltage source we get the circuit as shown in Figure E2.10(*a*). Here 1 A current source with 4 Ω parallel resistance has been replaced by 4 V source with 4 Ω series resistance, and 1 A current source with 20 Ω parallel resistance has been replaced by 20 V source with 20 Ω series resistance.

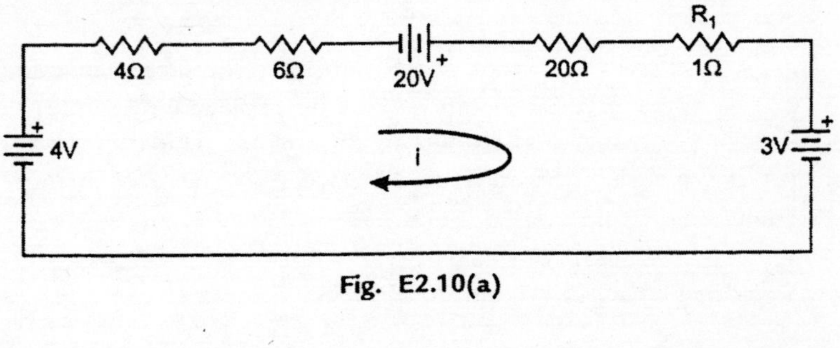

Fig. E2.10(a)

From Figure E2.10(*a*) current i is given by,

$$i = \frac{20+4-3}{4+6+20+1} = 0.677 \text{ A}$$

Then the power across R_1 is

$$P = i^2 R_1 = 0.677^2 \times 1 = 0.458 \text{ W}$$

2.3 STAR (Y)-DELTA (Δ) AND DELTA (Δ)-STAR (Y)

In a circuit analysis, the topology of some network is too complicated to permit a straightforward reduction by series-parallel combinations. By means of star-delta or delta-star transformation, most of such network can be transformed into an equivalent structure that will permit series parallel reduction.

A star network is shown in Figure 2.2(*a*) and Figure 2.2(*b*). A delta network is shown in Figure 2.3(*a*) and Figure 2.3(*b*).

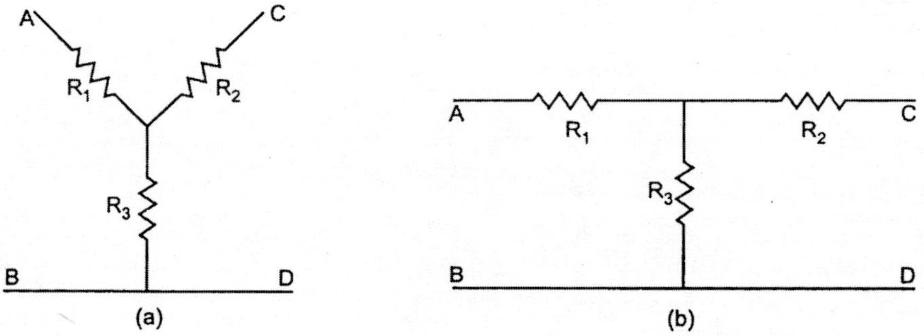

Fig. 2.2

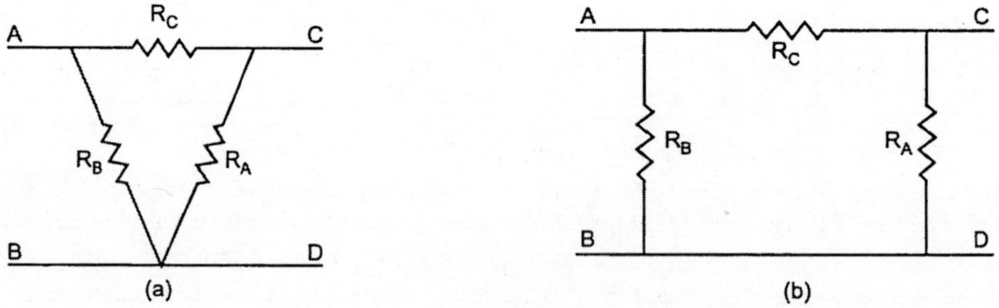

Fig. 2.3

Suppose we are given three resistances R_A, R_B and R_C connected in delta (Δ) fashion as shown in Figure 2.3, the three given resistances can be replaced by three resistances R_1, R_2 and R_3 connected in star (Y) fashion as shown in Figure 2.2. The two arrangements will be equivalent if and only if the resistances measured between any pair of terminals are the same in both the arrangements.

2.3.1 Delta (Δ) to Star (Y) Transformation

From Figure 2.2,

The equivalent resistance as seen from the terminal *A-B* is

$$R_{AB}(Y) = R_1 + R_3 \qquad ...(2.8)$$

The equivalent resistance as seen from the terminal *A-C* is

$$R_{AC}(Y) = R_1 + R_2 \qquad ...(2.9)$$

The equivalent resistance as seen from the terminal *C-D* is

$$R_{CD}(Y) = R_2 + R_3 \qquad ...(2.10)$$

From Figure 2.3

The equivalent resistance as seen from the terminal *A-B* is

$$R_{AB}(\Delta) = \frac{R_B \times (R_A + R_C)}{R_A + R_B + R_C} \qquad ...(2.11)$$

The equivalent resistance as seen from the terminal *A-C* is

$$R_{AC}(\Delta) = \frac{R_C \times (R_A + R_B)}{R_A + R_B + R_C} \qquad ...(2.12)$$

The equivalent resistance as seen from the terminal *C-D* is

$$R_{CD}(\Delta) = \frac{R_A \times (R_B + R_C)}{R_A + R_B + R_C} \qquad ...(2.13)$$

Thus for the two circuit to be identical

$$R_{AB}(Y) = R_{AB}(\Delta)$$

$$R_{AC}(Y) = R_{AC}(\Delta)$$

$$R_{CD}(Y) = R_{CD}(\Delta)$$

Substituting equation 2.8 through equation 2.13 in the above three relations we get

$$R_1 + R_3 = \frac{R_B \times (R_A + R_C)}{R_A + R_B + R_C} \qquad ...(2.14)$$

$$R_1 + R_2 = \frac{R_C \times (R_A + R_B)}{R_A + R_B + R_C} \qquad ...(2.15)$$

$$R_2 + R_3 = \frac{R_A \times (R_B + R_C)}{R_A + R_B + R_C} \qquad ...(2.16)$$

Solving equations 2.14, 2.15 and 2.16 we get

$$R_1 = \frac{R_B R_C}{R_A + R_B + R_C}\ ;\ R_2 = \frac{R_A R_C}{R_A + R_B + R_C}\ ;\ R_3 = \frac{R_A R_B}{R_A + R_B + R_C} \qquad ...(2.17)$$

Thus each resistor in the star network in the product of the resistors in the two adjacent delta branches divided by the sum of the three delta resistors, as shown in Figure 2.4.

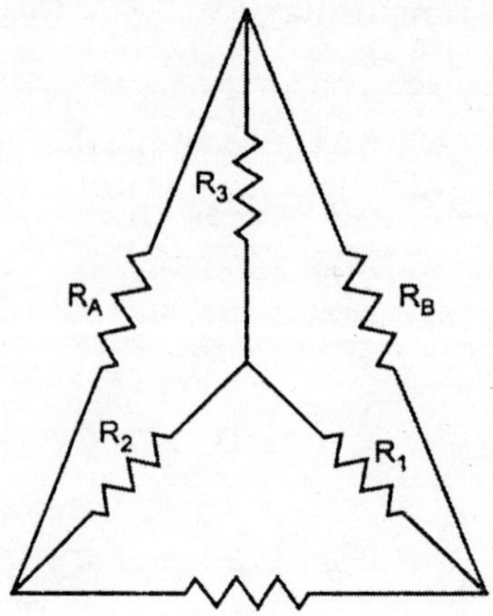

Fig. 2.4

2.3.2 Star (Y) to Delta (Δ) Transformation

Multiplying equation (2.14) and (2.15), equation (2.15) and (2.16) and equation (2.16) and (2.14) and then adding them together and then simplifying we get the following equation.

$$R_A = \frac{R_1R_2 + R_2R_3 + R_3R_1}{R_1} \quad ...(2.18)$$

$$R_B = \frac{R_1R_2 + R_2R_3 + R_3R_1}{R_2} \quad ...(2.19)$$

$$R_C = \frac{R_1R_2 + R_2R_3 + R_3R_1}{R_3} \quad ...(2.20)$$

SOLVED EXAMPLES

EXAMPLE 2.11. *For the circuit shown in Figure E2.11, find the equivalent resistance across X-Y.*

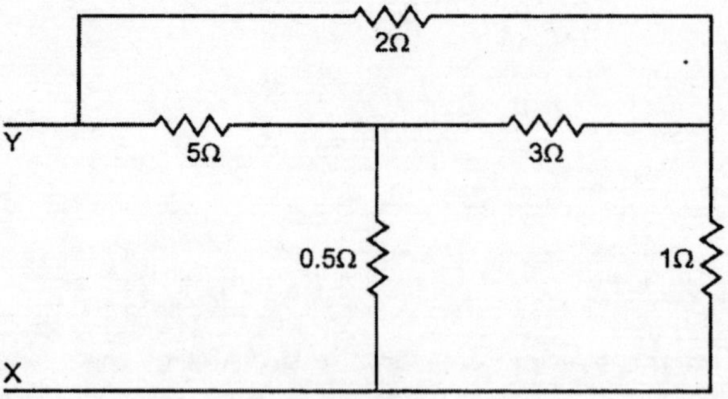

Fig. E2.11

SOLUTION. Redrawing the circuit we get Figure E2.11(*a*). Converting the delta into star we get the circuit shown in Figure E2.11(*b*), where,

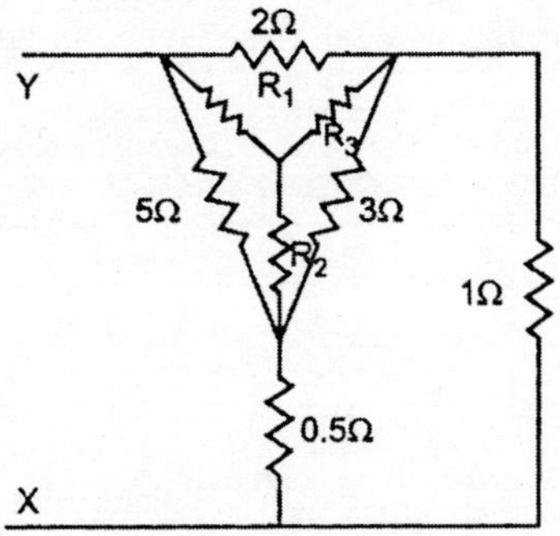

Fig. E2.11(a)

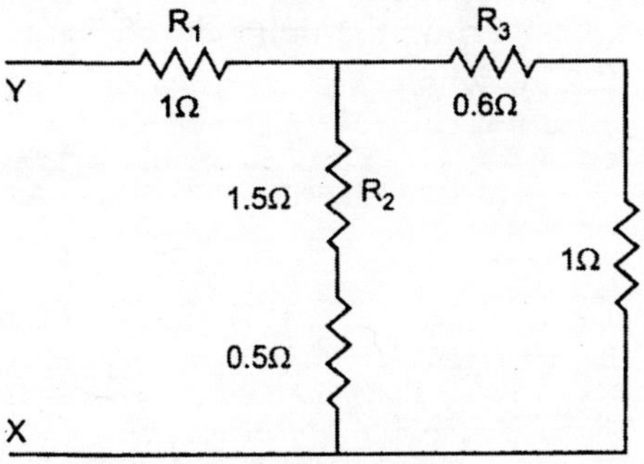

Fig. E2.11(b)

$$R_1 = \frac{5\times 2}{5+3+2} = 1\ \Omega$$

$$R_2 = \frac{5\times 3}{5+3+2} = 1.5\ \Omega$$

$$R_3 = \frac{3\times 2}{5+3+2} = 0.6\ \Omega$$

Adding the resistances in series we get the circuit shown in Figure E2.11(*c*).

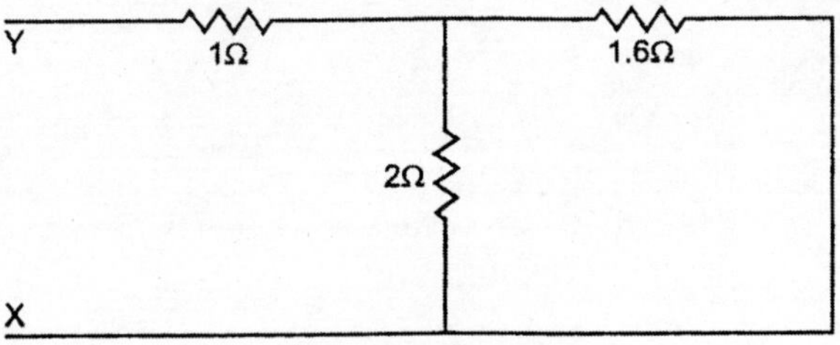

Fig. E2.11(c)

So the equivalent resistance as seen from the terminal X-Y is,

$$R_{eq} = 1+\frac{1.6\times 2}{1.6+2} = 1.87\ \Omega$$

EXAMPLE 2.12. *For the circuit shown in Figure E2.12, find the equivalent resistance across X-Y.*

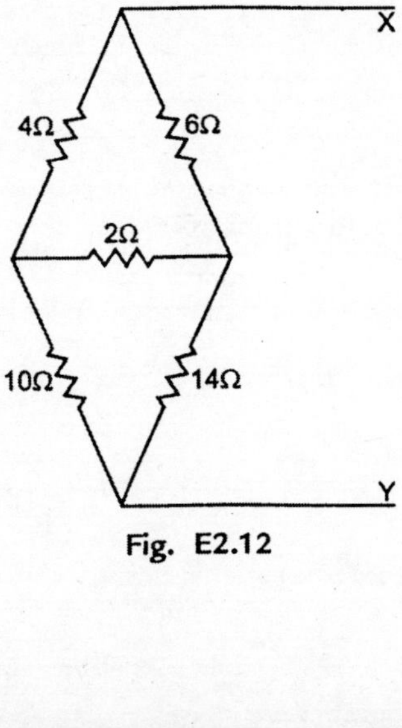

Fig. E2.12

SOLUTION. Converting the delta *abc* of Figure E2.12(*a*), into star we get the circuit as shown in Figure E2.12(*b*), where,

$$R_1 = \frac{4\times6}{4+6+2} = 2\ \Omega$$

$$R_2 = \frac{4\times2}{4+6+2} = 0.667\ \Omega$$

$$R_3 = \frac{6\times2}{4+6+2} = 1\ \Omega$$

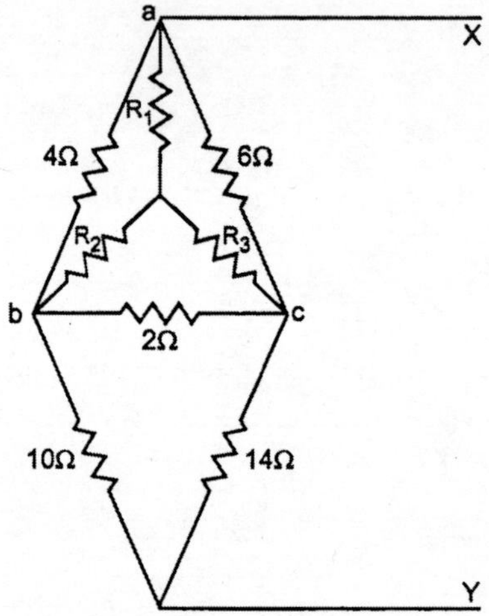

Fig. E2.12(a)

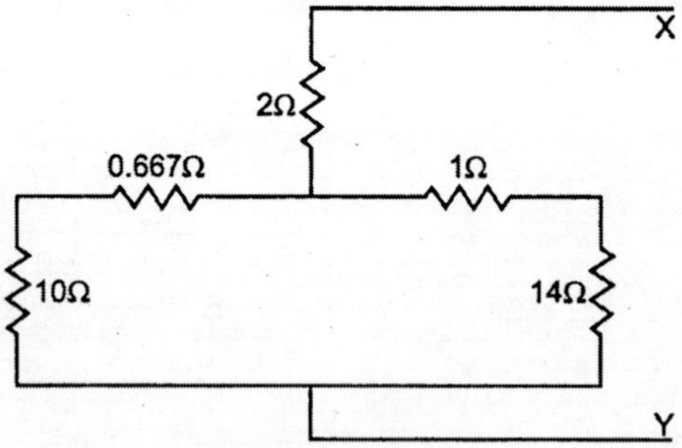

Fig. E2.12(b)

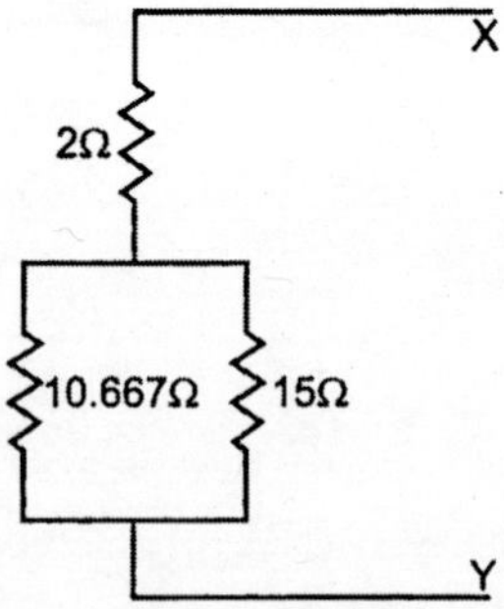

Fig. E2.12(c)

The circuit of Figure E2.12(*b*) can be simplified as shown in Figure E2.12(*c*). Then the equivalent resistance as seen from the terminal *X-Y* is

$$R_{XY} = 2+\frac{10.667\times15}{15+10.667} = 8.233\ \Omega$$

EXAMPLE 2.13. *For the circuit shown in Figure E2.13, find R_{ab}, R_{bc} and R_{ca} using delta-star transformation.*

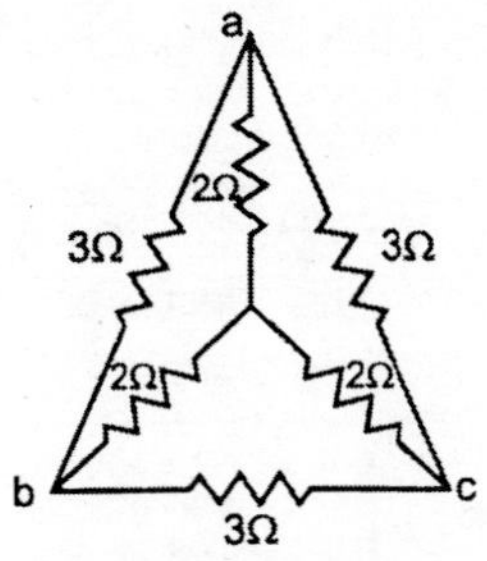

Fig. E2.13

SOLUTION. Converting the delta of Figure E2.13 into star we get the circuit as shown in Figure E2.13(*a*), where,

$$R_1 = R_2 = R_3 = \frac{3\times 3}{3+3+3} = 1\ \Omega$$

Simplifying the circuit of Figure E2.13(*a*), we get the circuit shown in Figure E2.13(*b*).

$$R_{ab} = R_{bc} = R_{ca} = 0.667 + 0.667 = 1.334\ \Omega$$

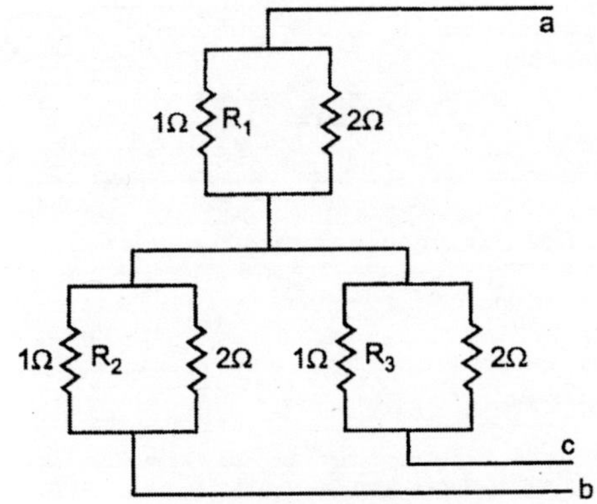

Fig. E2.13(a)

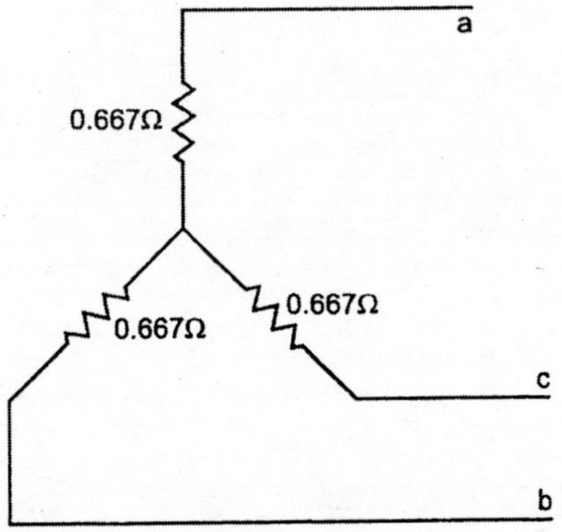

Fig. E2.13(b)

EXAMPLE 2.14. *Find the current delivered by the source for the circuit shown in Figure E2.14.*

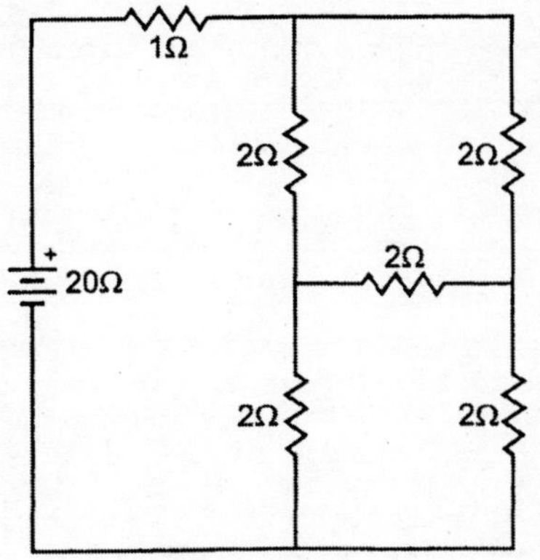

Fig. E2.14

SOLUTION. Redrawing the circuit of Figure E2.14, we get the circuit of Figure E2.14(*a*). Here the upper delta *abc* is converted into a star.

Where $$R_1 = R_2 = R_3 = \frac{2\times 2}{2+2+2} = 0.667\ \Omega$$

Then the circuit after converting into star becomes as shown in Figure E2.14(*b*).

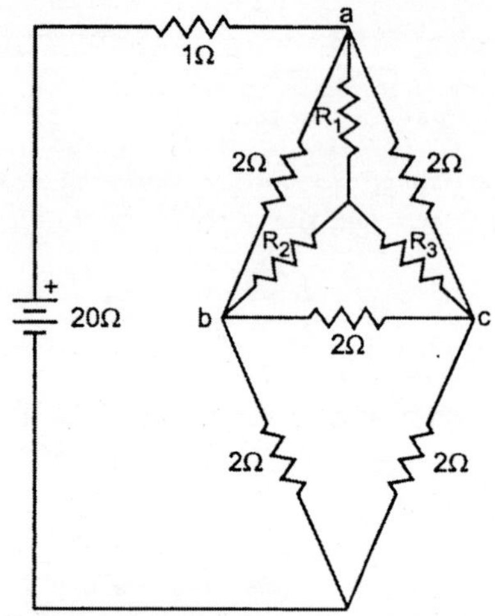

Fig. E2.14(a)

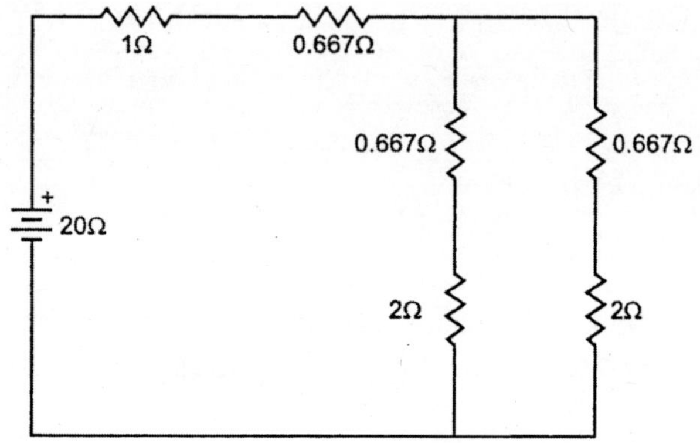

Fig. E2.14(b)

From Figure E2.14(*b*),

$$R_{eq} = (1+0.667)+\frac{(2+0.667)\times(2+0.667)}{2+0.667+2+0.667} = 3\ \Omega$$

Then the source current is

$$I = \frac{V}{R_{eq}} = \frac{20}{3} = 6.667\ \text{A}$$

EXAMPLE 2.15. *Find R_{XY} for the circuit shown in Figure E2.15.*

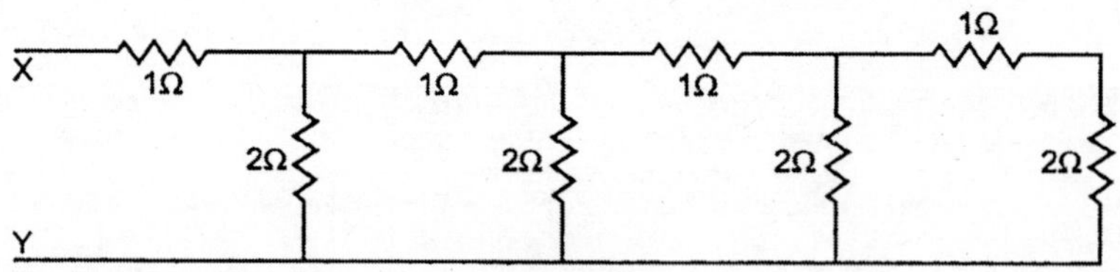

Fig. E2.15

SOLUTION. Converting the delta *abc* of Figure E2.15(*a*), into star we get the circuit shown in Figure E2.15(*b*).

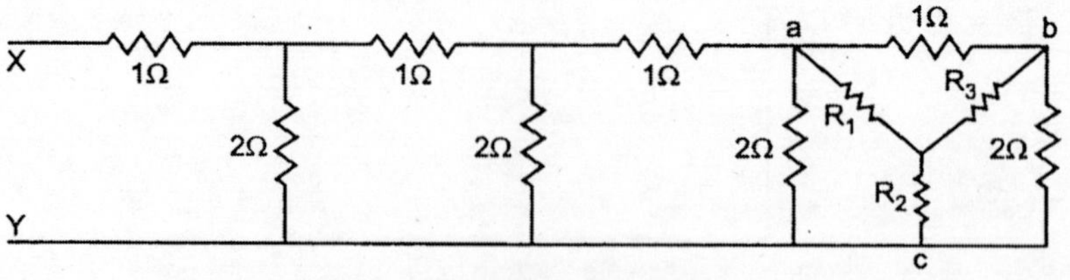

Fig. E2.15(a)

Where $R_1 = R_3 = \dfrac{2\times 1}{2+2+1} = 0.4\ \Omega$

$$R_2 = \frac{2\times 2}{2+2+1} = 0.8\ \Omega$$

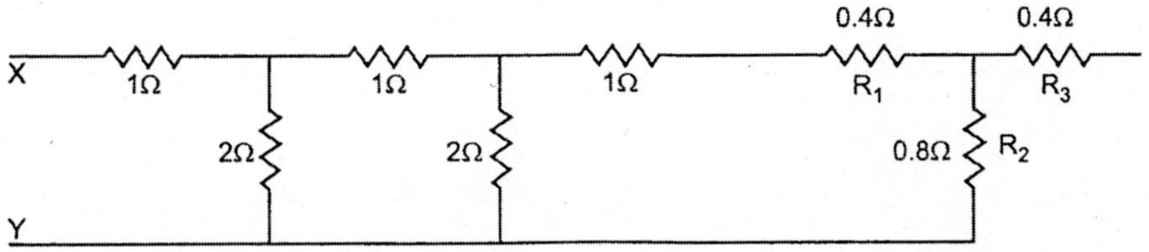

Fig. E2.15(b)

Redrawing the circuit of Figure E2.15(*b*), we get the circuit shown in Figure E2.15(*c*).

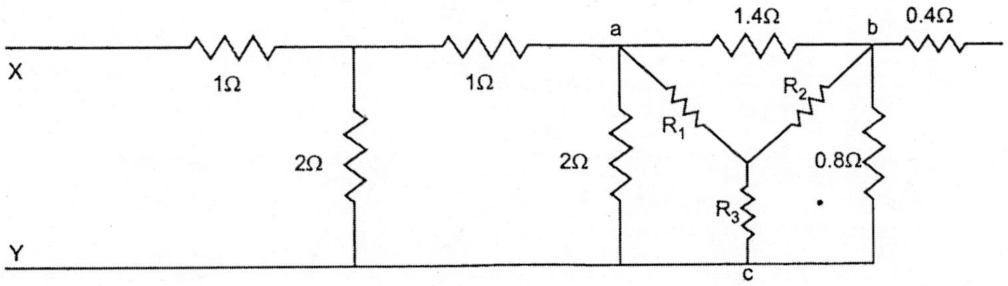

Fig. E2.15(c)

Converting the delta *abc* of Figure E2.15(*c*), into star we get the circuit shown in Figure E2.15(*d*).

Where $R_1 = \dfrac{2\times 1.4}{2+1.4+0.8} = 0.667\ \Omega$ $\quad R_2 = \dfrac{0.8\times 1.4}{2+1.4+0.8} = 0.267\ \Omega$

$$R_3 = \frac{2\times 0.8}{2+1.4+0.8} = 0.381\ \Omega$$

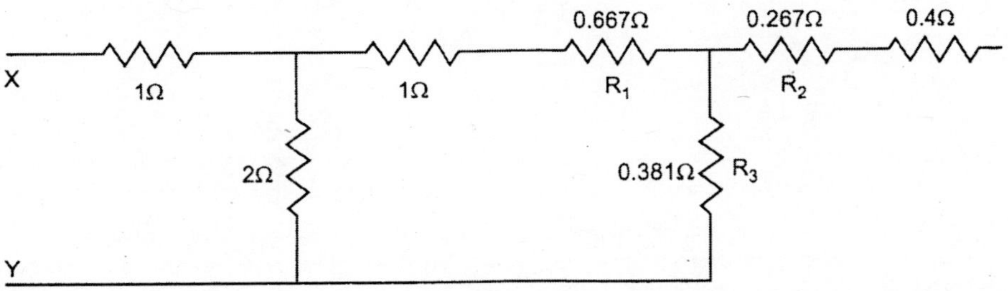

Fig. E2.15(d)

Redrawing the circuit of Figure E2.15(*d*), we get the circuit shown in Figure E2.15(*e*).

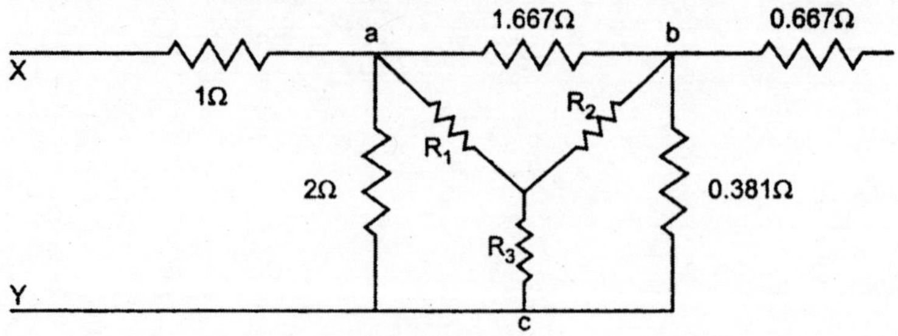

Fig. E2.15(e)

Converting the delta *abc* of Figure E2.15(*e*), into star we get the circuit shown in Figure E2.15(*f*), where,

$$R_1 = \frac{2 \times 1.667}{2 + 1.667 + 0.381} = 0.824\ \Omega$$

$$R_2 = \frac{0.381 \times 1.667}{2 + 1.667 + 0.381} = 0.157\ \Omega$$

$$R_3 = \frac{2 \times 0.381}{2 + 1.667 + 0.381} = 0.188\ \Omega$$

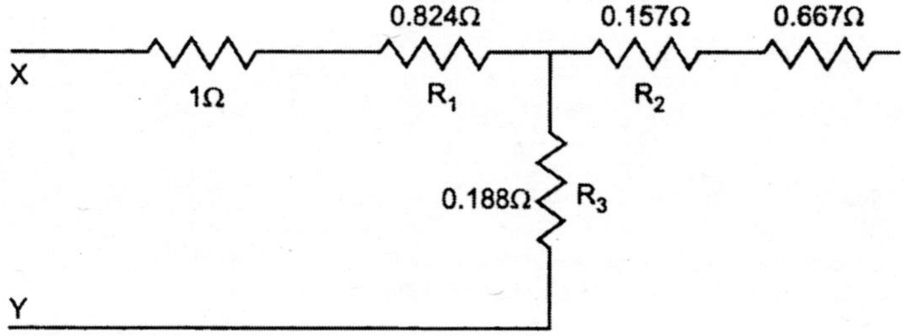

Fig. E2.15(f)

Then from Figure E2.15(*f*),

$$R_{XY} = 1 + 0.824 + 0.188 = 2.012\ \Omega$$

EXAMPLE 2.16. *Find R_{XY} for the circuit shown in Figure E2.16.*

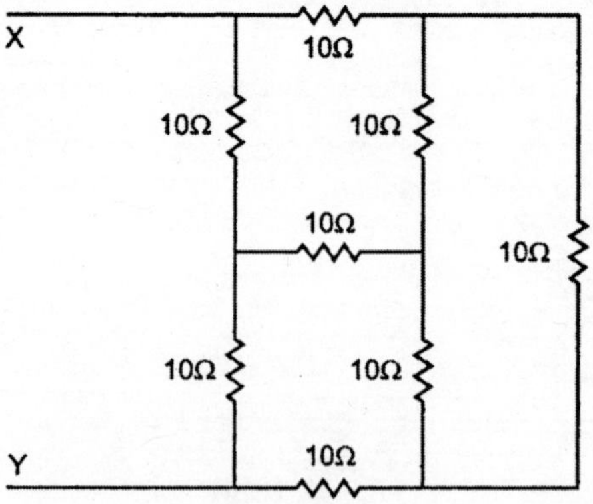

Fig. E2.16

SOLUTION. Redrawing the circuit of Figure E2.16, we get the circuit shown on Figure E2.16(*a*).

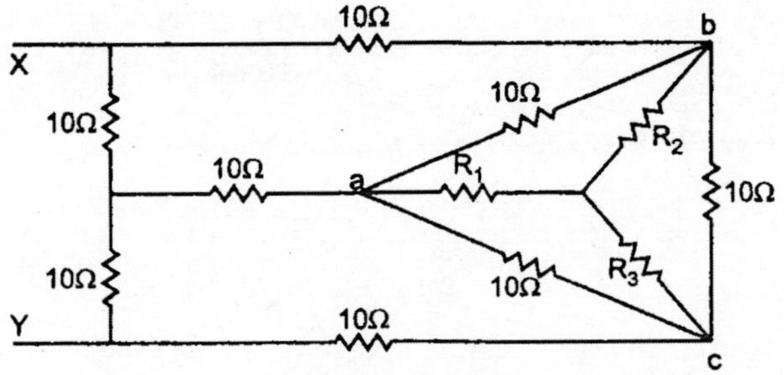

Fig. E2.16(a)

Now converting the delta *abc* into star for the circuit shown in Figure E2.16(*a*), we get the circuit shown in Figure E2.16(*b*).

Where $$R_1 = R_2 = R_3 = \frac{10 \times 10}{10+10+10} = 3.33\ \Omega$$

Redrawing the circuit of Figure E2.16(*b*), we get the circuit shown on Figure E2.16(*c*).

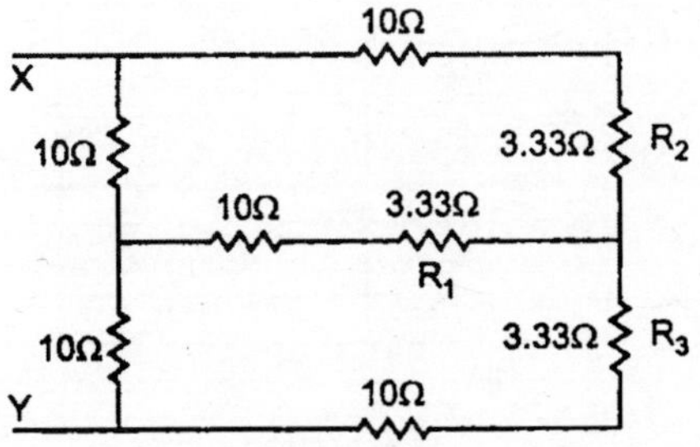

Fig. E2.16(b)

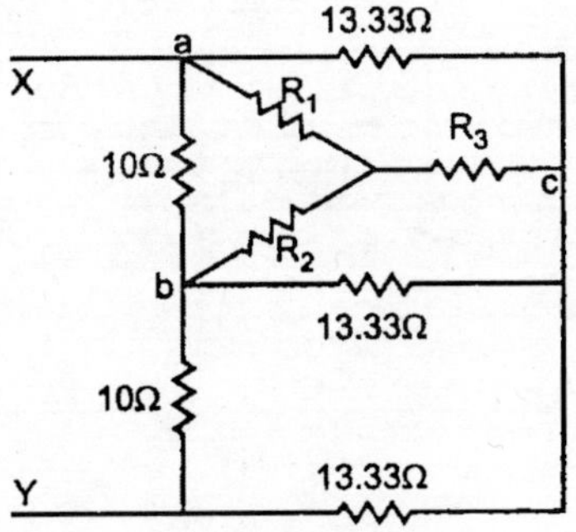

Fig. E2.16(c)

Now converting the delta *abc* into star for the circuit shown in Figure E2.16(*c*), we get the circuit shown in Figure E2.16(*d*).

Where $$R_1 = R_2 = \frac{10 \times 13.33}{10+13.33+13.33} = 3.64\ \Omega$$

$$R_3 = \frac{13.33 \times 13.33}{10+13.33+13.33} = 4.85\ \Omega$$

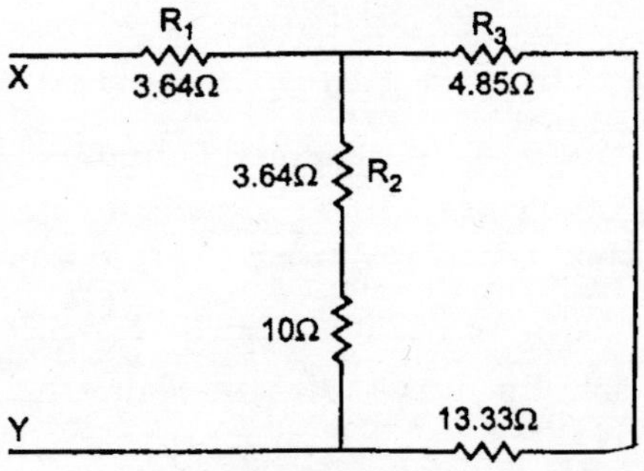

Fig. E2,16(d)

From Figure E2.16(*d*),

$$R_{eq} = 3.64 + \frac{(3.64+10)\times(4.85+13.33)}{3.64+10+4.85+13.33} = 11.433\ \Omega$$

EXAMPLE 2.17. *Find R_{XY} for the circuit shown in Figure E2.17.*

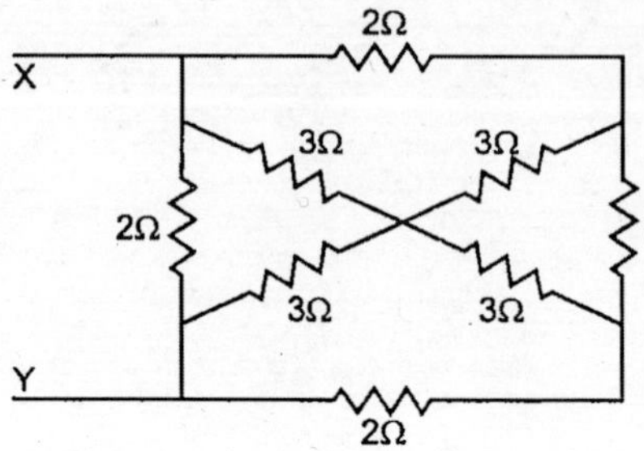

Fig. E2.17

SOLUTION. The circuit is redrawn again in Figure E2.17(*a*).

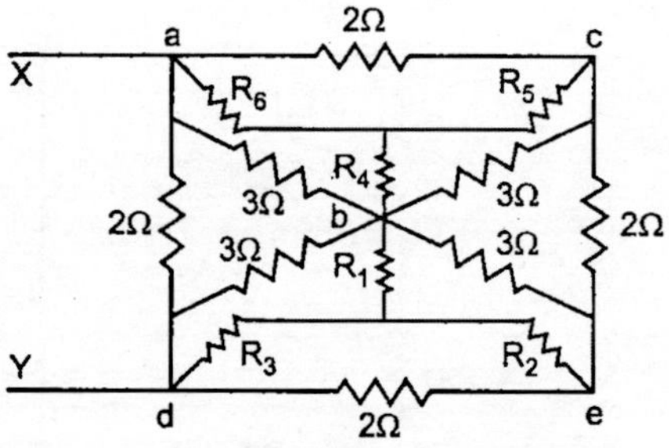

Fig. E2.17(a)

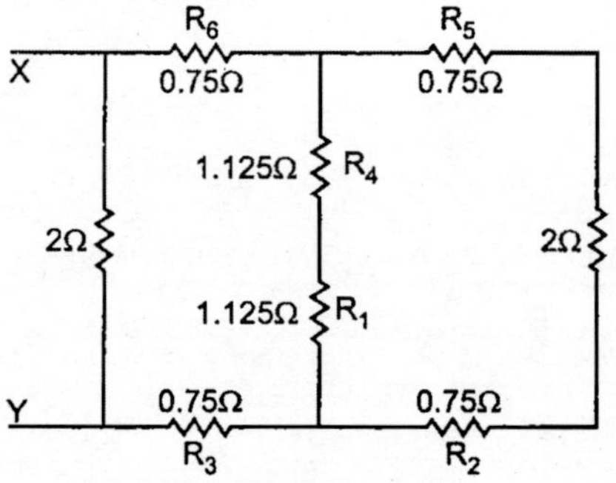

Fig. E2.17(b)

Converting the delta *abc* and delta *dbe* of Figure E2.17(*a*) into star we get the circuit shown in Figure E2.17(*b*).

Where

$$R_1 = \frac{3\times3}{3+3+2} = 1.125\ \Omega$$

$$R_2 = R_3 = \frac{3\times2}{3+3+2} = 0.75\ \Omega$$

$$R_4 = \frac{3\times3}{3+3+2} = 1.125\ \Omega$$

$$R_5 = R_6 = \frac{3\times2}{3+3+2} = 0.75\ \Omega$$

Adding all the series resistances of the circuit of Figure E2.17(*b*), we get a simplified circuit shown in Figure E2.17(*c*), where,

$$R_{eq1} = 0.75 + 0.75 + 2 = 3.5\ \Omega$$

and

$$R_{eq2} = 1.125 + 1.125 = 2.25\ \Omega$$

The circuit can be further simplified as shown in Figure E2.17(*d*), where,

$$R_{eq3} = \frac{2.25\times3.5}{2.25+3.5} = 1.369\ \Omega$$

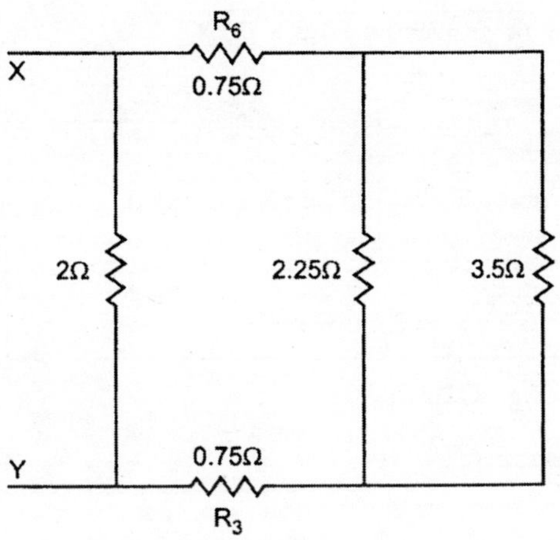

Fig. E2.17(c)

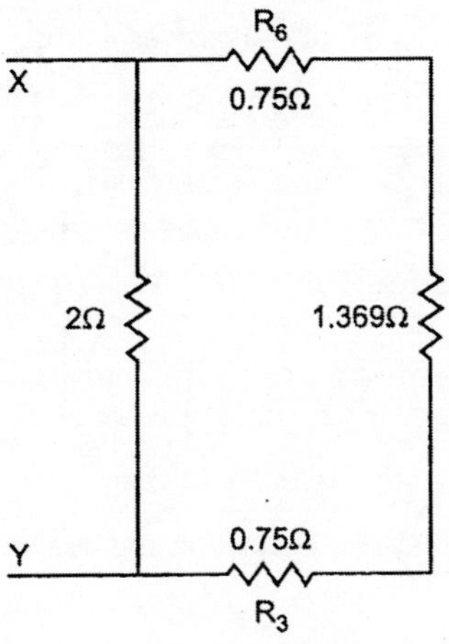

Fig. E2.17(d)

Then from Figure E2.17(*d*),

$$R_{eq} = \frac{(0.75+1.369+0.75)\times 2}{0.75+1.369+0.75+2} = 1.178\ \Omega$$

EXERCISES

1. Using source transformations find the current through 3 Ω resistor for the circuit shown in Figure Q2.1.

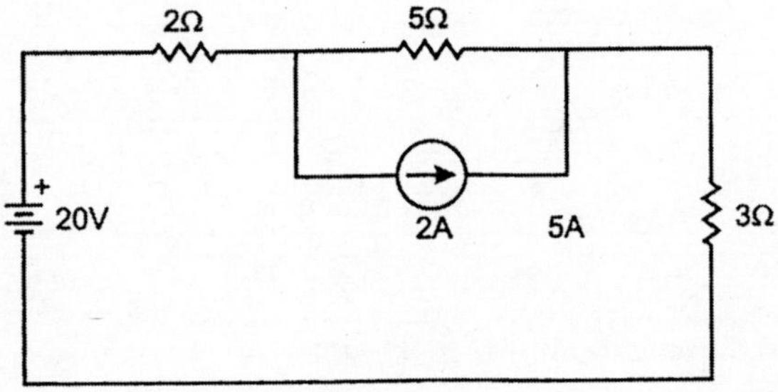

Fig. Q2.1

2. Using source transformations find the current through R_1, for the circuit shown in Figure Q2.2.

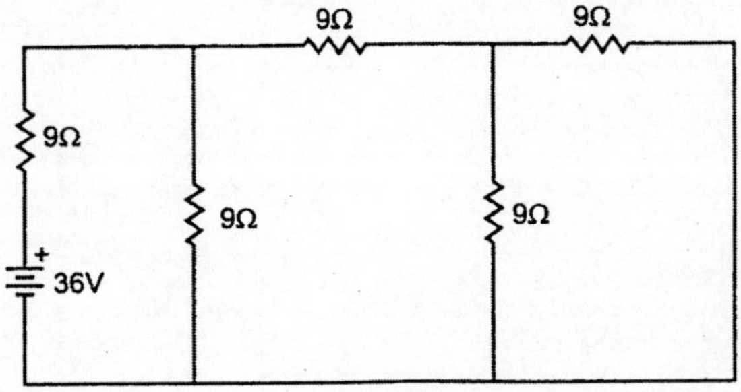

Fig. Q2.2

3. Find the power across R_1 for the circuit shown in Figure Q2.3.

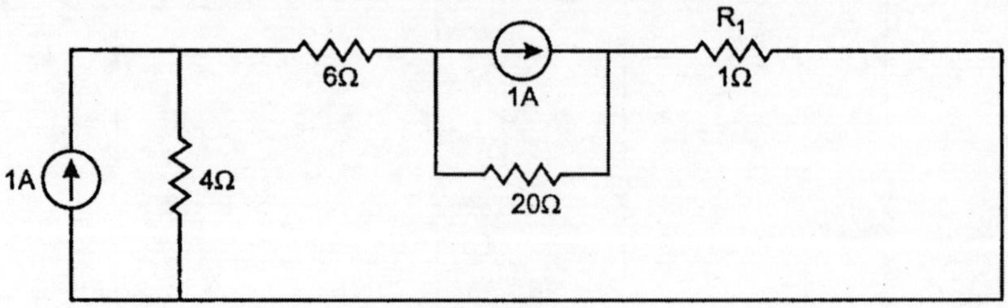

Fig. Q2.3

4. Find R_{XY} for the circuit shown in Figure Q2.4.

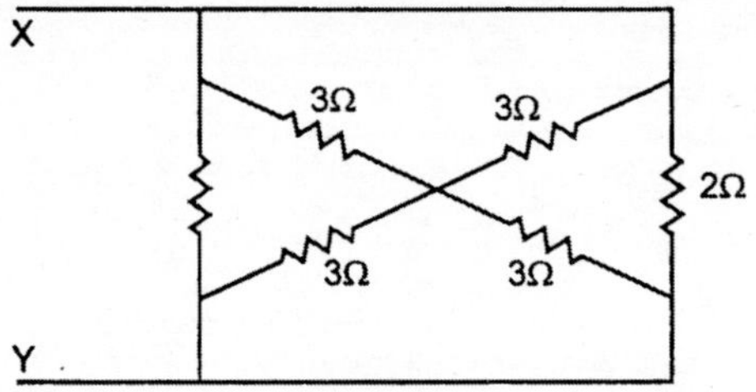

Fig. Q2.4

5. For the circuit shown in Figure Q2.5, find the equivalent resistance across *X-Y*.

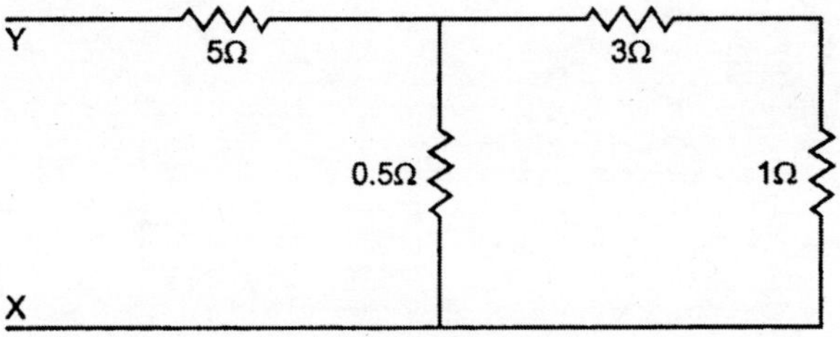

Fig. Q2.5

6. Find R_{XY} for the circuit shown in Figure Q2.6.

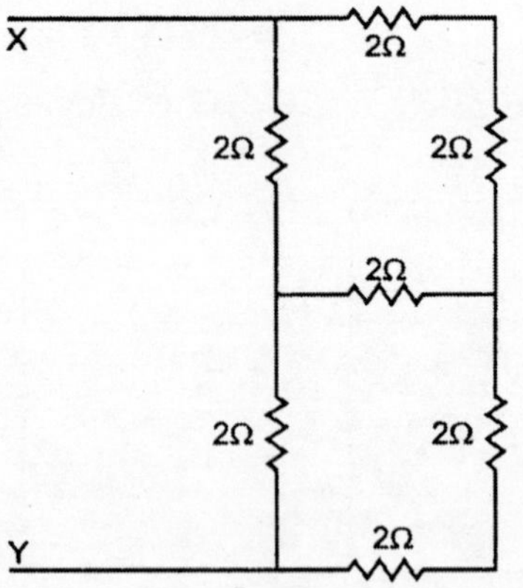

Fig. Q2.6

Chapter 3

MESH AND LOOP ANALYSIS

3.1 INTRODUCTION

The mesh/loop and node analysis are the two simplest methods of solving a particular network/circuits. In these methods the actual circuit is changed during the analysis process. To solve a circuit using mesh analysis we make use of KVL (Kirchhoff's Voltage Law). And to solve a circuit using node analysis we make use of KCL (Kirchhoff's Current Law).

3.2 MESH ANALYSIS

Steps to Solve a Circuit Using Mesh Analysis

Type I : The Network with Only Voltage Source (Figure 3.1)

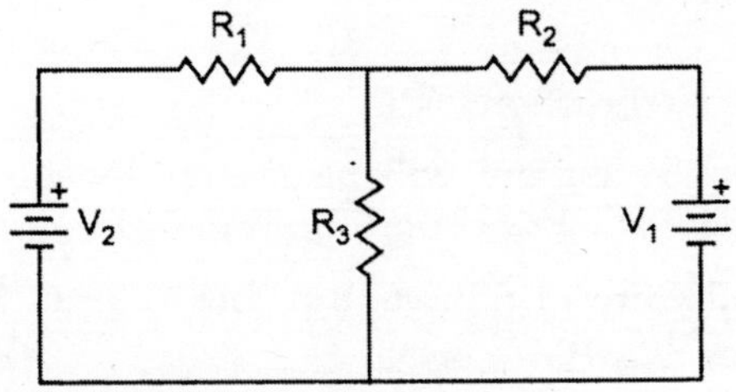

Fig. 3.1

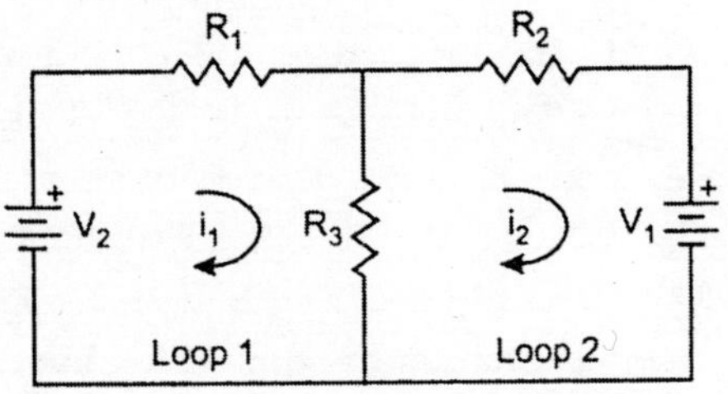

Fig.3.1(a)

Step 1. Identify the number of mesh present in the given circuit. In the circuit shown in Figure 3.1 there are two mesh/loop, as indicated in Figure 3.1(*a*)

Step 2. For each mesh assign a mesh current i_1, i_2, ... i_n, in a particular direction of your choice (anticlockwise or clockwise) as indicated in Figure 3.1(*a*). In Figure 3.1(*a*) two loop currents i_1 and i_2 has been taken for loop 1 and loop 2 respectively in clockwise direction.

Step 3. Write the KVL equation for each mesh/loop. For the circuit shown in Figure 3.1(*a*), the KVL equation will be,

Loop 1 : $(R_1 + R_3)i_1 - R_3 i_2 = V_2$

Loop 2 : $(R_2 + R_3)i_2 - R_3 i_1 = -V_1$

Step 4. Step 3 shows that if there are n mesh then there will be n simultaneous KVL equations. Solve the simultaneous equations to get the mesh/loop currents.

Type II : The Network with Voltage Source and Current Source, But The Current Source Exists Only in One Mesh (Figure 3.2)

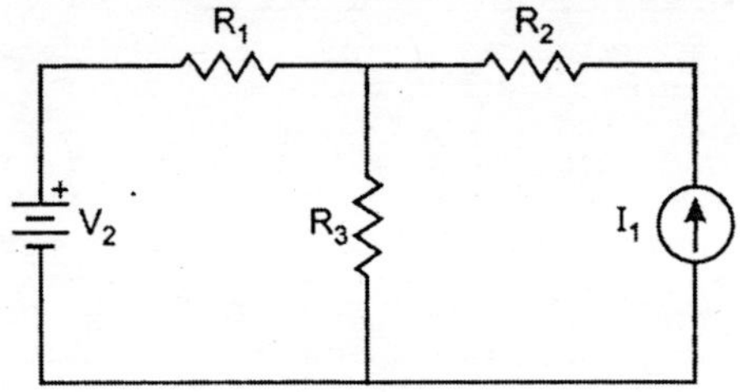

Fig. 3.2

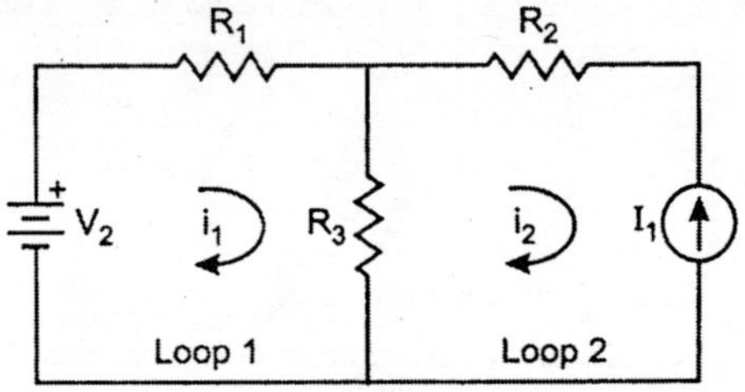

Fig. 3.2(a)

Step 1. Identify the number of mesh present in the given circuit. In the circuit shown in Figure 3.2, there are two mesh/loop, as indicated in Figure 3.2(*a*)

Step 2. For each mesh assign a mesh current $i_1, i_2, \ldots i_n$, in a particular direction of your choice (anticlockwise or clockwise) as indicated in Figure 3.2(*a*). In Figure 3.2(*a*) two loop currents i_1 and i_2 have been taken for loop 1 and loop 2 respectively in clockwise direction.

Step 3. Write the KVL equation for each mesh/loop containing the voltage source. That is for Figure 3.2(*a*) write the KVL equation for loop 1.

Loop 1 : $(R_1 + R_3)i_1 - R_3 i_2 = V_2$

For the loop containing the current source assign the value of the current source to the loop current. That is for Figure 3.2(*a*) assign $i_2 = -I_1$. The negative sign is due to because i_2 is flowing in opposite direction of I_1.

Step 4. Step 3 shows that if there are n mesh then there will be n simultaneous KVL equations. Solve the simultaneous equations to get the mesh/loop currents.

Type III : The Network with Voltage Source and Current Source, But The Current Source is Shared by Both The Loops (Figure 3.3)

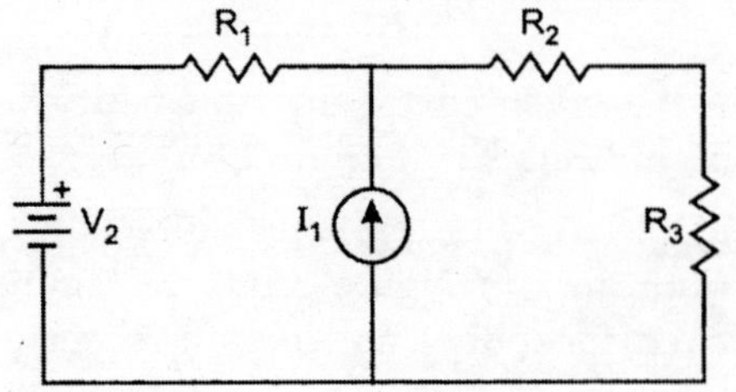

Fig. 3.3

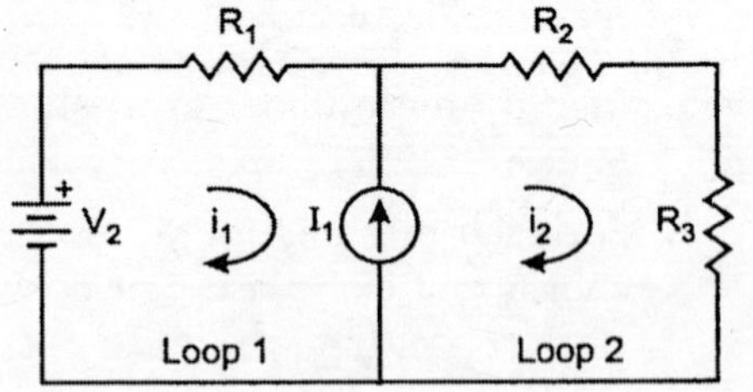

Fig. 3.3(a)

Step 1. Identify the number of mesh present in the given circuit. In the circuit shown in Figure 3.3 there are two mesh/loop, as indicated in Figure 3.3(*a*)

Step 2. For each mesh assign a mesh current $i_1, i_2, \ldots i_n$, in a particular direction of your choice (anticlockwise or clockwise) as indicated in Figure 3.3(*a*). In Figure 3.3(*a*) two loop currents i_1 and i_2 have been taken for loop 1 and loop 2 respectively in clockwise direction.

Step 3. Write the KVL equation for the modified circuit that will be obtained after removing the current source. That is if we remove the current source from the circuit shown in Figure 3.3(*a*), we get a new circuit shown in Figure 3.3(*b*). We will have to write the KVL equation for the circuit shown in Figure 3.3(*b*). The KVL equation will be,

$$R_1 i_1 + R_2 i_2 + R_3 i_2 = V_2$$

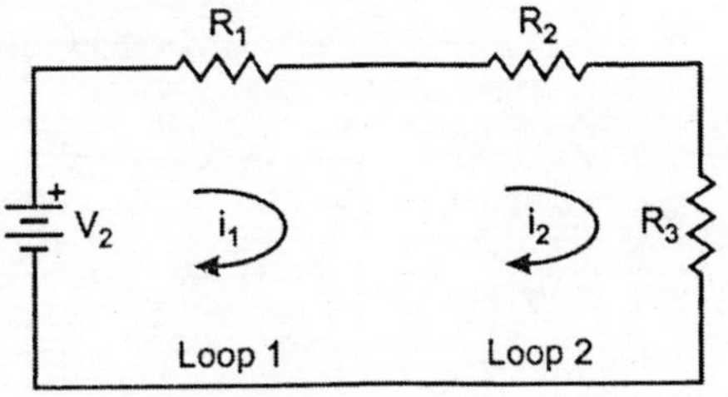

Fig. 3.3(b)

Step 4. Express the current delivered by the current source by loop currents. That is for the circuit shown in Figure 3.3(*a*),

$$I_1 = i_2 - i_1$$

Step 5. Solve all the simultaneous equations, to get the value of loop currents.

SOLVED EXAMPLES

EXAMPLE 3.1. *For the circuit shown in Figure E3.1, find the loop currents.*

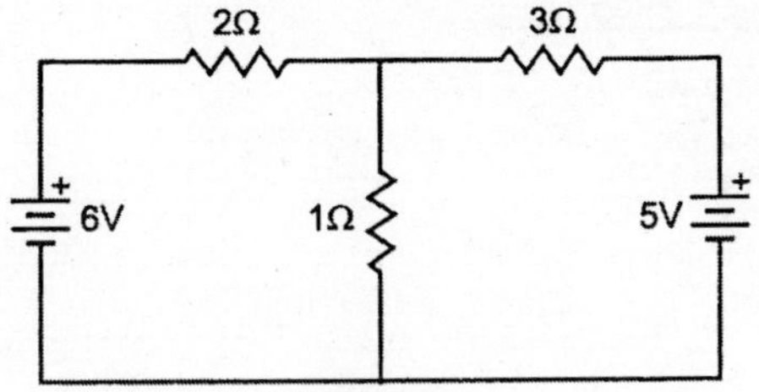

Fig.E3.1

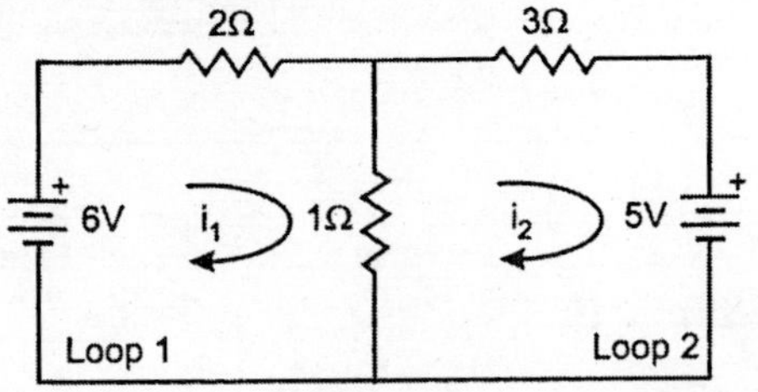

Fig.E3.1(a)

SOLUTION. The loop currents are indicated in Figure E3.1(*a*). Writing the loop equations we get

Loop 1 : $3i_1 - i_2 = 6$ ...(*i*)

Loop 2 : $4i_2 - i_1 = -5$ ...(*ii*)

Writing equations (*i*) and (*ii*) in matrix form we get

$$\begin{bmatrix} 3 & -1 \\ -1 & 4 \end{bmatrix}\begin{bmatrix} i_1 \\ i_2 \end{bmatrix} = \begin{bmatrix} 6 \\ -5 \end{bmatrix}$$

Let $$\Delta = \begin{vmatrix} 3 & -1 \\ -1 & 4 \end{vmatrix} = 11$$

Then $$i_1 = \frac{\begin{vmatrix} 6 & -1 \\ -5 & 4 \end{vmatrix}}{\Delta} = 1.73 \text{ A} \qquad i_2 = \frac{\begin{vmatrix} 3 & 6 \\ -1 & -5 \end{vmatrix}}{\Delta} = -0.82 \text{ A}$$

EXAMPLE 3.2. *For the circuit shown in Figure E3.2, find the loop currents.*

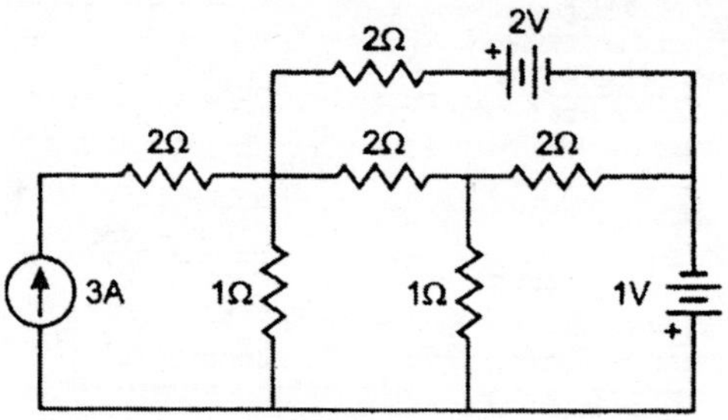

Fig. E3.2

SOLUTION. The loop currents are indicated in Figure E3.2(*a*). Writing the loop equations we get

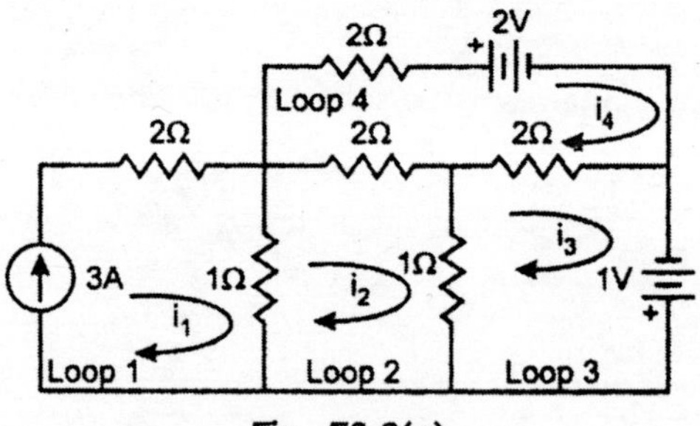

Fig. E3.2(a)

Loop 1 : $i_1 = 3$ A ...(*i*)

Loop 2 : $4i_2 - i_3 - i_1 - 2i_4 = 0$ ...(*ii*)

Substituting equation (*i*) in equation (*ii*) we get

$4i_2 - i_3 - 2i_4 = 3$ ...(*iii*)

Loop 3 : $3i_3 - i_2 - 2i_4 = 1$ ...(*iv*)

Loop 4 : $6i_4 - 2i_2 - 2i_3 = -2$ ...(*v*)

Writing equations (*iii*), (*iv*) and (*v*) in matrix form we get

$$\begin{bmatrix} 4 & -1 & -2 \\ -1 & 3 & -2 \\ -2 & -2 & 6 \end{bmatrix}\begin{bmatrix} i_2 \\ i_3 \\ i_4 \end{bmatrix} = \begin{bmatrix} 3 \\ 1 \\ -2 \end{bmatrix}$$

Let
$$\Delta = \begin{vmatrix} 4 & -1 & -2 \\ -1 & 3 & -2 \\ -2 & -2 & 6 \end{vmatrix} = 30$$

Then
$$i_2 = \frac{\begin{vmatrix} 3 & -1 & -2 \\ 1 & 3 & -2 \\ -2 & -2 & 6 \end{vmatrix}}{\Delta} = 1.2 \text{ A}$$

$$i_3 = \frac{\begin{vmatrix} 4 & 3 & -2 \\ -1 & 1 & -2 \\ -2 & -2 & 6 \end{vmatrix}}{\Delta} = 1 \text{ A}$$

$$i_4 = \frac{\begin{vmatrix} 4 & -1 & 3 \\ -1 & 3 & 1 \\ -2 & -2 & -2 \end{vmatrix}}{\Delta} = 0.4 \text{ A}$$

EXAMPLE 3.3. *For the circuit shown in Figure E3.3, find the loop currents. All resistances are of 1 Ω.*

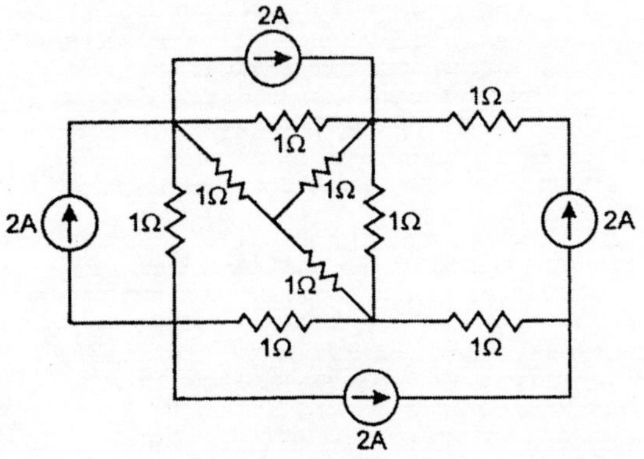

Fig. E3.3

SOLUTION. The loop currents are indicated in Figure E3.3(*a*). Writing the loop equations we get

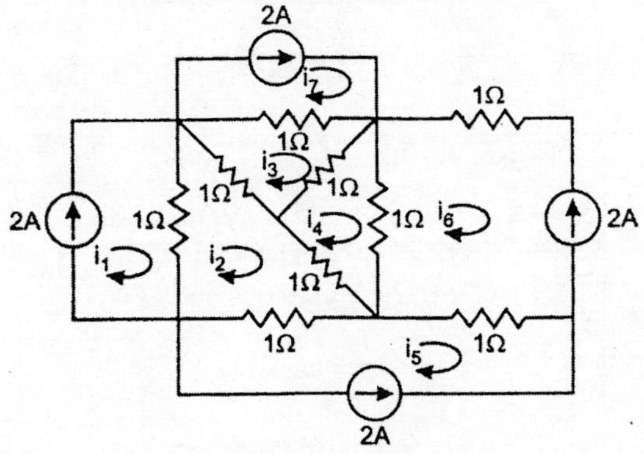

Fig. E3.3(a)

Loop 1 : $i_1 = 2\text{ A} \quad i_7 = 2\text{ A} \quad i_6 = -2\text{ A} \quad i_5 = -2\text{ A}$ ...(*i*)

Loop 2 : $4i_2 - i_1 - i_5 - i_3 - i_4 = 0$

Substituting equation (*i*) we get

$$4i_2 - i_3 - i_4 = 0 \qquad ...(ii)$$

Loop 3 : $3i_3 - i_7 - i_2 - i_4 = 0$

Substituting equation (*i*) we get

$$3i_3 - i_2 - i_4 = 2 \qquad ...(iii)$$

Loop 4 : $3i_4 - i_6 - i_2 - i_3 = 0$

Substituting equation (*i*) we get

$$3i_4 - i_2 - i_3 = -2 \qquad ...(iv)$$

Writing equations (*ii*), (*iii*) and (*iv*) in matrix form we get

$$\begin{bmatrix} 4 & -1 & -1 \\ -1 & 3 & -1 \\ -1 & -1 & 3 \end{bmatrix} \begin{bmatrix} i_2 \\ i_3 \\ i_4 \end{bmatrix} = \begin{bmatrix} 0 \\ 2 \\ -2 \end{bmatrix}$$

Let
$$\Delta = \begin{vmatrix} 4 & -1 & -1 \\ -1 & 3 & -1 \\ -1 & -1 & 3 \end{vmatrix} = 24$$

Then
$$i_2 = \frac{\begin{vmatrix} 0 & -1 & -1 \\ 2 & 3 & -1 \\ -2 & -1 & 3 \end{vmatrix}}{\Delta} = 0\text{ A}$$

$$i_3 = \frac{\begin{vmatrix} 4 & 0 & -1 \\ -1 & 2 & -1 \\ -1 & -2 & 3 \end{vmatrix}}{\Delta} = 0.5\text{ A}$$

$$i_4 = \frac{\begin{vmatrix} 4 & -1 & 0 \\ -1 & 3 & 2 \\ -1 & -1 & -2 \end{vmatrix}}{\Delta} = -0.5\text{ A}$$

EXAMPLE 3.4. *For the circuit shown in Figure E3.4, find the loop currents. All resistances are of 1 Ω*

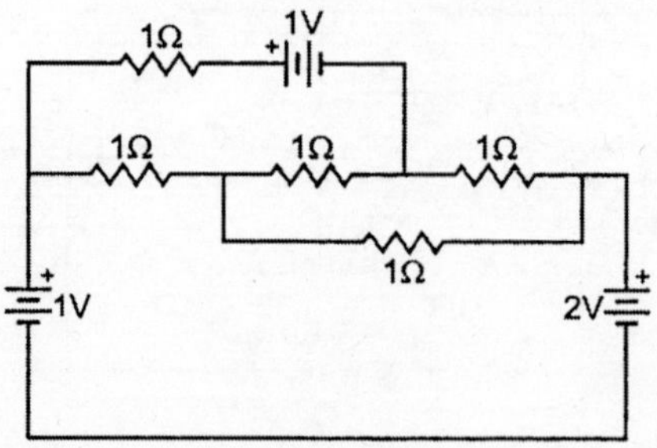

Fig. E3.4

SOLUTION. The loop currents are indicated in the Figure E3.4(*a*). Writing the loop equations we get

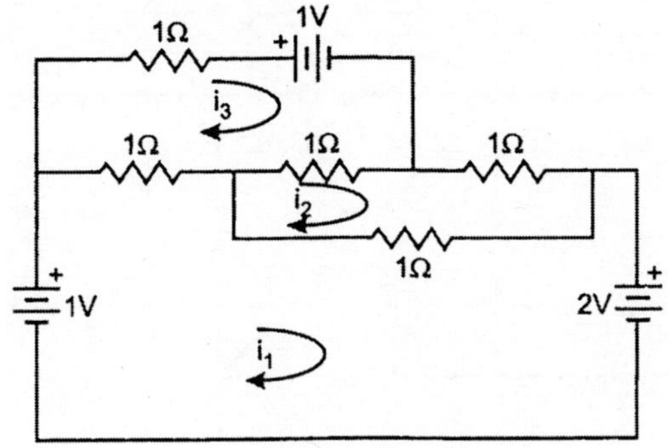

Fig. E3.4(a)

Loop 1 : $2i_1 - i_2 - i_3 = -1$ ...(*i*)

Loop 2 : $3i_2 - i_1 - i_3 = 0$ ...(*ii*)

Loop 3 : $3i_3 - i_2 - i_1 = -1$ ...(*iii*)

Writing equations (*ii*), (*iii*) and (*iv*) in matrix form we get

$$\begin{bmatrix} 2 & -1 & -1 \\ -1 & 3 & -1 \\ -1 & -1 & 3 \end{bmatrix} \begin{bmatrix} i_1 \\ i_2 \\ i_3 \end{bmatrix} = \begin{bmatrix} -1 \\ 0 \\ -1 \end{bmatrix}$$

Let
$$\Delta = \begin{vmatrix} 2 & -1 & -1 \\ -1 & 3 & -1 \\ -1 & -1 & 3 \end{vmatrix} = 8$$

Then
$$i_1 = \frac{\begin{vmatrix} -1 & -1 & -1 \\ 0 & 3 & -1 \\ -1 & -1 & 3 \end{vmatrix}}{\Delta} = -1.5 \text{ A}$$

$$i_2 = \frac{\begin{vmatrix} 2 & -1 & -1 \\ -1 & 0 & -1 \\ -1 & -1 & 3 \end{vmatrix}}{\Delta} = -0.875 \text{ A}$$

$$i_3 = \frac{\begin{vmatrix} 2 & -1 & -1 \\ -1 & 3 & 0 \\ -1 & -1 & -1 \end{vmatrix}}{\Delta} = 1.125 \text{ A}$$

EXAMPLE 3.5. *For the circuit shown in Figure E3.5, find the loop currents. All resistances are of 1 Ω*

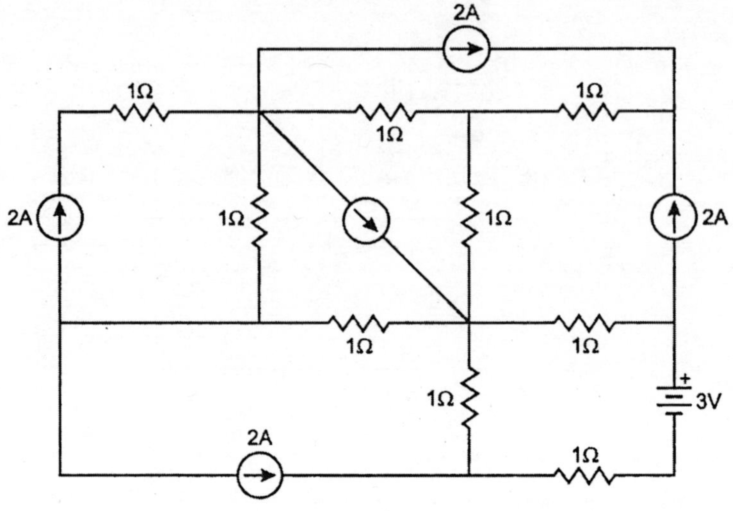

Fig. E3.5

SOLUTION. The loop currents are indicated in the Figure E3.5(*a*). Writing the loop equations we get

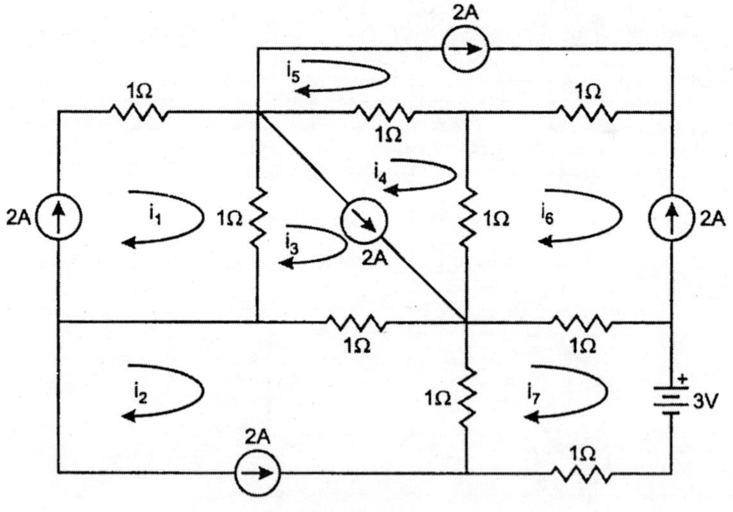

Fig. E3.5(a)

$$i_1 = 2\text{ A} \quad i_2 = -2\text{ A} \quad i_5 = 2\text{ A} \quad i_6 = -2\text{ A} \qquad ...(i)$$

Loop 3 and 4 :

$$i_3 - i_4 = 2 \qquad ...(ii)$$

$$2i_4 - i_5 - i_6 + 2i_3 - i_2 - i_1 = 0 \qquad ...(iii)$$

Substituting equation (*i*) in equation (*iii*) we get

$$i_4 + i_3 = 0 \qquad ...(iv)$$

Solving equations (*ii*) and (*iv*) we get

$$i_3 = 1\text{ A} \qquad i_4 = -1\text{ A}$$

Loop 7 : $3i_7 - i_2 - i_6 = -3$ ...(v)

Substituting equation (i) in equation (v) we get

$$3i_7 = -7$$

$$i_7 = -2.33 \text{ A} \quad \text{...(vi)}$$

EXAMPLE 3.6. *For the circuit shown in Figure E3.6, find the loop currents. All resistances are of 1 Ω.*

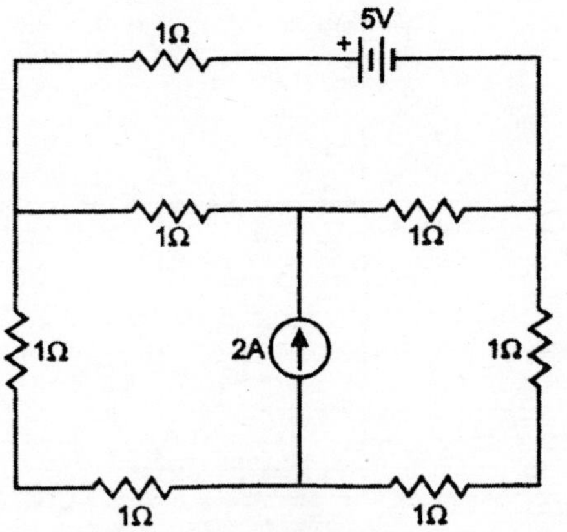

Fig. E3.6

SOLUTION. The loop currents are indicated in Figure E3.6(*a*). Writing the loop equations we get

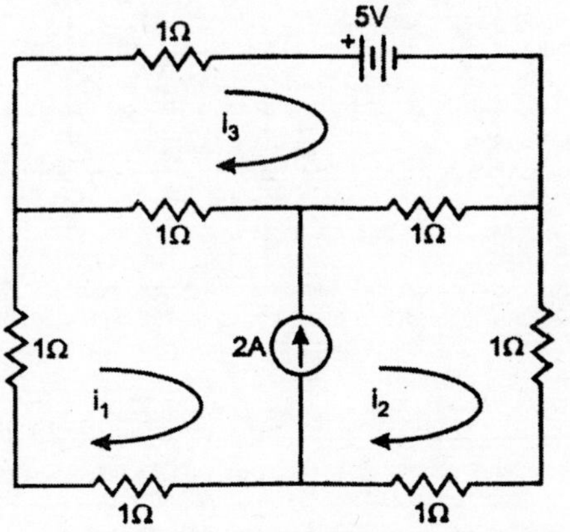

Fig. E3.6(a)

Loop 1 and 2 : $i_2 - i_1 = 2$ ...(i)

$$3i_1 - 2i_3 + 3i_2 = 0 \quad \text{...(ii)}$$

Loop 3 : $3i_3 - i_1 - i_2 = 5$ ...(iii)

Writing equations (i), (ii) and (iii) in matrix form we get

$$\begin{bmatrix} -1 & 1 & 0 \\ 3 & 3 & -2 \\ -1 & -1 & 3 \end{bmatrix} \begin{bmatrix} i_1 \\ i_2 \\ i_3 \end{bmatrix} = \begin{bmatrix} 2 \\ 0 \\ 5 \end{bmatrix}$$

Let $$\Delta = \begin{vmatrix} -1 & 1 & 0 \\ 3 & 3 & -2 \\ -1 & -1 & 3 \end{vmatrix} = -14$$

Then $$i_1 = \frac{\begin{vmatrix} 2 & 1 & 0 \\ 0 & 3 & -2 \\ 5 & -1 & 3 \end{vmatrix}}{\Delta} = -0.29 \text{ A}$$

$$i_2 = \frac{\begin{vmatrix} -1 & 2 & 0 \\ 3 & 0 & -2 \\ -1 & 5 & 3 \end{vmatrix}}{\Delta} = 1.71 \text{ A}$$

$$i_3 = \frac{\begin{vmatrix} -1 & 1 & 2 \\ 3 & 3 & 0 \\ -1 & -1 & 5 \end{vmatrix}}{\Delta} = 2.14 \text{ A}$$

EXAMPLE 3.7. *For the circuit shown in Figure E3.7, find the loop currents. All resistances are of 1 Ω.*

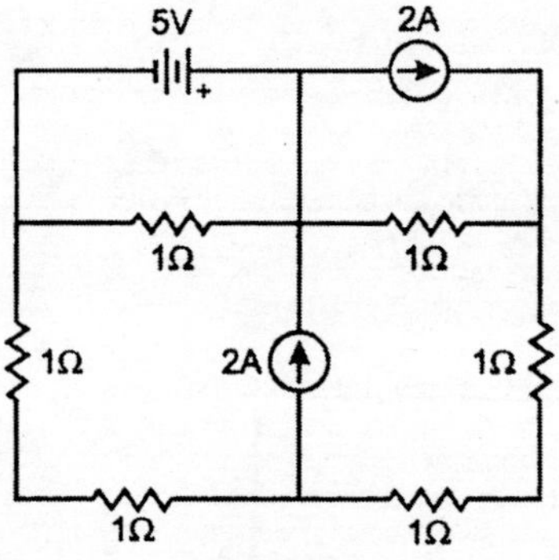

Fig. E3.7

SOLUTION. The loop currents are indicated in Figure E3.7(*a*). Writing the loop equations we get

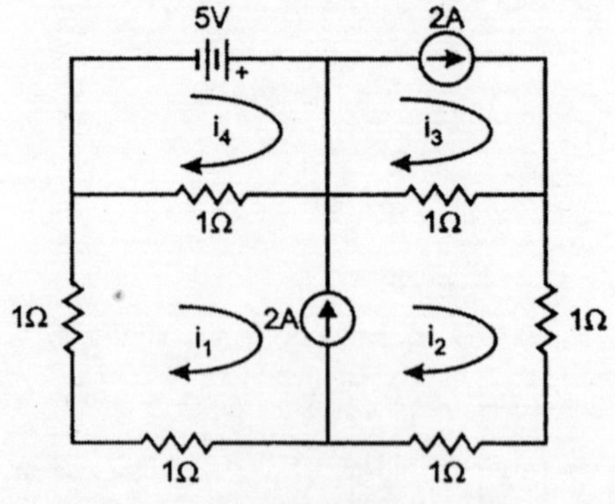

Fig. E3.7(a)

$$i_3 = 2 \text{ A} \quad \text{...}(i)$$

Loop 1 and 2 : $$i_2 - i_1 = 2 \quad \text{...}(ii)$$

$$3i_1 + 3i_2 - i_3 - i_4 = 0 \quad \text{...}(iii)$$

Substituting equation (*i*) in equation (*iii*) we get

$$3i_1 + 3i_2 - i_4 = 2 \quad \text{...}(iv)$$

Loop 4 : $$i_4 - i_1 = 5 \quad \text{...}(v)$$

Writing equations (*ii*), (*iv*) and (*v*) in matrix form we get

$$\begin{bmatrix} -1 & 1 & 0 \\ 3 & 3 & -1 \\ -1 & 0 & 1 \end{bmatrix} \begin{bmatrix} i_1 \\ i_2 \\ i_4 \end{bmatrix} = \begin{bmatrix} 2 \\ 2 \\ 5 \end{bmatrix}$$

Let $$\Delta = \begin{vmatrix} -1 & 1 & 0 \\ 3 & 3 & -1 \\ -1 & 0 & 1 \end{vmatrix} = -5$$

Then $$i_1 = \frac{\begin{vmatrix} 2 & 1 & 0 \\ 2 & 3 & -1 \\ 5 & 0 & 1 \end{vmatrix}}{\Delta} = 0.2 \text{ A}$$

$$i_2 = \frac{\begin{vmatrix} -1 & 2 & 0 \\ 3 & 2 & -1 \\ -1 & 5 & 1 \end{vmatrix}}{\Delta} = 2.2 \text{ A}$$

$$i_4 = \frac{\begin{vmatrix} -1 & 1 & 2 \\ 3 & 3 & 2 \\ -1 & 0 & 5 \end{vmatrix}}{\Delta} = 5.2 \text{ A}$$

EXAMPLE 3.8. *For the circuit shown in Figure E3.8, find the loop currents.*

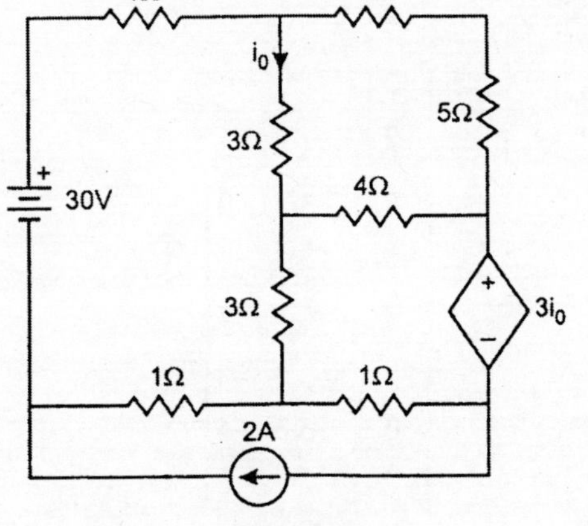

Fig. E3.8

SOLUTION. The loop currents are indicated in Figure E3.8(*a*). Writing the loop equations we get

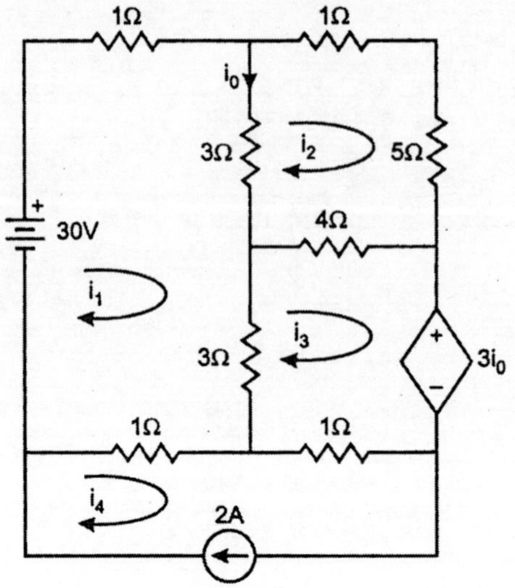

Fig. E3.8(a)

Loop 4 : $i_4 = 2$ A ...(*i*)

Loop 1 : $8i_1 - 3i_2 - 3i_3 - i_4 = 30$ ...(*ii*)

Substituting equation (*i*) in equation (*ii*) we get

$$8i_1 - 3i_2 - 3i_3 = 32 \quad ...(iii)$$

Loop 2 : $13i_2 - 3i_1 - 4i_3 = 0$ ...(*iv*)

Loop 3 : $8i_3 - 3i_1 - 4i_2 - i_4 = -3i_0$ ...(*v*)

From Figure E3.8,

$$i_0 = i_1 - i_2 \quad ...(vi)$$

Substituting equations (*i*) and (*vi*) in equation (*v*) we get

$$8i_3 - 7i_2 = 2 \quad ...(vii)$$

Writing equations (*iii*), (*iv*) and (*vii*) in matrix form we get

$$\begin{bmatrix} 8 & -3 & -3 \\ -3 & 13 & -4 \\ 0 & -7 & 8 \end{bmatrix} \begin{bmatrix} i_1 \\ i_2 \\ i_3 \end{bmatrix} = \begin{bmatrix} 32 \\ 0 \\ 2 \end{bmatrix}$$

Let
$$\Delta = \begin{vmatrix} 8 & -3 & -3 \\ -3 & 13 & -4 \\ 0 & -7 & 8 \end{vmatrix} = 473$$

Then
$$i_1 = \frac{\begin{vmatrix} 32 & -3 & -3 \\ 0 & 13 & -4 \\ 2 & -7 & 8 \end{vmatrix}}{\Delta} = 5.36 \text{ A}$$

$$i_2 = \frac{\begin{vmatrix} 8 & 32 & -3 \\ -3 & 0 & -4 \\ 0 & 2 & 8 \end{vmatrix}}{\Delta} = 1.79 \text{ A}$$

$$i_3 = \frac{\begin{vmatrix} 8 & -3 & 32 \\ -3 & 13 & 0 \\ 0 & -7 & 2 \end{vmatrix}}{\Delta} = 1.82 \text{ A}$$

Thus $$i_0 = i_1 - i_2 = 3.57 \text{ A}$$

EXAMPLE 3.9. *For the circuit shown in Figure E3.9, find the loop currents.*

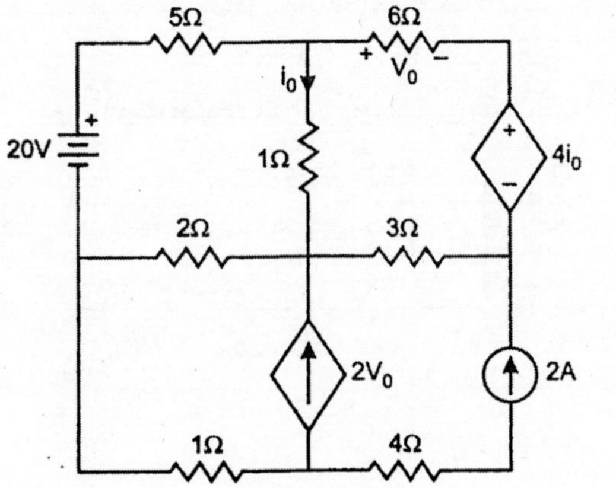

Fig. E3.9

SOLUTION. The loop currents are indicated in Figure E3.9(*a*). Writing the loop equations we get

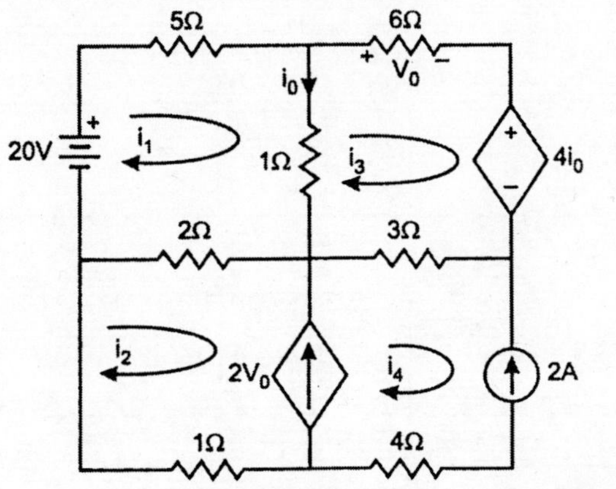

Fig. E3.9(a)

Loop 4 : $i_4 = -2$ A ...(*i*)

Loop 1 : $8i_1 - i_3 - 2i_2 = 20$ ...(*ii*)

Loop 3 : $10i_3 - i_1 - 3i_4 = -4i_0$ ...(*iii*)

From Figure E3.9(*a*)

$$i_0 = i_1 - i_3 \quad ...(iv)$$

Substituting equations (*i*) and (*iv*) in equation (*iii*) we get

$$6i_3 + 3i_1 = -6 \quad ...(v)$$

Loop 2 and Loop 4 : $i_4 - i_2 = 2V_0$ ...(*vi*)

From Figure E3.9(*a*),

$$V_0 = 6i_3 \quad ...(vii)$$

Substituting equations (*i*) and (*vii*) in equation (*vi*) we get

$$12i_3 + i_2 = -2 \quad ...(viii)$$

Writing equations (*ii*), (*v*) and (*viii*) in matrix form we get

$$\begin{bmatrix} 8 & -1 & -2 \\ 3 & 0 & 6 \\ 0 & 1 & 12 \end{bmatrix} \begin{bmatrix} i_1 \\ i_2 \\ i_3 \end{bmatrix} = \begin{bmatrix} 20 \\ -6 \\ -2 \end{bmatrix}$$

Let $$\Delta = \begin{vmatrix} 8 & -1 & -2 \\ 3 & 0 & 6 \\ 0 & 1 & 12 \end{vmatrix} = -18$$

Then $$i_1 = \frac{\begin{vmatrix} 20 & -1 & -2 \\ -6 & 0 & 6 \\ -2 & 1 & 12 \end{vmatrix}}{\Delta} = 9.33 \text{ A}$$

$$i_2 = \frac{\begin{vmatrix} 8 & 20 & -2 \\ 3 & -6 & 6 \\ 0 & -2 & 12 \end{vmatrix}}{\Delta} = 66 \text{ A}$$

$$i_3 = \frac{\begin{vmatrix} 8 & -1 & 20 \\ 3 & 0 & -6 \\ 0 & 1 & -2 \end{vmatrix}}{\Delta} = -5.67 \text{ A}$$

Thus $$i_0 = i_1 - i_3 = 15 \text{ A}$$

$$V_0 = 6i_3 = -34.02 \text{ V}$$

EXAMPLE 3.10. *For the circuit shown in Figure E3.10, find the ratio* $\frac{V_0}{V_i}$.

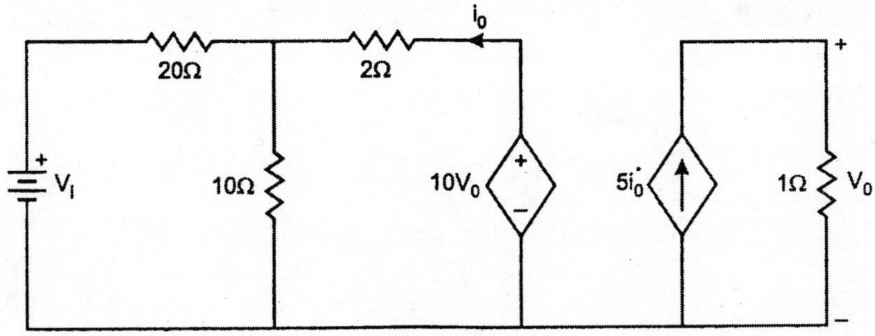

Fig. E3.10

SOLUTION. The loop currents are shown in Figure E3.10(*a*). Writing the loop equations we get

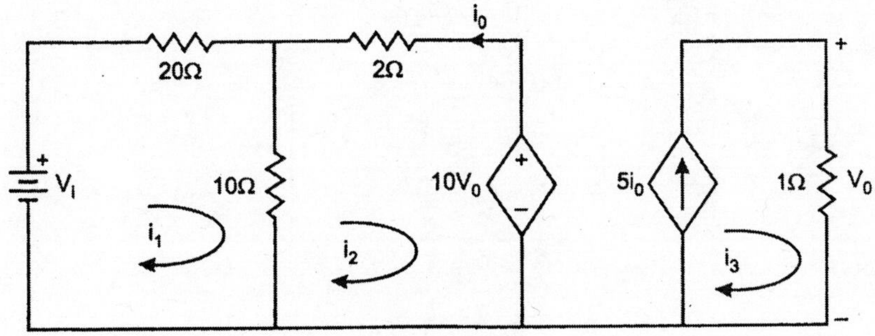

Fig. E3.10(a)

Loop 1 : $30i_1 - 10i_2 = V_i$ ...(*i*)

Loop 2 : $12i_2 - 10i_1 = -10V_0$ ...(*ii*)

Loop 3 : $V_0 = 1 \times i_3$ ...(*iii*)

$i_3 = 5i_0$ ...(*iv*)

From Figure E3.10(*a*), $i_0 = -i_2$ ...(*v*)

Substituting equations (*iii*), (*iv*) and (*v*) in equation (*ii*) we get

$$38i_2 + 10i_1 = 0 \qquad \text{...(vi)}$$

Writing equations (*i*) and (*vi*) in matrix form we get

$$\begin{bmatrix} 30 & -10 \\ 10 & 38 \end{bmatrix} \begin{bmatrix} i_1 \\ i_2 \end{bmatrix} = \begin{bmatrix} V_i \\ 0 \end{bmatrix}$$

Let $$\Delta = \begin{vmatrix} 30 & -10 \\ 10 & 38 \end{vmatrix} = 1240$$

Then $$i_2 = \frac{\begin{vmatrix} 30 & V_i \\ 10 & 0 \end{vmatrix}}{\Delta} = -\frac{10V_i}{1240} \qquad \text{...(vii)}$$

From equations (*iii*), (*iv*) and (*v*) we get

$$V_0 = -5i_2 \quad \text{...(viii)}$$

Substituting equation (*vii*) in equation (*viii*) we get

$$V_0 = \frac{50V_i}{1240}$$

or

$$\frac{V_0}{V_i} = \frac{50}{1240} = 0.04$$

EXAMPLE 3.11. *For the circuit shown in Figure E3.11, find the current i_0.*

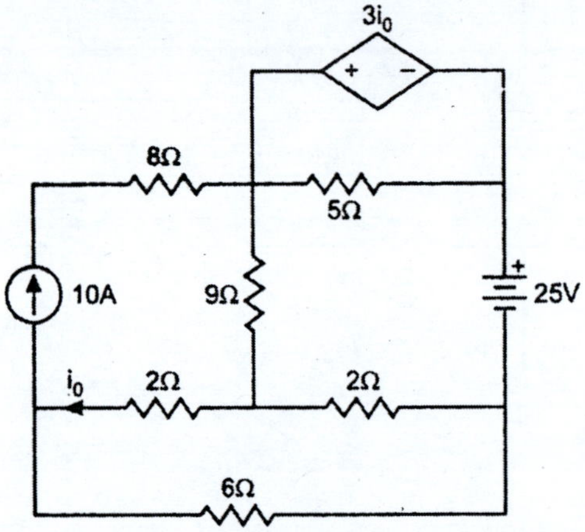

Fig. E3.11

SOLUTION. The loop currents are shown in Figure E3.11(*a*). Writing the loop equations we get

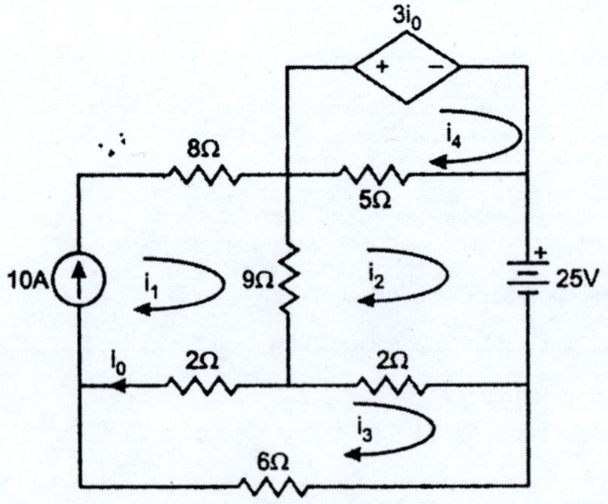

Fig. E3.11(a)

Loop 1 : $i_1 = 10\text{ A}$ ...(*i*)

Loop 2 : $14i_2 - 9i_1 - 2i_3 - 5i_4 = -25$ ...(*ii*)

Substituting equation (*i*) in equation (*ii*) we get

$$14i_2 - 2i_3 - 5i_4 = 65 \quad \text{...(iii)}$$

Loop 3 : $$10i_3 - 2i_1 - 2i_2 = 0 \quad \text{...(iv)}$$

Substituting equation (*i*) in equation (*iv*) we get

$$10i_3 - 2i_2 = 20 \quad \text{...(v)}$$

Loop 4 : $$5i_4 - 5i_2 = -3i_0 \quad \text{...(vi)}$$

Where $$i_0 = i_1 - i_3 \quad \text{...(vii)}$$

Substituting equation (*vii*) in equation (*vi*) we get

$$5i_4 - 5i_2 = -3(i_1 - i_3) \quad \text{...(viii)}$$

$$5i_4 - 5i_2 - 3i_3 = -30 \quad \text{...(ix)}$$

Writing equations (*iii*), (*v*) and (*ix*) we get

$$\begin{bmatrix} 14 & -2 & -5 \\ -2 & 10 & 0 \\ -5 & -3 & 5 \end{bmatrix} \begin{bmatrix} i_2 \\ i_3 \\ i_4 \end{bmatrix} = \begin{bmatrix} 65 \\ 20 \\ -30 \end{bmatrix}$$

Let $$\Delta = \begin{vmatrix} 14 & -2 & -5 \\ -2 & 10 & 0 \\ -5 & -3 & 5 \end{vmatrix} = 400$$

Then $$i_3 = \frac{\begin{vmatrix} 14 & 65 & -5 \\ -2 & 20 & 0 \\ -5 & -30 & 5 \end{vmatrix}}{\Delta} = 3.12 \text{ A}$$

$$i_0 = i_1 - i_3 = 6.88 \text{ A}$$

EXAMPLE 3.12. *For the circuit shown in Figure E3.12, find the current delivered by 5∠30 V source and 10∠90 V source.*

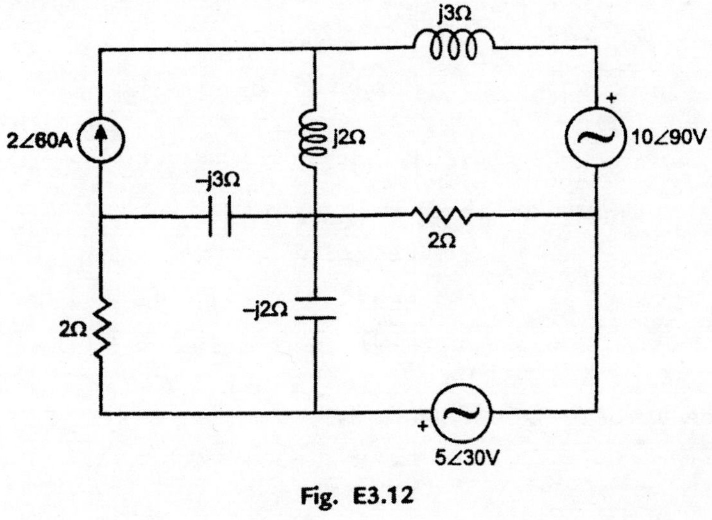

Fig. E3.12

SOLUTION. The loop currents are indicated in Figure E3.12(*a*). Writing the loop equations we get

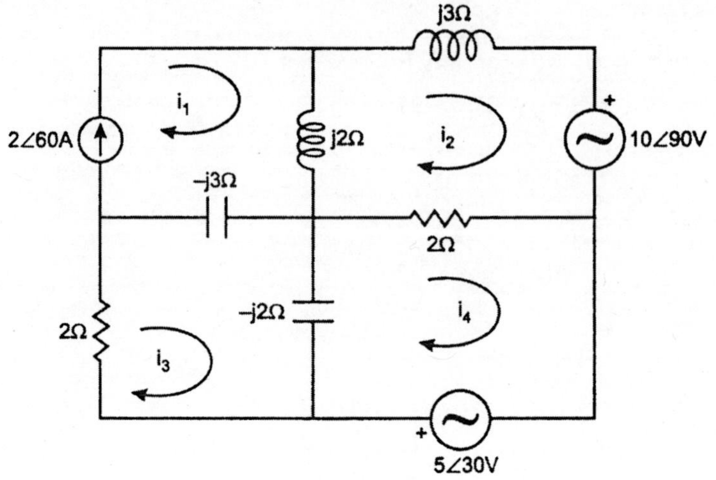

Fig. E3,12(a)

Loop 1 : $i_1 = 2\angle 60 \text{ A} = 1 + j1.73 \text{ A}$ ...(*i*)

Loop 2 : $(2 + j5)i_2 - 2i_4 - j2i_1 = -j10$ ...(*ii*)

Substituting the value of equation (*i*) in equation (*ii*) we get

$$(2 + j5)i_2 - 2i_4 = -3.46 - j8 \quad \text{...(iii)}$$

Loop 3 : $(2 - j5)i_3 + j3i_1 + j2i_4 = 0$ ...(*iv*)

Substituting the value of equation (*i*) in equation (*iv*) we get

$$(2 - j5)i_3 + j2i_4 = -5.19 + j3 \quad \text{...(v)}$$

Loop 4 : $(2 - j2)i_4 - 2i_2 + j2i_3 = 4.33 + j2.5$ ...(*vi*)

Writing equations (*iii*), (*v*) and (*vi*) in matrix form we get

$$\begin{bmatrix} 2+j5 & 0 & -2 \\ 0 & 2-j5 & j2 \\ -2 & j2 & 2-j2 \end{bmatrix} \begin{bmatrix} i_2 \\ i_3 \\ i_4 \end{bmatrix} = \begin{bmatrix} -3.46-j8 \\ -5.19+j3 \\ 4.33+j2.5 \end{bmatrix}$$

Let
$$\Delta = \begin{vmatrix} 2+j5 & 0 & -2 \\ 0 & 2-j5 & j2 \\ -2 & j2 & 2-j2 \end{vmatrix} = 58 - j18$$

Then the current delivered by 10∠90 V source is

$$i_2 = \frac{\begin{vmatrix} -3.46-j8 & 0 & -2 \\ -5.19+j3 & 2-j5 & j2 \\ 4.33+j2.5 & j2 & 2-j2 \end{vmatrix}}{\Delta} = (-1.05 + j0.57) \text{ A}$$

and the current delivered by 5∠30 V source is

$$i_3 = \frac{\begin{vmatrix} 2+j5 & -3.46-j8 & -2 \\ 0 & -5.19+j3 & j2 \\ -2 & 4.33+j2.5 & 2-j2 \end{vmatrix}}{\Delta} = (-0.86 + j0.083) \text{ A}$$

EXAMPLE 3.13. *For the circuit shown in Figure E3.13, find the loop currents.*

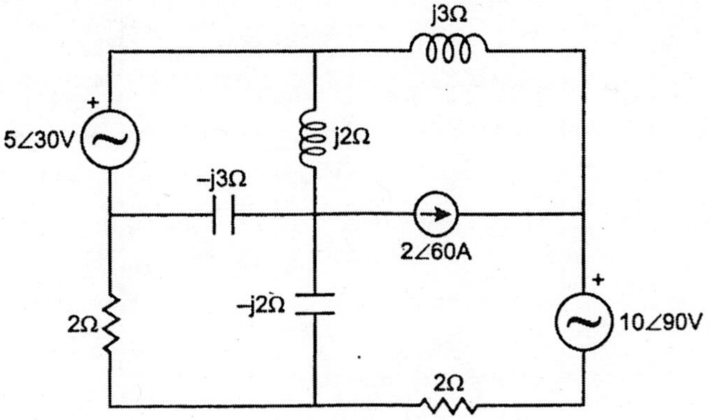

Fig. E3.13

SOLUTION. The loop currents are indicated in Figure E3.13(*a*). Writing the loop equations we get

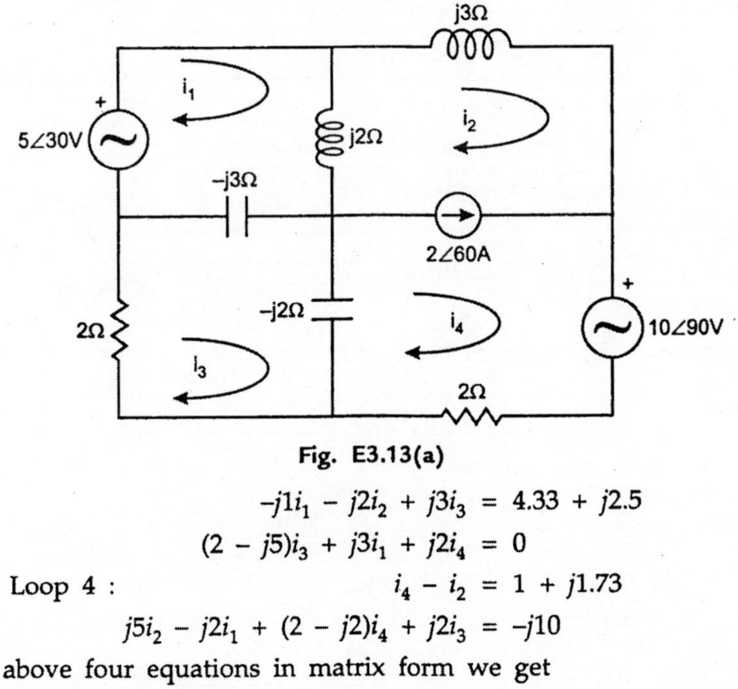

Fig. E3.13(a)

Loop 1 : $-j1i_1 - j2i_2 + j3i_3 = 4.33 + j2.5$ ...(*i*)

Loop 3 : $(2 - j5)i_3 + j3i_1 + j2i_4 = 0$ ...(*ii*)

Loop 2 and Loop 4 : $i_4 - i_2 = 1 + j1.73$ ...(*iii*)

And $j5i_2 - j2i_1 + (2 - j2)i_4 + j2i_3 = -j10$ ...(*iv*)

Writing the above four equations in matrix form we get

$$\begin{bmatrix} -j1 & -j2 & j3 & 0 \\ j3 & 0 & 2-j5 & j2 \\ 0 & -1 & 0 & 1 \\ -j2 & j5 & j2 & 2-j2 \end{bmatrix} \begin{bmatrix} i_1 \\ i_2 \\ i_3 \\ i_4 \end{bmatrix} = \begin{bmatrix} 4.33+j2.5 \\ 0 \\ 1+j1.73 \\ -j10 \end{bmatrix}$$

Let $$\Delta = \begin{vmatrix} -j1 & -j2 & j3 & 0 \\ j3 & 0 & 2-j5 & j2 \\ 0 & -1 & 0 & 1 \\ -j2 & j5 & j2 & 2-j2 \end{vmatrix} = 22 + j8$$

Then

$$i_1 = \frac{\begin{vmatrix} 4.33+j2.5 & -j2 & j3 & 0 \\ 0 & 0 & 2-j5 & j2 \\ 1+j1.73 & -1 & 0 & 1 \\ -j10 & j5 & j2 & 2-j2 \end{vmatrix}}{\Delta} = (6.59 - j2.73)\text{ A}$$

$$i_2 = \frac{\begin{vmatrix} -j1 & 4.33+j2.5 & j3 & 0 \\ j3 & 0 & 2-j5 & j2 \\ 0 & 1+j1.73 & 0 & 1 \\ -j2 & -j10 & j2 & 2-j2 \end{vmatrix}}{\Delta} = (-1.18 - j0.28)\text{ A}$$

$$i_3 = \frac{\begin{vmatrix} -j1 & -j2 & 4.33+j2.5 & 0 \\ j3 & 0 & 0 & j2 \\ 0 & -1 & 1+j1.73 & 1 \\ -j2 & j5 & -j10 & 2-j2 \end{vmatrix}}{\Delta} = (1.92 - j3.44)\text{ A}$$

$$i_2 = \frac{\begin{vmatrix} -j1 & -j2 & j3 & 4.33+j2.5 \\ j3 & 0 & 2-j5 & 0 \\ 0 & -1 & 0 & 1+j1.73 \\ -j2 & j5 & j2 & -j10 \end{vmatrix}}{\Delta} = (-0.18 + j1.45)\text{ A}$$

EXAMPLE 3.14. *For the circuit shown in Figure E3.14, find V_0.*

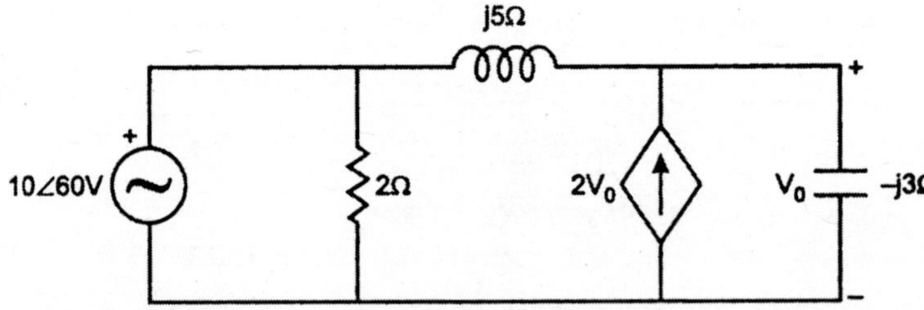

Fig. E3.14

SOLUTION. The loop currents are indicated in Figure E3.14(*a*). Writing the loop equations we get

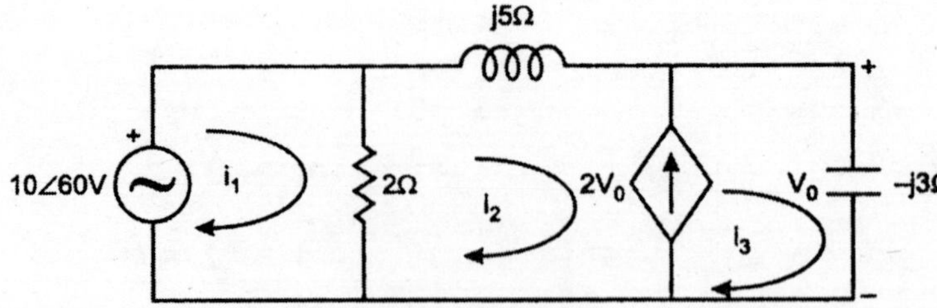

Fig. E3.14(a)

Loop 1 : $2i_1 - 2i_2 = 5 + j8.66$ ...(*i*)

Loop 2 and Loop 3 : $i_3 - i_2 = 2V_0$ ...(*ii*)

Again $V_0 = -j3 \times i_3$ ...(*iii*)

Substituting equation (*iii*) in equation (*ii*) we get

$$(1 + j6)i_3 - i_2 = 0 \quad ...(iv)$$

Again $(2 + j5)i_2 - 2i_1 - j3i_3 = 0$ ...(*v*)

Writing equations (*i*), (*iv*) and (*v*) in matrix form we get

$$\begin{bmatrix} 2 & -2 & 0 \\ 0 & -1 & 1+j6 \\ -2 & 2+j5 & -j3 \end{bmatrix} \begin{bmatrix} i_1 \\ i_2 \\ i_3 \end{bmatrix} = \begin{bmatrix} 5+j8.66 \\ 0 \\ 0 \end{bmatrix}$$

Let
$$\Delta = \begin{vmatrix} 2 & -2 & 0 \\ 0 & -1 & 1+j6 \\ -2 & 2+j5 & -j3 \end{vmatrix} = (60 - j4)$$

Then
$$i_3 = \frac{\begin{vmatrix} 2 & -2 & 5+j8.66 \\ 0 & -1 & 0 \\ -2 & 2+j5 & 0 \end{vmatrix}}{\Delta} = (-0.15 - j0.29) \text{ A}$$

And $V_0 = -j3 \times i_3 = (-0.87 + j0.45)$

EXAMPLE 3.15. *For the circuit shown in Figure E3.15, find i_0 and V_0.*

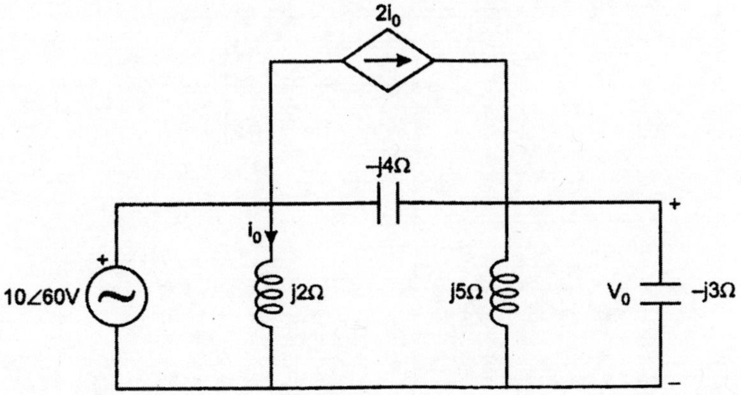

Fig. E3.15

SOLUTION. The loop currents are indicated in Figure E3.15(*a*). Writing the loop equations we get

Loop 1 : $j2i_1 - j2i_2 = 5 + j8.66$ ...(*i*)

Loop 2 : $j3i_2 - j2i_1 - j5i_3 + j4i_4 = 0$ ...(*ii*)

Loop 3 : $j2i_3 - j5i_2 = 0$ ...(*iii*)

Loop 4 : $i_4 = 2i_0$ ...(*iv*)

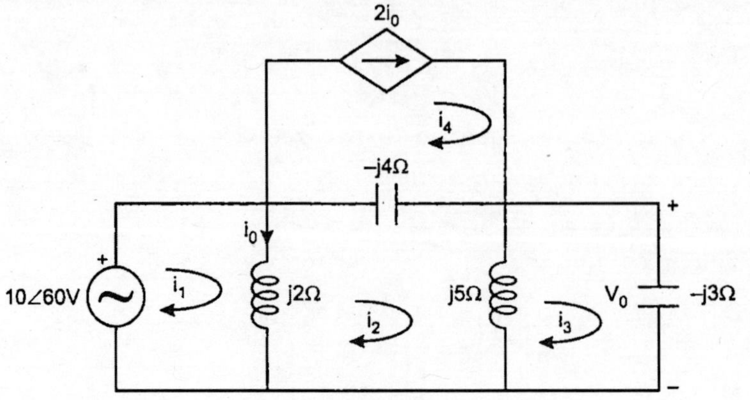

Fig. E3.15(a)

From Figure E3.15(*a*),

$$i_0 = i_1 - i_2 \quad ...(v)$$

Substituting equation (*v*) in equation (*iv*) we get

$$i_4 = 2i_1 - 2i_2 \quad ...(vi)$$

Substituting equation (*v*) in equation (*ii*) we get

$$j6i_1 - j5i_2 - j5i_3 = 0 \quad ...(vii)$$

Writing equations (*i*), (*iii*) and (*vii*) in matrix form we get

$$\begin{bmatrix} j2 & -j2 & 0 \\ 0 & -j5 & j2 \\ j6 & -j5 & -j5 \end{bmatrix}\begin{bmatrix} i_1 \\ i_2 \\ i_3 \end{bmatrix} = \begin{bmatrix} 5+j8.66 \\ 0 \\ 0 \end{bmatrix}$$

Let
$$\Delta = \begin{vmatrix} j2 & -j2 & 0 \\ 0 & -j5 & j2 \\ j6 & -j5 & -j5 \end{vmatrix} = -j46$$

Then
$$i_1 = \frac{\begin{vmatrix} 5+j8.66 & -j2 & 0 \\ 0 & -j5 & j2 \\ 0 & -j5 & -j5 \end{vmatrix}}{\Delta} = (6.59 - j3.8)\text{ A}$$

$$i_2 = \frac{\begin{vmatrix} j2 & 5+j8.66 & 0 \\ 0 & 0 & j2 \\ j6 & 0 & -j5 \end{vmatrix}}{\Delta} = (2.26 - j1.3)\text{ A}$$

$$i_3 = \frac{\begin{vmatrix} j2 & -j2 & 5+j8.66 \\ 0 & -j5 & 0 \\ j6 & -j5 & 0 \end{vmatrix}}{\Delta} = (5.65 - j3.26)\text{ A}$$

Thus
$$i_0 = i_1 - i_2 = (4.33 - j2.5)\text{ A}$$

And
$$V_0 = -j3i_3 = (-9.78 - ji6.95)\text{ V}$$

EXAMPLE 3.16. *For the circuit shown in Figure E3.16, find i_0 and V_0.*

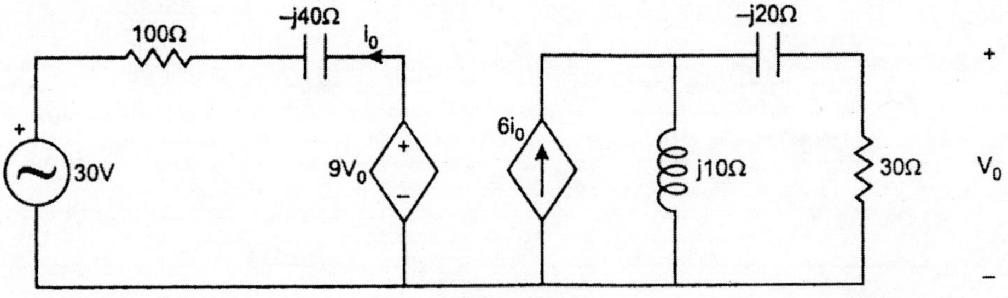

Fig. E3.16

SOLUTION. The loop currents are indicated in Figure E3.16(*a*). Writing the loop equations we get

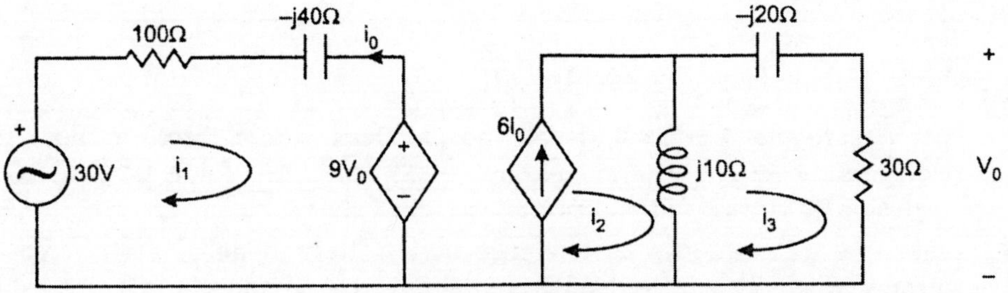

Fig. E3.16(a)

Loop 1 : $(100 - j40)i_1 + 9V_0 = 30$ ...(*i*)

Loop 2 : $i_2 = 6i_0$ ...(*ii*)

Loop 3 : $(30 - j10)i_3 - j10i_2 = 0$ ...(*iii*)

From Figure E3.16(*a*),

$$V_0 = 30 \times i_3 \qquad \text{...}(iv)$$

and

$$i_0 = -i_1 \qquad \text{...}(v)$$

Substituting equation (*iv*) in equation (*i*) we get

$$(100 - j40)i_1 + 270i_3 = 30 \qquad \text{...}(vi)$$

Substituting equation (*v*) in equation (*ii*) we get

$$i_2 = -6i_1 \qquad \text{...}(vii)$$

Writing equations (*iii*), (*vi*) and (*vii*) in matrix form we get

$$\begin{bmatrix} 0 & -j10 & 30-j10 \\ 100-j40 & 0 & 270 \\ 6 & 1 & 0 \end{bmatrix} \begin{bmatrix} i_1 \\ i_2 \\ i_3 \end{bmatrix} = \begin{bmatrix} 0 \\ 30 \\ 0 \end{bmatrix}$$

Let

$$\Delta = \begin{vmatrix} 0 & -j10 & 30-j10 \\ 100-j40 & 0 & 270 \\ 6 & 1 & 0 \end{vmatrix} = (2.6 \times 10^3 - j1.84 \times 10^4)$$

Then
$$i_1 = \frac{\begin{vmatrix} 0 & -j10 & 30-j10 \\ 30 & 0 & 270 \\ 0 & 1 & 0 \end{vmatrix}}{\Delta} = (0.02 + j0.04) \text{ A}$$

And
$$i_3 = \frac{\begin{vmatrix} 0 & -j10 & 0 \\ 100-j40 & 0 & 30 \\ 6 & 1 & 0 \end{vmatrix}}{\Delta} = (j0.11) \text{ A}$$

Thus
$$i_0 = -i_1 = -(0.02 + j0.04) \text{ A}$$

And
$$V_0 = 30 \times i_3 = j3.42 \text{ V}$$

3.3 NODE ANALYSIS

Steps to Solve a Circuit Using Mesh Analysis

Step 1. For a particular circuit finds all the junctions where three or more than three elements meet. These junctions are called nodes. So for a particular circuit the first step is to find all the nodes.

Step 2. Take one of the nodes as reference node. The voltage of the reference node is always taken as *zero* volts. For remaining $n - 1$ nodes the voltages V_1, V_2, ... V_{n-1} are assigned.

Step 3. KCL is applied to all the $n - 1$ nodes.

Step 4. The $n - 1$ simultaneous equations are solved to get the node voltages.

Step 5. If a voltage source is present between the reference node and a non reference node. Then the non reference node voltage is set equal to the value of the voltage source.

Step 6. If a voltage source is present between the two non reference node, then those two nodes are termed as supernode. While finding such node voltages, both KCL and KVL is applied to the nodes.

SOLVED EXAMPLES

EXAMPLE 3.17. *For the circuit shown in Figure E3.17, find*

(i) The number of nodes *(ii) The node voltage.*

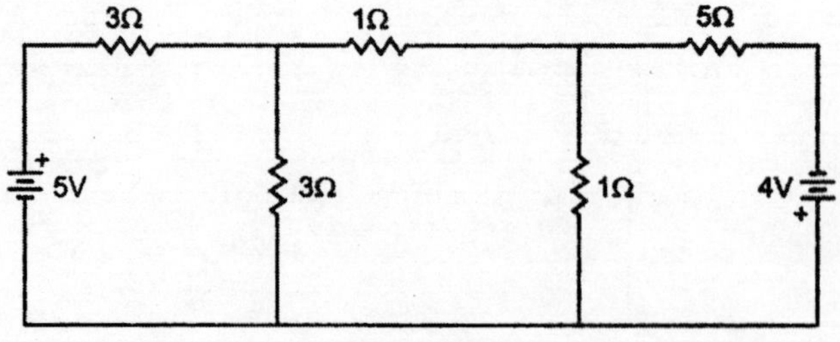

Fig. E3.17

SOLUTION. The circuit shown in Figure E3.17, can be redrawn as shown in Figure E3.17(*a*)

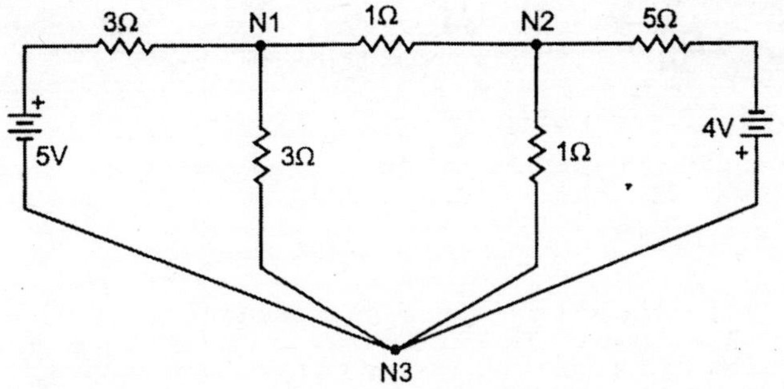

Fig. E3.17(a)

(*i*) From the circuit shown Figure E3.17(*a*). There are three nodes, Node N1, node N2 and node N3.

(*ii*) Let the voltage at node N1 be V_1 and the voltage at node N2 be V_2. Node N3 is taken as reference node and its voltage is taken as 0 V.

The node equations are,

$$\text{Node N1 :} \quad \frac{(V_1-5)}{2}+\frac{V_1-V_2}{1}+\frac{V_1}{3} = 0$$

or $$1.67V_1 - V_2 = 1.67 \qquad ...(i)$$

$$\text{Node N2 :} \quad \frac{(V_2+4)}{5}+\frac{V_2-V_1}{1}+\frac{V_2}{1} = 0$$

or $$2.2V_2 - V_1 = -0.8 \qquad ...(ii)$$

Writing equations (*i*) and (*ii*) in matrix form we get

$$\begin{bmatrix} 1.67 & -1 \\ -1 & 2.2 \end{bmatrix} \begin{bmatrix} V_1 \\ V_2 \end{bmatrix} = \begin{bmatrix} 1.67 \\ -0.8 \end{bmatrix}$$

Let $$\Delta = \begin{vmatrix} 1.67 & -1 \\ -1 & 2.2 \end{vmatrix} = 2.67$$

Then $$V_1 = \frac{\begin{vmatrix} 1.67 & -1 \\ -0.8 & 2.2 \end{vmatrix}}{\Delta} = 1.07 \text{ V}$$

$$V_2 = \frac{\begin{vmatrix} 1.67 & 1.67 \\ -1 & -0.8 \end{vmatrix}}{\Delta} = 0.13 \text{ V}$$

EXAMPLE 3.18. *For the circuit shown in Figure E3.18, find the node voltages.*

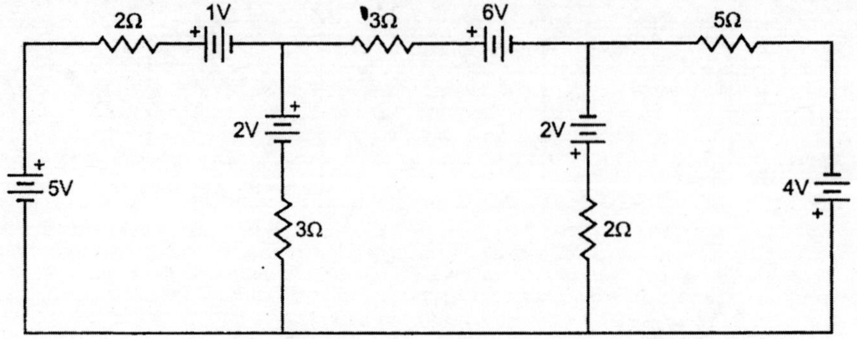

Fig. E3.18

SOLUTION. The circuit shown in Figure E3.18, can be redrawn as shown in Figure E3.18(*a*).

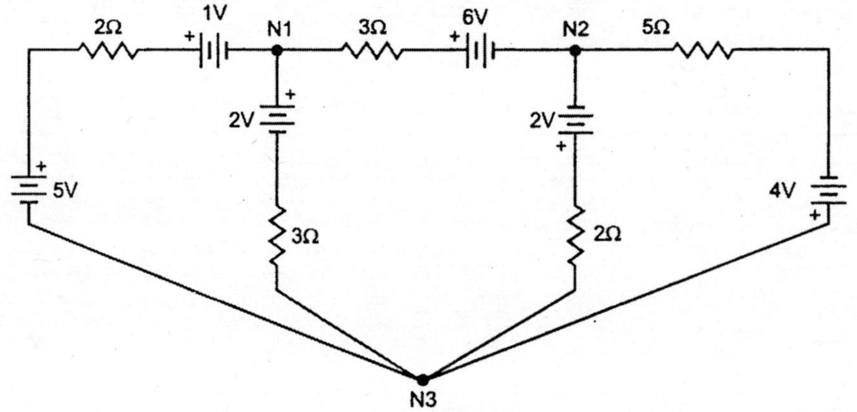

Fig. E3.18(a)

From the circuit shown Figure E3.18(*a*). There are three nodes, Node N1, node N2 and node N3. Let the voltage at node N1 be V_1 and the voltage at node N2 be V_2. Node N3 is taken as reference node and its voltage is taken as 0 V.

The node equations are,

$$\text{Node N1 :}\quad \frac{(V_1+1-5)}{2}+\frac{V_1-6-V_2}{3}+\frac{V_1-2}{3} = 0$$

or $$1.67V_1 - 0.33V_2 = 4.67 \qquad \text{...}(i)$$

$$\text{Node N2 :}\quad \frac{V_2+2}{2}+\frac{V_2+4}{5}+\frac{V_2+6-V_1}{3} = 0$$

or $$1.03V_2 - 0.33V_1 = -3.8 \qquad \text{...}(ii)$$

Writing equations (*i*) and (*ii*) in matrix form we get

$$\begin{bmatrix} 1.67 & -0.33 \\ -0.33 & 1.03 \end{bmatrix}\begin{bmatrix} V_1 \\ V_2 \end{bmatrix} = \begin{bmatrix} 4.67 \\ -3.8 \end{bmatrix}$$

Let
$$\Delta = \begin{vmatrix} 1.67 & -0.33 \\ -0.33 & 1.03 \end{vmatrix} = 1.61$$

Then
$$V_1 = \frac{\begin{vmatrix} 4.67 & -0.33 \\ -3.8 & 1.03 \end{vmatrix}}{\Delta} = 2.2 \text{ V}$$

$$V_2 = \frac{\begin{vmatrix} 1.67 & 4.67 \\ -0.33 & -3.8 \end{vmatrix}}{\Delta} = -3 \text{ V}$$

EXAMPLE 3.19. *For the circuit shown in Figure E3.19, find the current through 4 Ω.*

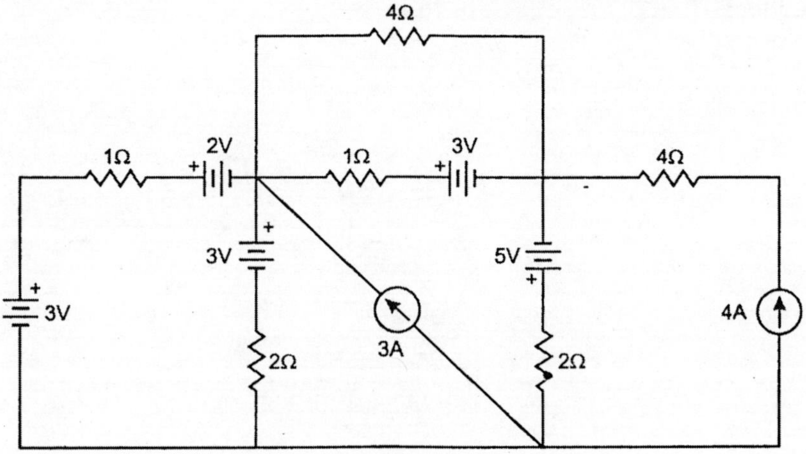

Fig. E3.19

SOLUTION. The circuit shown in Figure E3.19, can be redrawn as shown in Figure E3.19(*a*).

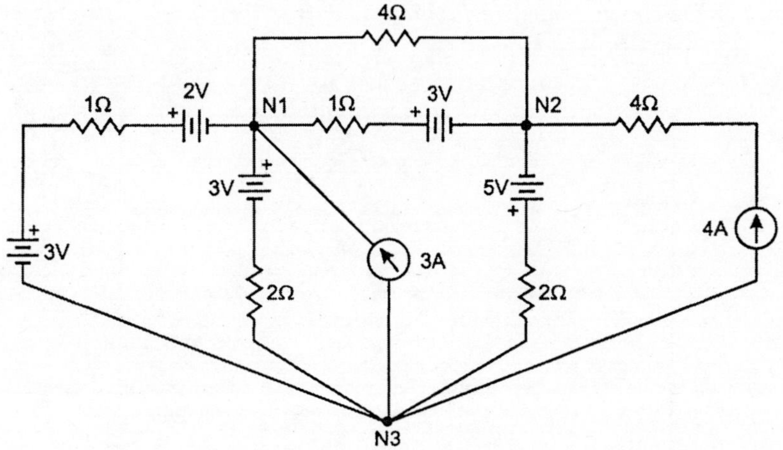

Fig. E3.19(a)

From the circuit shown Figure E3.19(*a*). There are three nodes, Node N1, node N2 and node N3. Let the voltage at node N1 be V_1 and the voltage at node N2 be V_2. Node N3 is taken as reference node and its voltage is taken as 0 V.

The node equations are,

Node N1 :

$$\frac{(V_1+2-3)}{1}+\frac{V_1-3}{2}+\frac{V_1-3-V_2}{1}+\frac{V_1-V_2}{4}-3 = 0$$

or $$2.75V_1 - 1.25V_2 = 8.5 \qquad ...(i)$$

Node N2 :

$$\frac{V_2+3-V_1}{1}+\frac{V_2-V_1}{4}+\frac{V_2+5}{2}-4 = 0$$

or $$1.75V_2 - 1.25V_1 = -1.5 \qquad ...(ii)$$

Writing equations (*i*) and (*ii*) in matrix form we get

$$\begin{bmatrix} 2.75 & -1.25 \\ -1.25 & 1.75 \end{bmatrix}\begin{bmatrix} V_1 \\ V_2 \end{bmatrix} = \begin{bmatrix} 8.5 \\ -1.5 \end{bmatrix}$$

Let $$\Delta = \begin{vmatrix} 2.75 & -1.25 \\ -1.25 & 1.75 \end{vmatrix} = 3.25$$

Then $$V_1 = \frac{\begin{vmatrix} 8.5 & -1.25 \\ -1.5 & 1.75 \end{vmatrix}}{\Delta} = 4 \text{ V}$$

$$V_2 = \frac{\begin{vmatrix} 2.75 & 8.5 \\ -1.25 & -1.5 \end{vmatrix}}{\Delta} = 2 \text{ V}$$

So the current through 4 Ω $= \dfrac{V_1-V_2}{4} = 0.5$ A

EXAMPLE 3.20. *For the circuit shown in Figure E3.20, find the current delivered by all the voltage sources.*

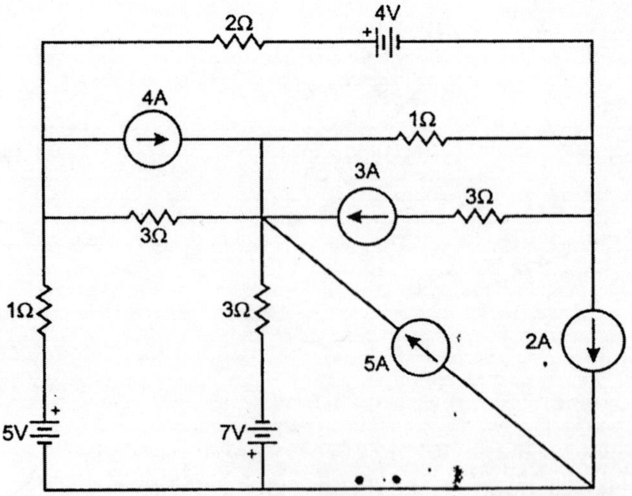

Fig. E3.20

SOLUTION. In the circuit shown in Figure E3.20, there are all together four nodes. The nodes N1, N2, N3 and N4 are indicated in the circuit shown in Figure E3.20(*a*). Let the voltage at node N1 be V_1, at node N2 be V_2 and at node N3 be V_3. Node N4 is taken as reference node and its voltage is taken as 0 V.

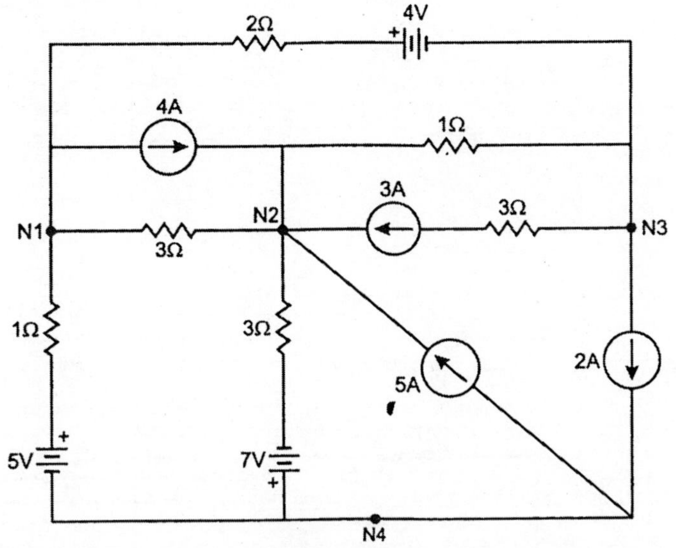

Fig. E3.20(a)

The node equations are,

Node N1 :
$$\frac{V_1-5}{1}+\frac{V_1-V_2}{3}+4+\frac{V_1-4-V_3}{2} = 0$$

or
$$1.83V_1 - 0.33V_2 - 0.5V_3 = 3 \qquad ...(i)$$

Node N2 :
$$\frac{V_2-V_1}{3}+\frac{V_2+7}{3}-3-5-4+\frac{V_2-V_3}{1} = 0$$

or
$$1.67V_2 - 0.33V_1 - V_3 = 9.76 \qquad ...(ii)$$

Node N3 :
$$2+3+\frac{V_3-V_2}{1}+\frac{V_3+4-V_1}{2} = 0$$

or
$$1.5V_3 - V_2 - 0.5V_1 = -6.33 \qquad ...(iii)$$

Writing equations (*i*), (*ii*) and (*iii*) in matrix form we get

$$\begin{bmatrix} 1.83 & -0.33 & -0.5 \\ -0.33 & 1.67 & -1 \\ -0.5 & -1 & 1.5 \end{bmatrix}\begin{bmatrix} V_1 \\ V_2 \\ V_3 \end{bmatrix} = \begin{bmatrix} 3 \\ 9.67 \\ -6.33 \end{bmatrix}$$

Let
$$\Delta = \begin{vmatrix} 1.83 & -0.33 & -0.5 \\ -0.33 & 1.67 & -1 \\ -0.5 & -1 & 1.5 \end{vmatrix} = 1.84$$

Then

$$V_1 = \frac{\begin{vmatrix} 3 & -0.33 & -0.5 \\ 9.67 & 1.67 & -1 \\ -6.33 & -1 & 1.5 \end{vmatrix}}{\Delta} = 3.67 \text{ V}$$

$$V_2 = \frac{\begin{vmatrix} 1.83 & 3 & -0.5 \\ -0.33 & 9.67 & -1 \\ -0.5 & -6.33 & 1.5 \end{vmatrix}}{\Delta} = 7.86 \text{ V}$$

$$V_3 = \frac{\begin{vmatrix} 1.83 & -0.33 & 3 \\ -0.33 & 1.67 & 9.67 \\ -0.5 & -1 & -6.33 \end{vmatrix}}{\Delta} = 2.24 \text{ V}$$

EXAMPLE 3.21. *For the circuit shown in Figure E3.21, find the current delivered by the voltage source.*

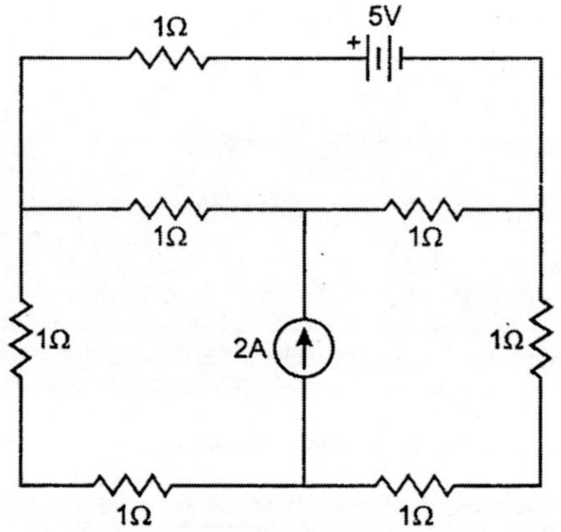

Fig. E3.21

SOLUTION. The circuit shown in Figure E3.21, can be redrawn as shown in Figure E3.21(*a*).

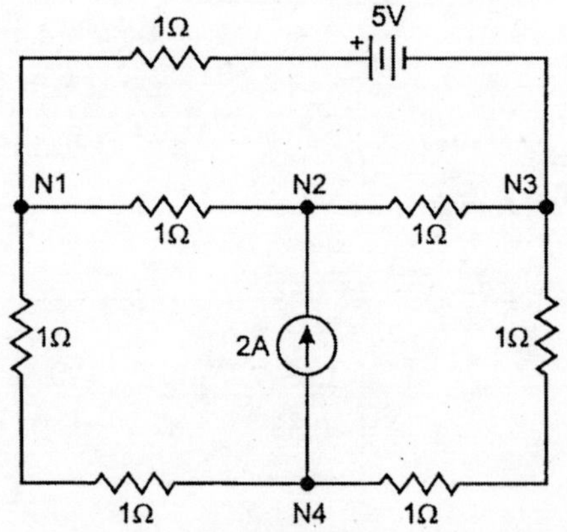

Fig. E3.21(a)

In the circuit shown in Figure E3.21, there are all together four nodes. The nodes N1, N2, N3 and N4 are indicated in the circuit shown in Figure E3.21(*a*). Let the voltage at node N1 be V_1, at node N2 be V_2 and at node N3 be V_3. Node N4 is taken as reference node and its voltage is taken as 0 V.

The node equations are,

Node N1 : $$\frac{V_1 - V_2}{1} + \frac{V_1 - 5 - V_3}{1} + \frac{V_1}{2} = 0$$

or $$2.5V_1 - V_2 - V_3 = 5 \qquad ...(i)$$

Node N2 : $$\frac{V_2 - V_1}{1} + \frac{V_2 - V_3}{1} - 2 = 0$$

or $$2V_2 - V_1 - V_3 = 2 \qquad ...(ii)$$

Node N3 : $$\frac{V_3 - V_2}{1} + \frac{V_3 + 5 - V_1}{1} + \frac{V_3}{2} = 0$$

or $$2.5V_3 - V_2 - V_1 = -5 \qquad ...(iii)$$

Writing equations (*i*), (*ii*) and (*iii*) in matrix form we get

$$\begin{bmatrix} 2.5 & -1 & -1 \\ -1 & 2 & -1 \\ -1 & -1 & 2.5 \end{bmatrix} \begin{bmatrix} V_1 \\ V_2 \\ V_3 \end{bmatrix} = \begin{bmatrix} 5 \\ 2 \\ -5 \end{bmatrix}$$

Let $$\Delta = \begin{vmatrix} 2.5 & -1 & -1 \\ -1 & 2 & -1 \\ -1 & -1 & 2.5 \end{vmatrix} = 3.5$$

Then $$V_1 = \frac{\begin{vmatrix} 5 & -1 & -1 \\ 2 & 2 & -1 \\ -5 & -1 & 2.5 \end{vmatrix}}{\Delta} = 3.43 \text{ V}$$

$$V_3 = \frac{\begin{vmatrix} 2.5 & -1 & 5 \\ -1 & 2 & 2 \\ -1 & -1 & -5 \end{vmatrix}}{\Delta} = 0.57 \text{ V}$$

$$i = \frac{V_1 - V_3}{1} = 2.86 \text{ A}$$

EXAMPLE 3.22. *For the circuit shown in Figure E3.22, find*

(*i*) *The number of nodes*

(*ii*) *The node voltage,* V_0 *and* i_0.

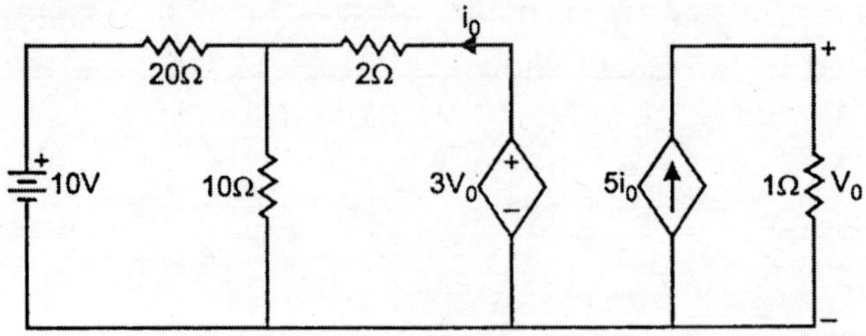

Fig. E3.22

SOLUTION. The circuit shown in Figure E3.22, can be redrawn as shown in Figure E3.22(*a*).

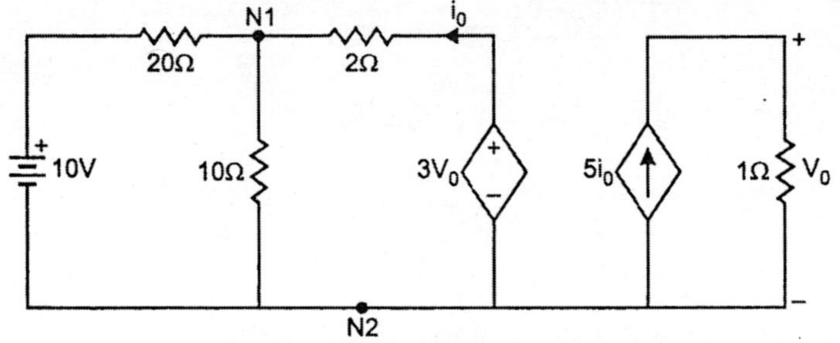

Fig. E3.22(a)

From the circuit shown Figure E3.22(*a*), there are only two nodes, node N1, node N2. Let the voltage at node N1 be V_1, node N2 is taken as reference node and its voltage is taken as 0 V.

The node equation is,

$$\frac{V_1 - 10}{20} + \frac{V_1 - 3V_0}{2} + \frac{V_1}{10} = 0$$

or $$0.65V_1 - 1.5V_0 = 0.5 \quad ...(i)$$

From Figure E3.22(*a*), $$V_0 = 1 \times 5i_0 \quad ...(ii)$$

Again $$i_0 = -\frac{V_1 - 3V_0}{2} \quad ...(iii)$$

Substituting equation (*iii*) in equation (*ii*) we get

$$V_0 = -\frac{5V_1 - 15V_0}{2}$$

or $$V_0 = \frac{5}{13}V_1 \quad ...(iv)$$

Substituting equation (*iv*) in equation (*i*) we get

$$0.65V_1 - \frac{1.5 \times 5}{13}V_1 = 0.5$$

or $$V_1 = 6.85 \text{ V} \quad ...(v)$$

Substituting equation (*v*) in equation (*iv*) we get

$$V_0 = \frac{5}{13} \times 6.85 = 2.63 \text{ V} \quad ...(vi)$$

Substituting equations (v) and (vi) in equation (iii) we get

$$i_0 = -\frac{V_1 - 3V_0}{2} = 0.52 \text{ A}$$

EXAMPLE 3.23. *For the circuit shown in Figure E3.23, find node voltages.*

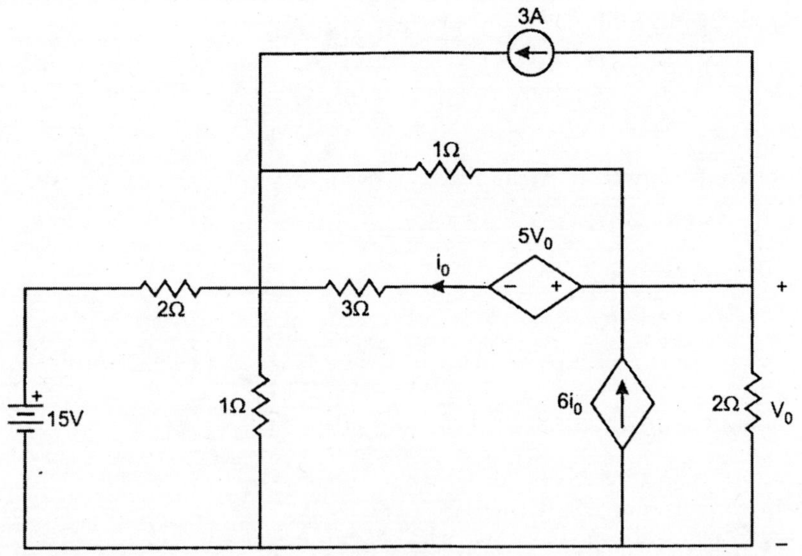

Fig. E3.23

SOLUTION. In the circuit shown in Figure E3.23, there are all together three nodes. The nodes N1, N2, and N3 are indicated in the circuit shown in Figure E3.23(a). Let the voltage at node N1 be V_1, and at node N2 be V_2. Node N3 is taken as reference node and its voltage is taken as 0 V.

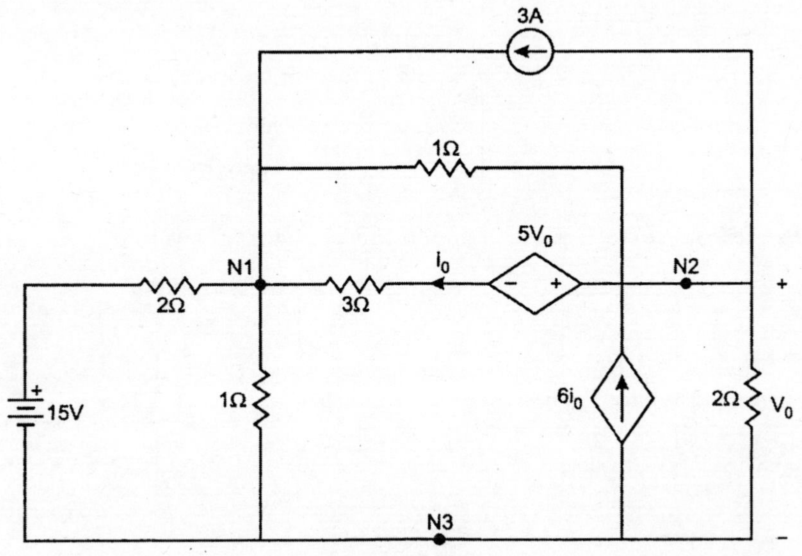

Fig. E3.23(a)

The node equations are,

Node N1 : $$\frac{V_1-15}{2}+\frac{V_1}{1}+\frac{V_1+5V_0-V_2}{3}+\frac{V_1-V_2}{1}-3 = 0$$

or $$2.83V_1 - 1.33V_2 + 1.67V_0 = 10.5 \qquad ...(i)$$

Node N2 : $$\frac{V_2}{2}+\frac{V_2-5V_0-V_1}{3}+\frac{V_2-V_1}{1}+3-6i_0 = 0$$

or $$1.83V_2 - 1.33V_1 - 1.67V_0 - 6i_0 = -3 \qquad ...(ii)$$

From Figure E3.23(*a*), $$V_0 = V_2 \qquad ...(iii)$$

Substituting equation (*iii*) in equation (*i*) and equation (*ii*) respectively we get

$$2.83V_1 + 0.34V_2 = 10.5 \qquad ...(iv)$$

$$0.16V_2 - 1.33V_1 - 6i_0 = -3 \qquad ...(v)$$

Again from Figure E3.23(*a*),

$$i_0 = \frac{V_2-5V_0-V_1}{3} = -\frac{4V_2-V_1}{3} \text{ (Substituting } V_0 = V_1) \qquad ...(vi)$$

Substituting equation (*v*) in equation (*iv*) we get

$$8.16V_2 - 3.33V_1 = -3 \qquad ...(vii)$$

Writing equations (*iv*) and (*vi*) in matrix form we get

$$\begin{bmatrix} 2.83 & 0.34 \\ -3.33 & 8.16 \end{bmatrix}\begin{bmatrix} V_1 \\ V_2 \end{bmatrix} = \begin{bmatrix} 10.5 \\ -3 \end{bmatrix}$$

Then $$\Delta = \begin{vmatrix} 2.83 & 0.34 \\ -3.33 & 8.16 \end{vmatrix} = 24.23$$

$$V_1 = \frac{\begin{vmatrix} 10.5 & 0.34 \\ -3 & 8.16 \end{vmatrix}}{\Delta} = 3.58 \text{ V}$$

$$V_2 = \frac{\begin{vmatrix} 2.83 & 10.5 \\ -3.33 & -3 \end{vmatrix}}{\Delta} = 1.09 \text{ V}$$

EXAMPLE 3.24. *For the circuit shown in Figure E3.24, find node voltages.*

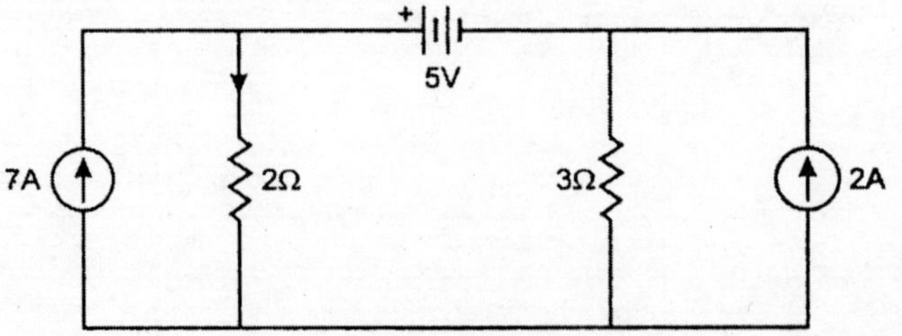

Fig. E3.24

SOLUTION. In the circuit shown in Figure E3.24, there are all together three nodes. The nodes N1, N2, and N3 are indicated in the circuit shown in Figure E3.24(*a*). Let the voltage at node N1 be V_1, and at node N2 be V_2. Node N3 is taken as reference node and its voltage is taken as 0 V. The node N1 and N2 are supernode.

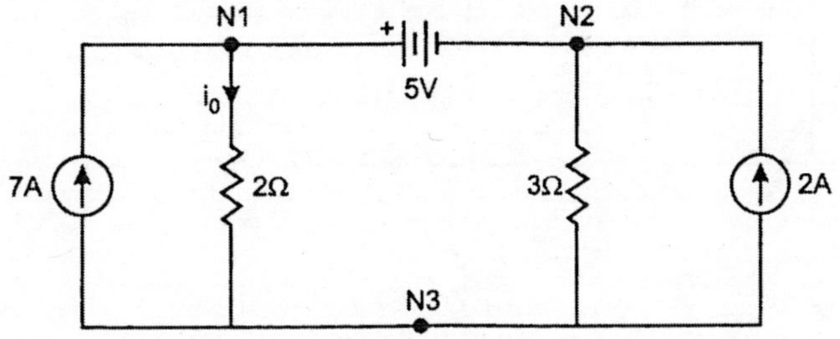

Fig. E3.24(a)

The KCL equation of supernode N1 and N2 are,

$$\frac{V_1}{2}+\frac{V_2}{3}-7-2 = 0$$

or
$$0.5V_1 + 0.33V_2 = 9 \qquad ...(i)$$

From Figure E3.24(*a*),

$$V_1 - V_2 = 5 \qquad ...(ii)$$

Solving equations (*i*) and (*ii*) we get

$$V_1 = 12.83\text{V}$$
$$V_2 = 7.83\text{V}$$

EXAMPLE 3.25. *For the circuit shown in Figure E3.25, find i_0.*

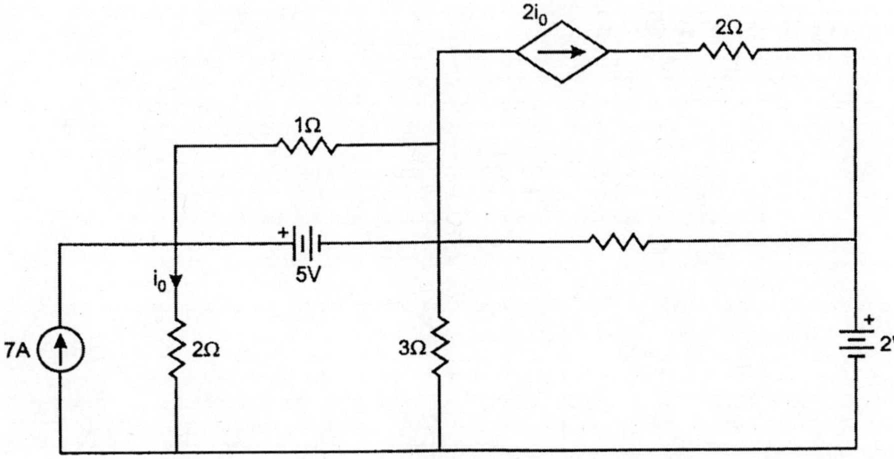

Fig. E3.25

SOLUTION. In the circuit shown in Figure E3.25, there are all together four nodes. The nodes N1, N2, N3 and N4 are indicated in the circuit shown in Figure E3.25(*a*). Let the voltage at node N1 be V_1, node N2 be V_2 and node N3 be V_3. Node N4 is taken as reference node and its voltage is taken as 0 V. The node N1 and N2 are supernode.

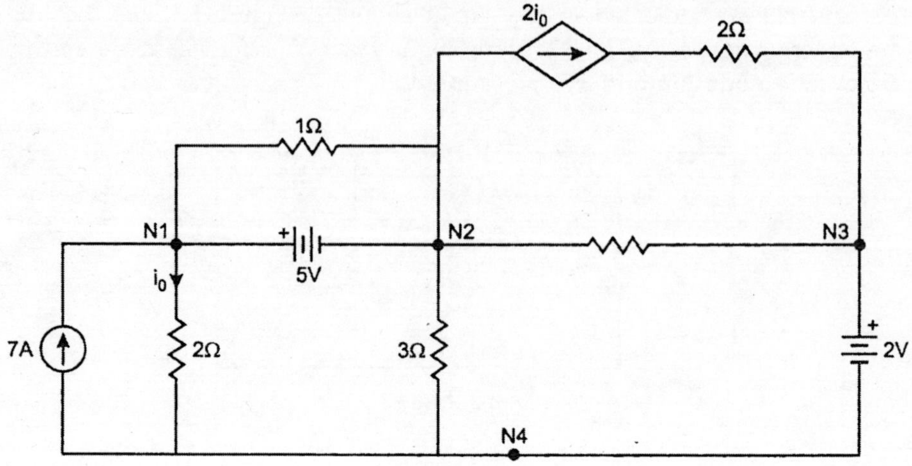

Fig. E3.25(a)

The KCL equation of supernode N1 and N2 are,

$$\frac{V_1}{2} - 7 + \frac{V_2}{3} + \frac{V_2 - V_3}{1} + 2i_0 = 0 \quad ...(i)$$

and $$V_1 - V_2 = 5 \quad ...(ii)$$

From Figure E3.25(*a*),

$$V_3 = 2 \text{ V} \quad ...(iii)$$

And $$i_0 = \frac{V_1}{2} \quad ...(iv)$$

Substituting equations (*iii*) and (*iv*) in equation (*i*) we get

$$1.5V_1 + 1.33V_2 = 9 \quad ...(v)$$

Solving equations (*ii*) and (*v*) we get

$$V_1 = 5.53\text{V}$$

$$V_2 = 0.53\text{V}$$

$$i_0 = \frac{V_1}{2} = 2.77\text{A}$$

EXAMPLE 3.26. *For the circuit shown in Figure E3.26, find i_0 and V_0.*

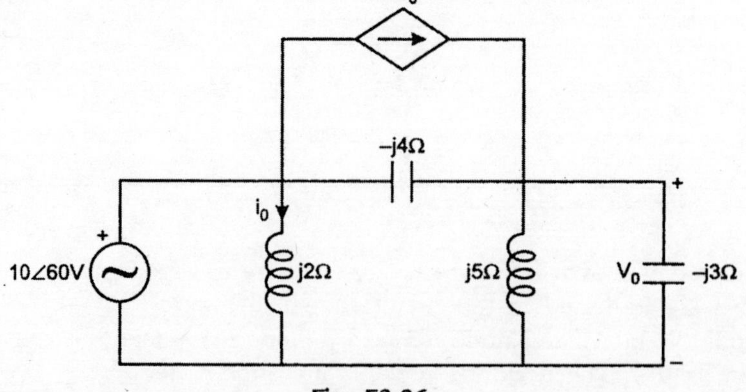

Fig. E3.26

SOLUTION. In the circuit shown in Figure E3.26, there are all together three nodes. The nodes N1, N2, and N3 are indicated in the circuit shown in Figure E3.26(*a*). Let the voltage at node N1 be V_1, and at node N2 be V_2. Node N3 is taken as reference node and its voltage is taken as 0 V.

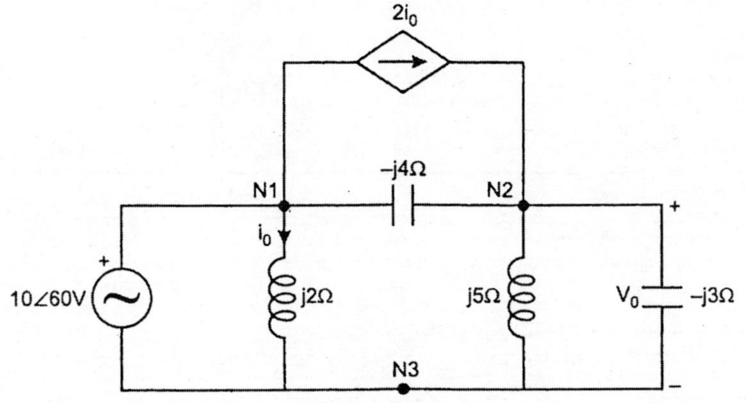

Fig. E3.26(a)

From Figure E3.26(*a*),

$$V_1 = 10\angle 60 \text{ V} = 5 + j8.66 \text{ V} \qquad ...(i)$$

The node equation of node N2,

$$\text{Node N2 :} \quad \frac{V_2}{j5} + \frac{V_2}{-j3} + \frac{V_2 - V_1}{-j4} - 2i_0 = 0 \qquad ...(ii)$$

Again

$$i_0 = \frac{V_1}{j2} = \frac{5 + j8.66}{j2} = 4.33 - j2.5 \text{ amp} \qquad ...(iii)$$

Substituting equations (*i*) and (*iii*) in equation (*ii*) we get

$$\frac{V_2}{j5} - \frac{V_2}{j3} - \frac{V_2 - 5 - j8.66}{j4} = 8.66 - j5$$

or

$$V_0 = V_2 = -3.88 - j0.167 \text{ V}$$

EXAMPLE 3.27. *For the circuit shown in Figure E3.27, find node voltages.*

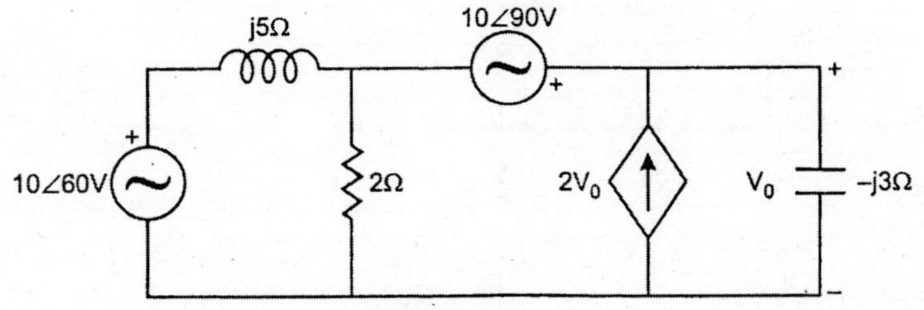

Fig. E3.27

SOLUTION. In the circuit shown in Figure E3.27, there are all together three nodes. The nodes N1, N2, and N3 are indicated in the circuit shown in Figure E3.27(*a*). Let the voltage at node N1 be V_1, and at node N2 be V_2. Node N3 is taken as reference node and its voltage is taken as 0V. The node N1 and N2 are supernode.

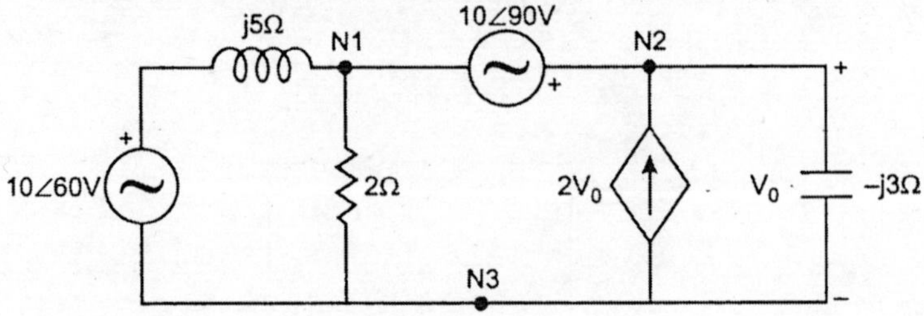

Fig. E3.27(a)

The KCL equation of supernode are,

$$\frac{V_1}{2}+\frac{V_1-10\angle 60}{j5}+\frac{V_2}{-j3}-2V_0 = 0 \qquad ...(i)$$

From Figure E3.26(*a*) $\qquad V_0 = V_2 \qquad ...(ii)$

Substituting equation (*ii*) in equation (*i*) we get

$$V_1(0.5 - j0.2) - V_2(2 - j0.33) = 1.73 - j1 \qquad ...(iii)$$

Again from Figure E3.26(*a*),

$$V_2 - V_1 = 10\angle 60 = 5 + j8.66 \qquad ...(iv)$$

Solving equations (*iii*) and (*iv*) we get

$$V_1 = (-8.81 - j10.54)\text{ V}$$
$$V_2 = (-3.81 - j1.88)\text{ V}$$

EXERCISES

1. For the circuit shown in Figure Q3.1, find the loop currents.

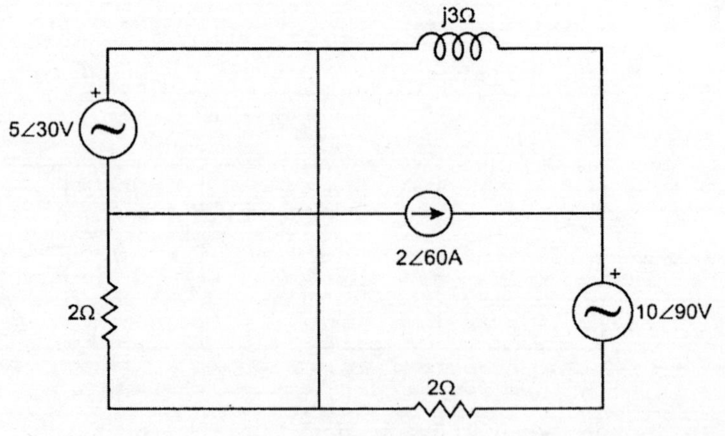

Fig Q3.1

2. For the circuit shown in Figure Q3.2, find the loop currents. All resistances are of 1 Ω.

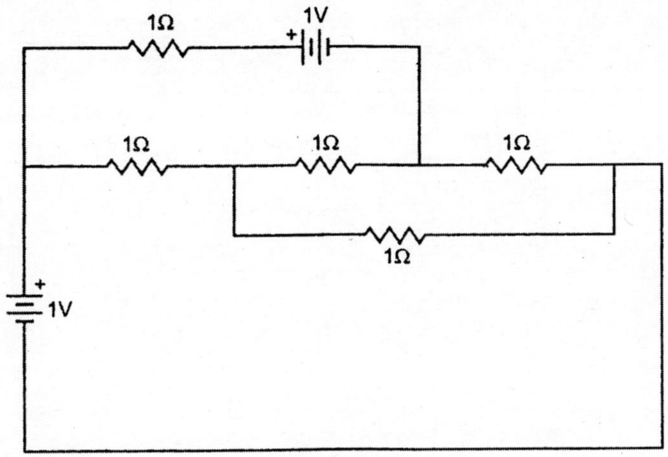

Fig. Q3.2

3. For the circuit shown in Figure Q3.3, find the current i_0.

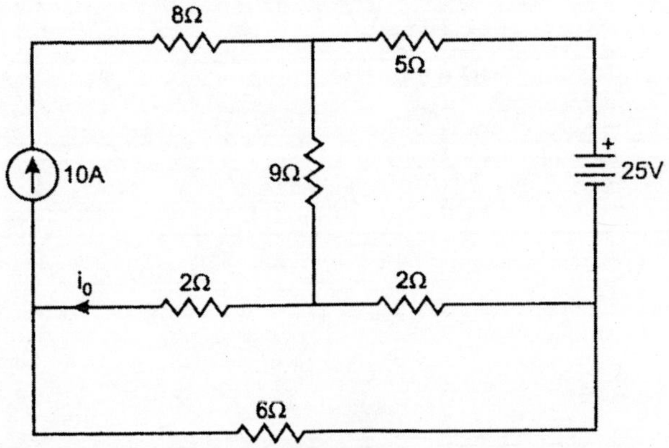

Fig. Q3.3

4. For the circuit shown in Figure Q3.4, find the current through 4 Ω using node analysis.

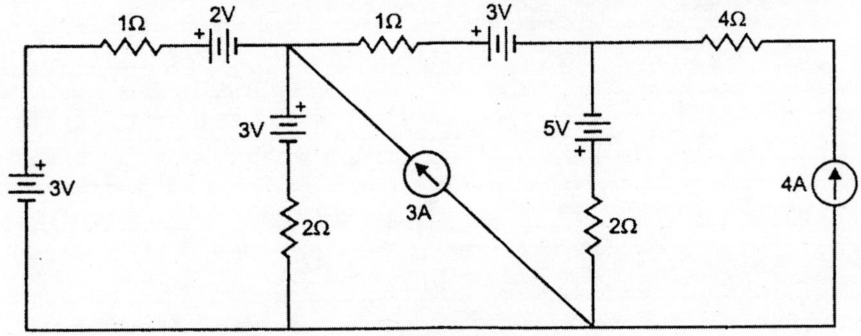

Fig. Q3.4

5. For the circuit shown in Figure Q3.5, find node voltages.

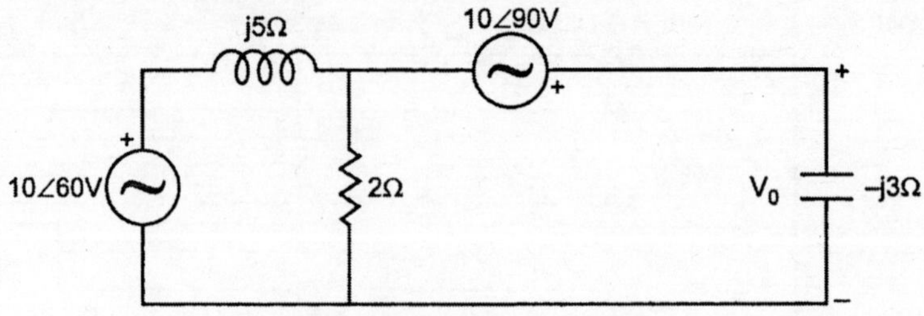

Fig. Q3.5

6. For the circuit shown in Figure Q3.6, find node voltages.

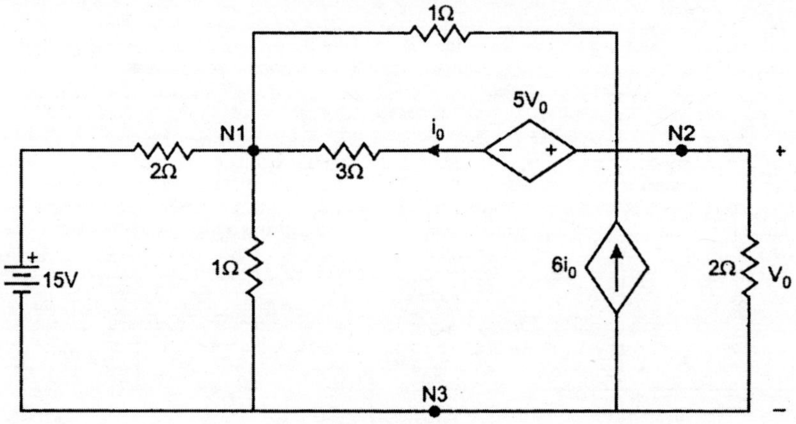

Fig. Q3.6

Chapter 4

NETWORK THEOREMS (DC CIRCUITS)

4.1 INTRODUCTION

All electrical networks can be solved by the application of KCL and KVL. But if the network is large and complex, these methods become very lengthy, laborious and time consuming.

Network theorems provide a simple and short cut method to solve complex problems easily. These theorems are basically derived from basic laws

1. Ohm's law
2. Kirchhoff's Current Law (KCL)
3. Kirchhoff's Voltage Law (KVL).

Some of the most important theorems are:

(*i*) Superposition Theorem
(*ii*) Reciprocity Theorem
(*iii*) Compensation Theorem
(*iv*) Thevenin's Theorem
(*v*) Norton's Theorem
(*vi*) Maximum Power Transfer Theorem
(*vii*) Millman's Theorem.

4.2 SUPERPOSITION THEOREM

Statement : In a linear network containing several independent sources, the overall response at any point in the network, equals t he sum of response due to each independent source considered separately, with all the other independent sources made inoperative.

An independent voltage source can be made inoperative by replacing it by the internal resistance of the source. It is zero for an ideal voltage source.

An independent current source can be made inoperative by replacing it by the internal resistance of the source. It is infinity for an ideal current source.

Note that the dependent sources are not removed from the network at any point in time.

SOLVED EXAMPLES

EXAMPLE 4.1. *Using Superposition theorem, find the loop currents for the circuit shown in Figure E4.1.*

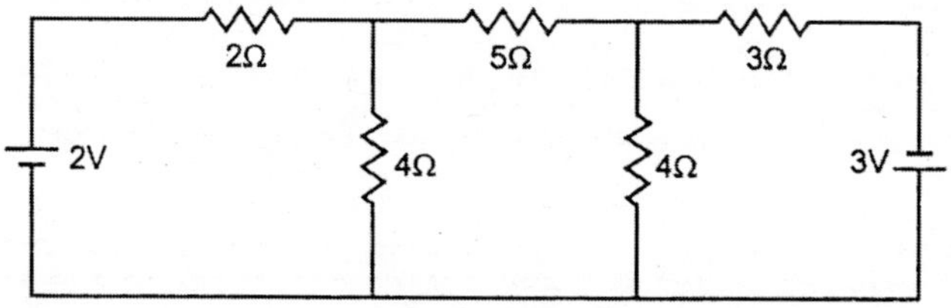

Fig. E4.1

SOLUTION. Considering only 2 V source and replacing the other source by short circuit (since the internal resistance of 3 V source is zero) we get Figure E4.1(*a*).

The loop currents are indicated in the circuit of Figure E4.1(*a*). Applying KVL in the loops we get.

For Loop 1 : $6i_1 - 4i_2 = 2$ ...(*i*)

For loop 2 : $9i_2 - 4i_1 - 4i_3 = 0$ ...(*ii*)

For loop 3 : $7i_3 - 4i_2 = 0$ ...(*iii*)

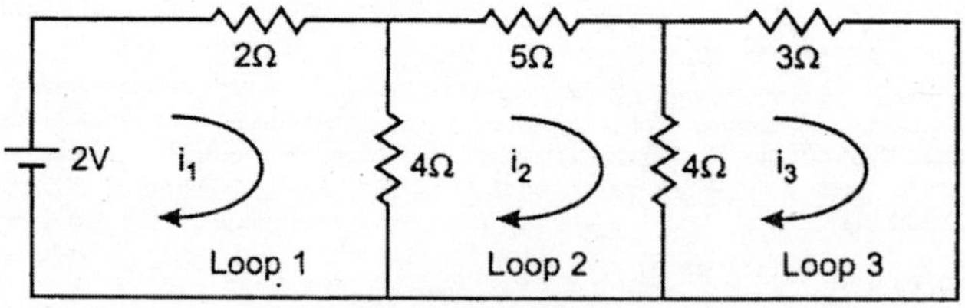

Fig. E4.1(a)

Writing the above three equations in matrix form we get

$$\begin{bmatrix} 6 & -4 & 0 \\ -4 & 9 & -4 \\ 0 & -4 & 7 \end{bmatrix} \times \begin{bmatrix} i_1 \\ i_2 \\ i_3 \end{bmatrix} = \begin{bmatrix} 2 \\ 0 \\ 0 \end{bmatrix}$$

Let
$$\Delta = \begin{vmatrix} 6 & -4 & 0 \\ -4 & 9 & -4 \\ 0 & -4 & 7 \end{vmatrix} = 170$$

Then the loop currents are,

$$i_1 = \frac{\begin{bmatrix} 2 & -4 & 0 \\ 0 & 9 & -4 \\ 0 & -4 & 7 \end{bmatrix}}{\Delta} = \frac{94}{170} = 0.553 \text{ Amp}$$

$$i_2 = \frac{\begin{vmatrix} 6 & 2 & 0 \\ -4 & 0 & -4a \\ 0 & 0 & 7 \end{vmatrix}}{\Delta} = \frac{56}{170} = 0.329 \text{ Amp}$$

and

$$i_3 = \frac{\begin{vmatrix} 6 & -4 & 2 \\ -4 & 9 & 0 \\ 0 & -4 & 0 \end{vmatrix}}{\Delta} = \frac{32}{170} = 0.188 \text{ Amp}$$

Considering only 3 V source and replacing the other source by short circuit (since the internal resistance of 2 V source is zero) we get Figure E4.1(*b*).

The loop currents are indicated in the circuit of Figure E4.1(*b*). Applying KVL in the loops we get.

For Loop 1 : $6i_1' - 4i_2' = 0$ ...(*iv*)

For loop 2 : $9i_2' - 4i_1' - 4i_3' = 0$ ...(*v*)

For loop 3 : $7i_3' - 4i_2' = 3$ ...(*vi*)

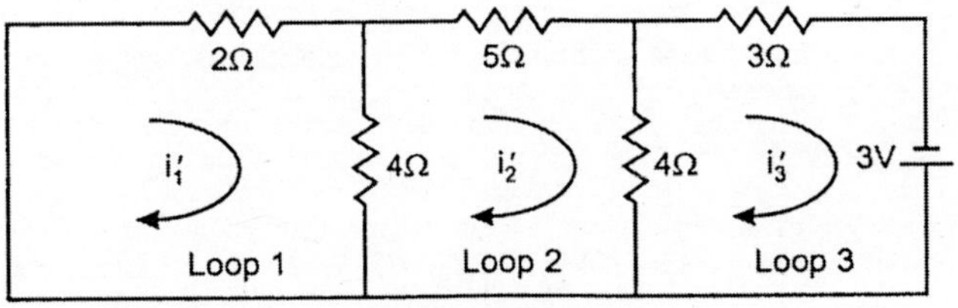

Fig. E4.1(b)

Writing the above three equations in matrix form we get,

$$\begin{bmatrix} 6 & -4 & 0 \\ -4 & 9 & -4 \\ 0 & -4 & 7 \end{bmatrix} \times \begin{bmatrix} i_1' \\ i_2' \\ i_3' \end{bmatrix} = \begin{bmatrix} 0 \\ 0 \\ 3 \end{bmatrix}$$

Let

$$\Delta = \begin{vmatrix} 6 & -4 & 0 \\ -4 & 9 & -4 \\ 0 & -4 & 7 \end{vmatrix} = 170$$

Then the loop currents are,

$$i_1' = \frac{\begin{bmatrix} 0 & -4 & 0 \\ 0 & 9 & -4 \\ 3 & -4 & 7 \end{bmatrix}}{\Delta} = \frac{48}{170} = 0.282 \text{ Amp}$$

$$i_2' = \frac{\begin{vmatrix} 6 & 0 & 0 \\ -4 & 0 & -4 \\ 0 & 3 & 7 \end{vmatrix}}{\Delta} = \frac{72}{170} = 0.424 \text{ Amp}$$

and $$i_3' = \frac{\begin{vmatrix} 6 & -4 & 0 \\ -4 & 9 & 0 \\ 0 & -4 & 3 \end{vmatrix}}{\Delta} = \frac{114}{170} = 0.671 \text{ Amp}$$

Thus, the total currents in the loops are

$$I_1 = 0.553 + 0.282 = 0.835 \text{ Amp}$$
$$I_2 = 0.329 + 0.424 = 0.753 \text{ Amp}$$
$$I_3 = 0.188 + 0.671 = 0.859 \text{ Amp}$$

EXAMPLE 4.2. *For the circuit shown in Figure E4.2, find the branch currents using superposition theorem.*

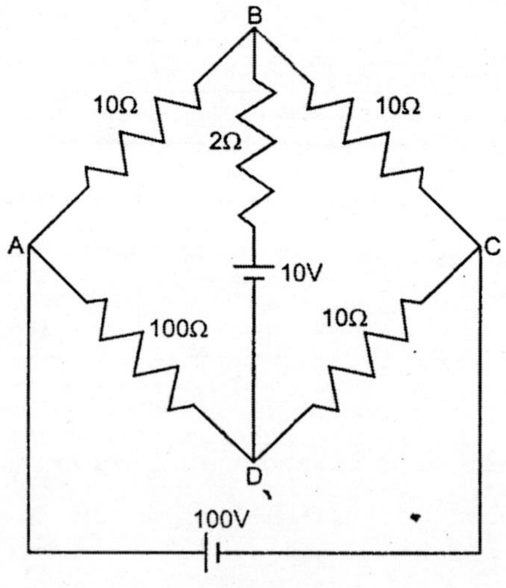

Fig. E4.2

SOLUTION. Considering only 100 V source and replacing the other source by short circuit (since the internal resistance of 10 V source is zero) we get Figure E4.2(*a*). The loop currents are indicated in the figure. The loop equations are:

Loop 1 : $112i_1 - 2i_2 - 100i_3 = 0$ ...(*i*)

Loop 2 : $22i_2 - 2i_1 - 10i_3 = 0$ ...(*ii*)

Loop 3 : $110i_3 - 100i_1 - 10i_2 = 100$ ...(*iii*)

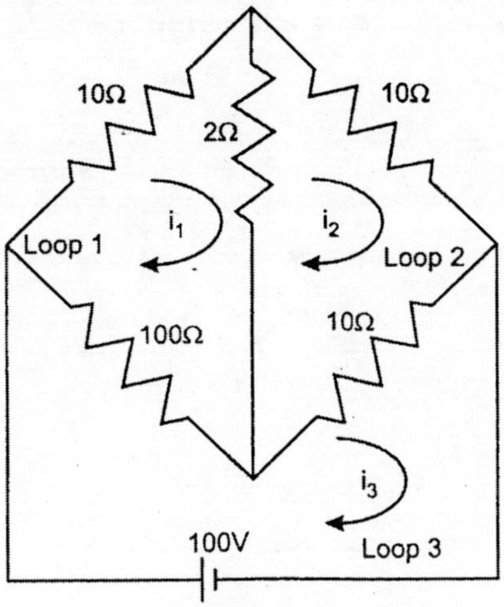

Fig. E4.2(a)

Writing equations (*i*), (*ii*) and (*iii*), in matrix form we get,

$$\begin{bmatrix} 112 & -2 & -100 \\ -2 & 22 & -10 \\ -100 & -10 & 110 \end{bmatrix} \begin{bmatrix} i_1 \\ i_2 \\ i_3 \end{bmatrix} = \begin{bmatrix} 0 \\ 0 \\ 100 \end{bmatrix}$$

Let
$$\Delta = \begin{vmatrix} 112 & -2 & -100 \\ -2 & 22 & -10 \\ -100 & -10 & 110 \end{vmatrix} = 35400$$

Then the loop currents are,

$$i_1 = \frac{\begin{vmatrix} 0 & -2 & -100 \\ 0 & 22 & -10 \\ 100 & -10 & 110 \end{vmatrix}}{\Delta} = \frac{222000}{35400} = 6.27 \text{ Amp}$$

$$i_2 = \frac{\begin{vmatrix} 112 & 0 & -100 \\ -2 & 0 & -10 \\ -100 & 100 & 110 \end{vmatrix}}{\Delta} = \frac{132000}{35400} = 3.73 \text{ Amp}$$

$$i_3 = \frac{\begin{vmatrix} 112 & -2 & 0 \\ -2 & 22 & 0 \\ -100 & -10 & 100 \end{vmatrix}}{\Delta} = \frac{246000}{35400} = 6.95 \text{ Amp}$$

Thus,

the current in the branch $AB = i_1 = 6.27$ Amp

the current in the branch $BC = i_2 = 3.73$ Amp

the current in the branch $CD = i_3 - i_1 = 3.22$ Amp

the current in the branch $AD = i_3 - i_1 = 0.68$ Amp

the current in the branch $BD = i_1 - i_2 = 2.54$ Amp

Source current $= i_3 = 6.95$ Amp

Considering only 10 V source and replacing the other source by short circuit (since the internal resistance of 100 V source is zero) we get Figure E4.2(*b*). The loop currents are indicated in the figure. The loop equations are :

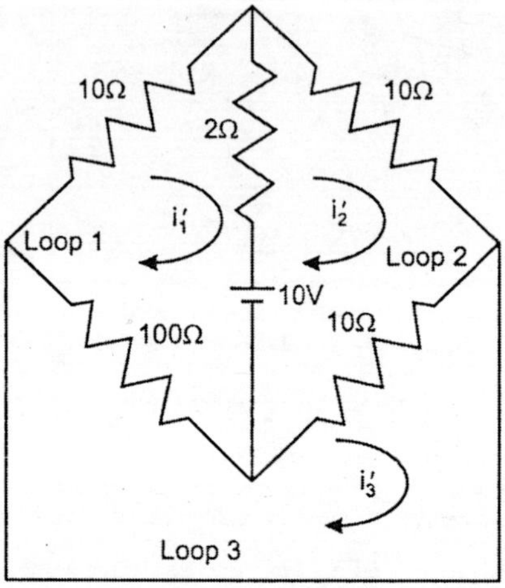

Fig. E4.2(b)

Loop 1 : $112i_1' - 2i_2' - 100i_3' = -10$ ...(*iv*)

Loop 2 : $22i_2' - 2i_1' - 10i_3' = 10$ ...(*v*)

Loop 3 : $110i_3' - 100i_1' - 10i_2' = 0$ ...(*vi*)

Writing equations (*iv*), (*v*) and (*vi*), in matrix form we get,

$$\begin{bmatrix} 112 & -2 & -100 \\ -2 & 22 & -10 \\ -100 & -10 & 110 \end{bmatrix} \begin{bmatrix} i_1' \\ i_2' \\ i_3' \end{bmatrix} = \begin{bmatrix} -10 \\ 10 \\ 0 \end{bmatrix}$$

Let
$$\Delta = \begin{vmatrix} 112 & -2 & -100 \\ -2 & 22 & -10 \\ -100 & -10 & 110 \end{vmatrix} = 35400$$

Then the loop currents are,

$$i_1' = \frac{\begin{vmatrix} -10 & -2 & -100 \\ 10 & 22 & -10 \\ 0 & -10 & 110 \end{vmatrix}}{\Delta} = \frac{-11000}{35400} = -0.3107 \text{ Amp}$$

$$i_2' = \frac{\begin{vmatrix} 112 & -10 & -100 \\ -2 & 10 & -10 \\ -100 & 0 & 110 \end{vmatrix}}{\Delta} = \frac{11000}{35400} = 0.3107 \text{ Amp}$$

$$i_3' = \frac{\begin{vmatrix} 112 & -2 & -10 \\ -2 & 22 & 10 \\ -100 & -10 & 0 \end{vmatrix}}{\Delta} = \frac{-9000}{35400} = -0.2542 \text{ Amp}$$

Thus,

the current in the branch $AB = i_1' = -0.3107$ Amp

the current in the branch $BC = i_2' = 0.3107$ Amp

the current in the branch $CD = i_3' - i_2' = -0.5649$ Amp

the current in the branch $AD = i_3' - i_2' = 0.0565$ Amp

the current in the branch $BD = i_1' - i_2' = -0.6214$ Amp

Source current $= i_3' = -0.2542$ Amp

Therefore the total currents through the branch $AB = i_1 + i_1' = 5.9593$ Amp

total currents through the branch $BC = i_2 + i_2' = 4.0407$ Amp

total currents through the branch $CD = (i_3 - i_2) + (i_3' - i_2') = 2.6551$ Amp

total currents through the branch $AD = (i_3 - i_1) + (i_3' - i_1') = 0.7365$ Amp

total currents through the branch $BD = (i_1 - i_2) + (i_1' - i_2') = 1.9186$ Amp

Source current $= i_3 + i_3' = 6.6958$ Amp.

EXAMPLE 4.3. *Using superposition theorem find the current through the branch AB for the circuit shown in Figure E4.3.*

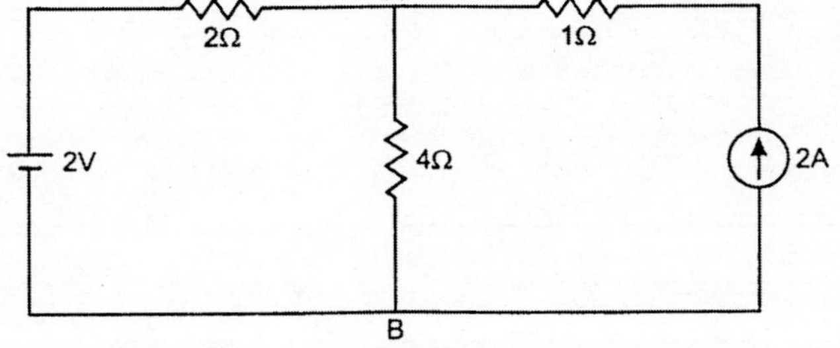

Fig. E4.3

SOLUTION. Considering only the voltage source and replacing the current source by open circuit (since the internal resistance of current source is infinity) we get Figure E4.3(*a*).

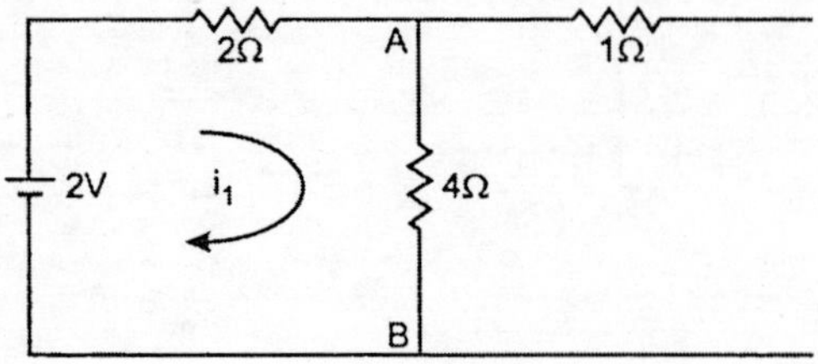

Fig. E4.3(a)

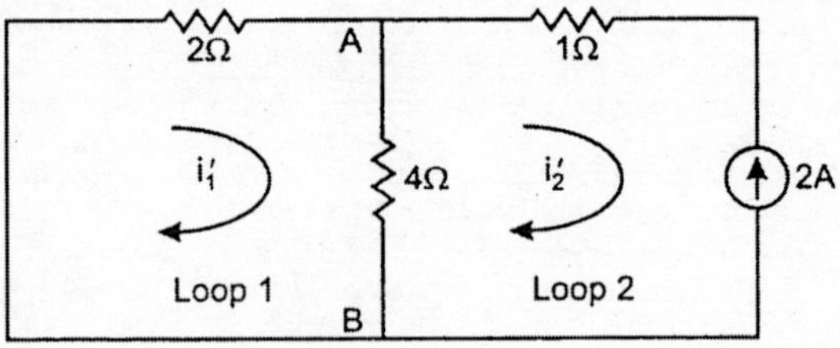

Fig. E4.3(b)

From Figure E4.3(*a*) the current through the branch *AB* is

$$i_1 = \frac{2}{2+4} = \frac{2}{6} = 0.33 \text{ Amp}$$

Now considering only the current source and replacing the voltage source by short circuit we get Figure E4.3(*b*). From loop 2 of Figure E4.3(*b*)

$$i_2' = -2 \text{ Amp}$$

Applying KVL in loop 1 of Figure E4.3(*b*) we get

$$6i_1' - 4i_2' = 0 \Rightarrow i_1' = \frac{4i_2'}{6} = \frac{-8}{6} = -1.33 \text{ Amp}$$

So the current through the branch *AB* is $i_1' - i_2' = -1.33 + 2 = 0.67$ Amp

So the total current through the branch *AB* is $i_1 + (i_1' - i_2') = 0.33 + 0.67 = 1$ Amp

EXAMPLE 4.4. *For the circuit shown in Figure E4.4, find the loop currents using Superposition theorem.*

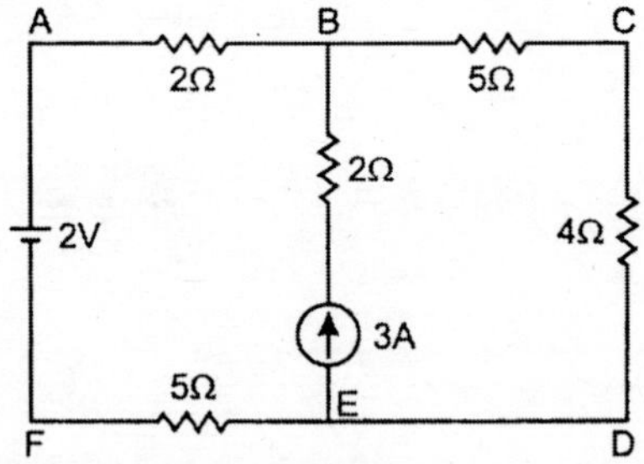

Fig. E4.4

SOLUTION. Considering only the voltage source and replacing the current source by open circuit we get the circuit shown in Figure E4.4(*a*)

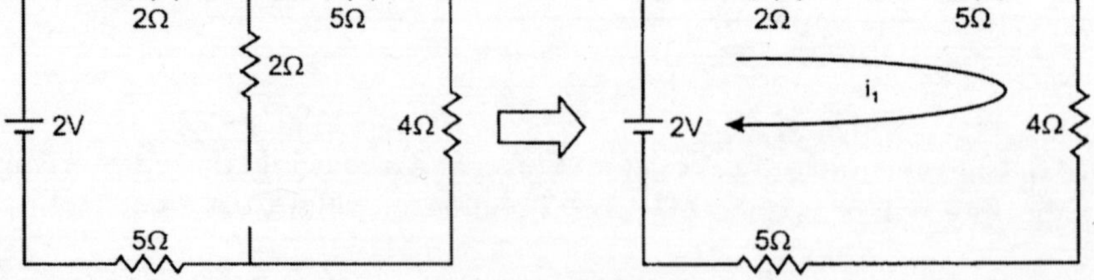

Fig. E4.4(a)

From Figure E4.4(*a*) we get $i_1 = \dfrac{2}{2+5+4+5} = 0.125$ Amp

Now considering only the Current source and replacing the voltage source by short circuit we get Figure E4.4(*b*).

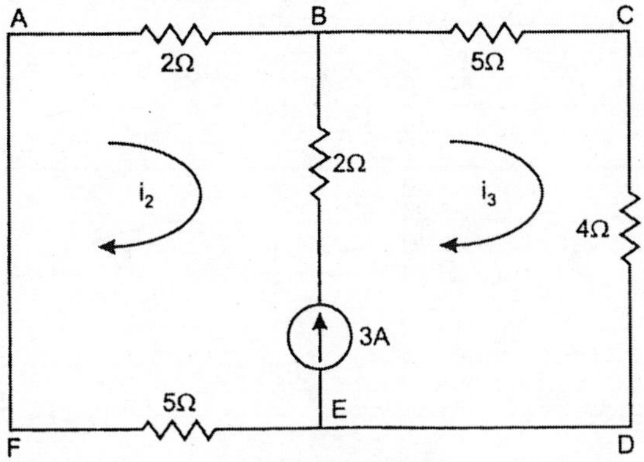

Fig. E4.4(b)

From Figure E4.4(*b*), we get

$$i_3 - i_2 = 3 \qquad \text{...(i)}$$

Writing the loop equation for the loop *ABCDEFA* we get

$$7i_2 + 9i_3 = 0 \qquad \text{...(ii)}$$

Solving the two equations we get

$$i_2 = -1.69 \text{ Amp}$$

$$i_3 = 1.31 \text{ Amp}$$

So the total current in the loops are

current in the loop *ABEFA* = 0.125 − 1.69 = −1.57 Amp

current in the loop *BCDEB* = 0.125 + 1.31 = 1.44 Amp

EXAMPLE 4.5. *For the circuit shown in Figure E4.5 find the current through the resistances using Superposition theorem.*

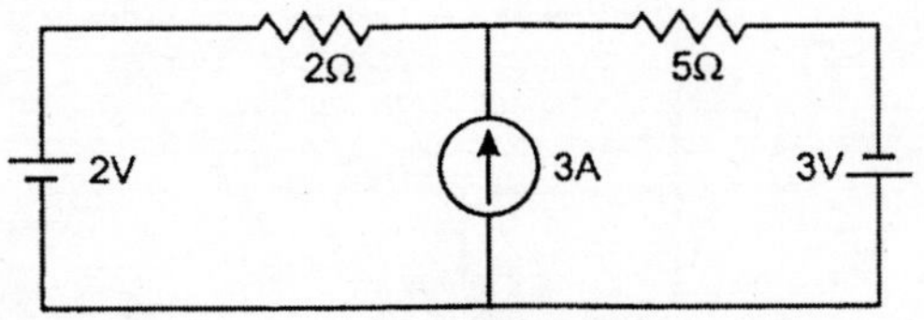

Fig. E4.5

SOLUTION. Considering only 2 V source and replacing the current source by open circuit and the voltage source by short circuit we get Figure E4.5(*a*).

From Figure E4.5(*a*) we get

$$i_2 = \frac{2}{2+5} = 0.286 \text{ Amp}$$

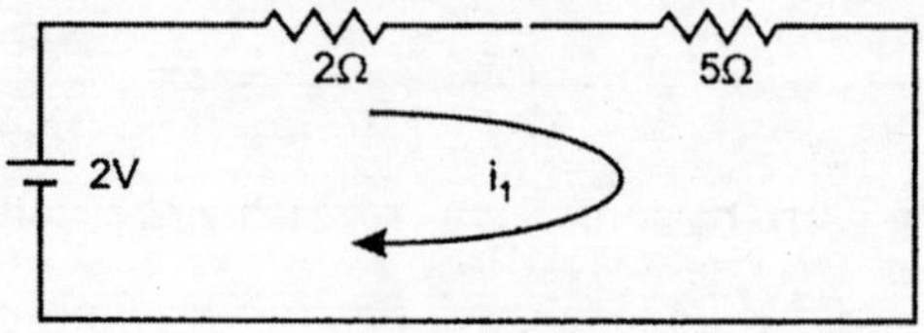

Fig. E4.5(a)

Considering only 3 V source and replacing the current source by open circuit and the voltage source by short circuit we get Figure E4.5(*b*).

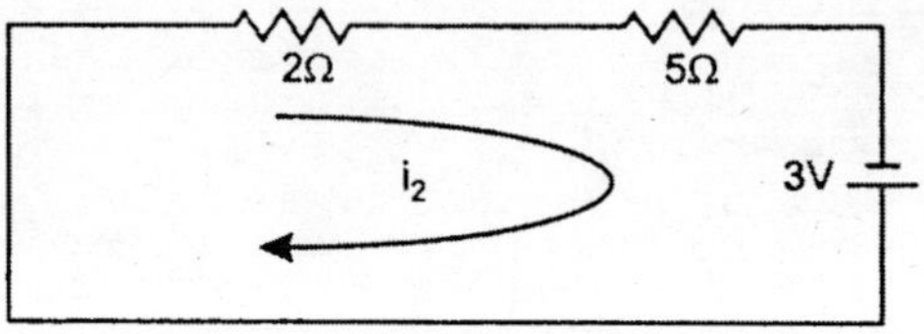

Fig. E4.5(b)

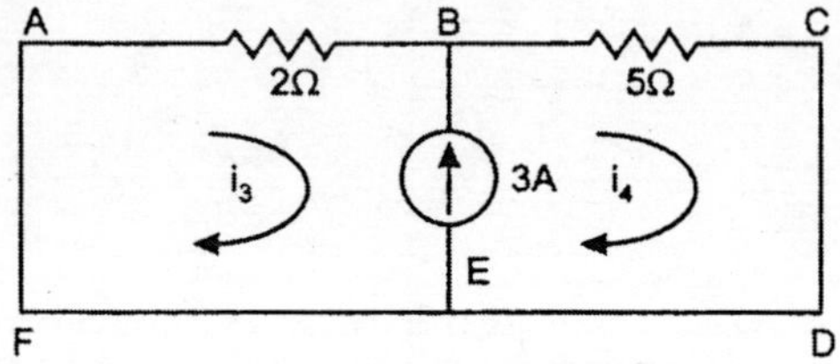

Fig. E4.5(c)

From Figure E4.5(*b*), we get

$$i_2 = \frac{-3}{2+5} = 0.429 \text{ Amp}$$

Considering only 3 A source and the voltage sources by short circuit we get Figure E4.5(*c*).

From Figure E4.5(*c*), we get

$$i_4 - i_3 = 3 \qquad ...(i)$$

Applying KVL in loop ABCDEFA we get,

$$2i_3 + 5i_4 = 0 \qquad ...(ii)$$

Solving the two equations we get

$$i_3 = -2.14 \text{ Amp} \quad \text{and} \quad i_4 = 0.857 \text{ Amp}$$

the current through 2 Ω = $i_1 + i_2 + i_3$ = 0.286 + 0.429 – 2.14 = –1.425 Amp

the current through 5 Ω = $i_1 + i_2 + i_4$ = 0.286 + 0.429 + 0.857 = 1.572 Amp

EXAMPLE 4.6. *For the circuit shown in Figure E4.6, find the open circuit voltage V_{AB} using Superposition theorem.*

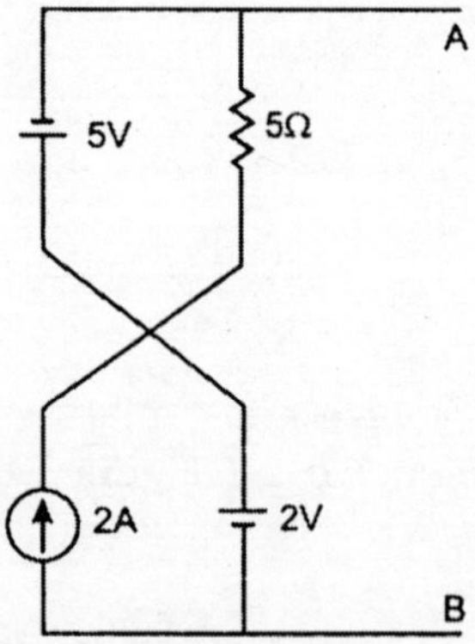

Fig. E4.6

SOLUTION. Considering only the 5 V source and replacing the 2 V source by short circuit and the 2 A current source by open circuit we get Figure E4.6(*a*).

From Figure E4.6(*a*) we get

$$i = \frac{5}{5} = 1 \text{ Amp} \qquad V_{AB1} = 5 \times 1 = 5 \text{ V}$$

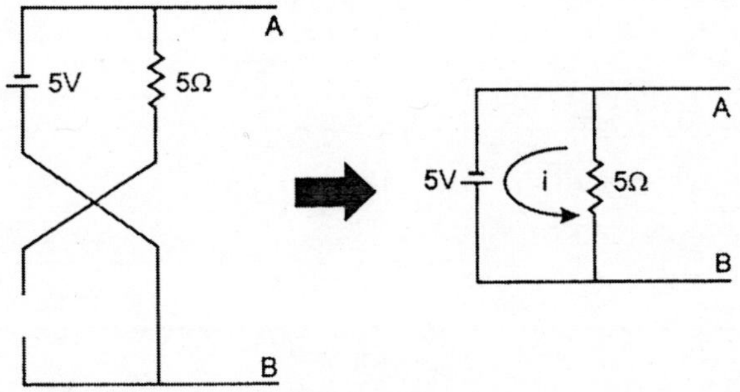

Fig. E4.6(a)

Now considering only the 2 V source and replacing the 5 V source by short circuit and the 2 A current source by open circuit we get Figure E4.6(*b*).

From Figure E4.6(*b*) we get

$$V_{AB2} = 2 \text{ V}$$

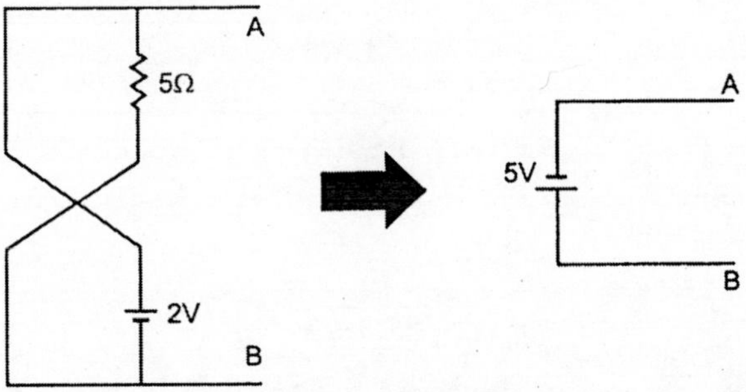

Fig. E4.6(b)

Now considering only the 2 A source and replacing the 5 V source by short circuit and the 2 V source by short circuit we get Figure E4.6(*c*).

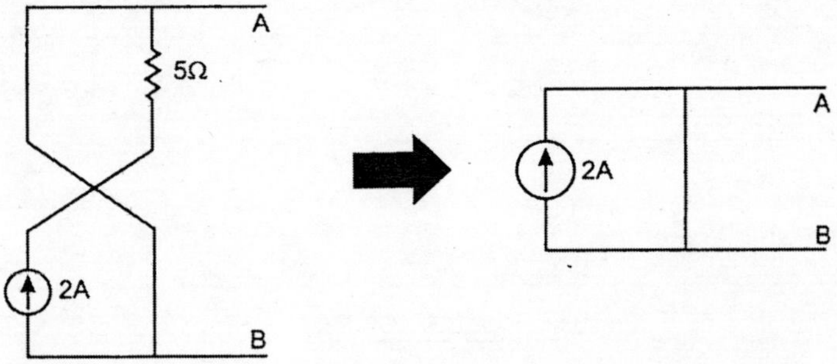

Fig. E4.6(c)

From Figure E4.6(c) we get

$$V_{AB3} = .0 \text{ V}$$

Therefore the open circuit voltage

$$V_{AB} = V_{AB1} + V_{AB2} + V_{AB3} = 5 - 2 + 0 = 3 \text{ V}$$

EXAMPLE 4.7. *For the circuit shown in Figure E4.7, find the drop across 3 Ω resistance using Superposition theorem.*

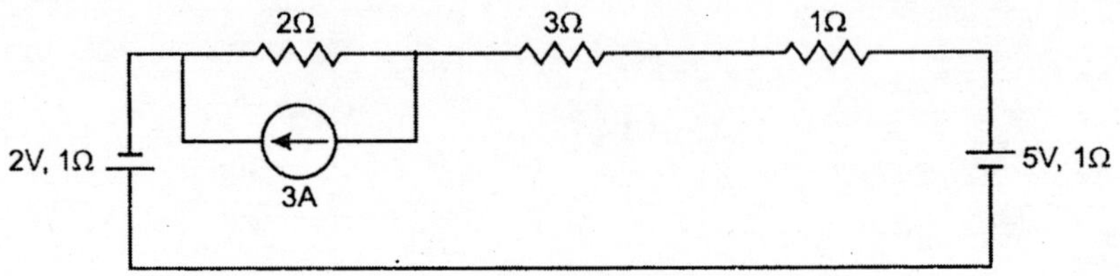

Fig. E4.7

SOLUTION. Considering only 2 V source and replacing the other voltage source by short circuit and the current source by open circuit we get Figure E4.7(a)

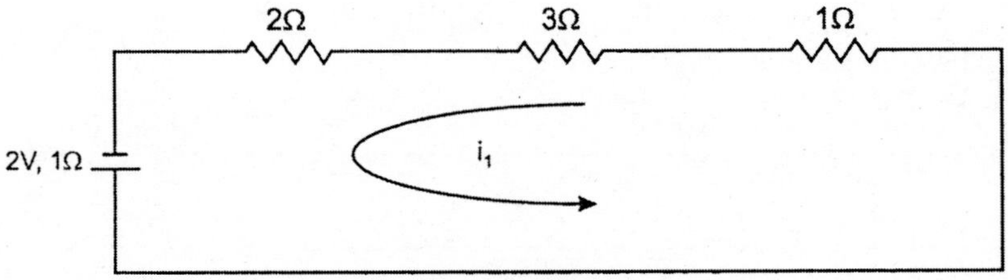

Fig. E4.7(a)

From Figure E4.7(a) we get

$$i_1 = \frac{2}{2+3+1+2} = 0.25 \text{ Amp}$$

Considering only 5 V source and replacing the other voltage source by short circuit and the current source by open circuit we get Figure E4.7(b)

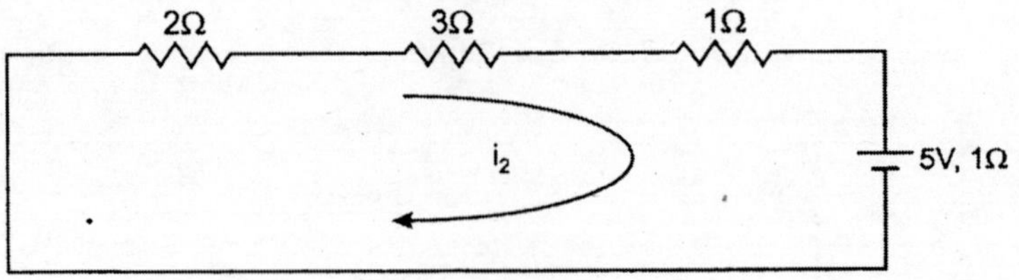

Fig. E4.7(b)

From Figure E4.7(b) we get

$$i_2 = \frac{-5}{2+3+1+1+1} = -0.625 \text{ Amp}$$

Considering only the 3 A current source and replacing the voltage sources by the internal resistances we get Figure E4.7(*c*)

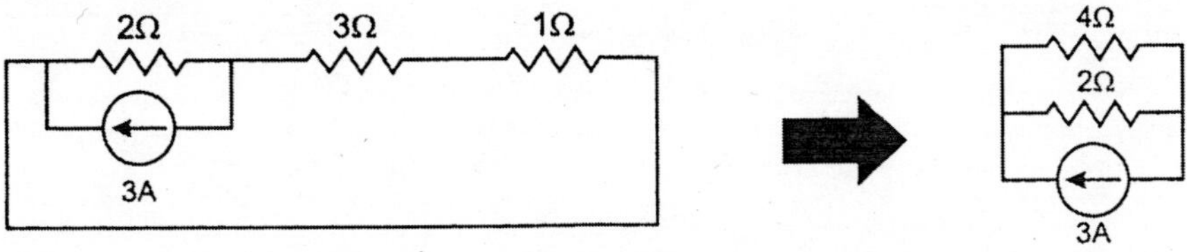

Fig. E4.7(c)

From Figure E4.7(*c*), current through 2 Ω resistor $= 3\times\dfrac{4}{4+2} = 2$ Amp

Current through 4 Ω $= 3\times\dfrac{2}{4+2} = 1$ Amp

So the current through 3 Ω is 1 Amp

Thus, the total current through 3 Ω = 0.25 − (−0.625) − 1 = −0.125Amp

EXAMPLE 4.8. *For the circuit shown in Figure E4.8, find the current through 2 Ω resistance using Superposition theorem.*

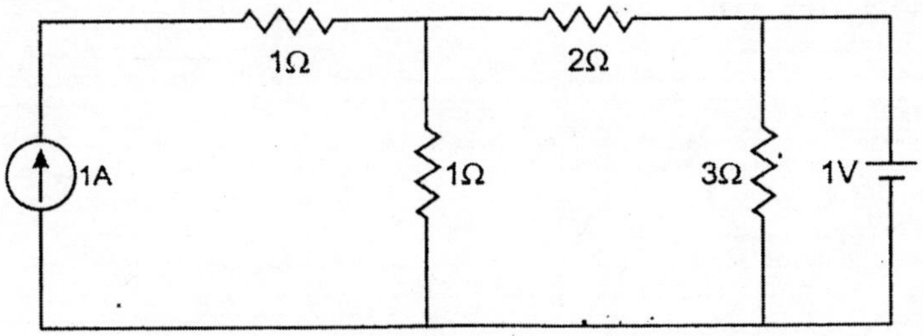

Fig. E4.8

SOLUTION. Considering only 1 V source and replacing the current source by open circuit we get Figure E4.8(*a*).

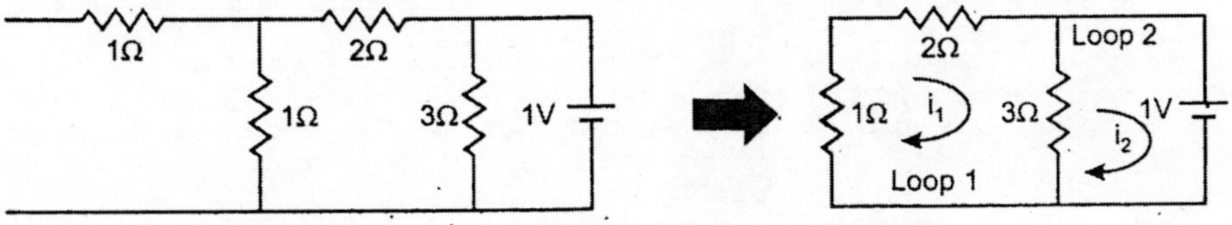

Fig. E4.8(a)

The loop currents are indicated in the Figure E4.8(*a*)

Writing the KVL equations we get

Loop 1 : $6i_1 - 3i_2 = 0$ ...(*i*)

Loop 2 : $3i_2 - 3i_1 = -1$ ...(*ii*)

Writing the above equation in matrix form we get

$$\begin{bmatrix} 6 & -3 \\ -3 & 3 \end{bmatrix} \times \begin{bmatrix} i_1 \\ i_2 \end{bmatrix} = \begin{bmatrix} 0 \\ -1 \end{bmatrix}$$

Let
$$\Delta = \begin{vmatrix} 6 & -3 \\ -3 & 3 \end{vmatrix} = 9$$

Then the current through the resistance 2 Ω = $i_1 = \dfrac{\begin{vmatrix} 0 & -3 \\ -1 & 3 \end{vmatrix}}{\Delta} = \dfrac{-3}{9} = -0.333$ Amp

Now considering only the current source and replacing the voltage source by short circuit we get Figure E4.8(*b*)

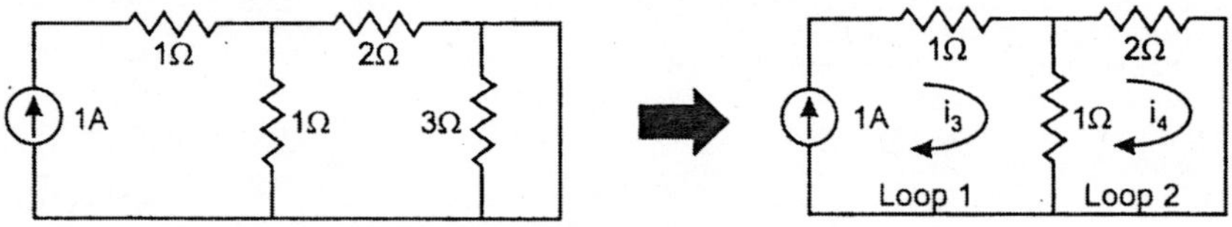

Fig. E4.8(b)

The loop currents are indicated in the Figure E4.8(*b*).

From Figure E4.8(*b*) we get

$$i_3 = 1 \text{ Amp} \quad \text{(From loop 1)} \qquad ...(iii)$$

Writing the KVL equation for loop 2 we get

$$3i_4 - i_3 = 0 \qquad ...(iv)$$

Solving equation (*iii*) and (*iv*) we get

current through 2 Ω = $i_4 = 0.33$ Amp

So the total current through 2 Ω = $i_1 + i_4 = 0$ Amp

EXAMPLE 4.9. *For the circuit shown in Figure E4.9 find i_0 using Superposition theorem.*

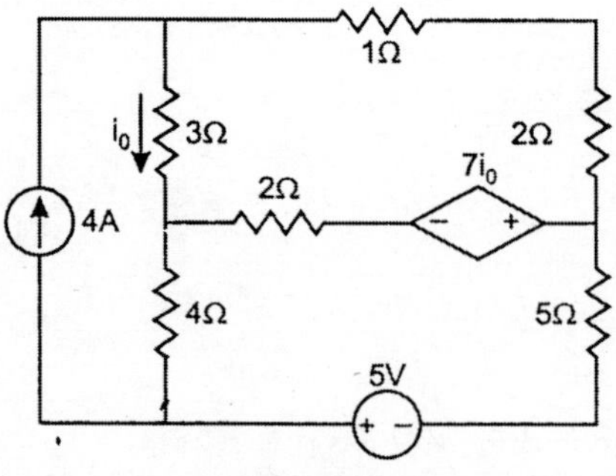

Fig. E4.9

SOLUTION. Considering only 4 A source and replacing the voltage source by short circuit we get Figure E4.9(*a*). Note that the independent source should not be removed from the circuit at all.

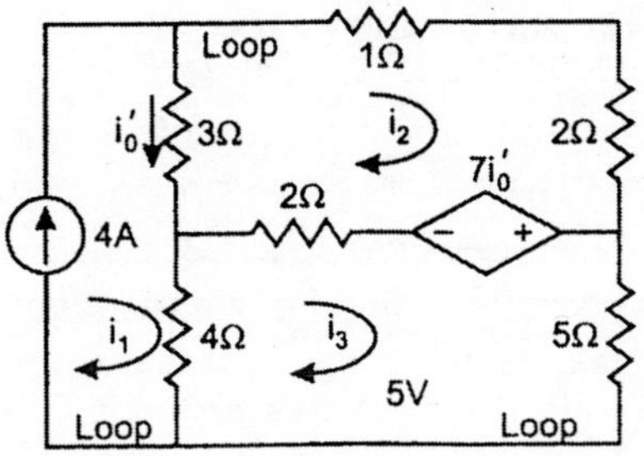

Fig. E4.9(a)

Writing the loop equations we get

Loop 1 : $i_1 = 4$ A ...(*i*)

Loop 2 : $8i_2 - 2i_3 - 3i_1 = -7i_0$ ...(*ii*)

Loop 3 : $11i_3 - 2i_2 - 4i_1 = 7i_0$ ...(*iii*)

Further from the circuit we can say $i_0' = i_1 - i_2$ ...(*iv*)

Substituting equation (*i*) and equation (*iv*) in equation (*ii*) and equation (*iii*) we get

$$i_2 - 2i_3 = -16 \quad ...(v)$$

$$5i_2 + 11i_3 = 44 \quad ...(vi)$$

Writing the equations in matrix form we get

$$\begin{bmatrix} 1 & -2 \\ 5 & 11 \end{bmatrix} \times \begin{bmatrix} i_2 \\ i_3 \end{bmatrix} = \begin{bmatrix} -16 \\ 44 \end{bmatrix}$$

Let $$\Delta = \begin{vmatrix} 1 & -2 \\ 5 & 11 \end{vmatrix} = 21$$

Then $$i_2 = \frac{\begin{vmatrix} -16 & -2 \\ 44 & 11 \end{vmatrix}}{\Delta} = 4.2 \text{ A}$$

Thus from equation (*iv*) $i_0' = i_1 - i_2 = -0.2$ A

Now considering only 5 V source and replacing the current source by open circuit we get Figure E4.9(*b*). Note that the independent source should not be removed from the circuit at all.

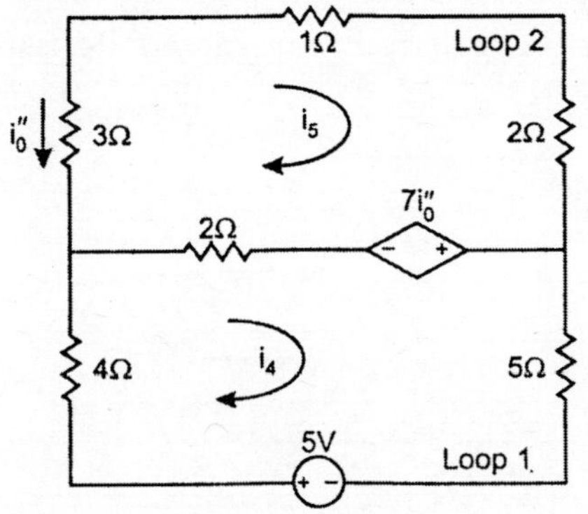

Fig. E4.9(b)

Writing the KVL equations for each loop we get

Loop 1 : $11i_4 - 2i_5 = 5 + 7i_0''$ ...(*vii*)

Loop 2 : $8i_5 - 2i_4 = -7i_0''$ ...(*viii*)

Again from the Figure E4.9(*b*) we can say $i_0'' = -i_5$ ...(*ix*)

Then substituting equation (*ix*) in equation (*vii*) and equation (*viii*) we get

$$11i_4 + 5i_5 = 5 \qquad ...(x)$$

$$i_5 - 2i_4 = 0 \qquad ...(xi)$$

Writing equations (*x*) and (*xi*) in matrix form we get

$$\begin{bmatrix} 11 & 5 \\ -2 & 1 \end{bmatrix} \times \begin{bmatrix} i_4 \\ i_5 \end{bmatrix} = \begin{bmatrix} 5 \\ 0 \end{bmatrix}$$

Let $\Delta = \begin{vmatrix} 11 & 5 \\ -2 & 1 \end{vmatrix} = 21$

Then $i_0'' = i_5 = \dfrac{\begin{vmatrix} 11 & 5 \\ -2 & 0 \end{vmatrix}}{\Delta} = 0.48 \text{ A}$

So the total current $i_0 = i_0' + i_0'' = -0.68 \text{ A}$

EXAMPLE 4.10. *For the circuit shown in Figure E4.10, find the current through 5 Ω using Superposition theorem.*

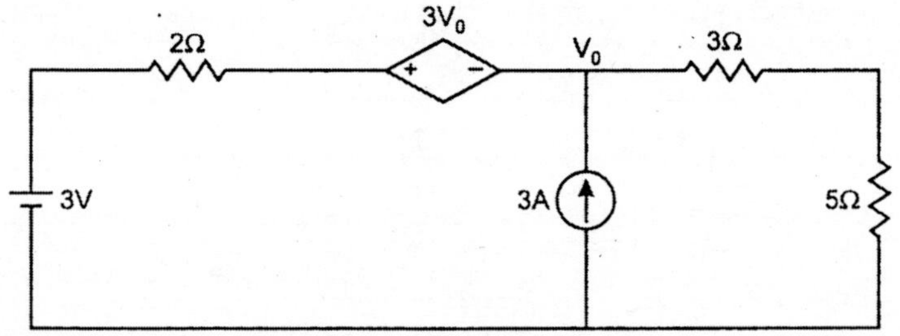

Fig. E4.10

SOLUTION. Considering only 3 V source and replacing the current source by open circuit, the dependent source is not removed at all, we get Figure E4.10(*a*)

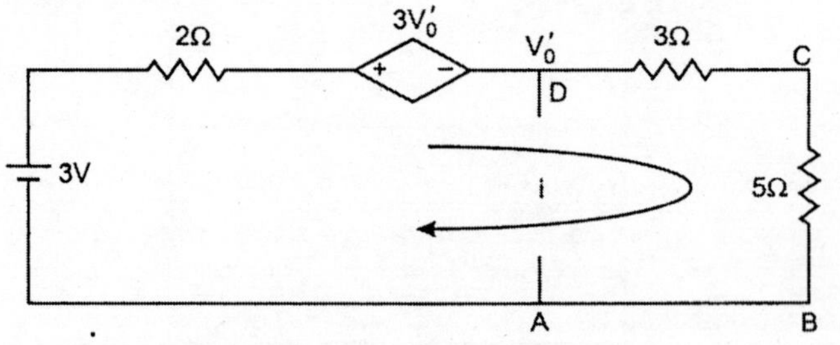

Fig. E4.10(a)

Applying KVL for the circuit in Figure E4.10(*a*) we get

$$10i + 3V_0' - 3 = 0 \quad ...(i)$$

Again applying KVL in the loop *ABCDA* we get

$$8i = V_0' \quad ...(ii)$$

Substituting equation (*ii*) in equation (*i*) we get

$$i = 0.088 \text{ Amp} \quad ...(iii)$$

Considering only 3 A source and replacing the voltage source by short circuit, the dependent source is not removed at all, we get Figure E4.10(*b*)

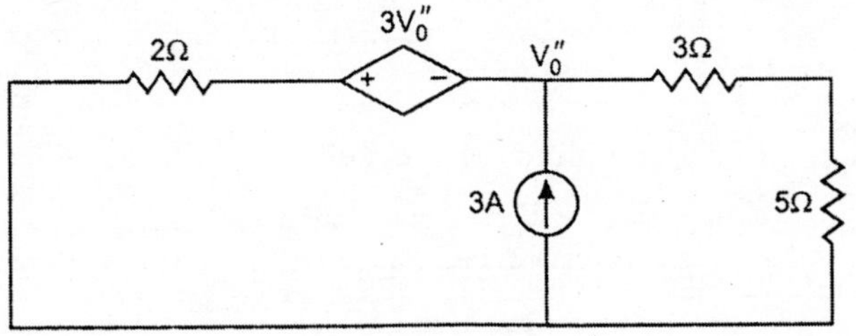

Fig. E4.10(b)

Applying node analysis for the circuit shown in Figure E4.10(*b*) we get

$$\frac{V_0''}{8} + \frac{V_0'' + 3V_0''}{2} - 3 = 0 \quad ...(iv)$$

Solving equation (*iv*) we get $V_0'' = 1.412$ volt

Then the current through 5 Ω = $\frac{V_0''}{8} = 0.176$ Amp

So the total current through 5 Ω = 0.088 + 0.176 = 0.264 Amp

EXAMPLE 4.11. *For the circuit shown in Figure E4.11, find V_0 using Superposition theorem.*

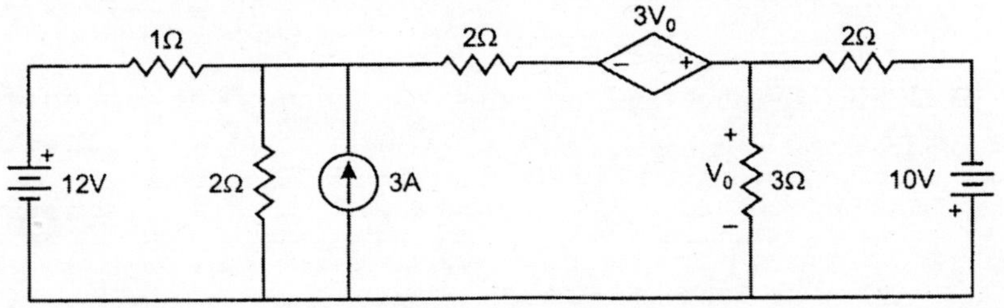

Fig. E4.11

SOLUTION. Considering only 12 V source and replacing the independent current source by open circuit and the independent 10 V voltage source by short circuit we get the circuit shown in Figure E4.11(*a*)

Writing the KVL equations for each loop we get

Loop 1 : $\quad 3i_1' - 2i_2' = 12 \quad ...(i)$

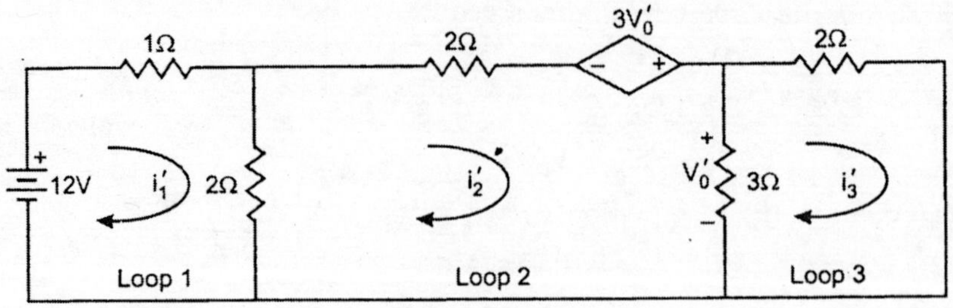

Fig. E4.11(a)

Loop 2 : $7i_2' - 2i_1' - 3i_3' = 3V_0'$ ...(*ii*)

Again $V_0' = 3 \times (i_2' - i_3')$ ...(*iii*)

Loop 3 : $5i_3' - 3i_2' = 0$ ...(*iv*)

Substituting equation (*iii*) in equation (*ii*) we get

$$4i_2 - 2i_1 = 0 \quad \text{...(v)}$$

Writing equations (*i*), (*iv*) and (*v*) in matrix form we get

$$\begin{bmatrix} 3 & -2 & 0 \\ 0 & -3 & 5 \\ -2 & 4 & 0 \end{bmatrix} \times \begin{bmatrix} i_1' \\ i_2' \\ i_3' \end{bmatrix} = \begin{bmatrix} 12 \\ 0 \\ 0 \end{bmatrix}$$

Let
$$\Delta = \begin{vmatrix} 3 & -2 & 0 \\ 0 & -3 & 5 \\ -2 & 4 & 0 \end{vmatrix} = -40$$

Then
$$i_2' = \frac{\begin{vmatrix} 3 & 12 & 0 \\ 0 & 0 & 5 \\ -2 & 0 & 6 \end{vmatrix}}{\Delta} = 30 \text{ A} \quad \text{and} \quad i_3' = \frac{\begin{vmatrix} 3 & -2 & 12 \\ 0 & -3 & 0 \\ -2 & -2 & 0 \end{vmatrix}}{\Delta} = 8 \text{ A}$$

Thus, $V_0' = 3 \times (i_2' - i_3') = 36$ V

Considering only 10 V source and replacing the independent current source by open circuit and the independent 12 V voltage source by short circuit we get the circuit shown in Figure E4.11(*b*).

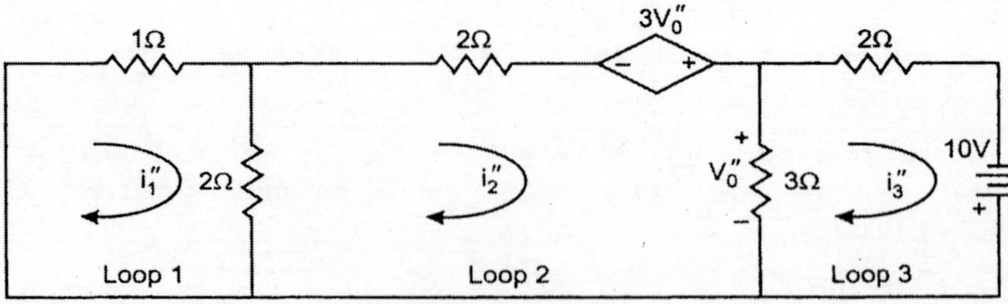

Fig. E4.11(b)

Writing the KVL equations for each loop we get

Loop 1 : $3i_1'' - 2i_2'' = 0$ ...(*vi*)

Loop 2 : $7i_2'' - 2i_1'' - 3i_3'' = 3V_0''$ ...(*vii*)

Again $V_0'' = 3 \times (i_2'' - i_3'')$ ...(*viii*)

Loop 3 : $5i_3'' - 3i_2'' = 10$ ...(*ix*)

Substituting equation (*iii*) in equation (*ii*) we get

$$-2i_1'' - 2i_2'' + 6i_3'' = 0 \quad ...(x)$$

Writing the equations (*vi*), (*ix*) and (*x*) in matrix form we get,

$$\begin{bmatrix} 3 & -2 & 0 \\ 0 & -3 & 5 \\ -2 & -2 & 6 \end{bmatrix} \times \begin{bmatrix} i_1'' \\ i_2'' \\ i_3'' \end{bmatrix} = \begin{bmatrix} 0 \\ 0 \\ 10 \end{bmatrix}$$

Let $$\Delta = \begin{vmatrix} 3 & -2 & 0 \\ 0 & -3 & 5 \\ -2 & -2 & 6 \end{vmatrix} = -4$$

Then $$i_2'' = \frac{\begin{vmatrix} 3 & 0 & 0 \\ 0 & 0 & 5 \\ -2 & 10 & 6 \end{vmatrix}}{\Delta} = 37.5 \text{ A and } i_3'' = \frac{\begin{vmatrix} 3 & -2 & 0 \\ 0 & -3 & 0 \\ -2 & -2 & 10 \end{vmatrix}}{\Delta} = 22.5 \text{ A}$$

Thus, $$V_0'' = 3 \times (i_2'' - i_3'') = 45 \text{ V}$$

Considering only 3 A source and replacing the independent voltage sources by short circuit we get the circuit shown in Figure E4.11(*c*).

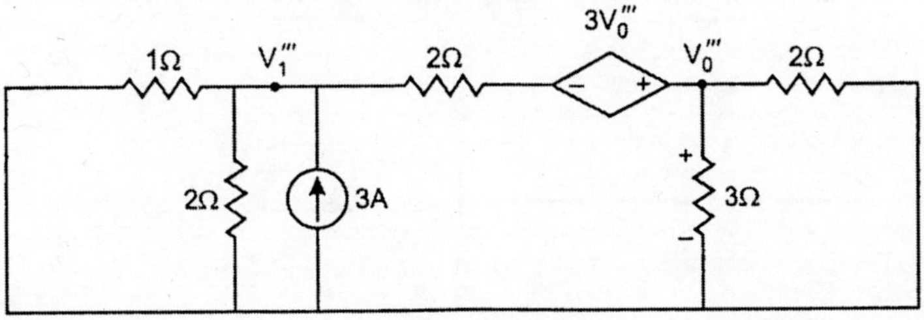

Fig. E4.11(c)

The two nodes with potential V_1''' and V_0''' are shown in the Figure E4.11(*c*). Writing the KCL equations for each node we get

$$\frac{V_1'''}{1} + \frac{V_1'''}{2} - 3 + \frac{V_1''' + 3V_0''' - V_0'''}{2} = 0$$

or $$2V_1''' + V_0''' = -3 \quad ...(xi)$$

$$\frac{V_0'''}{3} + \frac{V_0'''}{2} + \frac{V_0''' - 3V_0''' - V_1'''}{2} = 0$$

or $$-0.167V_0''' - 0.5V_1''' = 0 \quad \text{...(xii)}$$

Solving equations (*xi*) and (*xii*) we get

$$V_0''' = 9.04 \text{ V}$$

Then total $$V_0 = V_0' + V_0'' + V_0''' = 90.04 \text{ V}$$

EXAMPLE 4.12. *For the circuit shown in Figure E4.12, find V_0 and i_0 using Superposition theorem.*

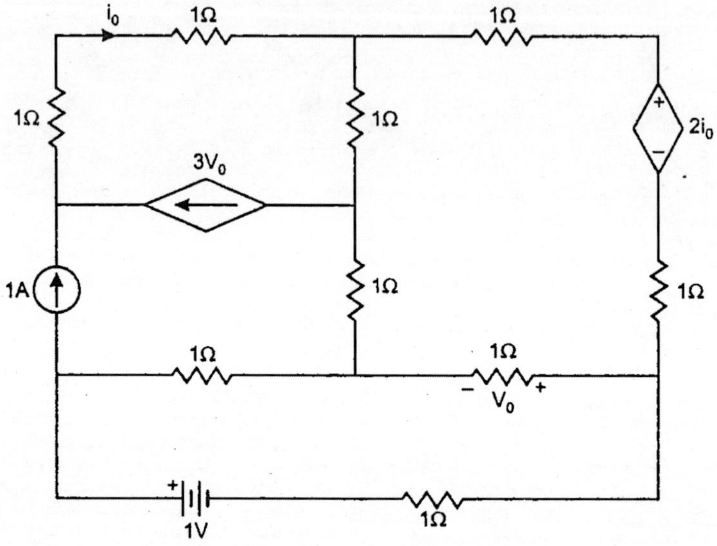

Fig. E4.12

SOLUTION. Considering only 1 V source and replacing the independent current source by open circuit we get Figure E4.12(*a*).

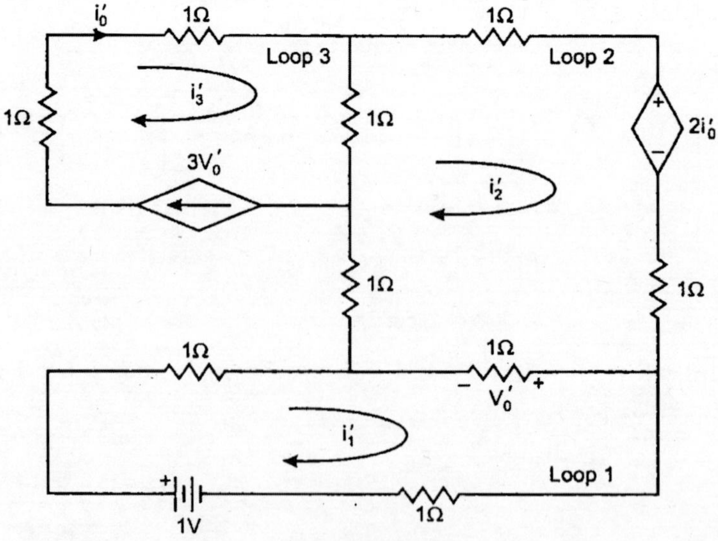

Fig. E4.12(a)

Writing the loop equations we get

Loop 1 : $3i_1' - i_2' = 1$ ...(*i*)

Loop 2 : $5i_2' - i_1' - i_3' = -2i_0'$ ...(*ii*)

$V_0' = 1 \times (i_2' - i_1')$ ...(*iii*)

Loop 3 : $i_3' = i_0' = 3V_0'$ ...(*iv*)

Substituting equation (*iv*) in equation (*ii*) and (*iii*) we get

$5i_2' - i_1' + i_3' = 0$ ...(*v*)

$0.333 \times i_3' = 1 \times (i_2' - i_1')$ ...(*vi*)

Writing equation (*i*), (*v*), and (*vi*) in matrix form we get

$$\begin{bmatrix} 3 & -1 & 0 \\ -1 & 5 & 1 \\ -1 & 1 & -0.333 \end{bmatrix} \times \begin{bmatrix} i_1' \\ i_2' \\ i_3' \end{bmatrix} = \begin{bmatrix} 1 \\ 0 \\ 0 \end{bmatrix}$$

Let
$$\Delta = \begin{vmatrix} 3 & -1 & 0 \\ -1 & 5 & 1 \\ -1 & 1 & -0.333 \end{vmatrix} = -6.62$$

Then
$$i_1' = \frac{\begin{vmatrix} 1 & -1 & 0 \\ 0 & 5 & 1 \\ 0 & 1 & -0.333 \end{vmatrix}}{\Delta} = 0.4 \text{ A}$$

$$i_2' = \frac{\begin{vmatrix} 3 & 1 & 0 \\ -1 & 0 & 1 \\ -1 & 0 & -0.333 \end{vmatrix}}{\Delta} = 0.2 \text{ A}$$

$$i_3' = \frac{\begin{vmatrix} 3 & -1 & 1 \\ -1 & 5 & 0 \\ -1 & 1 & 0 \end{vmatrix}}{\Delta} = -0.6 \text{ A}$$

Then $V_0' = 1 \times (i_2' - i_1') = -0.2$ V

and $i_3' = i_0' = -0.6$ A

Considering only 1 A source and replacing the independent voltage source by short circuit we get Figure E4.12(*b*).

Writing the loop equations, we get

Loop 1 : $3i_1'' - i_2'' - i_4'' = 0$ ...(*vii*)

Loop 2 : $5i_2'' - i_1'' - i_3'' - i_4'' = -2i_0''$ ...(*viii*)

$V_0'' = 1 \times (i_2'' - i_1'')$

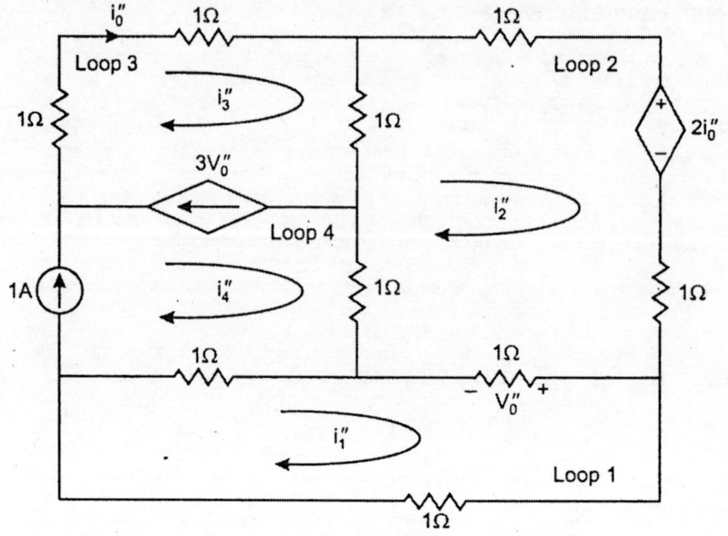

Fig. E4.12(b)

Loop 3 : $3i_3'' - i_2'' + 2i_4'' - i_2'' - i_1'' = 0$ ...(*ix*)

$$i_0'' = i_3'' \quad ...(x)$$

Loop 4 : $i_4'' = 1\text{ A}$ ...(*xi*)

Replacing equations (*x*) and (*xi*) in equation (*vii*), (*viii*) and (*ix*) we get

$$3i_1'' - i_2'' = 1 \quad ...(xii)$$

$$5i_2'' - i_1'' + i_3'' = 1 \quad ...(xiii)$$

$$3i_3'' - 2i_2'' - i_1'' = 2 \quad ...(xiv)$$

Writing equations (*xii*), (*xiii*) and (*xiv*) in matrix form we get

$$\begin{bmatrix} 3 & -1 & 0 \\ -1 & 5 & 1 \\ -1 & -2 & 3 \end{bmatrix} \times \begin{bmatrix} i_1'' \\ i_2'' \\ i_3'' \end{bmatrix} = \begin{bmatrix} 1 \\ 1 \\ -2 \end{bmatrix}$$

Let

$$\Delta = \begin{vmatrix} 3 & -1 & 0 \\ -1 & 5 & 1 \\ -1 & -2 & 3 \end{vmatrix} = 49$$

Then

$$i_1'' = \frac{\begin{vmatrix} 1 & -1 & 0 \\ 1 & 5 & 1 \\ -2 & -2 & 3 \end{vmatrix}}{\Delta} = 0.45\text{ A}$$

$$i_2'' = \frac{\begin{vmatrix} 3 & 1 & 0 \\ -1 & 1 & 1 \\ -1 & -2 & 3 \end{vmatrix}}{\Delta} = 0.35\text{ A}$$

$$i_3'' = \frac{\begin{vmatrix} 3 & -1 & 1 \\ -1 & 5 & 1 \\ -1 & -2 & -1 \end{vmatrix}}{\Delta} = 0 \text{ A}$$

Then $\quad V_0'' = 1 \times (i_2'' - i_1'') = -0.1 \text{ A}$

and $\quad i_3'' = i_0'' = 0 \text{ A}$

Thus $\quad V_0 = V_0' + V_0'' = -0.3 \text{ V}$

and $\quad i_0 = i_0' + i_0'' = -0.6 \text{ A}$

EXAMPLE 4.13. *For the circuit shown in Figure E4.13, find V_0 and i_0 using Superposition theorem.*

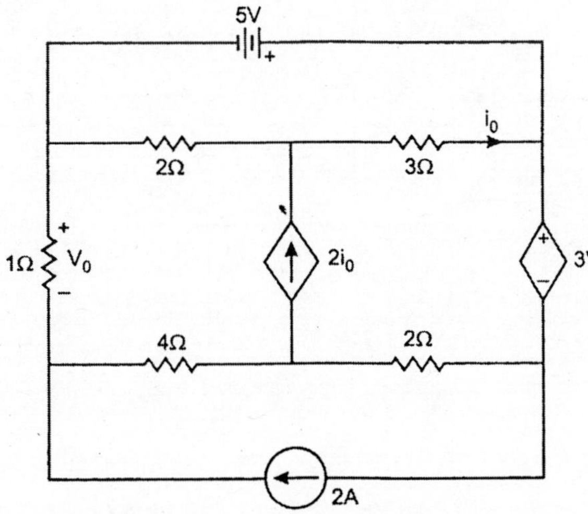

Fig. E4.13

SOLUTION. Considering only 5 V source and replacing the independent current source by open circuit we get Figure E4.13(*a*).

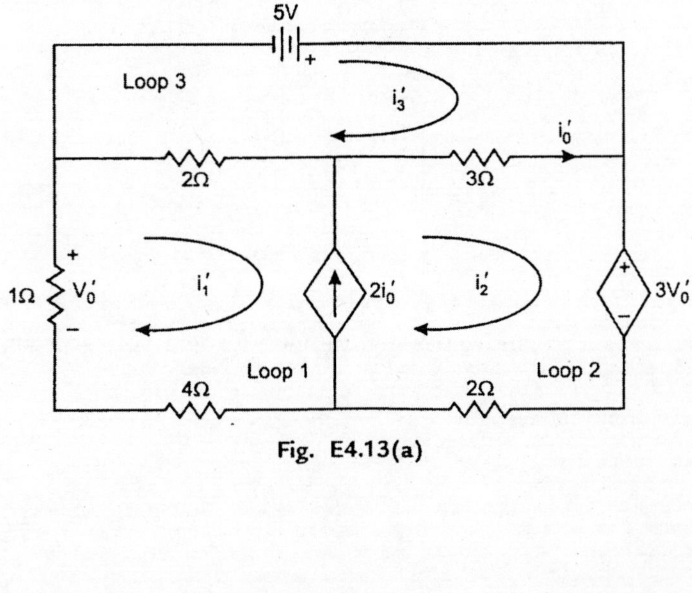

Fig. E4.13(a)

Writing the loop equations we get

Loop 1 and Loop 2 : $i_2' - i_1' = 2i_0'$ ...(*i*)

$$7i_1' + 5i_2' - 5i_3' = -3V_0' \quad ...(ii)$$

$$i_2' - i_3' = i_0' \quad ...(iii)$$

$$V_0' = -1 \times i_1' \quad ...(iv)$$

Loop 3 : $5i_3' - 2i_2' - 3i_1' = 5$ ...(*v*)

Substituting equation (*iii*) in equation (*i*) and equation (*iv*) in equation (*ii*), we get

$$-i_1' - i_2' + 2i_3' = 0 \quad ...(vi)$$

and $7i_1' + 2i_2' - 5i_3' = 0$ ...(*vii*)

Writing equations (*v*), (*vi*) and (*vii*) in matrix form we get

$$\begin{bmatrix} -2 & -3 & 5 \\ -1 & -1 & 2 \\ 7 & 2 & -5 \end{bmatrix} \times \begin{bmatrix} i_1' \\ i_2' \\ i_3' \end{bmatrix} = \begin{bmatrix} 5 \\ 0 \\ 0 \end{bmatrix}$$

Let
$$\Delta = \begin{vmatrix} -2 & -3 & 5 \\ -1 & -1 & 2 \\ 7 & 2 & -5 \end{vmatrix} = -4$$

Then
$$i_1' = \frac{\begin{vmatrix} 5 & -3 & 5 \\ 0 & -1 & 2 \\ 0 & 2 & -5 \end{vmatrix}}{\Delta} = -1.25 \text{ A}$$

$$i_2' = \frac{\begin{vmatrix} -2 & 5 & 5 \\ -1 & 0 & 2 \\ 7 & 0 & -5 \end{vmatrix}}{\Delta} = -11.25 \text{ A}$$

$$i_3' = \frac{\begin{vmatrix} -2 & -3 & 5 \\ -1 & -1 & 0 \\ 7 & 2 & 0 \end{vmatrix}}{\Delta} = -6.25 \text{ A}$$

Then $V_0' = 1 \times (-i_1') = 1.25$ V

and $i_0' = i_2' - i_3' = -5$ A

Considering only 2 A source and replacing the independent voltage source by short circuit we get Figure E4.13(*b*).

Writing the loop equations we get

From loop 1 and loop 2 : $i_2'' - i_1'' = 2i_0''$ ...(*viii*)

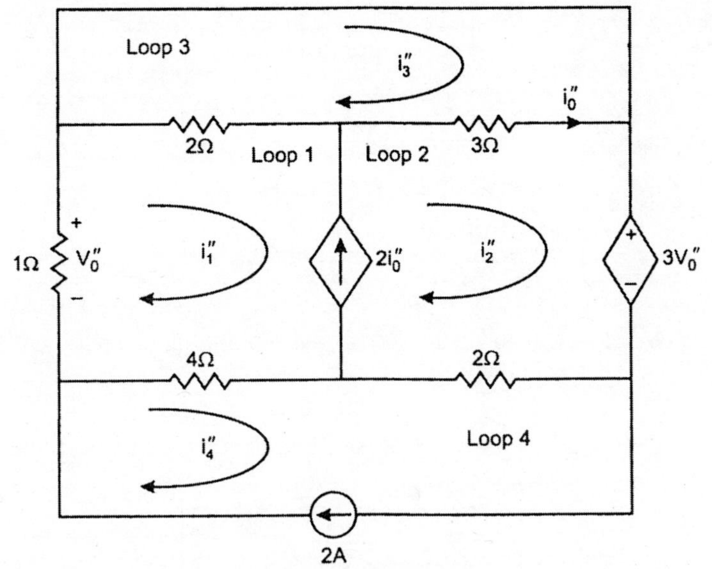

Fig. E4.13(b)

$$7i_1'' + 5i_2'' - 6i_4'' - 5i_3'' = -3V_0'' \quad \text{...(ix)}$$

$$V_0'' = 1 \times (-i_1'') \quad \text{...(x)}$$

Loop 3 :

$$5i_3'' - 2i_1'' - 3i_2'' = 0 \quad \text{...(xi)}$$

$$i_0'' = i_2'' - i_3'' \quad \text{...(xii)}$$

Loop 4 :

$$i_4'' = 2 \text{ A} \quad \text{...(xiii)}$$

Substituting equation (*xii*) in equation (*viii*) we get

$$-i_2'' - i_1'' + 2i_3'' = 0 \quad \text{...(xiv)}$$

Substituting equations (*x*) and (*xiii*) in equation (*ix*) we get

$$4i_1'' + 5i_2'' - 5i_3'' = 12 \quad \text{...(xv)}$$

Writing equations (*xi*), (*xiv*) and (*xv*) in matrix form we get

$$\begin{bmatrix} -2 & -3 & 5 \\ -1 & -1 & 2 \\ 4 & 5 & -5 \end{bmatrix} \times \begin{bmatrix} i_1'' \\ i_2'' \\ i_3'' \end{bmatrix} = \begin{bmatrix} 0 \\ 0 \\ 12 \end{bmatrix}$$

Let

$$\Delta = \begin{vmatrix} -2 & -3 & 5 \\ -1 & -1 & 2 \\ 4 & 5 & -5 \end{vmatrix} = -4$$

Then

$$i_1'' = \frac{\begin{vmatrix} 0 & -3 & 5 \\ 0 & -1 & 2 \\ 12 & 5 & -5 \end{vmatrix}}{\Delta} = 3 \text{ A}, \quad i_2'' = \frac{\begin{vmatrix} -2 & 0 & 5 \\ -1 & 0 & 2 \\ 4 & 12 & -5 \end{vmatrix}}{\Delta} = 3 \text{ A}$$

and $$i_3'' = \frac{\begin{vmatrix} -2 & -3 & 0 \\ -1 & -1 & 0 \\ 4 & 5 & 12 \end{vmatrix}}{\Delta} = 3 \text{ A}$$

Then $$V_0'' = 1 \times (-i_1'') = -3 \text{ A}$$

and $$i_0'' = i_2'' - i_3'' = 0 \text{ A}$$

Thus $$V_0 = V_0' + V_0'' = -1.75 \text{ V}$$

and $$i_0 = i_0' + i_0'' = -5 \text{ A}$$

EXAMPLE 4.14. *For the circuit shown in Figure E4.14 find the drop across 4 Ω using Superposition theorem.*

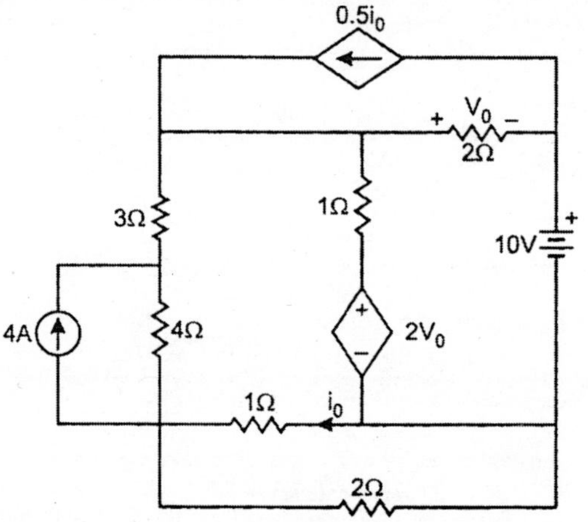

Fig. E4.14

SOLUTION. Considering only 10 V source and replacing the independent current source by open circuit we get Figure E4.14(*a*).

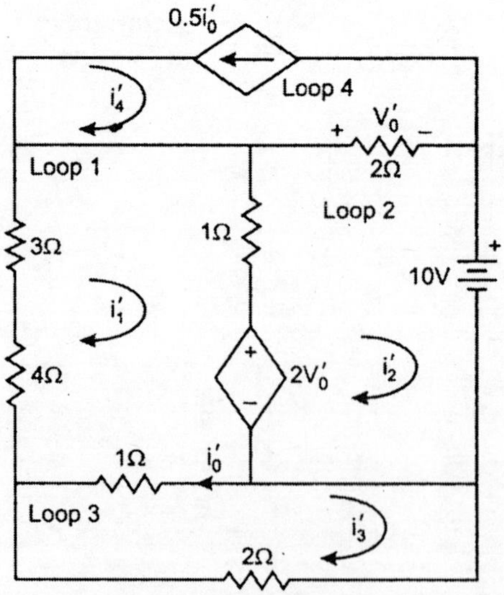

Fig. E4.14(a)

Writing the loop equations,

Loop 1 : $9i_1' - i_3' - i_2' = -2V_0'$ ...(*i*)

Loop 2 : $3i_2' - i_1' - i_4' - 2V_0' + 10 = 0$ ...(*ii*)

Loop 3 : $3i_3' - i_1' = 0$ ...(*iii*)

Loop 4 : $i_4' = -0.5i_0'$ ...(*iv*)

And $V_0' = 2(i_2' - i_4')$ ...(*v*)

$i_0' = i_1' - i_3'$ ...(*vi*)

Substituting equations (*iv*), (*v*) and (*vi*) in equation (*i*) we get

$$9.5i_1' - 1.5i_3' + 3i_2' = 0 \quad \text{...(vii)}$$

Substituting equations (*iv*), (*v*) and (*vi*) in equation (*ii*) we get

$$-2.5i_1' + 1.5i_3' - i_2' = -10 \quad \text{...(viii)}$$

Writing equations (*iii*), (*vii*) and (*viii*) in matrix form we get

$$\begin{bmatrix} -1 & 0 & 3 \\ 9.5 & 3 & -1.5 \\ -2.5 & -1 & 1.5 \end{bmatrix} \times \begin{bmatrix} i_1' \\ i_2' \\ i_3' \end{bmatrix} = \begin{bmatrix} 0 \\ 0 \\ -10 \end{bmatrix}$$

Let
$$\Delta = \begin{vmatrix} -1 & 0 & 3 \\ 9.5 & 3 & -1.5 \\ -2.5 & -1 & 1.5 \end{vmatrix} = -9$$

Then
$$i_1' = \frac{\begin{vmatrix} 0 & 0 & 3 \\ 0 & 3 & -1.5 \\ -10 & -1 & 1.5 \end{vmatrix}}{\Delta} = -10 \text{ A}$$

Thus, the drop across 4 Ω using only the voltage source is $4 \times i_1' = -40$ V

Considering only 4 A source and replacing the independent voltage source by short circuit we get Figure E4.14(*b*).

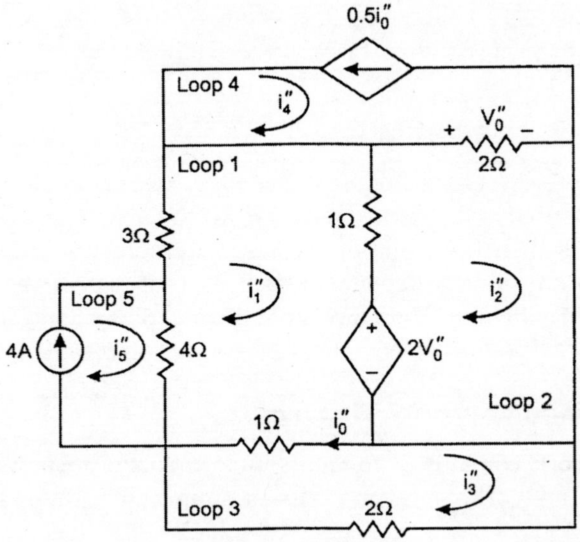

Fig. E4.14(b)

Writing the loop equations,

Loop 1 : $9i_1'' - i_3'' - i_2'' - 4i_5'' + 2V_0'' = 0$ ...(ix)

Loop 2 : $3i_2'' - i_1'' - i_4'' - 2V_0'' = 0$ ...(x)

Loop 3 : $3i_3'' - i_1'' = 0$ ...(xi)

Loop 4 : $i_4'' = -0.5i_0''$ ...(xii)

Loop 5 : $i_5'' = 4\text{ A}$ ...(xiii)

And $V_0'' = 2(i_2'' - i_4'')$ ...(xiv)

$i_0'' = i_1'' - i_3''$ ...(xv)

Substituting equations (xii), (xiii), (xiv) and (xv) in equation (i) we get

$$11i_1'' - 3i_3'' + 3i_2'' = 16 \quad \text{...(xvi)}$$

Substituting equations (xii), (xiii), (xiv) and (xv) in equation (ii) we get

$$-i_2'' - 2.5i_1'' + 1.5i_3'' = 0 \quad \text{...(xvii)}$$

Writing equations (xi), (xvi) and (xvii) in matrix form we get

$$\begin{bmatrix} -1 & 0 & 3 \\ 11 & 3 & -3 \\ -2.5 & -1 & 1.5 \end{bmatrix} \times \begin{bmatrix} i_1'' \\ i_2'' \\ i_3'' \end{bmatrix} = \begin{bmatrix} 0 \\ 16 \\ 0 \end{bmatrix}$$

Let

$$\Delta = \begin{vmatrix} -1 & 0 & 3 \\ 11 & 3 & -3 \\ -2.5 & -1 & 1.5 \end{vmatrix} = -12$$

Then

$$i_1' = \frac{\begin{vmatrix} 0 & 0 & 3 \\ 16 & 3 & -3 \\ 0 & -1 & 1.5 \end{vmatrix}}{\Delta} = 4\text{ A}$$

Thus, the drop across 4 Ω using only the voltage source is $4 \times (i_1' - i_5'') = 0\text{ V}$

Thus, the total drop is −40 V.

4.3 RECIPROCITY THEOREM

Statement : In a linear active bilateral network, if the excitation and the corresponding response in interchanged, then the value of the response does not change at all. This is valid if the network contains only one independent source. That is this theorem fails for a multi-source network. Also this theorem does not hold if any dependent source is present in the network.

4.3.1 Steps for Solving Reciprocity Theorem

Step 1 : If the network contains a voltage source then the response measured is current. And if the network contains a current source then the response measured is the voltage.

Step 2 : Identify clearly the position of the source and the response from the circuit.

Step 3 : from the first circuit calculate the value of the response.

Step 4 : Then interchange the position of the source and response and find the value of the response again. For the Reciprocity to be valid the value of the two responses should be the same

SOLVED EXAMPLES

EXAMPLE 4.15. *For the circuit shown in Figure E4.15(a) and Figure E4.15(b), verify Reciprocity theorem.*

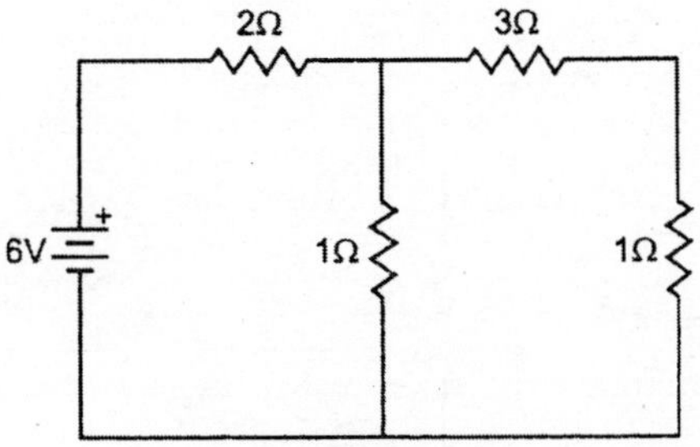

Fig. E4.15(a)

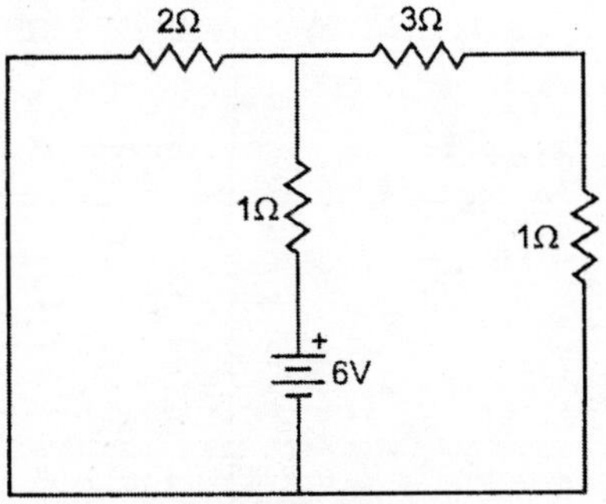

Fig. E4.15(b)

SOLUTION. The circuit shown in Figure E4.15(*a*), is redrawn in Figure E4.15(*c*) with loop currents.

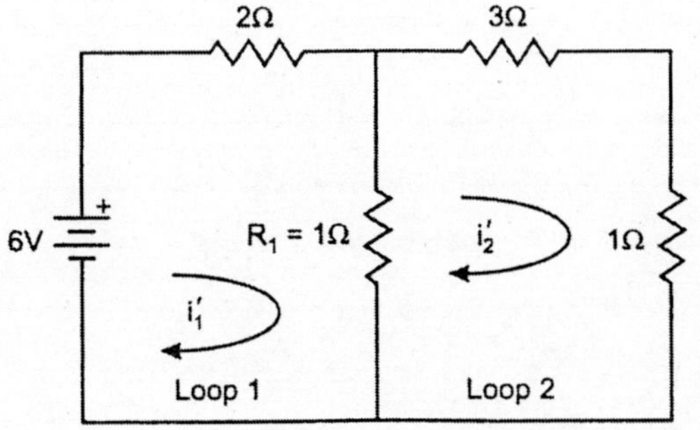

Fig. E4.15(c)

The loop equations are

Loop 1 : $3i_1' - i_2' = 6$ ...(*i*)

Loop 2 : $5i_2' - i_1' = 0$ ...(*ii*)

Writing equations (*i*) and (*ii*) in matrix form we get

$$\begin{bmatrix} 3 & -1 \\ -1 & 5 \end{bmatrix} \begin{bmatrix} i_1' \\ i_2' \end{bmatrix} = \begin{bmatrix} 6 \\ 0 \end{bmatrix}$$

Let, $$\Delta = \begin{vmatrix} 3 & -1 \\ -1 & 5 \end{vmatrix} = 14$$

Then $$i_1' = \frac{\begin{vmatrix} 6 & -1 \\ 0 & 5 \end{vmatrix}}{\Delta} = 2.12 \text{ A}$$

$$i_2' = \frac{\begin{vmatrix} 3 & 6 \\ -1 & 0 \end{vmatrix}}{\Delta} = 0.43 \text{ A}$$

The current through $R_1 = I' = i_1' - i_2' = 1.69$ A

The circuit shown in Figure E4.15(*b*), is redrawn in Figure E4.15(*d*) with loop currents.

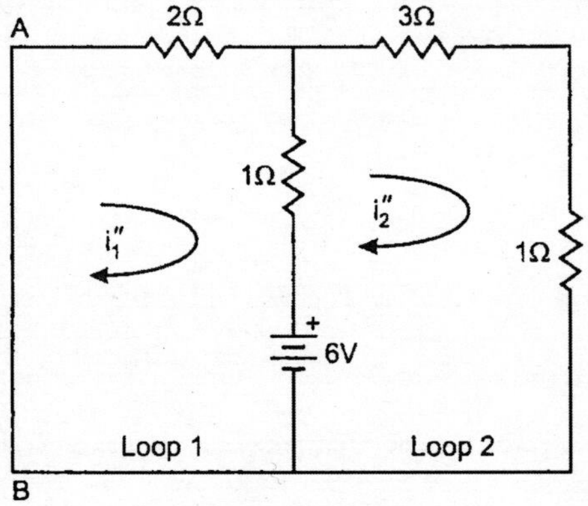

Fig. E4.15(d)

The loop equations are

Loop 1 : $3i_1'' - i_2'' = -6$ ...(*iii*)

Loop 2 : $5i_2'' - i_1'' = 6$ ...(*iv*)

Writing equations (*iii*) and (*iv*) in matrix form we get

$$\begin{bmatrix} 3 & -1 \\ -1 & 5 \end{bmatrix}\begin{bmatrix} i_1'' \\ i_2'' \end{bmatrix} = \begin{bmatrix} -6 \\ 6 \end{bmatrix}$$

Let, $$\Delta = \begin{vmatrix} 3 & -1 \\ -1 & 5 \end{vmatrix} = 14$$

Then $$i_1'' = \frac{\begin{vmatrix} -6 & -1 \\ 6 & 5 \end{vmatrix}}{\Delta} = -1.71 \text{ A}$$

$$i_2'' = \frac{\begin{vmatrix} 3 & -6 \\ -1 & 6 \end{vmatrix}}{\Delta} = 0.86 \text{ A}$$

Since $I' = i''$ the Reciprocity theorem is verified.

EXAMPLE 4.16. *For the circuit shown in Figure E4.16(a) and Figure E4.16(b), verify Reciprocity theorem.*

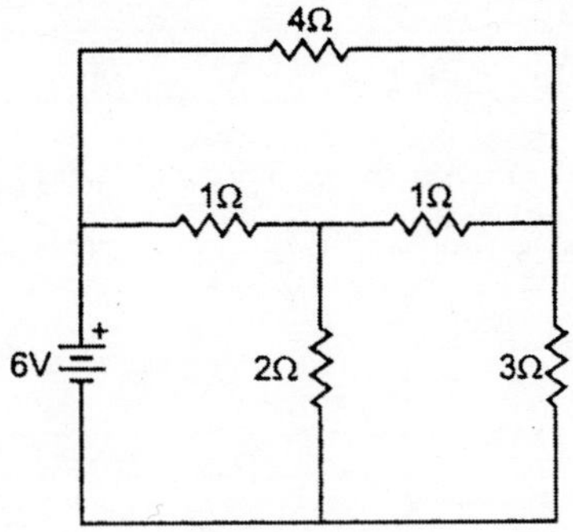

Fig. E4.16(a)

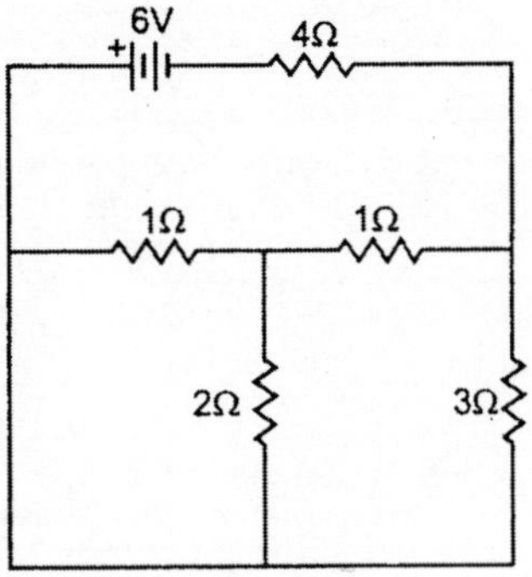

Fig. E4.16(b)

SOLUTION. The circuit shown in Figure E4.16(a), is redrawn in Figure E4.16(c) with loop currents.

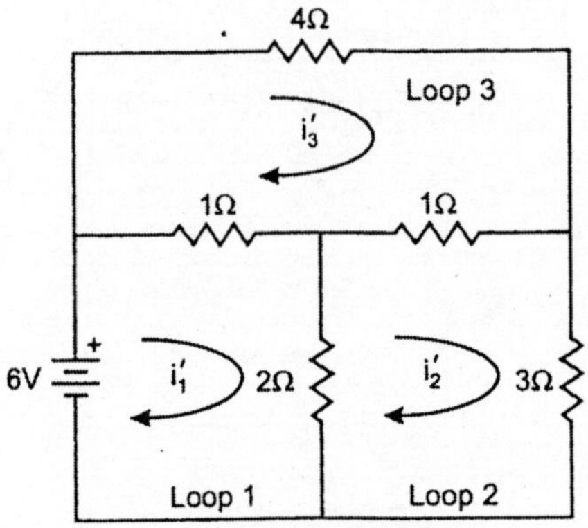

Fig. E4.16(c)

The loop equations are

Loop 1 : $3i_1' - 2i_2' - i_3' = 6$ ...(i)

Loop 2 : $6i_2' - 2i_1' - i_3' = 0$ ...(ii)

Loop 3 : $6i_3' - i_1' - i_2' = 0$ ...(iii)

Writing equations (i), (ii) and (iii) in matrix form we get

$$\begin{bmatrix} 3 & -2 & -1 \\ -2 & 6 & -1 \\ -1 & -1 & 6 \end{bmatrix} \begin{bmatrix} i_1' \\ i_2' \\ i_3' \end{bmatrix} = \begin{bmatrix} 6 \\ 0 \\ 0 \end{bmatrix}$$

Let $$\Delta = \begin{vmatrix} 3 & -2 & -1 \\ -2 & 6 & -1 \\ -1 & -1 & 6 \end{vmatrix} = 37$$

Then $$i_3' = \frac{\begin{vmatrix} 3 & -2 & 6 \\ -2 & 6 & 0 \\ -1 & -1 & 0 \end{vmatrix}}{\Delta} = 0.68 \text{ A}$$

The circuit shown in Figure E4.16(*b*), is redrawn in Figure E4.16(*d*) with loop currents.

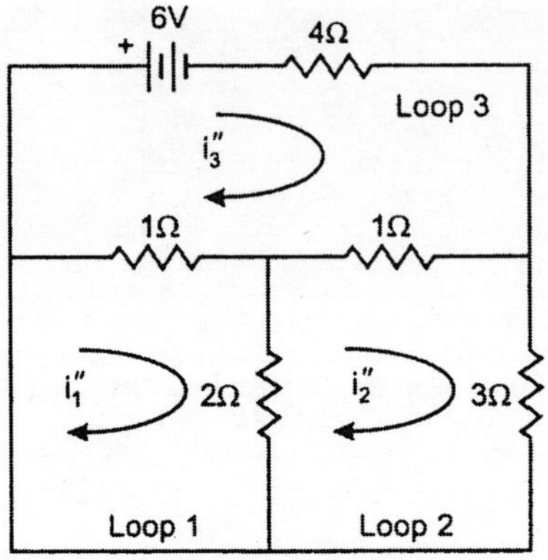

Fig. E4.16(d)

The loop equations are

Loop 1 : $3i_1'' - 2i_2'' - i_3'' = 0$ ...(*i*)

Loop 2 : $6i_2'' - 2i_1'' - i_3'' = 0$ ...(*ii*)

Loop 3 : $6i_3'' - i_1'' - i_2'' = -6$ ...(*iii*)

Writing equations (*i*), (*ii*) and (*iii*) in matrix form we get

$$\begin{bmatrix} 3 & -2 & -1 \\ -2 & 6 & -1 \\ -1 & -1 & 6 \end{bmatrix} \begin{bmatrix} i_1'' \\ i_2'' \\ i_3'' \end{bmatrix} = \begin{bmatrix} 0 \\ 0 \\ -6 \end{bmatrix}$$

Let $$\Delta = \begin{vmatrix} 3 & -2 & -1 \\ -2 & 6 & -1 \\ -1 & -1 & 6 \end{vmatrix} = 71$$

Then $$i_1'' = \frac{\begin{vmatrix} 0 & -2 & -1 \\ 0 & 6 & -1 \\ -6 & -1 & 6 \end{vmatrix}}{\Delta} = -0.68 \text{ A}$$

Since $i_3' = i_1''$ the Reciprocity theorem is verified.

EXAMPLE 4.17. *For the circuit shown in Figure E4.17(a) and Figure E4.17(b), verify Reciprocity theorem.*

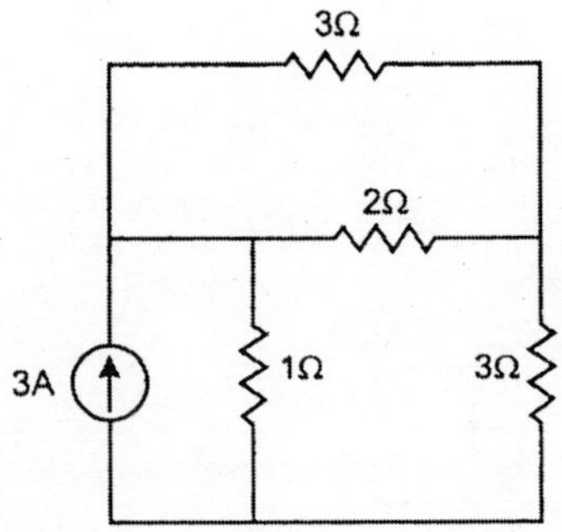

Fig. E4.17(a)

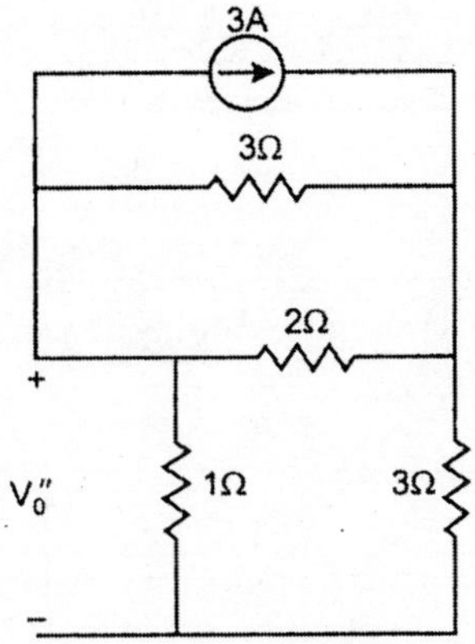

Fig. E4.17(b)

SOLUTION. The circuit shown in Figure E4.17(*a*), is redrawn in Figure E4.17(*c*) with loop currents.

The loop equations are

Loop 1 : $i_1' = 3$ A ...(*i*)

Loop 2 : $6i_2' - i_1' - 2i_3' = 0$ ...(*ii*)

Substituting equation (*i*) in equation (*ii*) we get

$$6i_2' - 2i_3' = 3 \quad ...(iii)$$

Loop 3 : $5i_3' - 2i_2' = 0$ ...(*iv*)

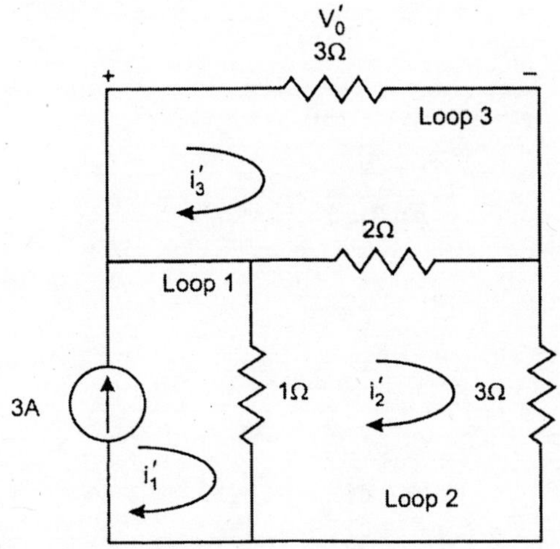

Fig. E4.17(c)

Writing equations (*iii*) and (*iv*) in matrix form we get

$$\begin{bmatrix} 6 & -2 \\ -2 & 5 \end{bmatrix} \begin{bmatrix} i_2' \\ i_3' \end{bmatrix} = \begin{bmatrix} 3 \\ 0 \end{bmatrix}$$

Let $$\Delta = \begin{vmatrix} 6 & -2 \\ -2 & 5 \end{vmatrix} = 26$$

Then $$i_3' = \frac{\begin{vmatrix} 6 & 3 \\ -2 & 0 \end{vmatrix}}{\Delta} = 0.23 \text{ A}$$

Then $$V_0' = 3i_3' = 0.69 \text{ V}$$

The circuit shown in Figure E4.17(*b*), is redrawn in Figure E4.17(*d*) with loop currents.

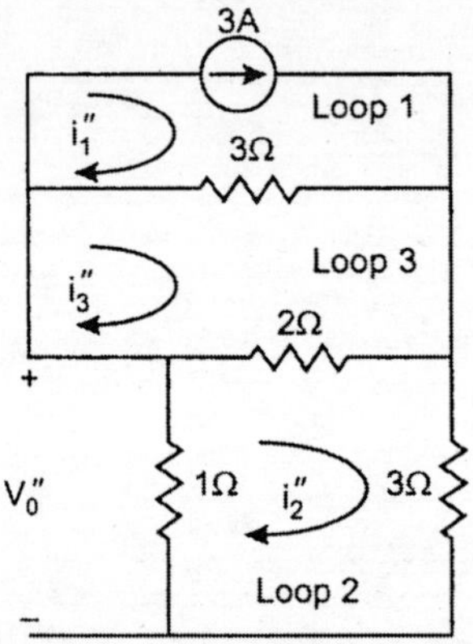

Fig. E4.17(d)

The loop equations are

Loop 1 : $$i_1'' = 3 \text{ A} \qquad ...(v)$$

Loop 2 : $$6i_2'' - 2i_3'' = 0 \qquad ...(vi)$$

Loop 3 : $$5i_3'' - 2i_2'' - 3i_1'' = 0 \qquad ...(vii)$$

Substituting equation (*v*) in equation (*vii*) we get

$$5i_3'' - 2i_2'' = 9 \qquad ...(viii)$$

Writing equations (*vi*) and (*viii*) in matrix form we get

$$\begin{bmatrix} 6 & -2 \\ -2 & 5 \end{bmatrix} \begin{bmatrix} i_2'' \\ i_3'' \end{bmatrix} = \begin{bmatrix} 0 \\ 9 \end{bmatrix}$$

Let $$\Delta = \begin{vmatrix} 6 & -2 \\ -2 & 5 \end{vmatrix} = 26$$

Then $$i_2'' = \frac{\begin{vmatrix} 0 & -2 \\ 9 & 5 \end{vmatrix}}{\Delta} = 0.69 \text{ A}$$

Th... $$V_0'' = 1i_2'' = 0.69 \text{ A}$$

Since $V_0' = V_0''$, Reciprocity Theorem is verified.

EXAMPLE 4.18. *For the circuit shown in Figure E4.18(a) and Figure E4.18(b), verify Reciprocity theorem.*

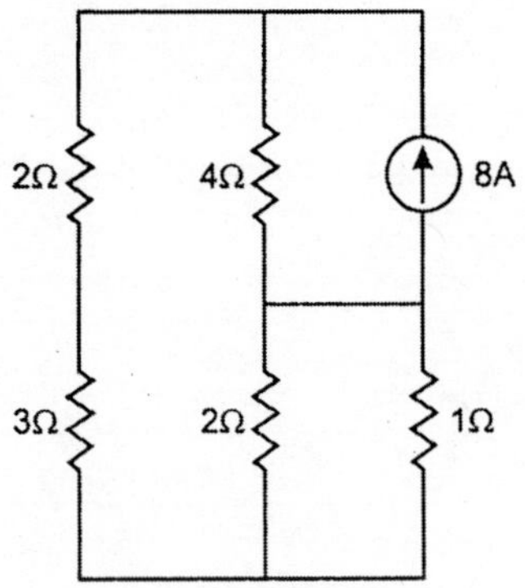

Fig. E4.18(a)

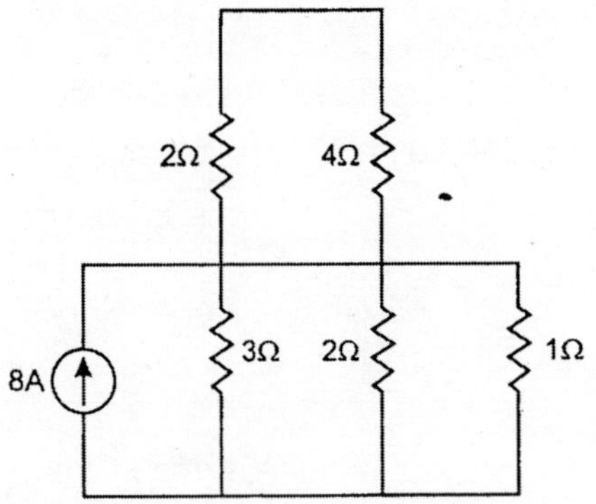

Fig. E4.18(b)

SOLUTION. The circuit shown in Figure E4.18(*a*), is redrawn in Figure E4.18(*c*) with loop currents.

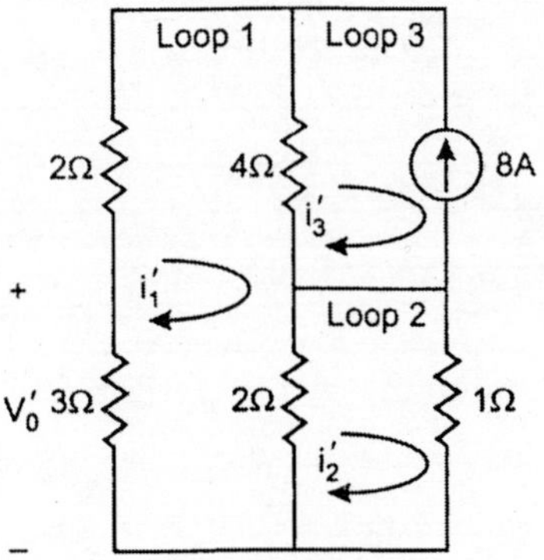

Fig. E4.18(c)

The loop equations are

Loop 1 : $11i_1' - 2i_2' - 4i_3' = 0$ ...(*i*)

Loop 2 : $3i_2' - 2i_1' = 0$ ...(*ii*)

Loop 3 : $i_3' = -8$ A ...(*iii*)

Substituting equation (*iii*) in equation (*i*) we get

$$11i_1' - 2i_2' = -32 \qquad ...(iv)$$

Writing equations (*ii*) and (*iv*) in matrix form we get

$$\begin{bmatrix} -2 & 3 \\ 11 & -2 \end{bmatrix} \begin{bmatrix} i_1' \\ i_2' \end{bmatrix} = \begin{bmatrix} 0 \\ -32 \end{bmatrix}$$

Let $$\Delta = \begin{vmatrix} -2 & 3 \\ 11 & -2 \end{vmatrix} = -29$$

Then $$i_1' = \frac{\begin{vmatrix} 0 & 3 \\ -32 & -2 \end{vmatrix}}{\Delta} = -3.31 \text{ A}$$

Then $$V_0' = 3i_1' = 9.93 \text{ V}$$

The circuit shown in Figure E4.18(*b*), is redrawn in Figure E4.18(*d*) with loop currents.

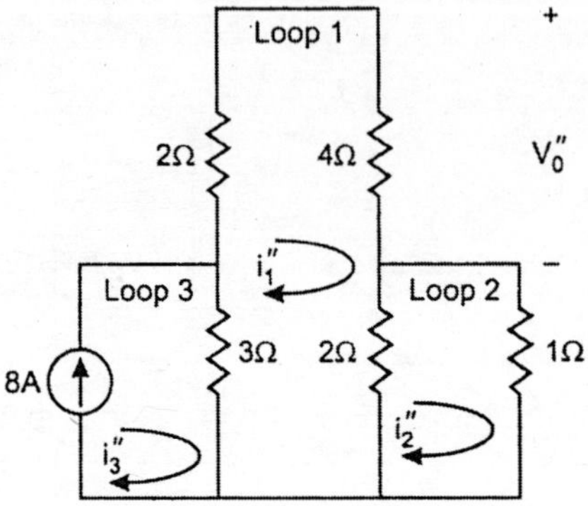

Fig. E4.18(d)

The loop equations are

Loop 1 : $11i_1'' - 2i_2'' - 3i_3'' = 0$ ...(*v*)

Loop 2 : $3i_2'' - 2i_1'' = 0$ ...(*vi*)

Loop 3 : $i_3'' = 8 \text{ A}$ ...(*vii*)

Substituting equation (*vii*) in equation (*v*) we get

$$11i_1'' - 2i_2'' = 24 \quad ...(viii)$$

Writing equations (*vi*) and (*viii*) in matrix form we get

$$\begin{bmatrix} -2 & 3 \\ 11 & -2 \end{bmatrix} \begin{bmatrix} i_1'' \\ i_2'' \end{bmatrix} = \begin{bmatrix} 0 \\ 20 \end{bmatrix}$$

Let $$\Delta = \begin{vmatrix} -2 & 3 \\ 11 & -2 \end{vmatrix} = -29$$

Then $$i_1'' = \frac{\begin{vmatrix} 0 & 3 \\ 20 & -2 \end{vmatrix}}{\Delta} = 2.07 \text{ A}$$

Then $$V_0'' = 4i_1'' = 8.28 \text{ V}$$

Since $V_0' = V_0''$, Reciprocity Theorem is verified.

4.4 COMPENSATION THEOREM

Statement : In a linear, active network if the resistance R carrying current I is changed by a small amount ΔR, then the change in the current in all branches can be obtained by adding a independent voltage source ΔV in series with $R \pm \Delta R$, and replacing all the independent voltage source and current source by their internal resistances. Where $\Delta V = I \times \Delta R$

4.4.1 Steps for Solving Compensation Theorem

Step 1 : Solve the original circuit to find the circuit current i.

Step 2 : Let the resistance R_1 of the circuit has been changed to a new value R_2. In the original circuit change R_1 to R_2 and connect an independent voltage source ΔV in series with R_2. Replace all other independent sources if any by their internal resistance. Where $\Delta V = i$ (the current calculated from step 1) $\times$ $(R_1 \sim R_2)$

Step 3 : Solve the circuit obtained from step 2. The current thus obtained will give the change in current Δi, in the circuit due to change in the value of resistance from R_1 to R_2.

SOLVED EXAMPLES

EXAMPLE 4.19. *If the 5 Ω resistance of the circuit shown in Figure E4.19 is changed to 8 Ω. Find the change in circuit current using compensation theorem.*

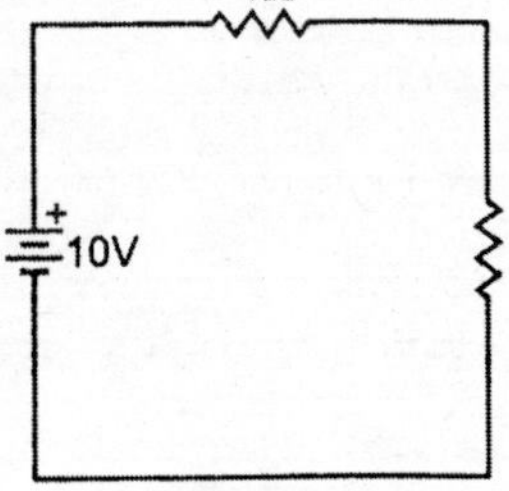

Fig. E4.19

SOLUTION. Let the current flowing through the circuit of Figure E4.19 be i_1.

Then $$i_1 = \frac{10}{6} = 1.67 \text{ A}$$

From compensation theorem we get the circuit as shown in Figure E4.19(*a*), to find the change in circuit current due to the change in resistance value from 5 Ω to 8 Ω.

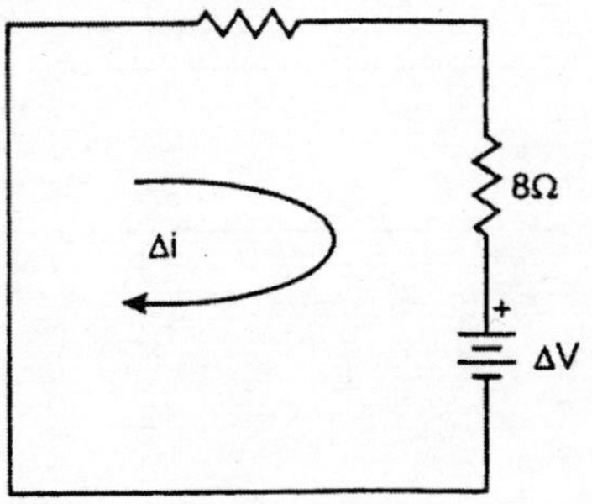

Fig. E4.19(a)

From Figure E4.19(*a*) we get

Let $$\Delta R = 8 - 5 = 3\ \Omega \qquad ...(i)$$

$$\Delta i = \frac{\Delta V}{9}$$

where $$\Delta V = i_1 \times \Delta R = 1.67 \times 3 = 5.01\ \text{V} \qquad ...(ii)$$

Substituting equation (*ii*) in equation (*i*) we get

The change in circuit current $\Delta i = 0.56$ A

EXAMPLE 4.20. *For the circuit shown in Figure E4.20, find the loop currents. Then find the change in loop current if the using compensation theorem if the resistance R_1 is changed from 2 Ω to 1 Ω.*

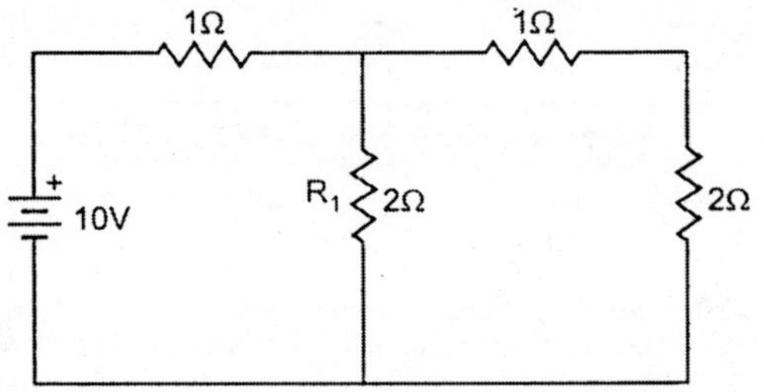

Fig. E4.20

SOLUTION. Figure E4.20, is redrawn in Figure E4.20(*a*) with loop currents.

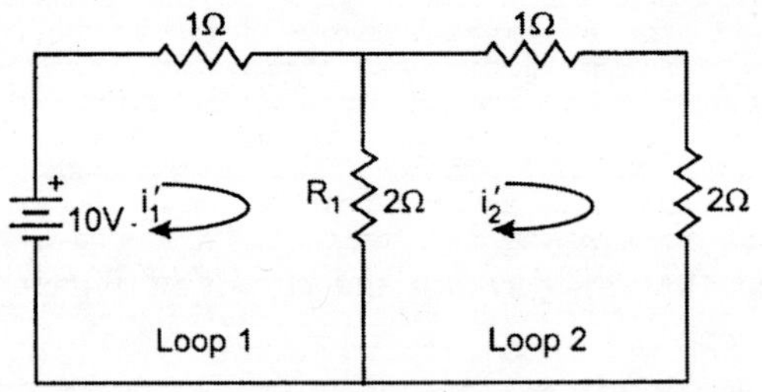

Fig. E4.20(a)

The loop equations are

Loop 1 : $$3i_1' - 2i_2' = 10 \qquad ...(i)$$

Loop 2 : $$5i_2' - 2i_1' = 0 \qquad ...(ii)$$

Solving equations (*i*) and (*ii*) we get

$$i_1' = 4.55\ \text{A} \quad \text{and} \quad i_2' = 1.82\ \text{A}$$

Current through $R_1 = 2.73$ A

To find the change in loop current using compensation theorem. The circuit needed for compensation theorem is shown in Figure E4.20(*b*).

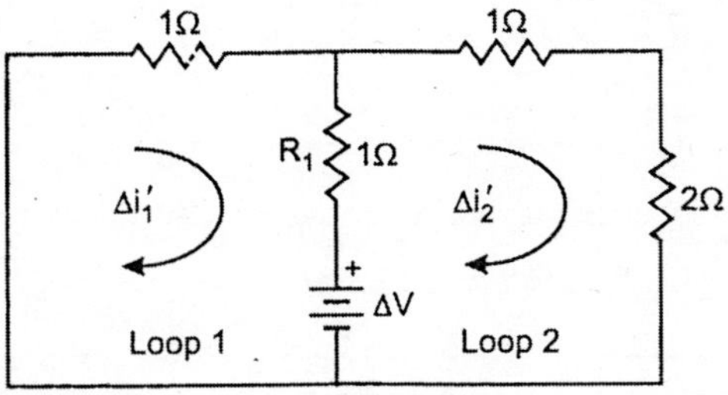

Fig. E4.20(b)

Here $\Delta R = 2 - 1 = 1\ \Omega$

$$\Delta V = 2.73 \times \Delta R = 2.73\ \text{V}$$

The loop equations are

Loop 1 : $2\Delta i_1' - 1\Delta i_2' = -2.73$ ...(*iii*)

Loop 2 : $4\Delta i_2' - 1\Delta i_1' = 2.73$ ...(*iv*)

Solving equations (*iii*) and (*iv*) we get

$$\Delta i_1' = -1.17\ \text{A} \quad \text{and} \quad \Delta i_2' = 0.39\ \text{A}$$

EXAMPLE 4.21. *For the circuit shown in Figure E4.21, find the loop currents. Then find the change in loop current if the using compensation theorem if the resistance R_1 is changed from 1 Ω to 2 Ω.*

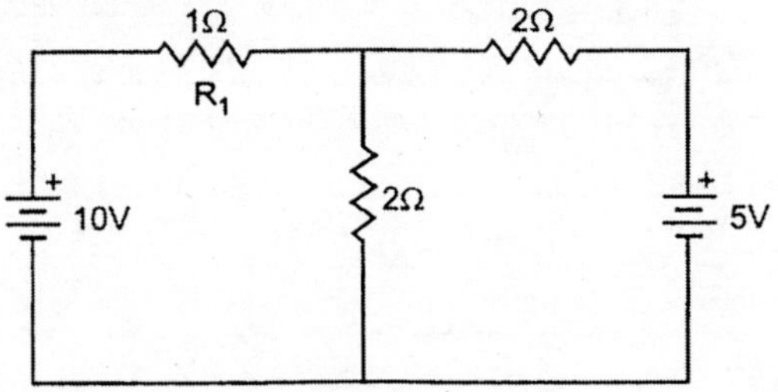

Fig. E4.21

SOLUTION. Figure E4.21, is redrawn in Figure E4.21(*a*) with loop currents.

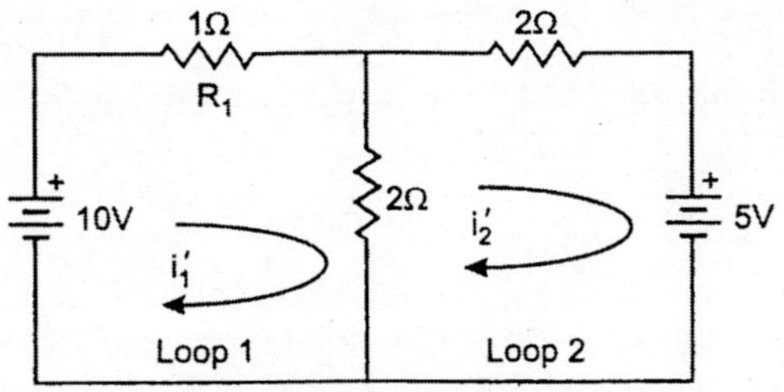

Fig. E.4.21(a)

The loop equations are

Loop 1 : $3i_1' - 2i_2' = 10$ ...(*i*)

Loop 2 : $4i_2' - 2i_1' = -5$ ...(*ii*)

Solving equations (*i*) and (*ii*) we get

$$i_1' = 3.75 \text{ A} \quad \text{and} \quad i_2' = 0.625 \text{ A}$$

Current through $R_1 = 3.75\text{A}$

To find the change in loop current using compensation theorem. The circuit needed for compensation theorem is shown in Figure E4.21(*b*).

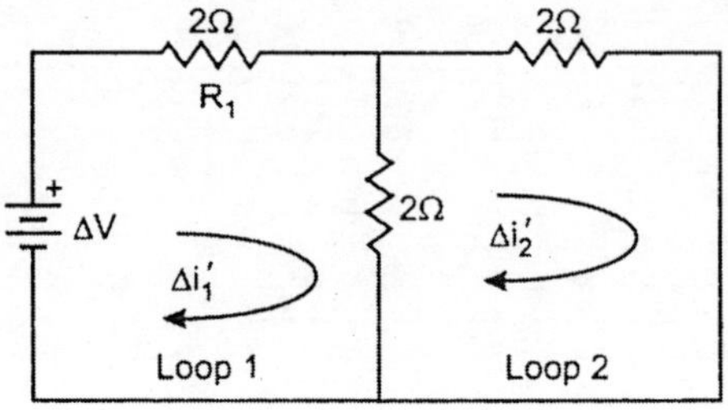

Fig. E4.21(b)

Here

$$\Delta R = 2 - 1 = 1 \ \Omega$$

$$\Delta V = 3.75 \times \Delta R = 3.75 \text{ V}$$

The loop equations are

Loop 1 : $4\Delta i_1' - 2\Delta i_2' = 3.75$ ...(*iii*)

Loop 2 : $4\Delta i_2' - 2\Delta i_1' = 0$ ...(*iv*)

Solving equations (*iii*) and (*iv*) we get

$$\Delta i_1' = 1.25 \text{ A} \quad \text{and} \quad \Delta i_2' = 0.625 \text{ A}$$

EXAMPLE 4.22. *For the circuit shown in Figure E4.22, find the loop currents. Then find the change in loop current using compensation theorem if the resistance R_1 is changed from 3 Ω to 5 Ω.*

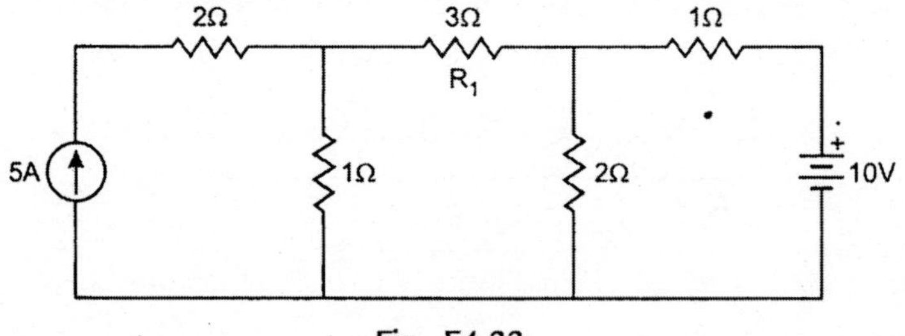

Fig. E4.22

SOLUTION. The circuit of Figure E4.22 is redrawn in Figure E4.22(*a*) with loop currents

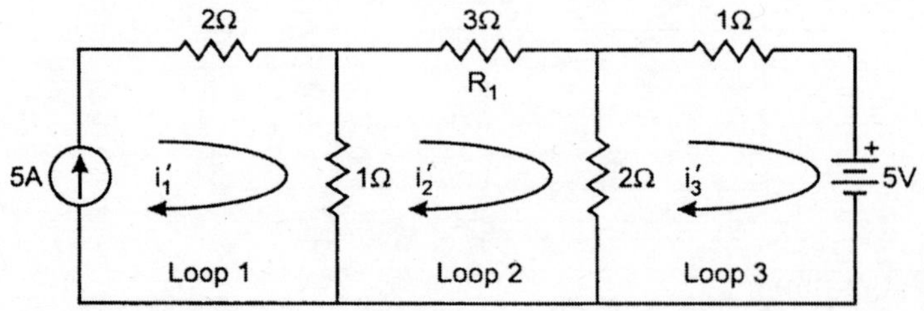

Fig. E4.22(a)

The loop equations are

Loop 1 : $i_1' = 5$ A ...(*i*)

Loop 2 : $6i_2' - i_1' - 2i_3' = 0$ ...(*ii*)

Substituting equation (*i*) in equation (*ii*) we get

$$6i_2' - 2i_3' = 5 \quad ...(iii)$$

Loop 3 : $3i_3' - 2i_2' = -5$ ...(*iv*)

Solving equations (*iii*) and (*iv*) we get

$$i_2' = 0.36 \text{ A} \quad \text{and} \quad i_3' = -1.43 \text{ A}$$

To find the change in loop current using compensation theorem. The circuit needed for compensation theorem is shown in Figure E4.22(*b*).

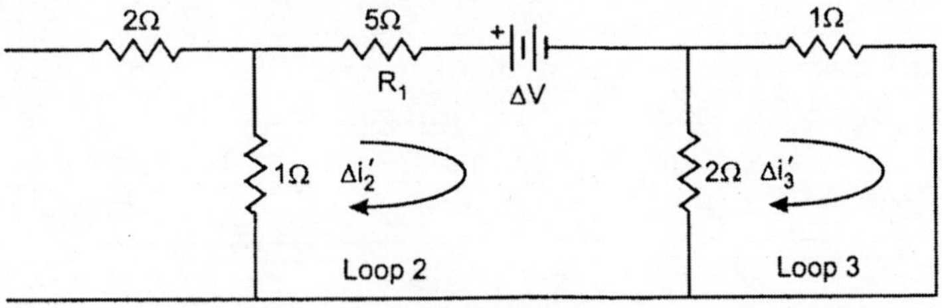

Fig. E4.22(b)

Here $\Delta R = 5 - 3 = 2\ \Omega$

$$\Delta V = i_2' \times \Delta R = 0.72 \text{ V}$$

The loop equations are

Loop 1 : $8\Delta i_2' - 2\Delta i_3' = 0.72$ ...(*v*)

Loop 2 : $3\Delta i_3' - 2\Delta i_2' = 0$ ...(*vi*)

Solving equations (*v*) and (*vi*) we get

$$\Delta i_2' = 0.11 \text{ A} \quad \text{and} \quad \Delta i_3' = 0.072 \text{ A}$$

4.5 THEVENIN'S THEOREM

Statement : Any two terminal linear, active, bilateral d.c. circuit can be replaced by an equivalent circuit consisting of a voltage source (V_{TH}) and a series resistance (R_{TH}).

Where V_{TH} is called the Thevenin`s voltage and $R_{TH}$ is called the Thevenin`s resistance.

4.5.1 Method to Find Thevenin's Voltage (V_{TH})

Remove the load resistance from the main circuit and then find the open circuit voltage. The open circuit voltage gives the Thevenin's voltage.

4.5.2 Methods to Find The Thevenin's Resistance (R_{TH})

1. Remove the load resistance from the main circuit. Replace an independent voltage source if present in the circuit by its internal resistance (by short circuit in case of an ideal voltage source). Replace an independent current source if present in the circuit by its

internal resistance (by open circuit in case of an ideal current source). Now find the equivalent resistance of the circuit as seen from the load terminal. This is the Thevenin's equivalent resistance of the circuit. This method is valid only if the circuit contains no dependent source. This method fails in case a dependent source is present in the circuit. For dependent source network the other two method discussed below holds.

2. Remove the load resistance from the main circuit. Short the load terminals and then find the short circuit current (i_{sc}). The Thevenin's resistance is given by the formula

$$R_{TH} = \frac{V_{TH}}{i_{sc}}$$

3. Remove the load resistance from the main circuit. Replace an independent voltage source if present in the circuit by its internal resistance (by short circuit in case of an ideal voltage source). Replace an independent current source if present in the circuit by its internal resistance (by open circuit in case of an ideal current source). Do not remove the dependent source if it is present in the circuit. Now add a dummy voltage source (v) across the load terminals and find the current (i) delivered by the dummy source. The Thevenin's resistance is given by the formula

$$R_{TH} = \frac{v}{i}$$

SOLVED EXAMPLES

EXAMPLE 4.23. *For the circuit shown in Figure E4.23, find the drop across the 4 Ω resistor, using Thevenin's theorem.*

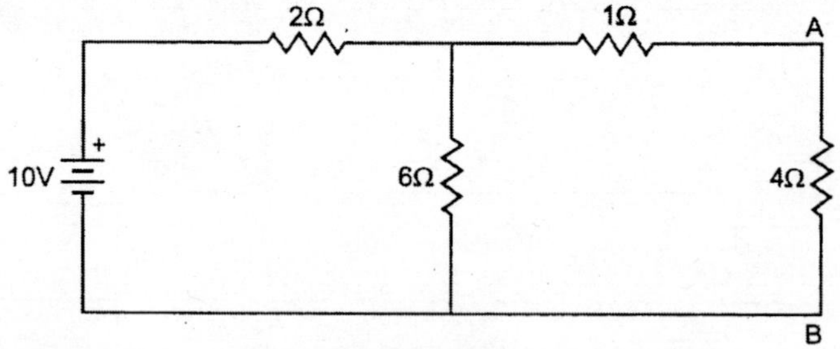

Fig. E4.23

SOLUTION. *To find* V_{TH} :

Removing the 4 Ω resistor form the circuit we get Figure E4.22(*a*)

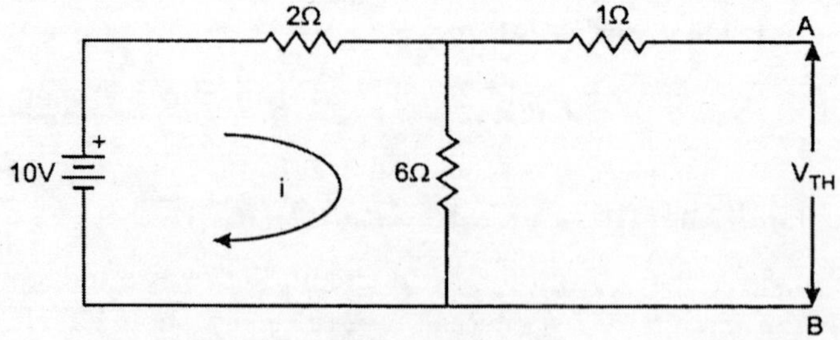

Fig. E4.23(a)

From Figure E4.23(*a*) we get

Current $$i = \frac{10}{6+2} = 1.25 \text{ A}$$

Then $$V_{TH} = 6 \times i = 6 \times 1.25 = 7.5 \text{ V}$$

To find R_{TH} :

Suppressing the independent voltage source by short circuit and removing the 4 Ω load resistance we get the circuit as shown in Figure E4.23(*b*)

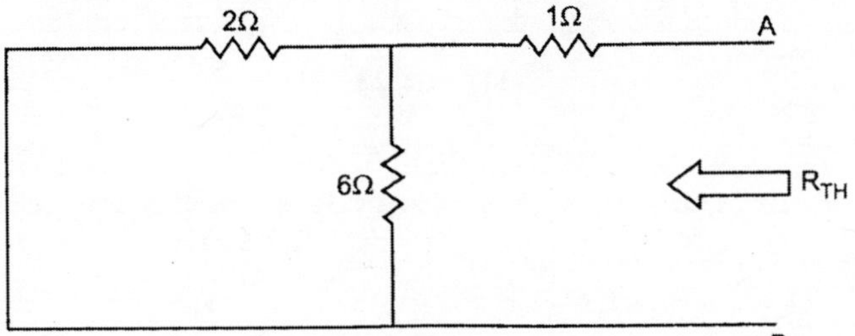

Fig. E4.23(b)

From Figure E4.23(*b*) we get $R_{TH} = \frac{6 \times 2}{6+2} + 1 = 2.5 \ \Omega$

Then the Thevenin's equivalent circuit is shown in Figure E4.23(*c*).

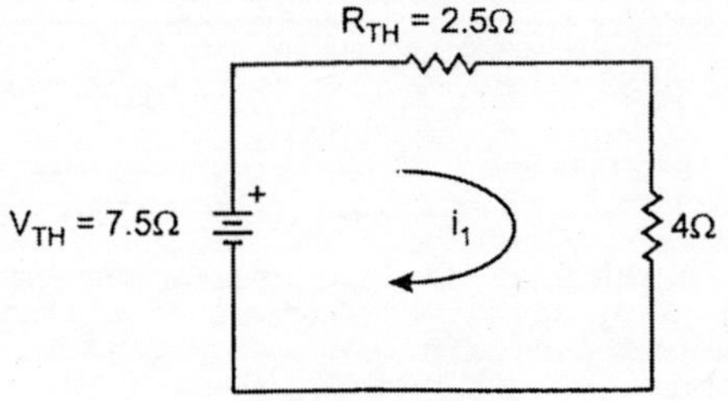

Fig. E4.23(c)

From Figure E4.23(*c*) we get $i_1 = \frac{7.5}{2.5+4} = 1.15 \text{ A}$

Then the drop across 4 Ω is

$$V = 4 \times i_1 = 4 \times 1.15 = 4.6 \text{ V}$$

EXAMPLE 4.24. *For the circuit shown in Figure E4.24 find the current through 12 Ω resistor, using Thevenin's theorem.*

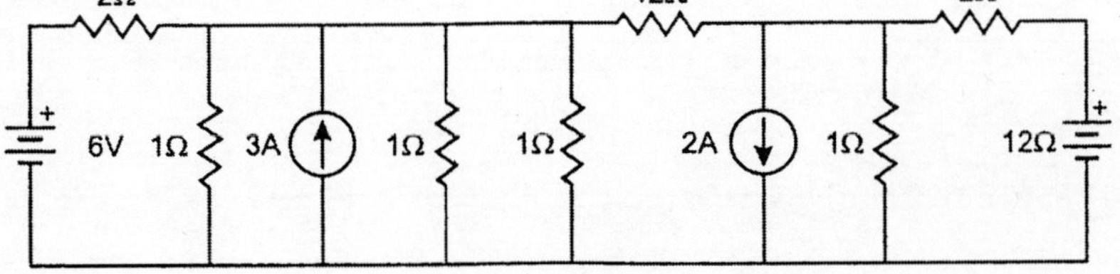

Fig. E4.24

SOLUTION. *To find V_{TH}* :

Removing the 12 Ω resistor from the circuit we get Figure E4.24(*a*).

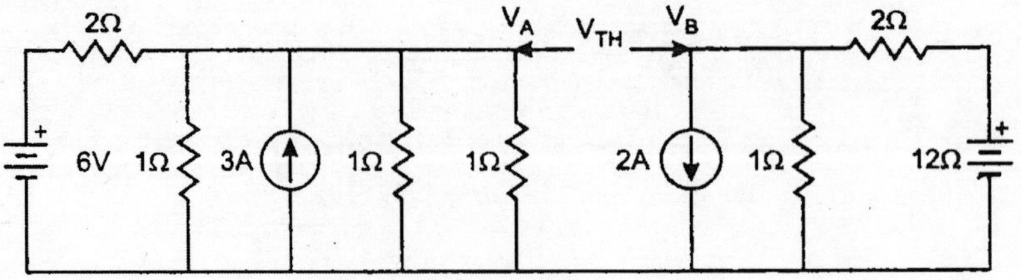

Fig. E4.24(a)

There are two nodes in the circuit, node *A* and node *B*. Let the corresponding node voltage be V_A and V_B respectively. writing the node equations we get.

Node *A* : $$\frac{V_A-6}{2}+\frac{V_A}{1}-3+\frac{V_A}{1}+\frac{V_A}{1} = 0$$

$$V_A = 1.71 \text{ V}$$

Node *B* : $$\frac{V_B-12}{2}+\frac{V_B}{1}+2 = 0$$

$$V_B = 2.67 \text{ V}$$

Then $$V_{TH} = V_B - V_A = 0.96 \text{ V}$$

To find R_{TH} :

Suppressing the independent voltage source by short circuit and independent current source by open citcuit and removing the 12 Ω load resistance we get the circuit as shown in Figure E4.24(*b*)

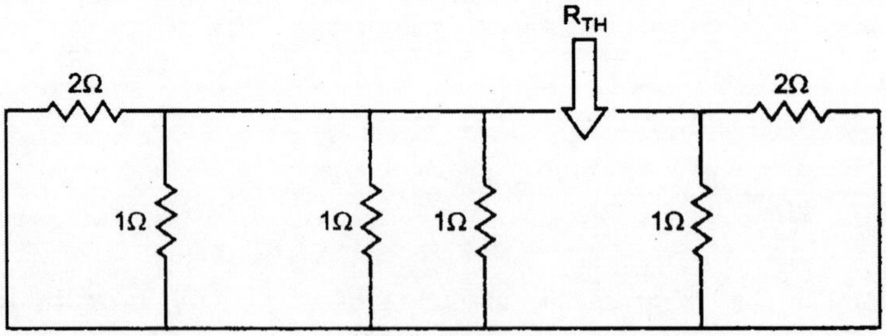

Fig. E4.24(b)

$$R_{th} = \frac{1}{\frac{1}{1}+\frac{1}{1}+\frac{1}{1}+\frac{1}{2}}+\frac{1}{\frac{1}{1}+\frac{1}{2}} = 0.95 \ \Omega$$

Then the Thevenin's equivalent circuit is shown in Figure E4.24(*c*).

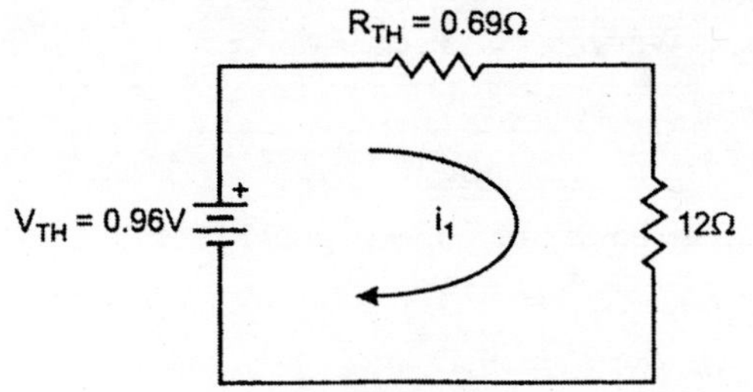

Fig. E4.24(c)

From Figure E4.24(*c*) we get $i_1 = \dfrac{0.96}{12+0.95} = 0.074$ A

EXAMPLE 4.25. *For the circuit shown in Figure E4.25, find the current through 3 Ω resistor, using Thevenin's theorem.*

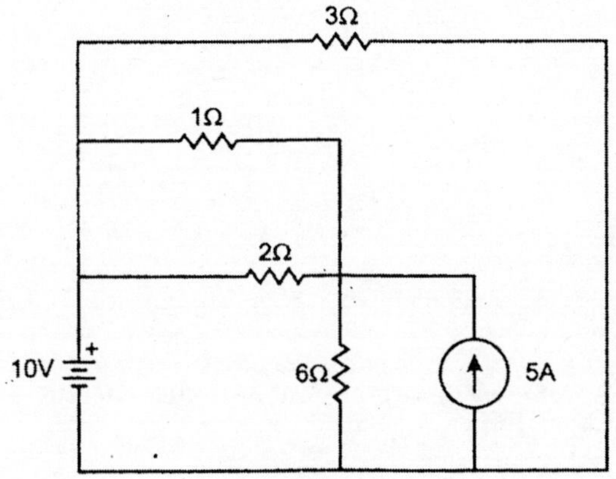

Fig. E4.25

SOLUTION. *To find* V_{TH} :

After removing the load resistance from the original circuit we get the circuit as shown in Figure E4.25(*a*). The loop currents are indicated in the Figure E4.25(*a*)

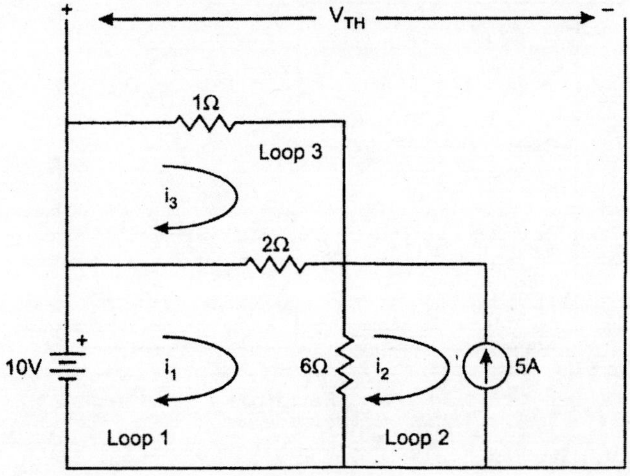

Fig. E4.25(a)

Writing the KVL equation we get,

Loop 1 : $8i_1 - 6i_2 - 2i_3 = 10$ ...(*i*)

Loop 2 : $i_2 = -5$ A ...(*ii*)

Substituting equation (*ii*) in equation (*i*), we get

$$8i_1 - 2i_3 = -20 \quad \text{...(iii)}$$

Loop 3 : $3i_3 - 2i_1 = 0$ ...(*iv*)

Writing equations (*iii*) and (*iv*) in matrix form we get

$$\begin{bmatrix} 8 & -2 \\ -2 & 3 \end{bmatrix} \begin{bmatrix} i_1 \\ i_3 \end{bmatrix} = \begin{bmatrix} -20 \\ 0 \end{bmatrix}$$

Let $$\Delta = \begin{vmatrix} 8 & -2 \\ -2 & 3 \end{vmatrix} = 20$$

Then $$i_3 = \frac{\begin{vmatrix} 8 & -20 \\ -2 & 0 \end{vmatrix}}{\Delta} = 2 \text{ A}$$

Then $$V_{TH} = i_3 \times 1 = 2 \text{ V}$$

To find R_{TH} :

Replacing the voltage source by short circuit and the current source by open circuit we get a circuit as shown in Figure E4.25(*b*)

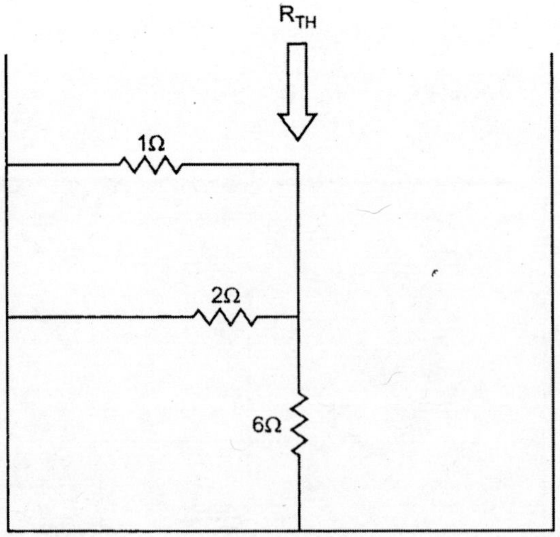

Fig. E4.25(b)

From Figure E4.25(*b*) we get $R_{TH} = 0\ \Omega$

Then the Thevenin's equivalent circuit will be as shown in Figure E4.25(*c*)

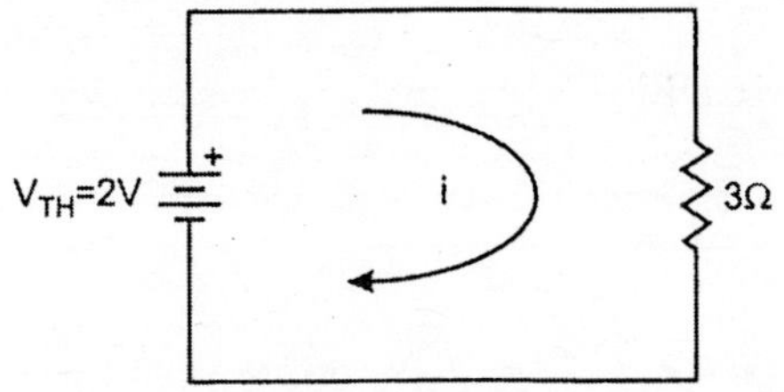

Fig. E4.25(c)

From Figure E4.25(*c*), $i = \frac{2}{3} = 0.67$ A

EXAMPLE 4.26. *Find the Thevenin's equivalent across the terminal AB.*

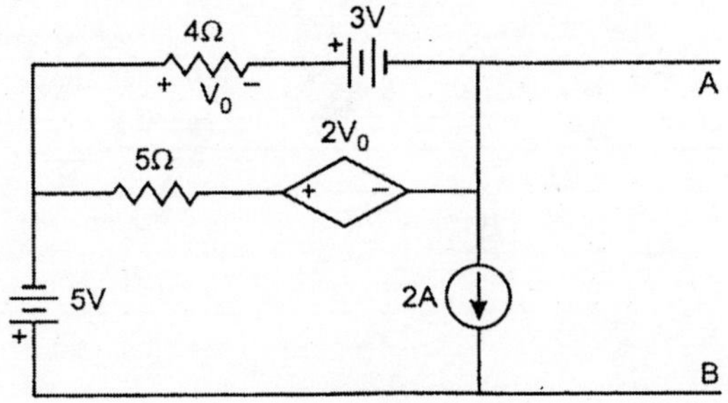

Fig. E4.26

SOLUTION. *To find* V_{TH} :

The circuit of Figure E4.26 is redrawn in Figure E4.26(*a*), with loop currents

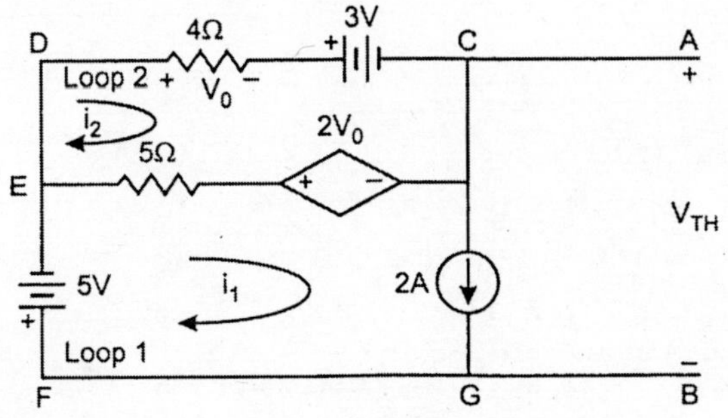

Fig. E4.26(a)

The loop equations are

Loop 1 : $i_1 = 2$ A ...(*i*)

Loop 2 : $9i_2 - 5i_1 = -3 + 2V_0$ ...(*ii*)

Substituting equation (*i*) in equation (*ii*) we get

$$9i_2 = 7 + 2V_0 \qquad ...(iii)$$

From Figure E4.26(*a*) we get $V_0 = 4i_2$ ...(*iv*)

Substituting equation (*iv*) in equation (*iii*) we get

$$i_2 = 7 \text{ A} \quad \text{...}(v)$$

Applying KVL in the path *ACDEFGB* we get

$$4i_2 + 3 + V_{TH} + 5 = 0$$

Thus $$V_{TH} = -36 \text{ V}$$

To find R_{TH} :

After shorting the open circuit terminal *A-B* we get the circuit as shown in Figure E4.26(*b*)

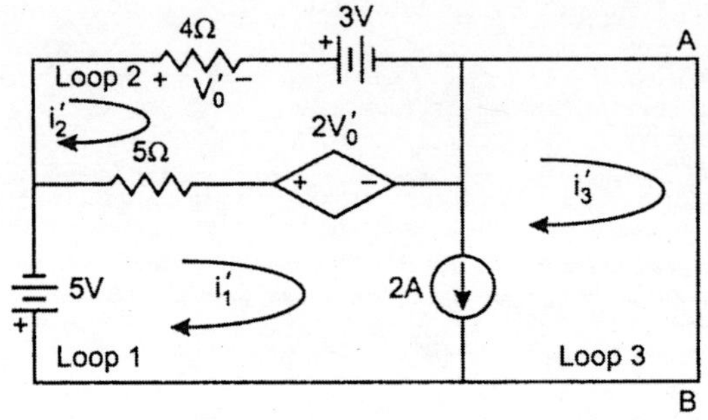

Fig. E4.26(b)

The Loop equations are

Loop 1 : $$5i_1' - 5i_2' = -5 - 2V_0' \quad \text{...}(i)$$

Loop 2 : $$9i_2' - 5i_1' = -3 + 2V_0' \quad \text{...}(ii)$$

And $$i_1' - i_3' = 2 \quad \text{...}(iii)$$

Again $$V_0' = 4i_2' \quad \text{...}(iv)$$

Substituting equation (*iv*) in equation (*i*) we get

$$5i_1' + 3i_2' = -5 \quad \text{...}(v)$$

Substituting equation (*iv*) in equation (*ii*) we get

$$i_2' - 5i_1' = -3 \quad \text{...}(vi)$$

Solving equations (*v*) and (*vi*) we get

$$i_1' = 1 \text{ A} \quad \text{and} \quad i_2' = 2 \text{ A}$$

Substituting the value $i_1' = 1$ A in equation (*iii*) we get

$$i_3' = 1 \text{ A} \quad \text{...}(vii)$$

Thus, $$R_{TH} = \frac{V_{TH}}{i_{sc}} = \frac{V_{TH}}{i_3'} = \frac{36}{1} = 36 \ \Omega$$

Then the Thevenin's equivalent circuit will be as shown in Figure E4.26(*c*)

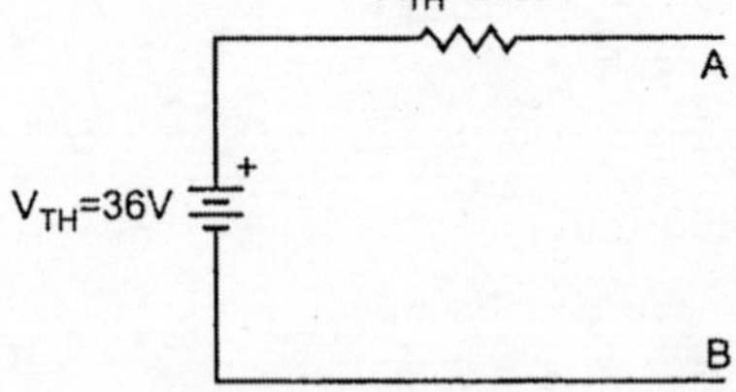

Fig. E4.26(c)

EXAMPLE 4.27. *Find the Thevenin's equivalent across the terminal AB.*

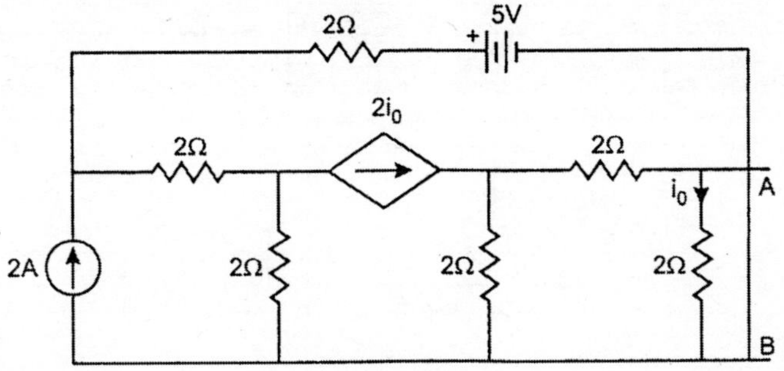

Fig. E4.27

SOLUTION. *To find* V_{TH} :

The circuit of Figure E4.27, is redrawn in Figure E4.27(*a*), with loop currents.

From Figure E4.27(*a*), the loop equations are

Loop 1 : $i_1' = 2\text{ A}$ ...(*i*)

Loop 3 : $6i_3' - 2i_2' - 2i_4' = 0$ ...(*ii*)

Loop 2 and Loop 4 : $i_2' - i_4' = 2i_0'$ ...(*iii*)

Again from Figure E4.27(*a*) we get

$$i_0' = i_3' \quad \text{...(iv)}$$

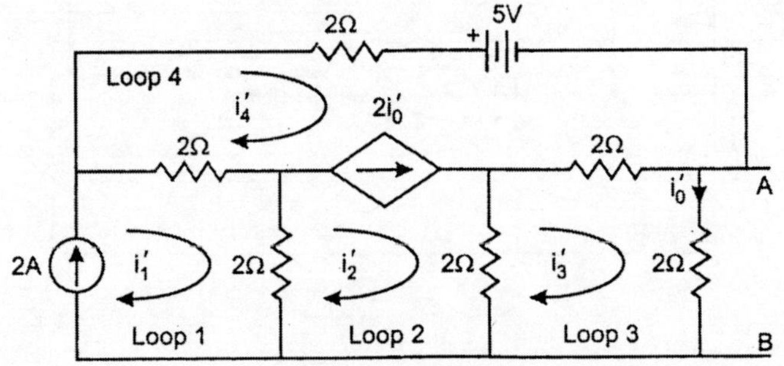

Fig. E4.27(a)

Substituting equation (*iv*) in equation (*iii*) we get

$$i_2' - i_4' - 2i_3' = 0 \quad \text{...(v)}$$

Again writing the KVL for loop 2 and loop 4 we get

$$4i_2' - 4i_1' - 4i_3' + 4i_4' = 5 \quad \text{...(vi)}$$

Substituting equation (*i*) in equation (*vi*) we get

$$4i_2' - 4i_3' + 4i_4' = 13 \quad \text{...(vii)}$$

Writing equations (*ii*), (*v*) and (*vii*) in matrix form we get

$$\begin{bmatrix} -2 & 6 & -1 \\ 1 & -2 & -1 \\ 4 & -4 & 4 \end{bmatrix} \begin{bmatrix} i_2' \\ i_3' \\ i_4' \end{bmatrix} = \begin{bmatrix} 0 \\ 0 \\ 13 \end{bmatrix}$$

Let
$$\Delta = \begin{vmatrix} -2 & 6 & -1 \\ 1 & -2 & -1 \\ 4 & -4 & 4 \end{vmatrix} = -28$$

Then
$$i_3' = \frac{\begin{vmatrix} -2 & 0 & -1 \\ 1 & 0 & -1 \\ 4 & 13 & 4 \end{vmatrix}}{\Delta} = \frac{79}{28} = 2.82 \text{ A}$$

Thus,
$$V_{TH} = 2 \times i_3' = 5.64 \text{ V}$$

To find R_{TH} :

Short circuiting the open terminals *A-B* we get the circuit as shown in Figure E4.27(*b*). Short circuiting the open terminals *A-B* will make $i_0 = 0$ A thus, the dependent current source will vanish from the circuit.

From Figure E4.27(*b*) we get

Loop 1 : $$i_1'' = 2 \text{ A} \qquad \text{...}(viii)$$

Loop 2 : $$10i_2'' - 4i_1'' - 4i_3'' = -5 \qquad \text{...}(ix)$$

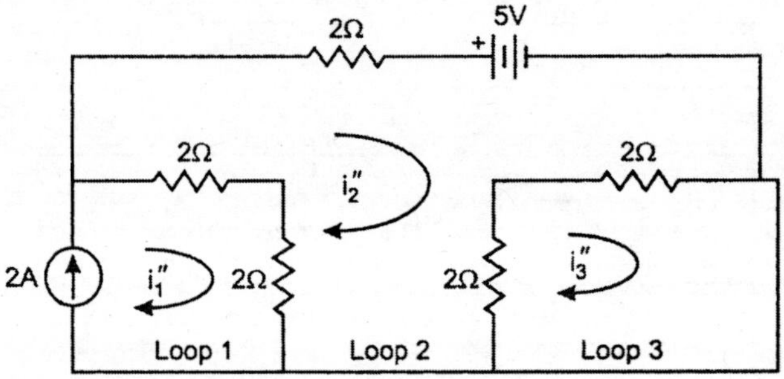

Fig. E4.27(b)

Substituting equation (*viii*) in equation (*ix*) we get

$$10i_2'' - 4i_3'' = 3 \qquad \text{...}(x)$$

Loop 3 : $$4i_3'' - 4i_2'' = 0 \qquad \text{...}(xi)$$

Writing equations (*x*) and (*xi*) in matrix form we get

$$\begin{bmatrix} 10 & -4 \\ -4 & 4 \end{bmatrix} \begin{bmatrix} i_2'' \\ i_3'' \end{bmatrix} = \begin{bmatrix} 3 \\ 0 \end{bmatrix}$$

Let
$$\Delta = \begin{vmatrix} 10 & -4 \\ -4 & 4 \end{vmatrix} = 24$$

Then
$$i_{sc} = i_3'' = \frac{\begin{vmatrix} 10 & 3 \\ -4 & 0 \end{vmatrix}}{\Delta} = \frac{12}{24} = 0.5 \text{ A}$$

Thus, $$R_{TH} = \frac{V_{TH}}{i_{sc}} = \frac{5.64}{0.5} = 11.28\ \Omega$$

Then the Thevenin`s equivalent circuit is shown in Figure E4.27(c)

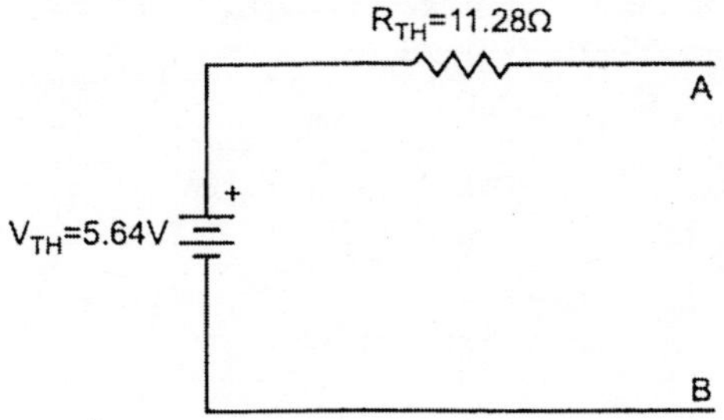

Fig. E4.27(c)

EXAMPLE 4.28. *Find the drop across A-B using Thevenin's theorem.*

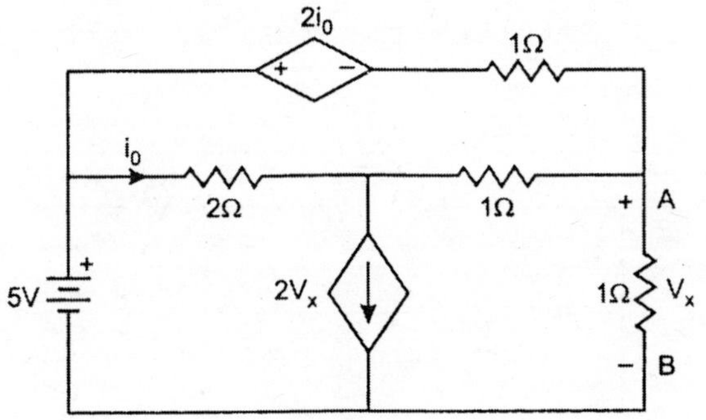

Fig. E4.28

SOLUTION. Removing the load resistance from the main circuit of Figure E4.28 we get a circuit as shown in Figure E4.28(*a*) The loop currents are also indicated in the circuit. The loop equations are

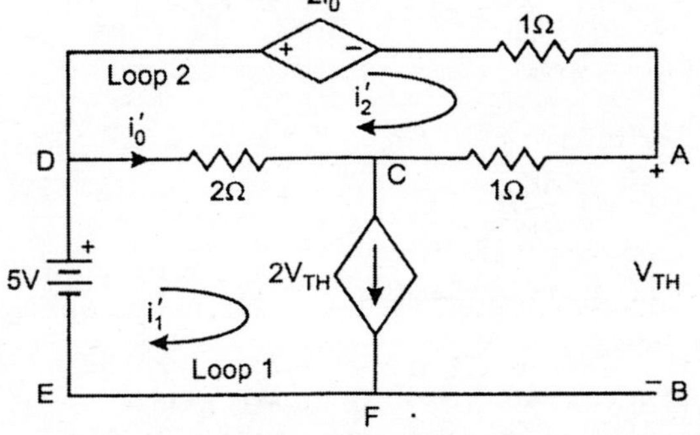

Fig. E4.28(a)

Loop 1 : $$i_1' = 2V_{TH}$$

or $$i_1' - 2V_{TH} = 0 \quad ...(i)$$

Loop 2 : $4i_2' - 2i_1' = -2i_0'$ ...(*ii*)

Again from Figure E4.28(*a*) $i_0' = i_1' - i_2'$ ...(*iii*)

Substituting equation (*iii*) in equation (*ii*) we get

$$i_2' = 0 \text{ A} \quad ...(iv)$$

Applying KVL in the loop *ACDEFB* we get

$$2i_1' + V_{TH} = 5 \quad ...(v)$$

Writing equations (*i*) and (*v*) in matrix form we get

$$\begin{bmatrix} 1 & -2 \\ 2 & 1 \end{bmatrix} \begin{bmatrix} i_1' \\ V_{TH} \end{bmatrix} = \begin{bmatrix} 0 \\ 5 \end{bmatrix}$$

Let $\Delta = \begin{vmatrix} 1 & -2 \\ 2 & 1 \end{vmatrix} = 5$

Then $V_{TH} = \dfrac{\begin{vmatrix} 1 & 0 \\ 2 & 5 \end{vmatrix}}{\Delta} = 1 \text{ V}$

To find R_{TH} :

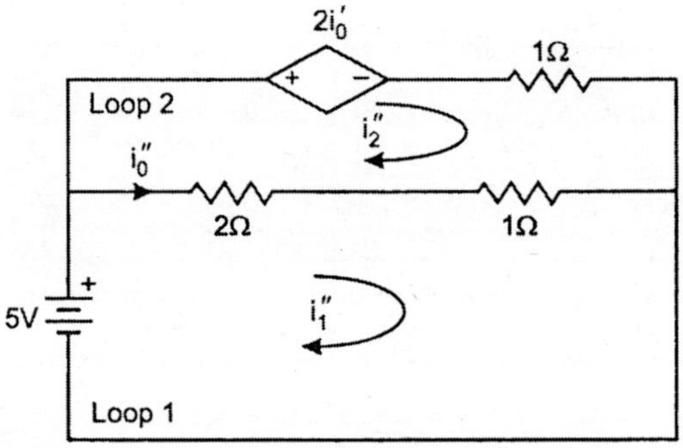

Fig. E4.28(b)

In Figure E4.28(*b*), the load terminal *AB* is shorted. This will make the Dependent current source an open circuit since $V_{TH} = 0$ V now. The loop currents are indicated in Figure E4.28(*b*) The loop equation are

Loop 1 : $3i_1'' - 3i_2'' = 5$ ...(*vi*)

Loop 2 : $4i_2'' - 3i_1'' = 2i_0''$ ...(*vii*)

Again $i_0'' = i_1'' - i_2''$ ...(*viii*)

Substituting equation (*viii*) in equation (*vii*) we get

$$6i_2'' - 5i_1'' = 0 \quad ...(ix)$$

Writing equations (*vi*) and (*ix*) in matrix form we get

$$\begin{bmatrix} 3 & -3 \\ -5 & 6 \end{bmatrix} \begin{bmatrix} i_1'' \\ i_2'' \end{bmatrix} = \begin{bmatrix} 5 \\ 0 \end{bmatrix}$$

Let $$\Delta = \begin{vmatrix} 3 & -3 \\ -5 & 6 \end{vmatrix} = 3$$

Then $$i_1'' = i_{sc} = \frac{\begin{vmatrix} 5 & -3 \\ 0 & 6 \end{vmatrix}}{\Delta} = 10 \text{ A}$$

Thus, $$R_{TH} = \frac{V_{TH}}{i_{sc}} = \frac{1}{10} = 0.1\ \Omega$$

The Thevenin's equivalent circuit is

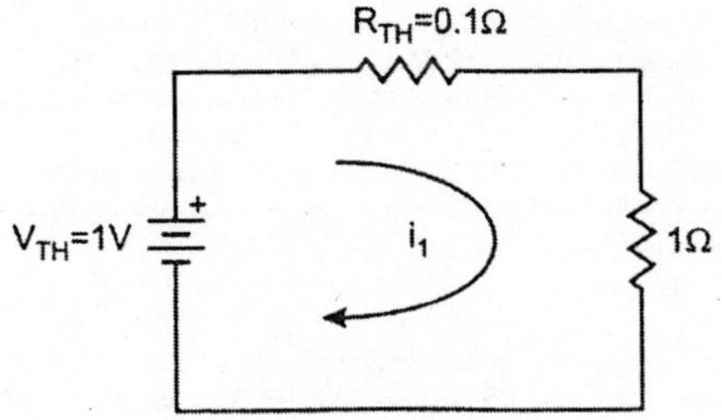

Fig. E4.28(c)

The current through 1 Ω is

$$\frac{1}{1+0.1} = 0.9 \text{ A}.$$

4.6 NORTON'S THEOREM

Statement : Any two terminal linear, active, bilateral d.c. circuit can be replaced by an equivalent circuit consisting of a current source (I_N) and a parallel resistance (R_N).

Where I_N is called the Norton's current and R_N is called the Norton's resistance.

4.6.1 Method to Find Norton's Current (I_N)

Remove the load resistance from the main circuit and short it. Find the short circuit current. This short circuit current is the Norton's current I_N.

4.6.2 Methods to Find The Norton's Resistance (R_N)

The method to find Norton's resistance is same as discussed in the section 5.5.2. That is

$$R_{TH} = R_N$$

SOLVED EXAMPLES

EXAMPLE 4.29. *For the circuit shown in Figure E4.29 find the drop across the 4 Ω resistor, using Thevenin's theorem.*

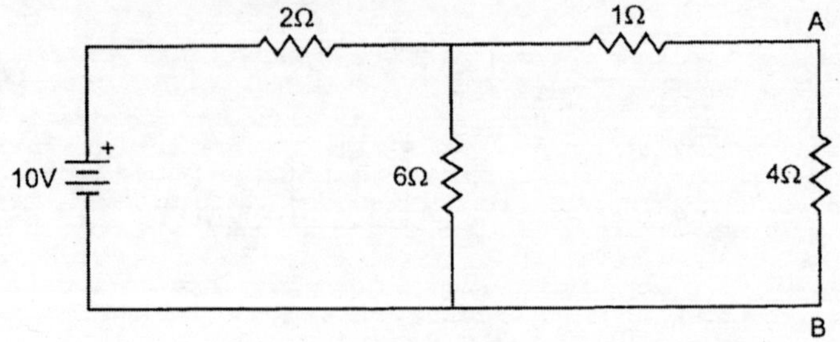

Fig. E4.29

SOLUTION. *To find* i_{sc} :

Removing the 4 Ω load resistance from the circuit and shorting the terminals *A-B* we get the Figure E4.29(*a*)

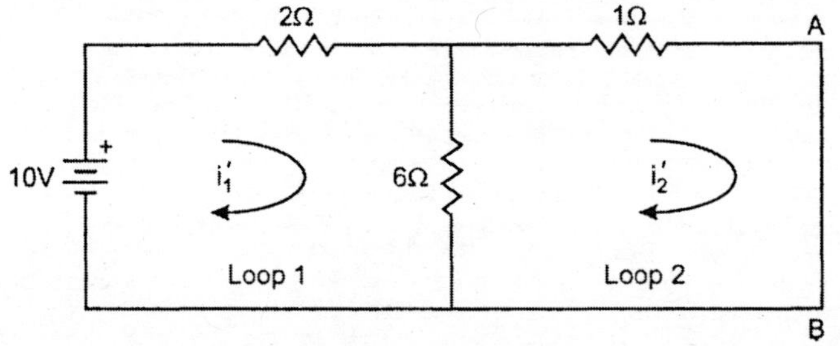

Fig. E4.29(a)

From Figure E4.29(*a*), the short circuit current is i_2'. Writing the loop equations we get,

Loop 1 : $8i_1' - 6i_2' = 10$ ...(*i*)

Loop 2 : $7i_2' - 6i_1' = 0$ ...(*ii*)

Solving equations (*i*) and (*ii*) we get the short circuit current is $i_{sc} = i_2' = 3$ A

To find R_n :

Replacing the voltage source by short circuit (the internal resistance), and removing the load resistance from the circuit we get Figure E4.29(*b*)

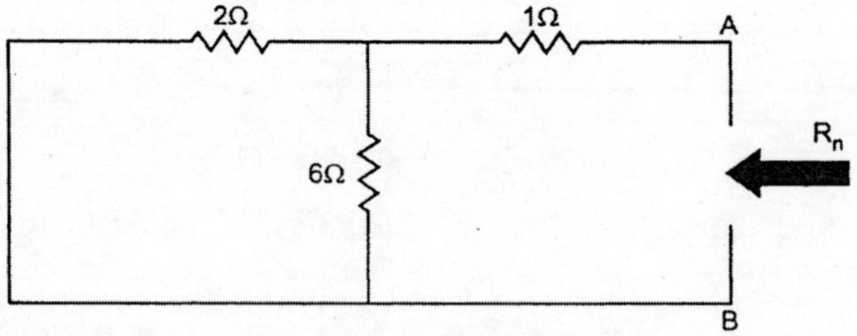

Fig. E4.29(b)

From Figure E4.29(*b*),

$$R_n = \frac{2 \times 6}{2+6} + 1 = 2.5\ \Omega$$

Then the Norton's circuit is shown in Figure E4.29(c),

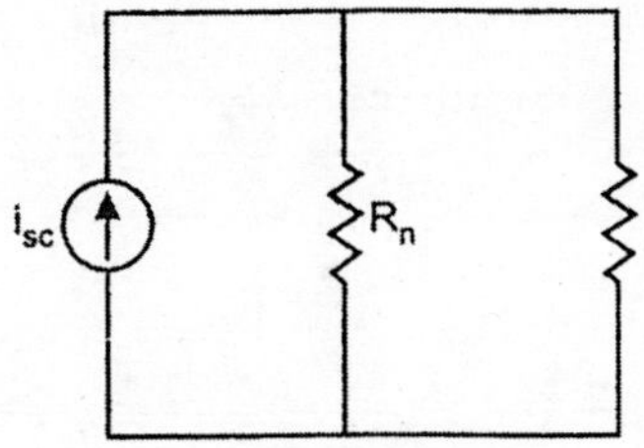

Fig. E4.29(c)

From Figure E4.29(c), current through 4 Ω is $= i_{sc} \times \dfrac{R_n}{R_n + 4} = 1.15$ A

EXAMPLE 4.30. *For the circuit shown in Figure E4.30, find the drop across the R1 resistor, using Norton's theorem.*

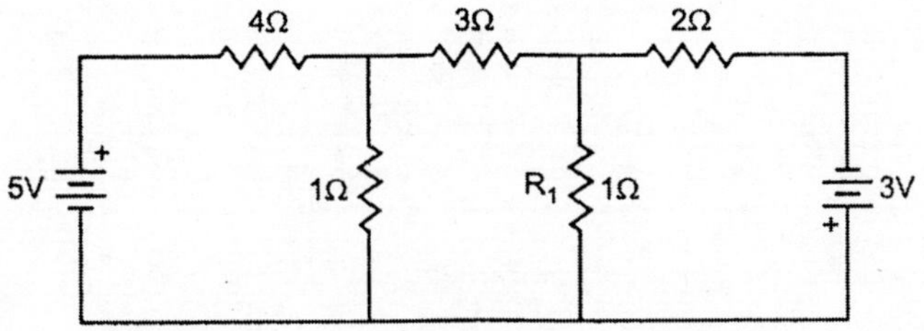

Fig. E4.30

SOLUTION. *To find* i_{sc} :

Removing the R_1 load resistance from the circuit and shorting the load terminals get the Figure E4.30(a).

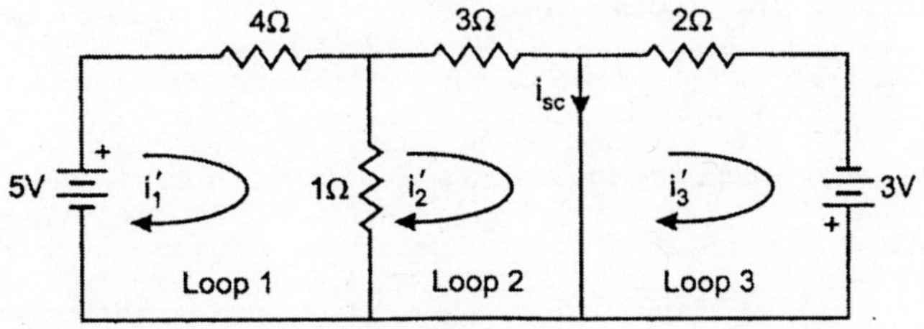

Fig. E4.30(a)

The loop equations are

Loop 1 : $5i_1' - i_2' = 5$ ...(i)

Loop 2 : $4i_2' - i_1' = 0$ ...(ii)

Loop 3 : $2i_3' = 3$ ...(iii)

From equation (iii), $i_3' = 1.5$ A

Solving equations (i) and (ii), we get

$$i_2' = 0.26 \text{ A}$$

Thus $i_{sc} = i_2' - i_3' = -1.24$ A

To find R_n :

Replacing the voltages source by short circuit (the internal resistance), and removing the load resistance from the circuit we get Figure E4.30(*b*)

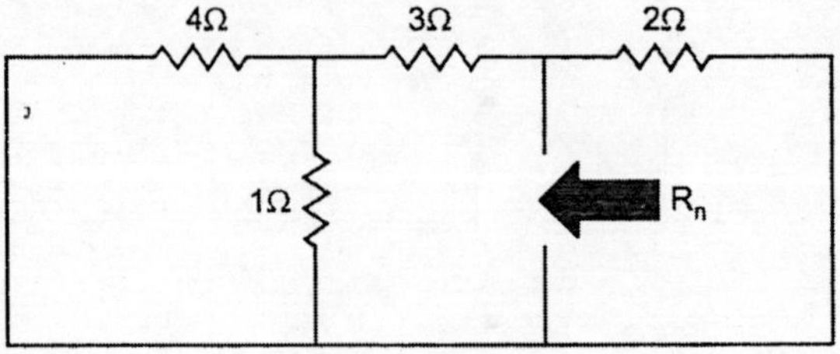

Fig. E4.30(b)

From Figure E4.30(*b*) we get

$$R_n = \frac{\left(\frac{4\times 1}{4+1}+3\right)\times 2}{\left(\frac{4\times 1}{4+1}+3\right)+2} = 1.31\ \Omega$$

Thus the Norton's equivalent circuit is shown in Figure E4.30(*c*).

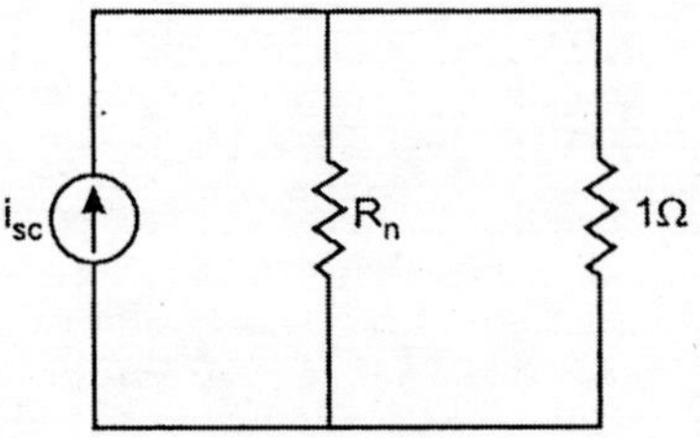

Fig. E4.30(c)

From Figure E4.30(*c*), current through 4 Ω is

$$= i_{sc} \times \frac{R_n}{R_n + 1} = 0.7\ \text{A}$$

EXAMPLE 4.31. *For the circuit shown in Figure E4.31, find the current through* R_1 *using Norton's theorem.*

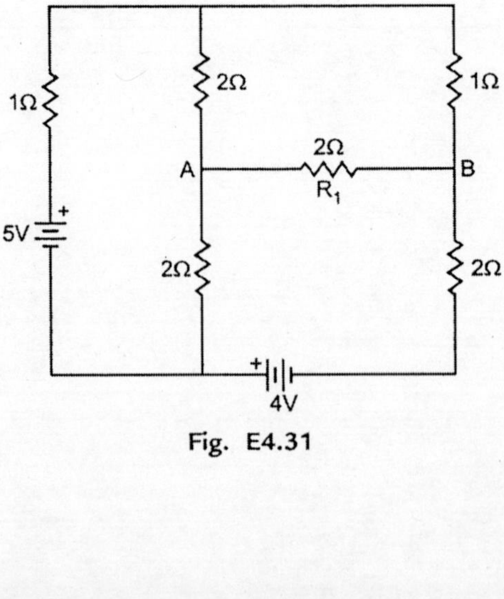

Fig. E4.31

SOLUTION. *To find* i_{sc} :

Removing the R_1, the load resistance, from the circuit and shorting the load terminals get the Figure E4.31(*a*). The loop equations are,

Loop 1 : $5i_1' - 2i_2' - 2i_3' = 5$ ...(*i*)

Loop 2 : $4i_2' - 2i_1' = 4$ ...(*ii*)

Loop 3 : $3i_3' - 2i_1' = 5$ ...(*iii*)

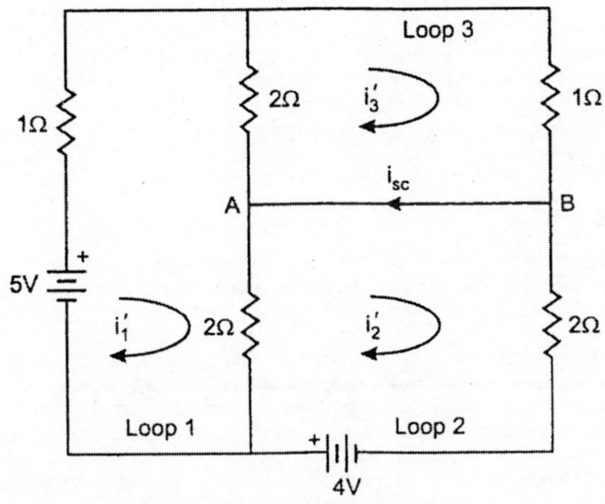

Fig. E4.31(a)

Solving equations (*i*), (*ii*) and (*iii*), we get

$$i_2' = 2.94 \text{ A} \quad \text{and} \quad i_3' = 4.25 \text{ A}$$

The short circuit current is

$$i_{sc} = i_3' - i_2' = 1.31 \text{ A}$$

To find R_n :

Replacing the voltages source by short circuit (the internal resistance), and removing the load resistance from the circuit we get Figure E4.31(*b*)

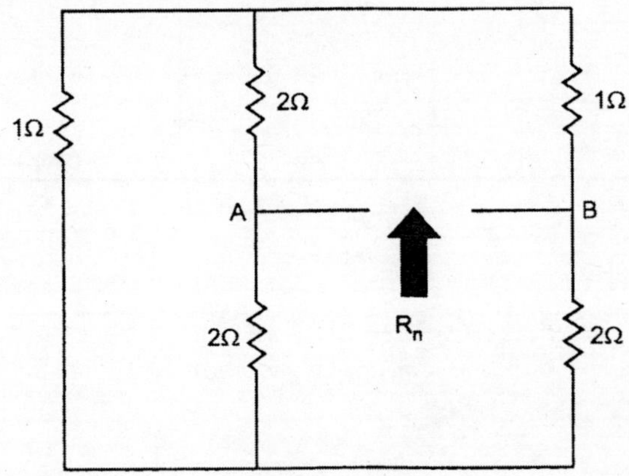

Fig. E4.31(b)

From Figure E4.31(b) we get

$$R_n = 2.9\ \Omega$$

Thus, the Norton's equivalent circuit is shown in Figure E4.31(c).

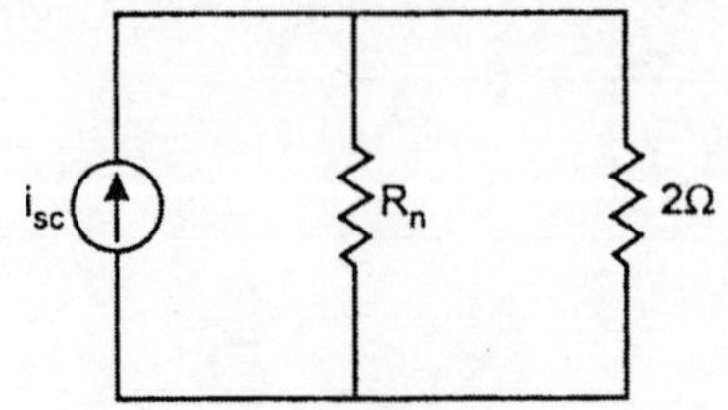

Fig. E4.31(c)

From Figure E4.31(c), current through 1 Ω is $= i_{sc} \times \dfrac{R_n}{R_n + 2} = 0.77$ A

EXAMPLE 4.32. *For the circuit shown in Figure E4.32 find the current through 3 Ω.*

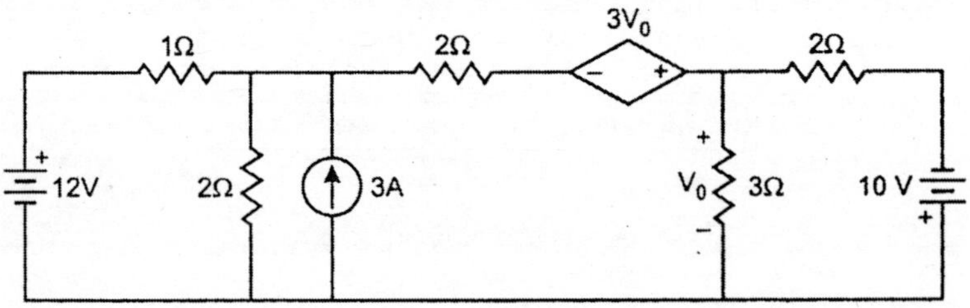

Fig. E4.32

SOLUTION. *To find* i_{sc} :

Removing the 3 Ω resisrance (the load resistance) from the circuit and shorting the load terminals get the Figure E4.32(a). Here after shorting the load terminal $V_0 = 0$ V, thus the dependent voltage source will also vanish from the circuit.

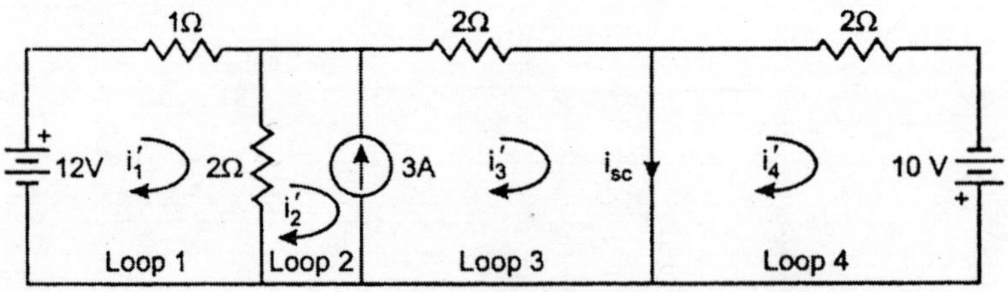

Fig. E4.32(a)

The loop equations are

Loop 1 : $3i'_1 - 2i'_2 = 12$ ...(i)

Loop 2 and loop 3 : $i'_3 - i'_2 = 3$ ...(ii)

And $2i'_2 - 2i'_3 - 2i'_1 = 0$ ...(iii)

Loop 4 : $i'_4 = \dfrac{10}{2} = 5$ A ...(iv)

Solving equations (*i*), (*ii*) and (*iii*), we get

$$i_3' = -7.5 \text{ A}$$

Thus, the short circuit current is

$$i_{sc} = i_3' - i_4' = -2.5 \text{ A}$$

To find V_{TH} :

The load has been removed from the network and the modified circuit is shown in Figure E4.32(*b*). The loop equations are

Loop 1 : $3i_1' - 2i_2' = 12$ ...(*v*)

Loop 2 and loop 3 : $i_3' - i_2' = 3$ ...(*vi*)

And $2i_2' + 4i_3' - 2i_1' = 10 + 3V_{TH}$ ...(*vii*)

And $V_{TH} = 10 - 2i_3'$ ...(*viii*)

Substituting equation (*viii*) in equation (*vii*) we get

$2i_2' + 10i_3' - 2i_1' = 10 + 30$ ...(*ix*)

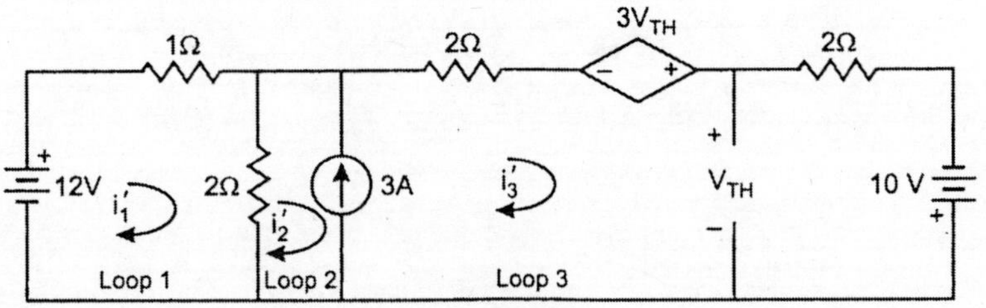

Fig. E4.32(b)

Solving equations (*v*), (*vi*) and (*ix*), we get

$$i_3' = 4.69 \text{ A}$$

and $V_{TH} = 10 - 2i_3' = 0.62 \text{ A}$

Thus, $R_n = \dfrac{V_{TH}}{i_{sc}} = 0.25\ \Omega$

Thus, the Norton's equivalent circuit is shown in Figure E4.32(*c*).

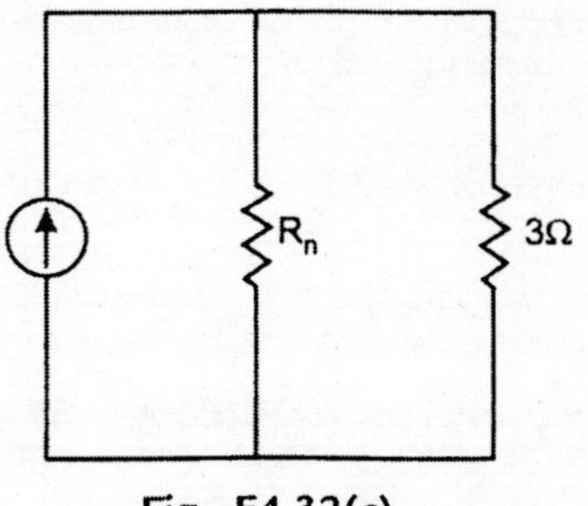

Fig. E4.32(c)

From Figure E4.32(*c*), current through 1 Ω is $= i_{sc} \times \dfrac{R_n}{R_n + 3} = 0.19 \text{ A}$

EXAMPLE 4.33. *Find the Norton's equivalent across the terminal AB.*

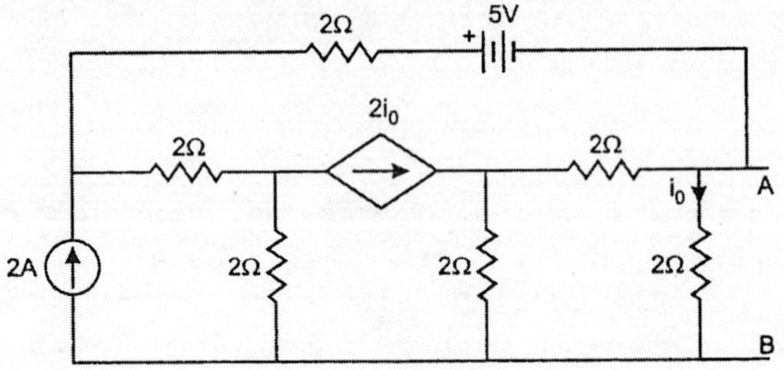

Fig. E4.33

SOLUTION. The circuit is same as in Example 27. From the solution of Example 27.

$$V_{TH} = 5.64 \text{ V}$$

$$i_{sc} = 0.5 \text{ A}$$

and
$$R_n = \frac{V_{TH}}{i_{sc}} = 11.28 \ \Omega$$

Thus, the Norton's equivalent circuit is shown in Figure E4.33(*a*)

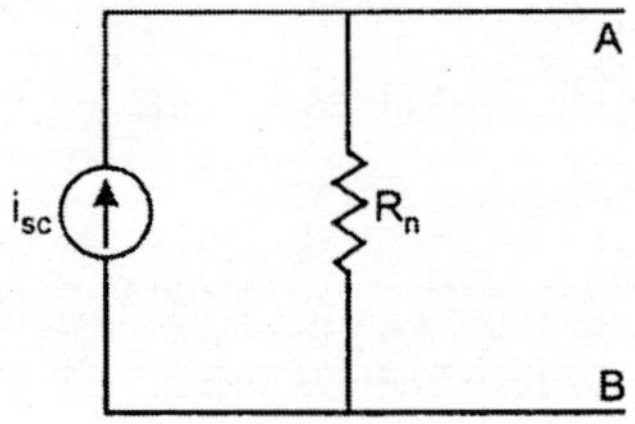

Fig. E4.33(a)

4.7 MAXIMUM POWER TRANSFER THEOREM

Statement : A load resistance in a linear, active, bilateral d.c. circuit will receive the maximum power if and only if the value of the load resistance becomes equal to the Thevenin's resistance (R_{TH}). And the maximum power will be

$$P_m = \frac{V_{TH}^2}{4 \times R_{TH}}$$

Where V_{TH} is the Thevenin's Voltage and R_{TH} is the Thevenin's resistance.

SOLVED EXAMPLES

EXAMPLE 4.34. *For the circuit shown in Figure E4.34, find the value of R to be connected across A-B for maximum power transfer, and thus find the power.*

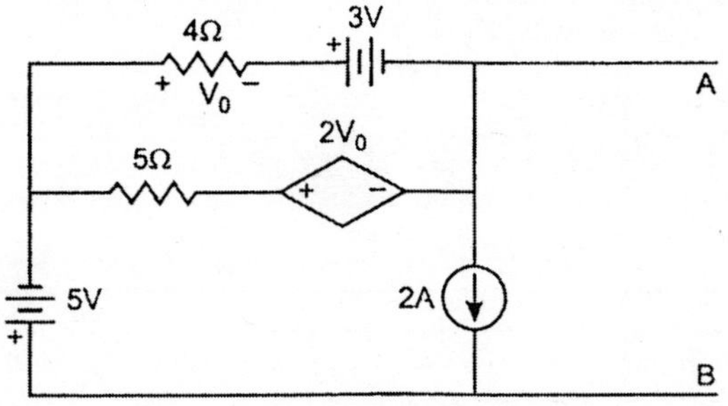

Fig. E4.34

SOLUTION. The circuit is same as of Example 26. From the solution of Example 26,

$$V_{TH} = -36 \text{ V}$$

$$R_{TH} = 36 \ \Omega$$

Thus, the value of R to be connected across A-B for maximum power transfer $R_{TH} = 36 \ \Omega$

And the maximum power is $\dfrac{V_{TH}^2}{4 \times R_{TH}} = 9 \text{ W}$

EXAMPLE 4.35. *For the circuit shown in Figure E4.35, find the value of R to be connected across A-B for maximum power transfer and thus find the power.*

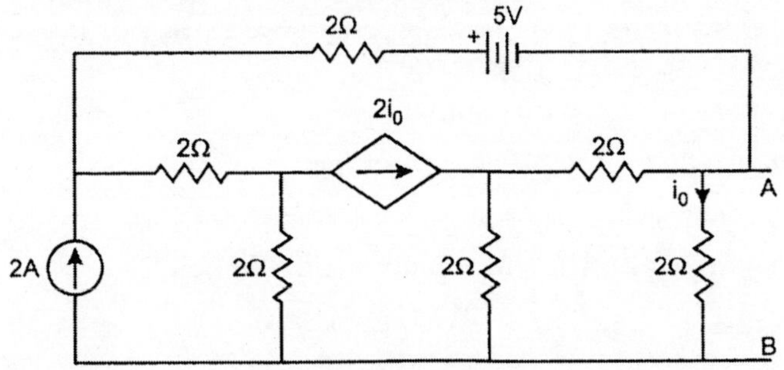

Fig. E4.35

SOLUTION. The circuit is same as of Example 27. From the solution of Example 27,

$$V_{TH} = 5.64 \text{ V}$$

$$R_{TH} = 11.18 \ \Omega$$

Thus, the value of R to be connected across A-B for maximum power transfer $R_{TH} = 11.18 \ \Omega$

And the maximum power is $\dfrac{V_{TH}^2}{4 \times R_{TH}} = 0.71 \text{ W}$

EXAMPLE 4.36. *For the circuit shown in Figure E4.36, find the value of R to be connected across A-B for maximum power transfer and thus find the power.*

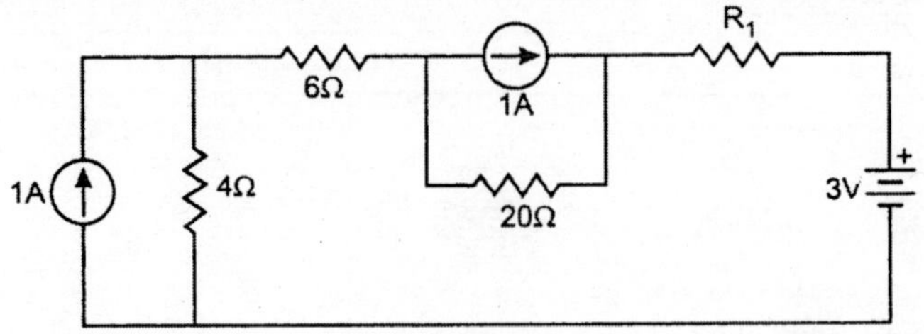

Fig. E4.36

SOLUTION. *To find* V_{TH} :

The circuit to find V_{TH} is shown in Figure E4.36(*a*), here the load resistance R_1 has been removed and we are calculating the open circuit voltage V_{TH}.

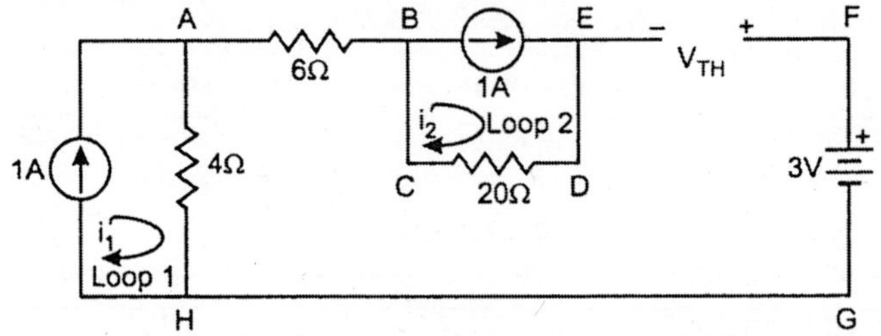

Fig. E4.36(a)

The loop equations are :

Loop 1 : $i_1' = 1$ A ...(*i*)

Loop 2 : $i_2' = 1$ A ...(*ii*)

Applying KVL in the loop *ABCDEFGHA* we get

$$-20i_2' - V_{TH} + 3 - 4i_1' = 0 \quad ...(iii)$$

Substituting equations (*i*) and (*ii*) in equation (*iii*) we get

$$V_{TH} = 21 \text{ V}$$

To find R_{TH} :

The circuit to find R_{TH} is shown in Figure E4.36(*b*). The voltage source has been replaced by short circuit (internal resistance) and the current source has been replaced by open circuit (internal resistance). From Figure E4.36(*b*)

$$R_{TH} = 30 \ \Omega$$

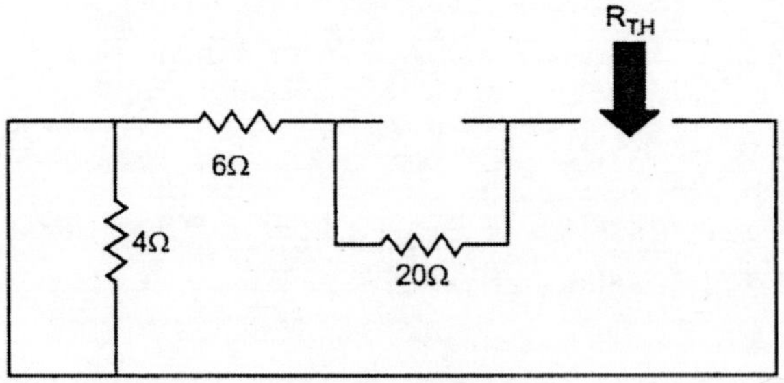

Fig. E4.36(b)

Thus, the value of R to be connected across A-B for maximum power transfer $R_{TH} = 30\ \Omega$

The maximum power $= \dfrac{V_{TH}^2}{4 \times R_{TH}} = 3.66$ W

EXAMPLE 4.37. *For the circuit shown in Figure E4.37, find the value of R to be connected across A-B for maximum power transfer and thus find the power.*

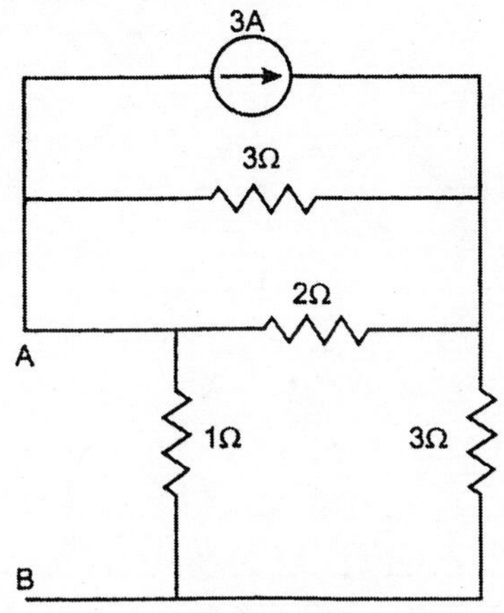

Fig. E4.37

SOLUTION. *To find* V_{TH} :

The circuit to find V_{TH} is shown in Figure E4.37(*a*),

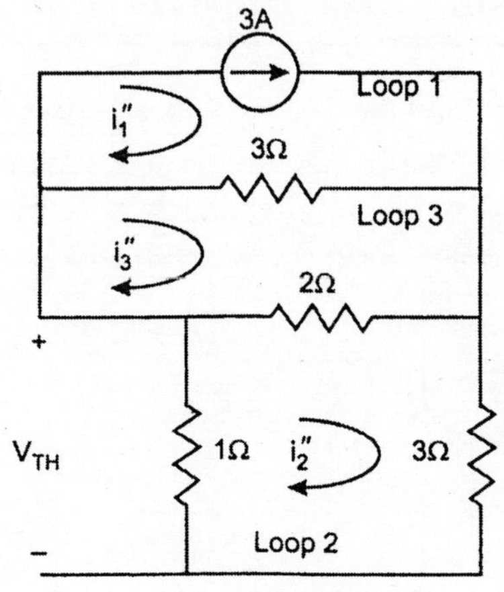

Fig. E4.37(a)

From the solution of Example 17, we get

$$V_{TH} = V_0'' = 0.69 \text{ V}$$

To find R_{TH} :

The circuit to find R_{TH} is shown in Figure E4.37(*b*),

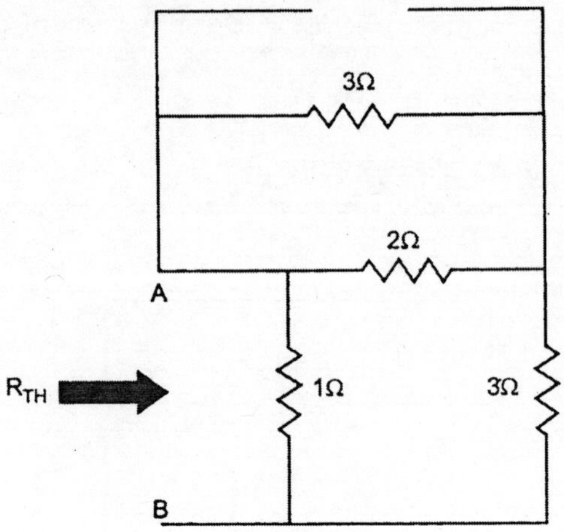

Fig. E4.37(b)

From Figure E4.37(*b*) we get

$$R_{TH} = 0.81\ \Omega$$

Thus, the value of R to be connected across A-B for maximum power transfer $R_{TH} = 0.81\ \Omega$

$$\text{The maximum power} = \frac{V_{TH}^2}{4 \times R_{TH}} = 0.15\ \Omega$$

4.8 MILLMAN'S THEOREM

Statement : If a linear active bilateral d.c. circuit consists of n voltage sources (V_1 to V_n) connected in parallel to each other, with their internal resistance (r_1 to r_n) connected in series with the source respectively (Figure 4.1). Then such a circuit can be replaced bt a single voltage source E_m with a series resistance R_m (Figure 4.2).

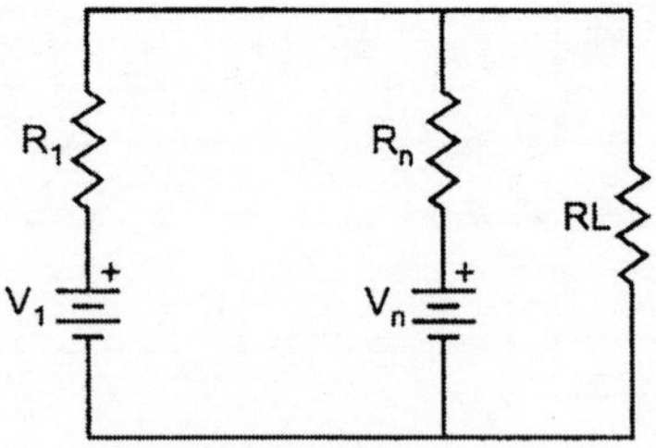

Fig. 4.1

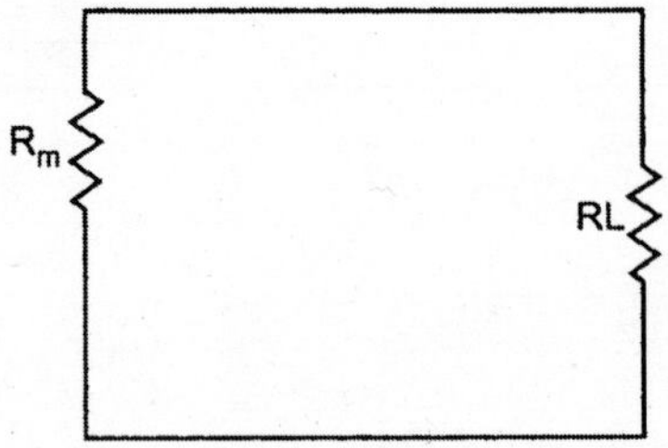

Fig. 4.2

Where,

$$E_m = \frac{V_1 \times \dfrac{1}{R_1} + \ldots\ldots\ldots\ldots + V_n \times \dfrac{1}{R_n}}{\dfrac{1}{R_1} + \ldots\ldots\ldots\ldots + \dfrac{1}{R_n}}$$

and

$$R_m = \frac{1}{\dfrac{1}{R_1} + \ldots\ldots\ldots\ldots + \dfrac{1}{R_n}}$$

SOLVED EXAMPLES

EXAMPLE 4.38. *For the circuit shown in Figure E4.38, find the current through 3 Ω, using Millman's theorem.*

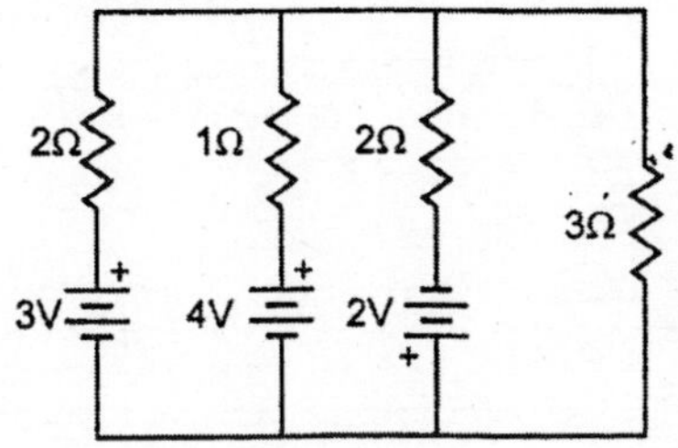

Fig. E4.38

SOLUTION. By applying Millman's theorem,

$$E_m = \frac{3\times\frac{1}{2}+4\times\frac{1}{1}-2\times\frac{1}{2}}{\frac{1}{2}+\frac{1}{1}+\frac{1}{2}} = 2.25 \text{ V}$$

$$Z_m = \frac{1}{\frac{1}{2}+\frac{1}{1}+\frac{1}{2}} = 0.5\ \Omega$$

Then current through 3 Ω $= \dfrac{E_m}{Z_m+3} = 0.64$ A

EXAMPLE 4.39. *For the circuit shown in Figure E4.39, find the current through 4 Ω, using Millman's theorem.*

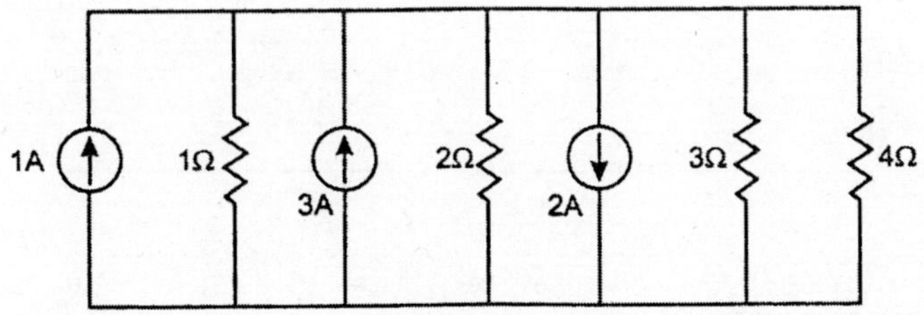

Fig. E4.39

SOLUTION. Converting all the current source into voltage source we get a new circuit as shown in Figure E4.39(*a*)

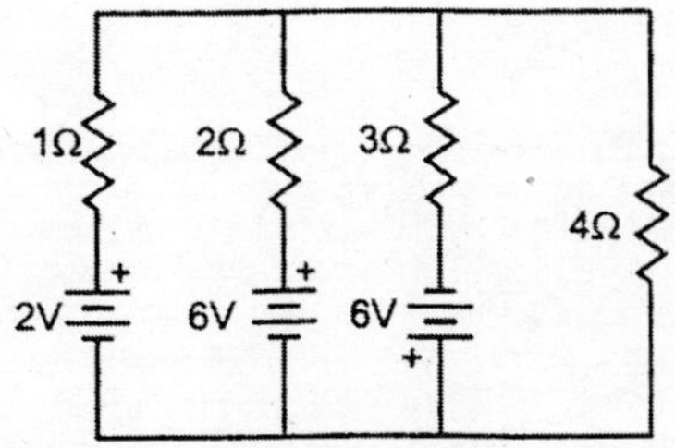

Fig. E4.39(a)

By applying Millman's theorem,

$$E_m = \frac{2\times\frac{1}{1}+6\times\frac{1}{2}-6\times\frac{1}{3}}{\frac{1}{1}+\frac{1}{2}+\frac{1}{3}} = 1.64 \text{ V}$$

$$Z_m = \frac{1}{\frac{1}{1}+\frac{1}{2}+\frac{1}{3}} = 0.55\ \Omega$$

Then current through 4 Ω $= \frac{E_m}{Z_m + 4} = 0.59$ A

EXERCISES

1. State and explain the following theorems :
   (*i*) Superposition Theorem
   (*ii*) Reciprocity Theorem
   (*iii*) Compensation Theorem
   (*iv*) Thevenin's Theorem
   (*v*) Norton's Theorem
   (*vi*) Maximum Power Transfer Theorem
   (*vii*) Millman's Theorem.
2. Find the drop across 5 Ω, using Thevenin's theorem for the circuit shown in Figure Q4.2

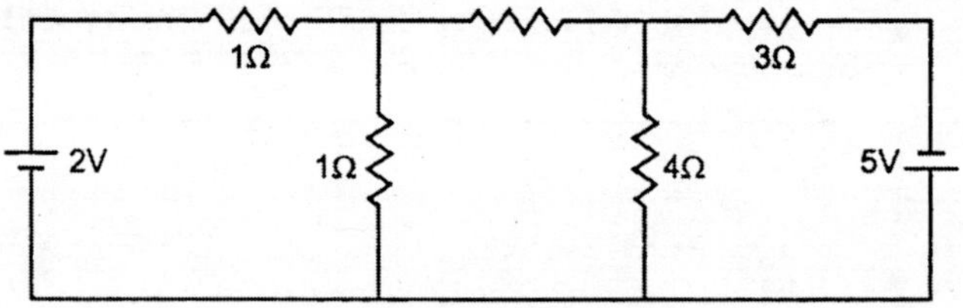

Fig. Q4.2

3. Find the value of resistance *R* to be connected across *A*-*B* for maximum power transfer, for the circuit shown in Figure Q4.3

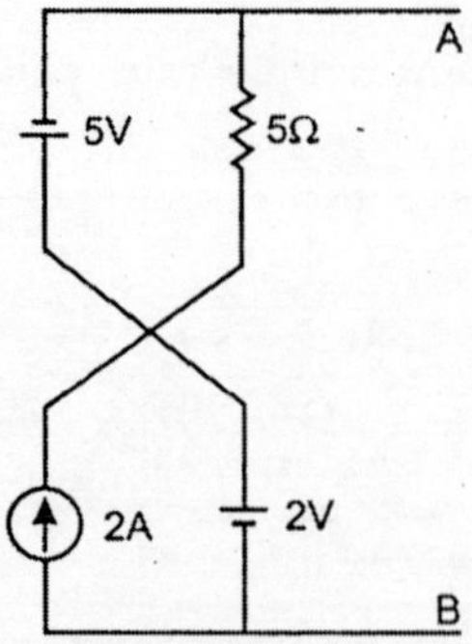

Fig. Q4.3

4. Find the current through 5 Ω, using Norton's theorem for the circuit shown in Figure Q4.4

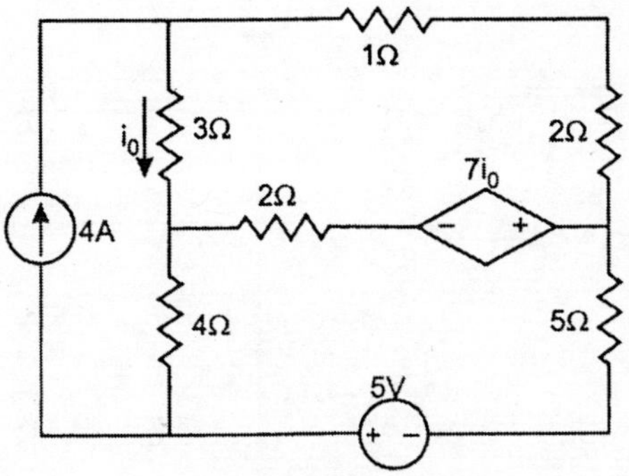

Fig. Q4.4

5. Find V_0, using Norton's theorem for the circuit shown in Figure Q4.5. Verify your result using Thevenin's theorem

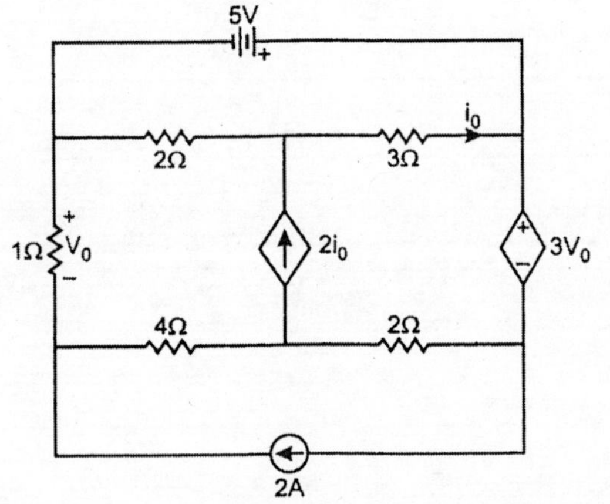

Fig. Q4.5

6. Find the loop currents using Superposition theorem for the circuit shown in Figure Q4.6

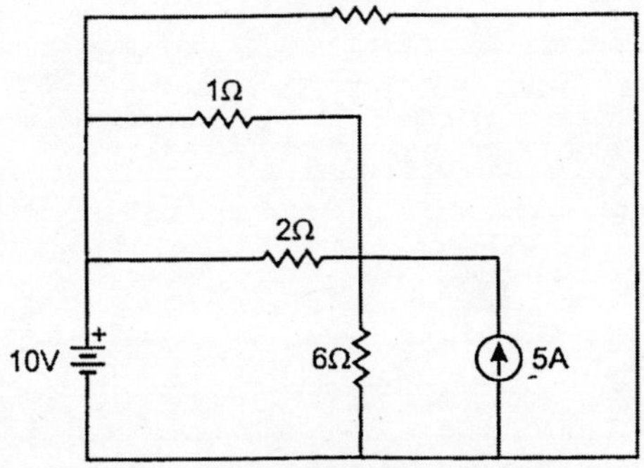

Fig. Q4.6

7. Find the value of resistance R to be connected across A-B for maximum power transfer, for the circuit shown in Figure Q4.7

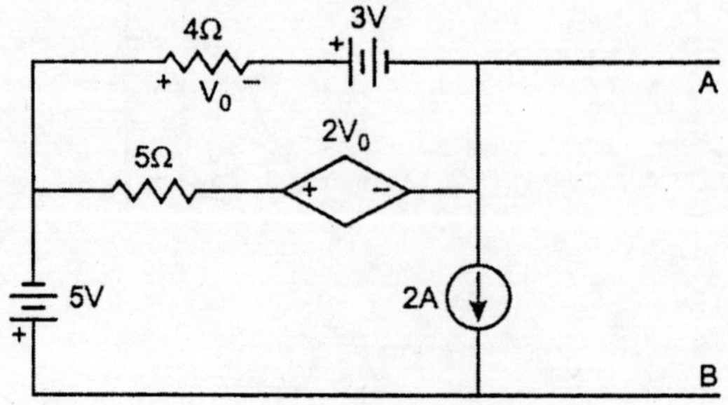

Fig. Q4.7

Chapter **5**

NETWORK THEOREMS (AC CIRCUITS)

5.1 INTRODUCTION

All the theorems that has been discussed in Chapter 4, for d.c. network is also applicable to a.c network. Here in this chapter we will apply all the theorems to a.c. circuit.

5.2 SUPERPOSITION THEOREM

Statement : In a linear network containing several independent sources, the overall response at any point in the network, equals the sum of response due to each independent source considered separately, with all the other independent sources made inoperative.

An independent voltage source can be made inoperative by replacing it by the internal impedance of the source. It is zero for an ideal voltage source.

An independent current source can be made inoperative by replacing it by the internal impedance of the source. It is infinity for an ideal current source.

Note that the dependent sources are not removed from the network at any point in time.

SOLVED EXAMPLES

EXAMPLE 5.1. *Using Superposition theorems find the drop across* $-j6\ \Omega$*, for the circuit shown in Figure E5.1.*

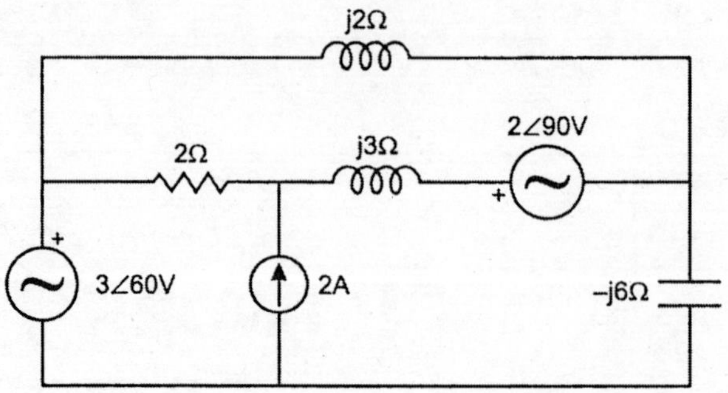

Fig. E5.1

SOLUTION. Considering only the 3∠60 V source and replacing the other voltage source by short circuit and the current source by open circuit we get the circuit shown in Figure E5.1(*a*)

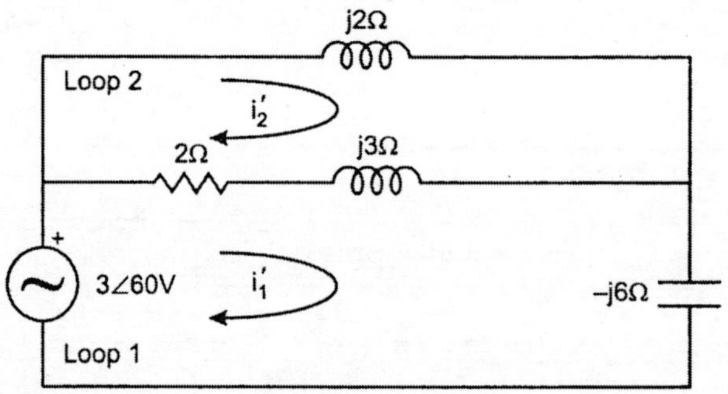

Fig. E5.1(a)

The loop equations are

Loop 1 : $(2 - j3)i'_1 - (2 + j3)i'_2 = 3\angle 60 = 1.5 + j2.6$ ...(*i*)

Loop 2 : $(2 + j5)i'_2 - (2 + j3)i'_1 = 0$ ...(*ii*)

Writing equations (*i*) and (*ii*) in matrix form we get

$$\begin{bmatrix} 2-j3 & -(2+j3) \\ -(2+j3) & 2+j5 \end{bmatrix}\begin{bmatrix} i'_1 \\ i'_2 \end{bmatrix} = \begin{bmatrix} 1.5+j2.6 \\ 0 \end{bmatrix}$$

Let $$\Delta = \begin{vmatrix} 2-j3 & -(2+j3) \\ -(2+j3) & 2+j5 \end{vmatrix} = 24 - j8$$

Then $$i'_1 = \frac{\begin{vmatrix} 1.5+j2.6 & -(2+j3) \\ 0 & 2+j5 \end{vmatrix}}{\Delta} = (-0.53 + j0.35)\text{ A}$$

So the drop across $-j6\ \Omega$ is $V' = -j6 \times i'_1 = (2.11 + j3.2)$ V

Considering only the 2∠90 V source and replacing the other voltage source by short circuit and the current source by open circuit we get the circuit shown in Figure E5.1(*b*)

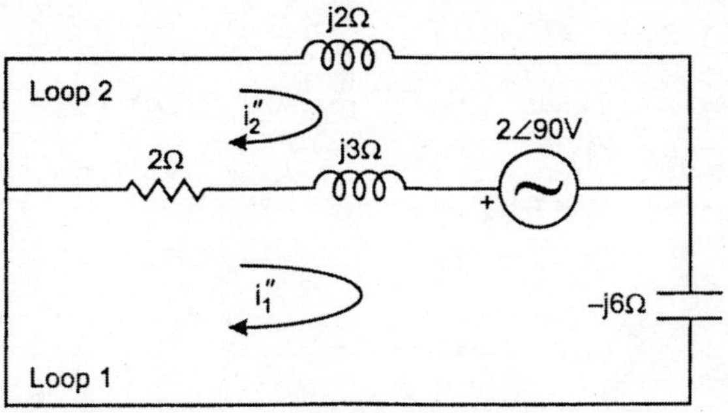

Fig. E5.1(b)

The loop equations are

Loop 1 : $(2 - j3)i_1'' - (2 + j3)i_2'' = -2\angle 90 = -j2$ ...(*iii*)

Loop 2 : $(2 + j5)i_2'' - (2 + j3)i_1'' = j2$ ...(*iv*)

Writing equations (*iii*) and (*iv*) in matrix form we get

$$\begin{bmatrix} 2-j3 & -(2+j3) \\ -(2+j3) & 2+j5 \end{bmatrix} \begin{bmatrix} i_1'' \\ i_2'' \end{bmatrix} = \begin{bmatrix} -j2 \\ j2 \end{bmatrix}$$

Let $$\Delta = \begin{vmatrix} 2-j3 & -(2+j3) \\ -(2+j3) & 2+j5 \end{vmatrix} = 24 - j8$$

Then $$i_1 = \frac{\begin{vmatrix} -j2 & -(2+j3) \\ j2 & 2+j5 \end{vmatrix}}{\Delta} = (0.15 + j0.05)\text{ A}$$

So the drop across $-j6\ \Omega$ is $V'' = -j6 \times i_1'' = (-0.6 - j4.2)$ A

Considering only the 2 A source and replacing the other voltage sources by short circuit we get the circuit shown in Figure E5.1(*c*)

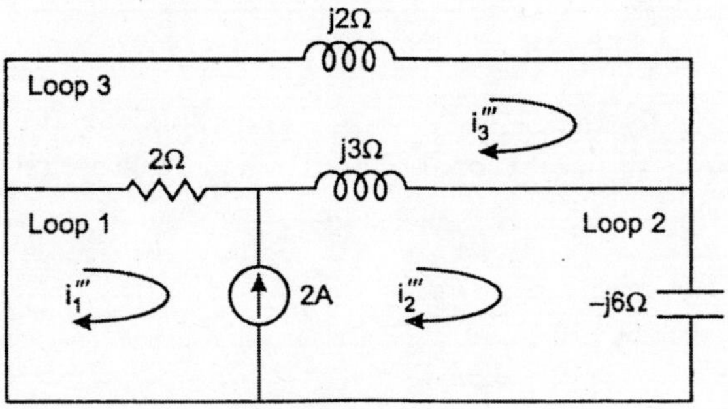

Fig. E5.1(c)

Writing the loop equations we get

Loop 1 & Loop 2 : $i_2''' - i_1''' = 2$ A ...(*v*)

$2i_1''' - j3i_2''' - (2 + j3)i_3''' = 0$ ...(*vi*)

Loop 3 : $(2 + j5)i_3''' - 2i_1''' - j3i_2''' = 0$ ...(*vii*)

Writing equations (*v*), (*vi*) and (*vii*) in matrix form we get

$$\begin{bmatrix} -1 & 1 & 0 \\ 2 & -j3 & -(2+j3) \\ -2 & -j3 & 2+j5 \end{bmatrix} \begin{bmatrix} i_1''' \\ i_2''' \\ i_3''' \end{bmatrix} = \begin{bmatrix} 2 \\ 0 \\ 0 \end{bmatrix}$$

Let
$$\Delta = \begin{vmatrix} -1 & 1 & 0 \\ 2 & -j3 & -(2+j3) \\ -2 & -j3 & 2+j5 \end{vmatrix} = -24 + j8$$

Then
$$i_2''' = \frac{\begin{vmatrix} -1 & 2 & 0 \\ 2 & 0 & -(2+j3) \\ -2 & 0 & 2+j5 \end{vmatrix}}{\Delta} = (-0.1 + j0.3)\text{ A}$$

So the drop across $-j6\ \Omega$ is $V''' = -j6 \times i_2''' = (1.8 + j0.6)$ V

So the total drop is $(3.31 - j0.4)$ V

EXAMPLE 5.2. *Using Superposition theorems find i_0 for the circuit shown in Figure E5.2.*

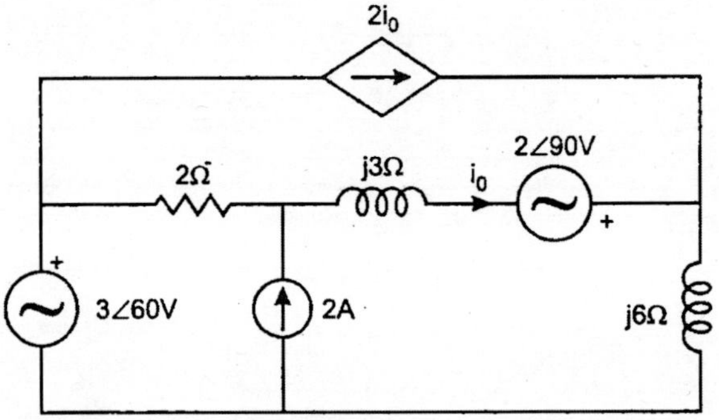

Fig. E5.2

SOLUTION. Considering only the 3∠60 V source and replacing the other voltage source by short circuit and the current source by open circuit we get the circuit shown in Figure E5.2(*a*)

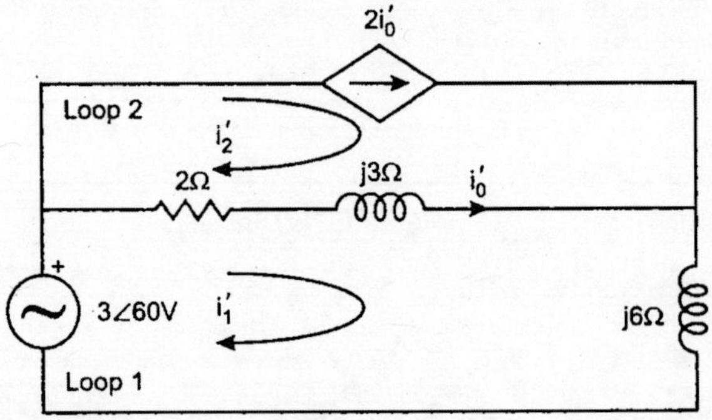

Fig. E5.2(a)

The loop equations are

Loop 1 : $(2 + j9)i_1' - (2 + j3)i_2' = 3\angle 60 = 1.5 + j2.59$ ...(*i*)

Loop 2 : $i_2' = 2i_0'$ ...(*ii*)

And $i_0' = i_1' - i_2'$ ...(*iii*)

Substituting equation (*ii*) in equation (*iii*) we get

$$i_1' - 1.5i_2' = 0 \qquad ...(iv)$$

Writing equations (*i*) and (*iv*) in matrix form we get

$$\begin{bmatrix} 2 + j9 & -(2 + j3) \\ 1 & -1.5 \end{bmatrix} \begin{bmatrix} i_1' \\ i_2' \end{bmatrix} = \begin{bmatrix} 1.5 + j2.59 \\ 0 \end{bmatrix}$$

Let $$\Delta = \begin{vmatrix} 2 + j9 & -(2 + j3) \\ 1 & -1.5 \end{vmatrix} = -1 - j10.5$$

Then $$i_1' = \frac{\begin{vmatrix} 1.5 + j2.59 & -(2 + j3) \\ 0 & -1.5 \end{vmatrix}}{\Delta} = (0.3869 - j0.77) \text{ A}$$

And $$i_2' = \frac{\begin{vmatrix} 2 + j9 & 1.5 + j2.59 \\ 1 & 0 \end{vmatrix}}{\Delta} = (0.26 - j0.12) \text{ A}$$

Then $i_0' = i_1' - i_2' = (0.13 - j0.66)$ A

Considering only the $2\angle 90$ V source and replacing the other voltage source by short circuit and the current source by open circuit we get the circuit shown in Figure E5.2(*b*)

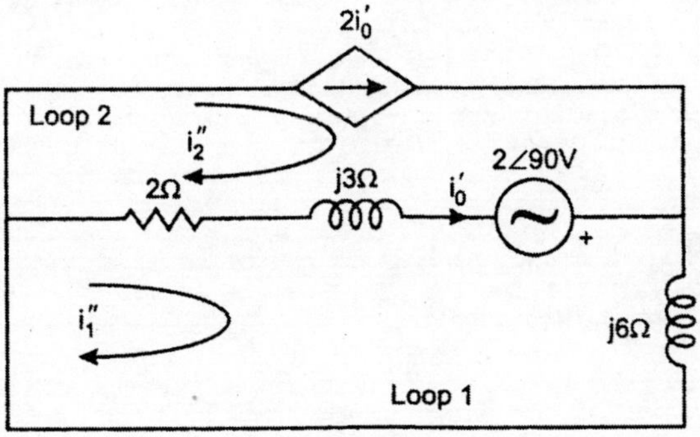

Fig. E5.2(b)

The loop equations are

Loop 1 : $(2 + j9)i_1'' - (2 + j3)i_2'' = 2\angle 90 = j2$ ...(*v*)

Loop 2 : $i_2'' = 2i_0''$ ...(*vi*)

And $i_0'' = i_1'' - i_2''$ ...(*vii*)

Substituting equation (*vi*) in equation (*vii*) we get

$$i_1'' - 1.5i_2'' = 0 \qquad ...(viii)$$

Writing equations (*i*) and (*iv*) in matrix form we get

$$\begin{bmatrix} 2+j9 & -(2+j3) \\ 1 & -1.5 \end{bmatrix}\begin{bmatrix} i_1'' \\ i_2'' \end{bmatrix} = \begin{bmatrix} j2 \\ 0 \end{bmatrix}$$

Let $$\Delta = \begin{vmatrix} 2+j9 & -(2+j3) \\ 1 & -1.5 \end{vmatrix} = (-1 - j10.5)$$

Then $$i_1'' = \frac{\begin{vmatrix} j2 & -(2+j3) \\ 0 & -1.5 \end{vmatrix}}{\Delta} = (0.28 - j0.27) \text{ A}$$

And $$i_2'' = \frac{\begin{vmatrix} 2+j9 & j2 \\ 1 & 0 \end{vmatrix}}{\Delta} = (0.19 + j0.018) \text{ A}$$

Then $$i_0'' = i_1'' - i_2'' = (0.094 - j0.25) \text{ A}$$

Considering only the 2 A source and replacing the other voltage sources by short circuit we get the circuit shown in Figure E5.2(*c*)

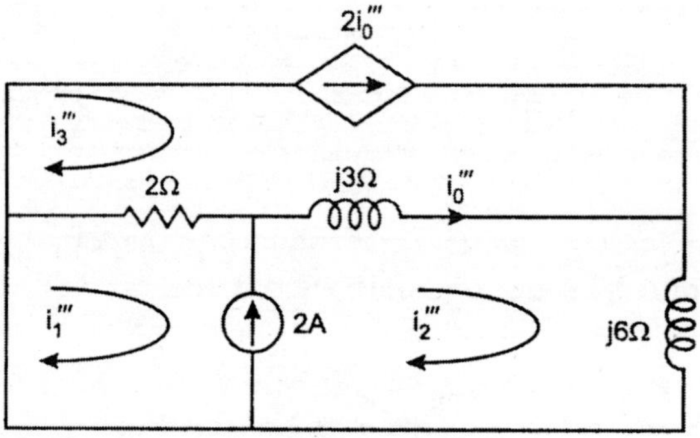

Fig. E5.2(c)

The loop equations are

Loop 1 and Loop 2 : $$i_2''' - i_1''' = 2 \quad \text{...(ix)}$$

$$2i_1''' + j9i_2''' - (2 + j3)i_3''' = 0 \quad \text{...(x)}$$

Loop 3 : $$i_3''' = 2i_0''' \quad \text{...(xi)}$$

Again $$i_0''' = i_1''' - i_2''' \quad \text{...(xii)}$$

Substituting equation (*xii*) in equation (*xi*) we get

$$i_3''' = 2i_1''' - 2i_2''' \quad \text{...(xiii)}$$

Writing equations (*ix*), (*x*) and (*xiii*) in matrix form we get

$$\begin{bmatrix} -1 & 1 & 0 \\ 2 & j9 & -(2+j3) \\ 2 & -2 & -1 \end{bmatrix}\begin{bmatrix} i_1''' \\ i_2''' \\ i_3''' \end{bmatrix} = \begin{bmatrix} 2 \\ 0 \\ 0 \end{bmatrix}$$

Let $$\Delta = \begin{vmatrix} -1 & 1 & 0 \\ 2 & j9 & -(2+j3) \\ 2 & -2 & -1 \end{vmatrix} = (2 + j9)$$

Then $$i_2''' = \frac{\begin{vmatrix} -1 & 2 & 0 \\ 2 & 0 & -(2+j3) \\ 2 & 0 & -1 \end{vmatrix}}{\Delta} = (-1.36 + j0.14)\text{ A}$$

And $$i_3''' = \frac{\begin{vmatrix} -1 & 1 & 2 \\ 2 & j9 & 0 \\ 2 & -2 & 0 \end{vmatrix}}{\Delta} = -4\text{ A}$$

Then $$i_0''' = i_2''' - i_3''' = (2.64 + j0.14)\text{ A}$$

Thus total $$i_0 = i_0' + i_0'' + i_0''' = (2.86 - j0.77)\text{ A}$$

5.3 RECIPROCITY THEOREM

Statement : In a linear active bilateral network, if the excitation and the corresponding response is interchanged, then the value of the response does not change at all. This is valid if the network contains only one independent source. That is this theorem fails for a multi-source network. Also this theorem does not hold if any dependent source is present in the network.

SOLVED EXAMPLES

EXAMPLE 5.1. *For the circuit shown in Figures E5.3(a) and E5.3(b), verify Reciprocity theorem.*

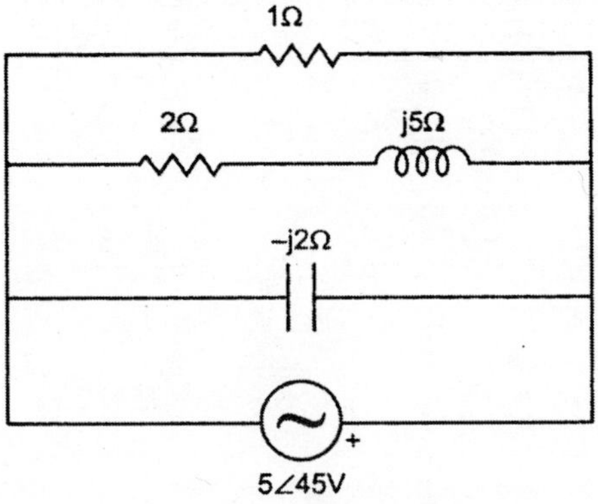

Fig. E5.3(a)

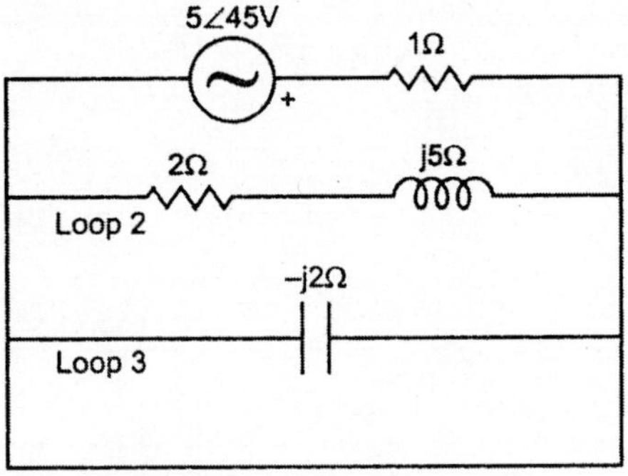

Fig. E5.3(b)

SOLUTION. The circuit shown in Figure E5.3(*a*), is redrawn in Figure E5.3(*c*) with loop currents.

The loop equations are

Loop 1 : $(3 + j5)i_1' - (2 + j5)i_2' = 0$ ...(*i*)

Loop 2 : $(2 + j3)i_2' - (2 + j5)i_1' + j2i_3' = 0$ ...(*ii*)

Loop 3 : $-j2i_3' + j2i_2' = -5\angle 45 = 3.53 + j3.53$ ...(*iii*)

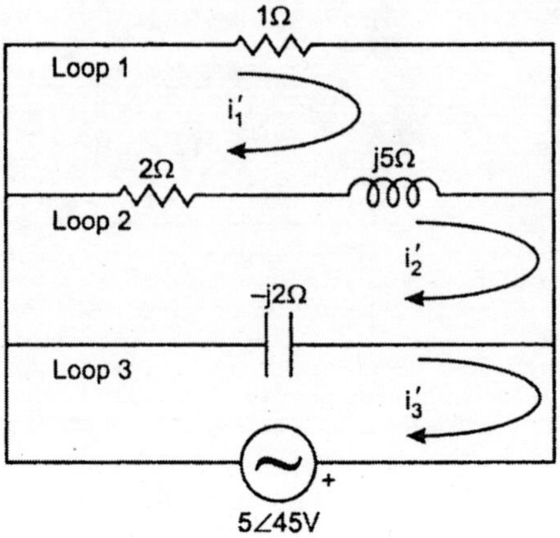

Fig. E5.3(c)

Writing equations (*i*), (*ii*) and (*iii*) in matrix form we get

$$\begin{bmatrix} 3+j5 & -(2+j5) & 0 \\ -(2+j5) & 2+j3 & j2 \\ 0 & j2 & -j2 \end{bmatrix}\begin{bmatrix} i_1' \\ i_2' \\ i_3' \end{bmatrix} = \begin{bmatrix} 0 \\ 0 \\ 3.53+j3.53 \end{bmatrix}$$

Let
$$\Delta = \begin{vmatrix} 3+j5 & -(2+j5) & 0 \\ -(2+j5) & 2+j3 & j2 \\ 0 & j2 & -j2 \end{vmatrix} = (10 - j4)$$

Then
$$i_1' = \frac{\begin{vmatrix} 0 & -(2+j5) & 0 \\ 0 & 2+j3 & j2 \\ 3.53+j3.53 & j2 & -j2 \end{vmatrix}}{\Delta} = (3.53 + j3.53)\text{ A}$$

The circuit shown in Figure E5.3(*b*), is redrawn in Figure E5.3(*d*) with loop currents.

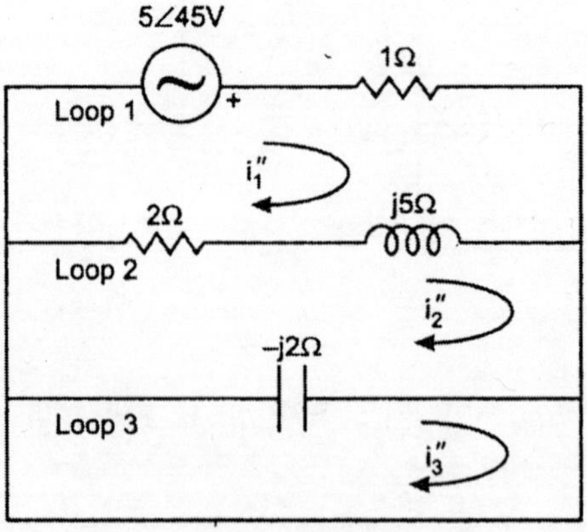

Fig. E5.3(d)

The loop equations are

Loop 1 : $(3 + j5)i_1'' - (2 + j5)i_2'' = 3.53 + j3.53$ ...(*iv*)

Loop 2 : $(2 + j3)i_2'' - (2 + j5)i_1'' + j2i_3'' = 0$ ...(*v*)

Loop 3 : $-j2i_3'' + j2i_2'' = 0$ ...(*vi*)

Writing equations (*iv*), (*v*) and (*vi*) in matrix form we get

$$\begin{bmatrix} 3+j5 & -(2+j5) & 0 \\ -(2+j5) & 2+j3 & j2 \\ 0 & j2 & -j2 \end{bmatrix} \begin{bmatrix} i_1'' \\ i_2'' \\ i_3'' \end{bmatrix} = \begin{bmatrix} 3.53+j3.53 \\ 0 \\ 0 \end{bmatrix}$$

Let $$\Delta = \begin{vmatrix} 3+j5 & -(2+j5) & 0 \\ -(2+j5) & 2+j3 & j2 \\ 0 & j2 & -j2 \end{vmatrix} = (10 - j4)$$

Then $$i_3'' = \frac{\begin{vmatrix} 3+j5 & -(2+j5) & 3.53+j3.53 \\ -(2+j5) & 2+j3 & 0 \\ 0 & j2 & 0 \end{vmatrix}}{\Delta} = (3.53 + j3.53)\text{ A}$$

Since $i_1' = i_3''$ the Reciprocity theorem is verified.

EXAMPLE 5.4. *For the circuit shown in Figures E5.4(a) and E5.4(b), verify Reciprocity theorem.*

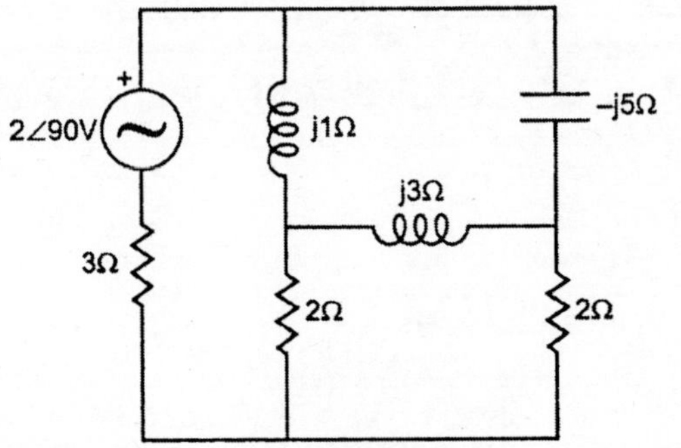

Fig. E5.4(a)

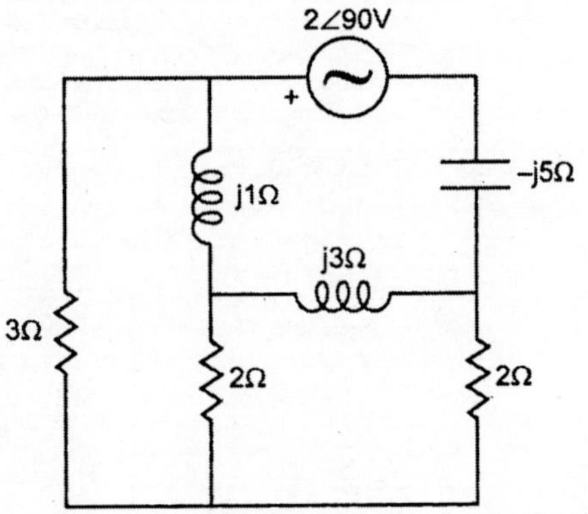

Fig. E5.4(b)

SOLUTION. The circuit shown in Figure E5.4(*a*), is redrawn in Figure E5.4(*c*) with loop currents.

The loop equations are

Loop 1 : $-j1i_1' - j1i_3' - j3i_2' = 0$ ...(*i*)

Loop 2 : $(4 + j3)i_2' - j3i_1' - 2i_3' = 0$ ...(*ii*)

Loop 3 : $(5 + j1)i_3' - j1i_1' - 2i_2' = j2$ ...(*iii*)

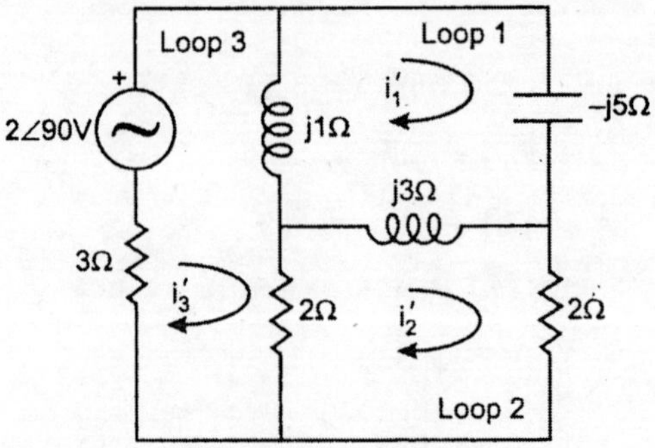

Fig. E5.4(c)

Writing equations (*i*), (*ii*) and (*iii*) in matrix form we get

$$\begin{bmatrix} -j1 & -j3 & -j1 \\ -j3 & (4+j3) & -2 \\ -j1 & -2 & (5+j1) \end{bmatrix} \begin{bmatrix} i_1' \\ i_2' \\ i_3' \end{bmatrix} = \begin{bmatrix} 0 \\ 0 \\ j2 \end{bmatrix}$$

Let $$\Delta = \begin{vmatrix} -j1 & -j3 & -j1 \\ -j3 & (4+j3) & -2 \\ -j1 & -2 & (5+j1) \end{vmatrix} = 80 - j1$$

Then $$i_1' = \frac{\begin{vmatrix} 0 & -j3 & -j1 \\ 0 & (4+j3) & -2 \\ j2 & -2 & (5+j1) \end{vmatrix}}{\Delta} = -0.25 - j0.08 \text{ A}$$

The circuit shown in Figure E5.4(*b*), is redrawn in Figure E5.4(*d*) with loop currents.

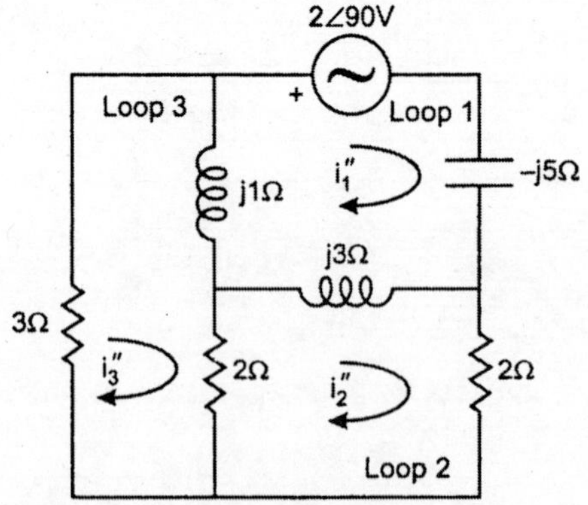

Fig. E5.4(d)

The loop equations are

Loop 1 : $-j1i_1'' - j1i_3'' - j3i_2'' = -j2$ ...(*iv*)

Loop 2 : $(4 + j3)i_2'' - j3i_1'' - 2i_3'' = 0$ ...(*v*)

Loop 3 : $(5 + j1)i_3'' - j1i_1'' - 2i_2'' = 0$ ...(*vi*)

Writing equations (*iv*), (*v*) and (*vi*) in matrix form we get

$$\begin{bmatrix} -j1 & -j3 & -j1 \\ -j3 & (4+j3) & -2 \\ -j1 & -2 & (5+j1) \end{bmatrix} \begin{bmatrix} i_1'' \\ i_2'' \\ i_3'' \end{bmatrix} = \begin{bmatrix} j2 \\ 0 \\ 0 \end{bmatrix}$$

Let
$$\Delta = \begin{vmatrix} -j1 & -j3 & -j1 \\ -j3 & (4+j3) & -2 \\ -j1 & -2 & (5+j1) \end{vmatrix} = 80 - j1$$

Then
$$i_3'' = \frac{\begin{vmatrix} -j1 & -j3 & j2 \\ -j3 & (4+j3) & 0 \\ -j1 & -2 & 0 \end{vmatrix}}{\Delta} = -0.25 - j0.08 \text{ A}$$

Since $i_1' = i_3''$ the Reciprocity theorem is verified.

EXAMPLE 5.5. *For the circuit shown in Figures E5.5(a) and E5.5(b), verify Reciprocity theorem.*

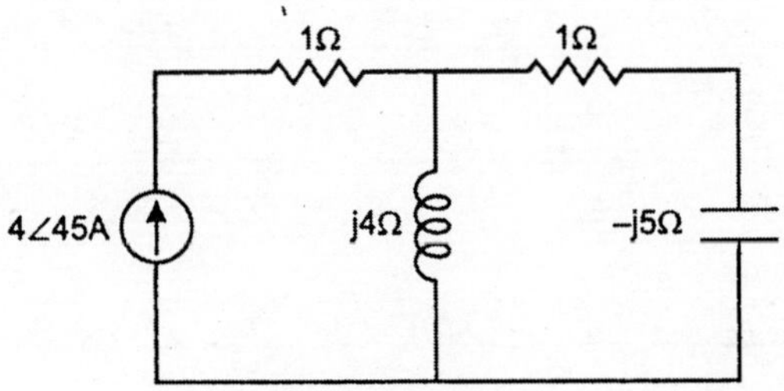

Fig. E5.5(a)

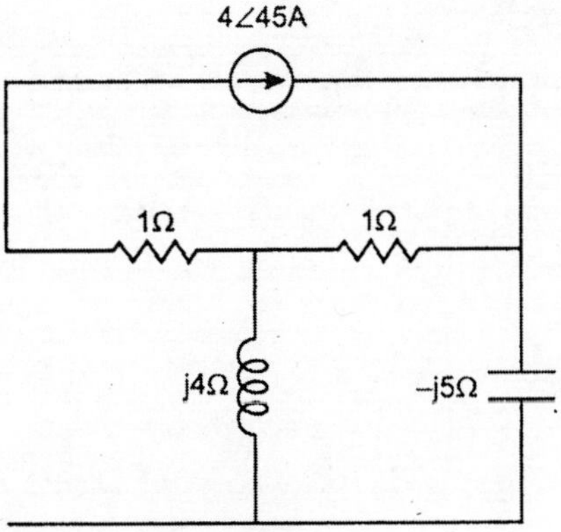

Fig. E5.5(b)

SOLUTION. The circuit shown in Figure E5.5(*a*), is redrawn in Figure E5.5(*c*) with loop currents.

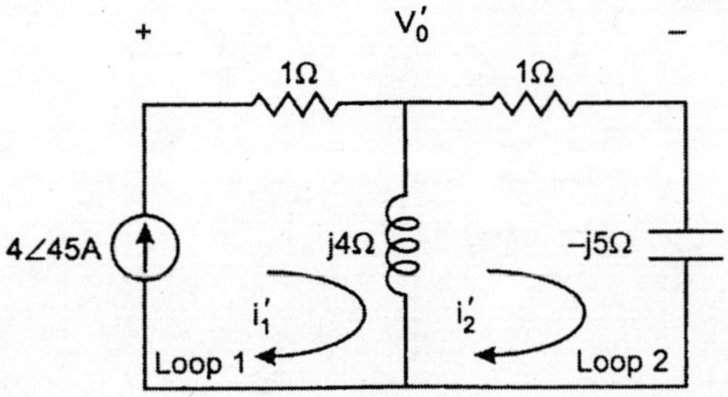

Fig. E5.5(c)

The loop equations are

Loop 1 : $i_1' = 4\angle 45 \text{ A} = 2.83 + j2.83 \text{ A}$ ...(*i*)

Loop 2 : $(1 - j1)i_2' - j4i_1' = 0$ ...(*ii*)

Substituting equation (*i*) in equation (*ii*) we get

$i_2' = -11.32 \text{ A}$ ...(*iii*)

Thus $V_0' = 1i_1' + 1i_2' = -8.49 + j2.83 \text{ V}$

The circuit shown in Figure E5.5(*b*), is redrawn in Figure E5.5(*d*) with loop currents.

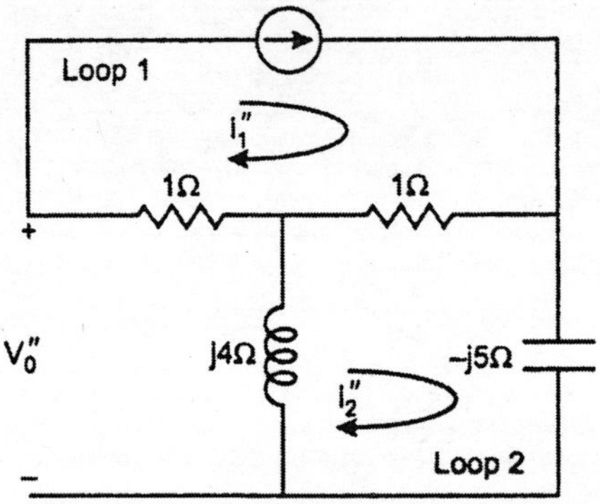

Fig. E5.5(d)

The loop equations are

Loop 1 : $i_1'' = 4\angle 45 \text{ A} = 2.83 + j2.83 \text{ A}$ ...(*iv*)

Loop 2 : $(1 - j1)i_2'' - i_1'' = 0$ ...(*v*)

Substituting equation (*i*) in equation (*v*) we get

$i_2'' = j2.83 \text{ A}$ ...(*vi*)

Thus $V_0'' = 1i_1'' + j4i_2'' = -8.49 + j2.83 \text{ V}$

Since $V_0' = V_0''$ Reciprocity theorem is verified.

EXAMPLE 5.6. *For the circuit shown in Figures E5.6(a) and E5.6(b), verify Reciprocity theorem.*

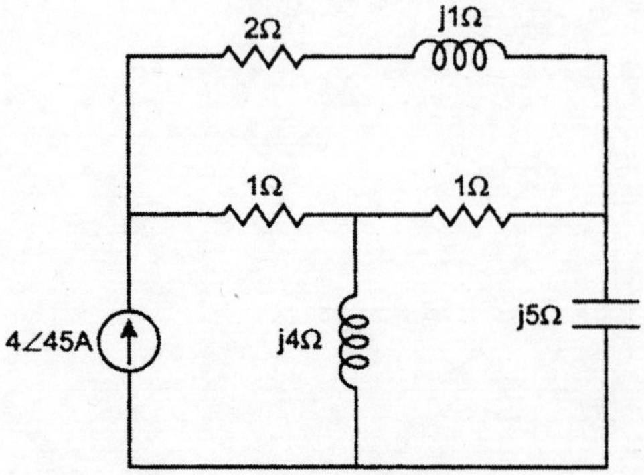

Fig. E5.6(a)

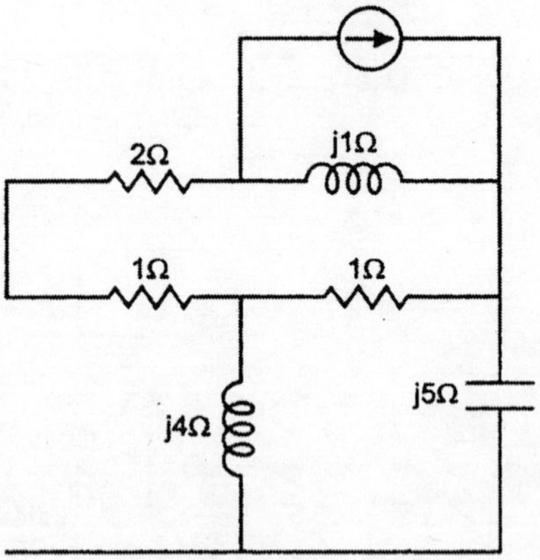

Fig. E5.6(b)

SOLUTION. The circuit shown in Figure E5.6(a), is redrawn in Figure E5.6(c) with loop currents.

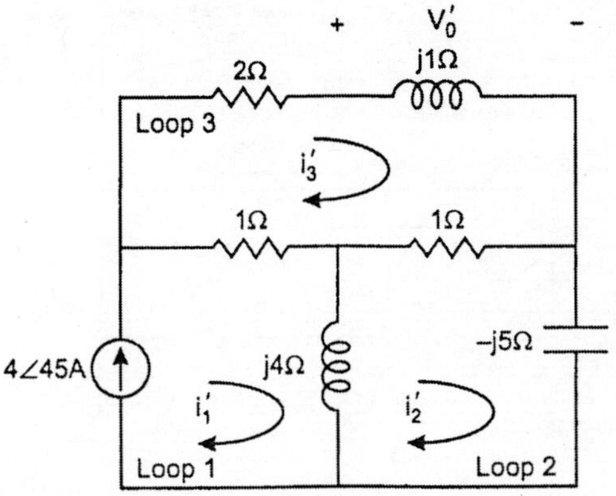

Fig. E5.6(c)

The Loop equations are

Loop 1 : $i_1' = 4\angle 45 \text{ A} = 2.83 + j2.83 \text{ A}$ ...(i)

Loop 2 : $(1 - j1)i_2' - j4i_1' - i_3' = 0$ ...(ii)

Substituting equation (i) in equation (ii) we get

$$(1 - j1)i_2' - i_3' = -11.32 + j11.32 \quad ...(iii)$$

Loop 3 : $(4 + j1)i_3' - i_1' - i_2' = 0$ ...(iv)

Substituting equation (i) in equation (iv) we get

$$(4 + j1)i_3' - i_2' = 2.83 + j2.83 \quad ...(v)$$

Writing equations (iii) and (v) in matrix form we get

$$\begin{bmatrix} 1-j1 & -1 \\ -1 & 4+j1 \end{bmatrix} \begin{bmatrix} i_2' \\ i_3' \end{bmatrix} = \begin{bmatrix} -11.32+j11.32 \\ 2.83+j2.83 \end{bmatrix}$$

Let $$\Delta = \begin{vmatrix} 1-j1 & -1 \\ -1 & 4+j1 \end{vmatrix} = 4 - j3$$

Then $$i_3' = \frac{\begin{vmatrix} 1-j1 & -11.32+j11.32 \\ -1 & 2.83+j2.83 \end{vmatrix}}{\Delta} = -2.26 + j1.13 \text{ A}$$

So $$V_0' = j1 \times i_3' = -1.13 - j2.26 \text{ V}$$

The circuit shown in Figure E5.6(b), is redrawn in Figure E5.6(d) with loop currents.

The Loop equations are

Loop 1 : $i_1'' = 4\angle 45 \text{ A} = 2.83 + j2.83 \text{ A}$ ...(vi)

Loop 2 : $(1 - j1)i_2'' - i_3'' = 0$ ...(vii)

Loop 3 : $(4 + j1)i_3'' - ji_1'' - i_2'' = 0$ ...($viii$)

Substituting equation (vi) in equation ($viii$) we get

$$(4 + j1)i_3'' - i_2'' = -2.83 + j2.83 \quad ...(ix)$$

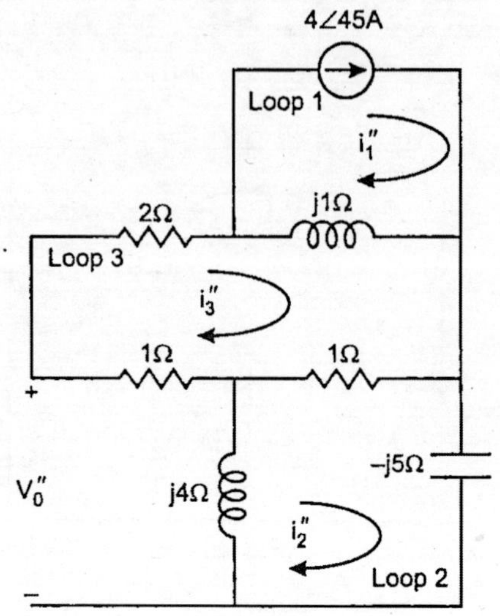

Fig. E5.6(d)

Writing equations (*vii*) and (*ix*) in matrix form we get

$$\begin{bmatrix} 1-j1 & -1 \\ -1 & 4+j1 \end{bmatrix}\begin{bmatrix} i_2'' \\ i_3'' \end{bmatrix} = \begin{bmatrix} 0 \\ -2.83+j2.83 \end{bmatrix}$$

Let $$\Delta = \begin{vmatrix} 1-j1 & -1 \\ -1 & 4+j1 \end{vmatrix} = 4 - j3$$

Then $$i_2'' = \frac{\begin{vmatrix} 0 & -1 \\ -2.83+j2.83 & 4+j1 \end{vmatrix}}{\Delta} = -0.79 + j0.11 \text{ A}$$

Then $$i_3'' = \frac{\begin{vmatrix} 1-j1 & 0 \\ -1 & -2.83+j2.83 \end{vmatrix}}{\Delta} = 0.91 + j0.68 \text{ A}$$

So $$V_0'' = j1 \times i_3'' + j4i_2'' = -1.13 - j2.26 \text{ V}$$

Since $V_0' = V_0''$ Reciprocity theorem is verified.

5.4 COMPENSATION THEOREM

Statement : In a linear, active network if the imredance Z carrying current I is changed by a small amount ΔZ, then the change in the current in all branches can be obtained by adding an independent voltage source ΔV in series with $Z \pm \Delta Z$, and replacing all the independent voltage source and current source by their internal impedance. Where

$$\Delta V = I \times \Delta Z$$

SOLVED EXAMPLES

EXAMPLE 5.7. *Consider the circuit shown in Figure E5.7. If resistance R_1 is changed from 2 Ω to 5 Ω, find the change in loop currents.*

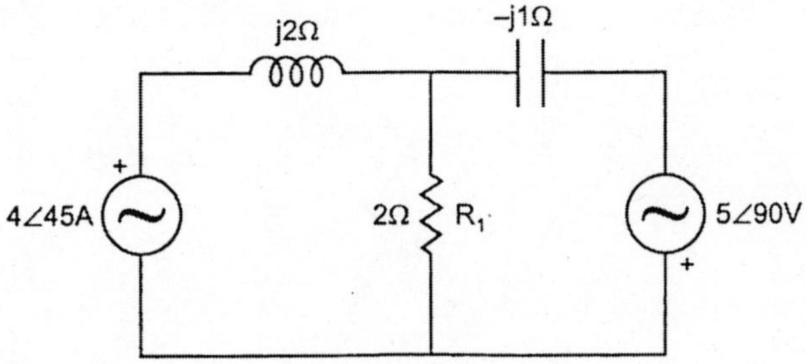

Fig. E5.7

SOLUTION. The circuit of Figure E5.7 is redrawn in Figure E5.7(*a*), with loop currents.

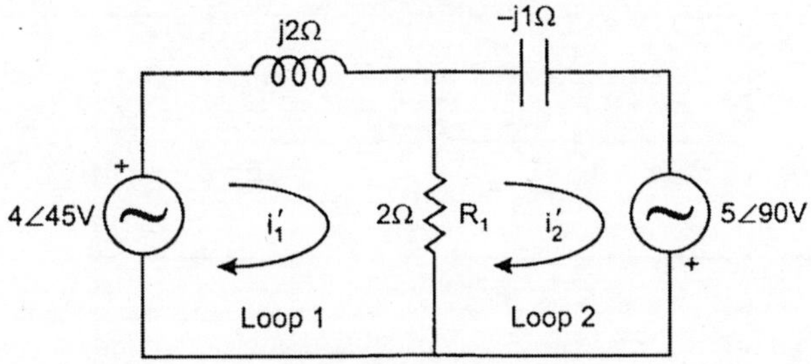

Fig. E5.7(a)

The loop equations are

Loop 1 : $(2 + j2)i_1' - 2i_2' = 4\angle 45$

or $(2 + j2)i_1' - 2i_2' = 2.83 + j2.83$ ...(*i*)

Loop 2 : $(2 - j1)i_2' - 2i_1' = 5\angle 90$

or $(2 - j1)i_2' - 2i_1' = j5$ ...(*ii*)

Solving equations (*i*) and (*ii*) we get

$$i_1' = 5.33 + j1.09 \text{ A} \quad \text{and} \quad i_2' = 2.83 + j5 \text{ A}$$

Then the current through R_1 is $= i_1' - i_2' = 2.50 - j3.91$ A

To find the change in loop current using compensation theorem. The circuit needed for compensation theorem is shown in Figure E5.7(*b*).

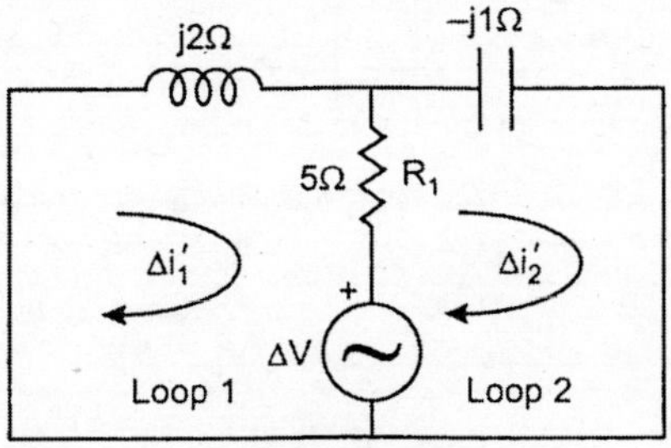

Fig. E5.7(b)

Here $\Delta R = 5 - 2 = 3\ \Omega$

$\Delta V = 7.50 - j11.73$ V

The loop equations are

Loop 1 : $(5 + j2)\Delta i_1' - 5\Delta i_2' = -\Delta V$

or $(5 + j2)\Delta i_1' - 5\Delta i_2' = -7.50 + j11.73$ ...(*iii*)

Loop 2 : $(5 - j1)\Delta i_2' - 5\Delta i_1' = \Delta V$

or $(5 - j1)\Delta i_2' - 5\Delta i_1' = 7.50 - j11.73$ ...(*iv*)

Solving equations (*iii*) and (*iv*) we get

$\Delta i_1' = 2.1 - j1.5$ A

and $\Delta i_2' = 4.2 - j3$ A

EXAMPLE 5.8. *Consider the circuit shown in Figure E5.8. If resistance R_1 is changed from 2 Ω to 5 Ω, find the change in current through R_1.*

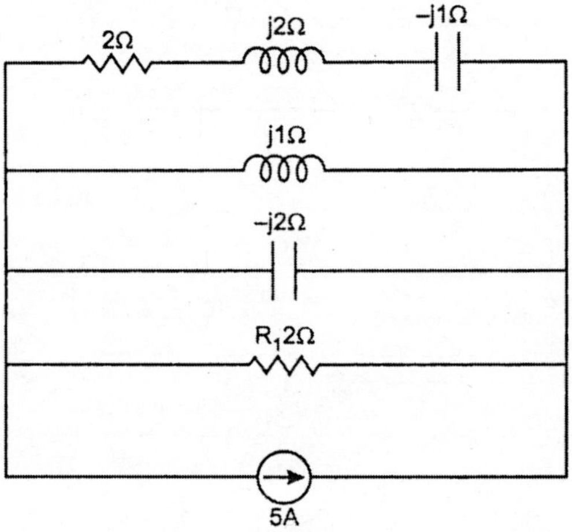

Fig. E5.8

SOLUTION. The circuit of Figure E5.8 is redrawn in Figure E5.8(*a*), with loop currents.

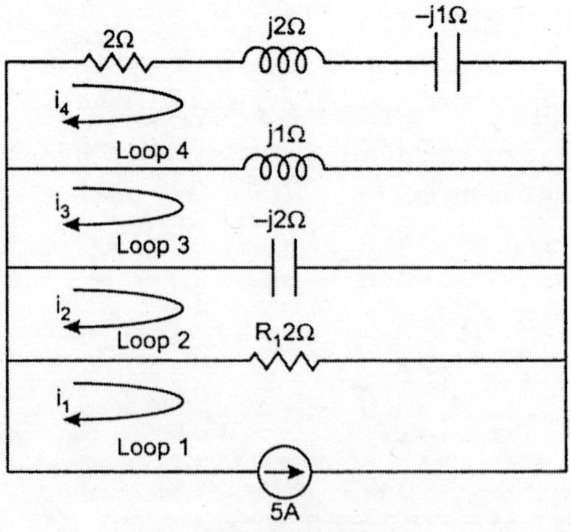

Fig. E5.8(a)

The loop equations are

Loop 1 : $i_1 = -5$ A ...(*i*)

Loop 2 : $(2 - j2)i_2 - 2i_1 + j2i_3 = 0$ ...(*ii*)

Substituting equation (*i*) in equation (*ii*) we get

$$(2 - j2)i_2 + j2i_3 = -10 \quad ...(iii)$$

Loop 3 : $(1 - j2)i_3 + j2i_2 - j1i_4 = 0$ ...(*iv*)

Loop 4 : $(2 + j2)i_4 - j1i_3 = 0$ ...(*v*)

Solving equations (*iii*), (*iv*) and (*v*) we get

$$i_2 = -3.44 - j0.8 \text{ A}$$

$$i_3 = -2.64 + j0.75 \text{ A}$$

and

$$i_4 = -0.85 - j0.47 \text{ A}$$

Then the current through R_1 is $= i_1 - i_2 = -1.56 + j0.84$

To find the change in loop current using compensation theorem. The circuit needed for compensation theorem is shown in Figure E5.8(*b*).

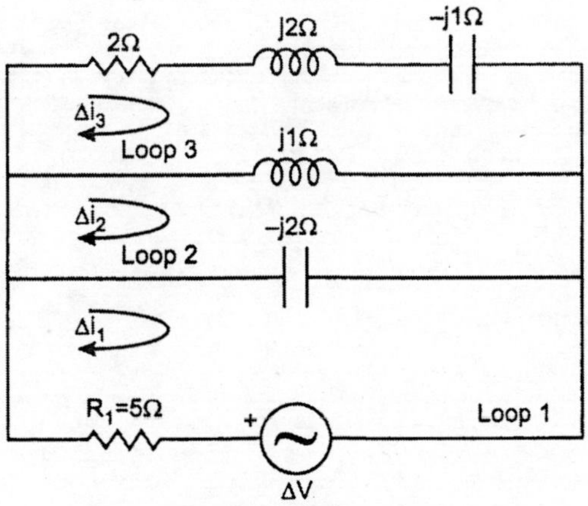

Fig. E5.8(b)

Here $\Delta R = 5 - 2 = 3\ \Omega$

$$\Delta V = -4.68 + j2.4 \text{ V}$$

The loop equations are

Loop 1 : $(5 - j2)\Delta i_1 - j2\Delta i_2 = \Delta V$ ...(*vi*)

Loop 2 : $(1 - j2)\Delta i_2 + j2\Delta i_1 - j1\Delta i_3 = 0$ ...(*vii*)

Loop 3 : $(2 + j2)\Delta i_3 - j1\Delta i_2 = 0$ ...(*viii*)

Solving equations (*vi*), (*vii*) and (*viii*) we get

$$\Delta i_1 = -0.95 - j0.19 \text{ A}$$

Thus, the change in the current of R_1 is $\Delta i_1 = -0.95 - j0.19$ A

EXAMPLE 5.9. *Consider the circuit shown in Figure E5.9. If $-j2\ \Omega$ is changed to $0\ \Omega$, find the change in loop currents.*

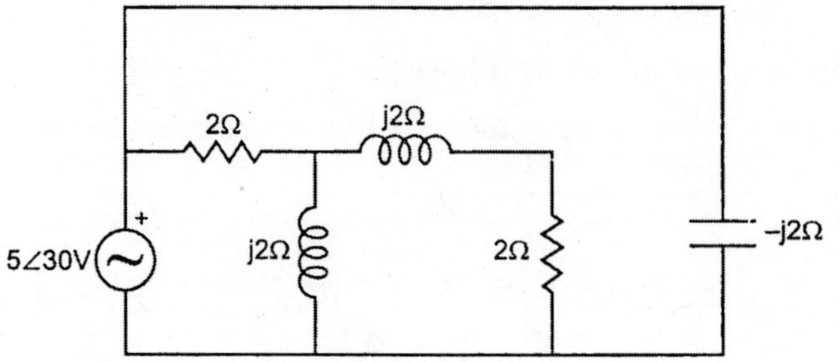

Fig. E5.9

SOLUTION. The circuit shown in Figure E5.9, is redrawn in Figure E5.9(*a*), with loop currents.

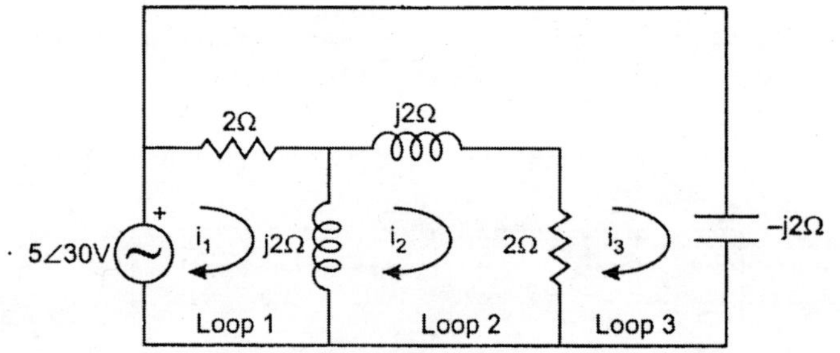

Fig. E5.9(a)

The loop equations are

Loop 1 : $(2 + j2)i_1 - j2i_2 - 2i_3 = 5\angle 30$ ...(*i*)

Loop 2 : $(2 + j4)i_2 - j2i_1 - (2 + j2)i_3 = 0$ ...(*ii*)

Loop 3 : $4i_3 - 2i_1 - (2 + j2)i_2 = 0$ ...(*iii*)

Solving equations (*i*), (*ii*) and (*iii*), we get

The current through $-j2\ \Omega = i_3 = -1.25 + j2.17$ A

To find the change in loop current using compensation theorem. The circuit needed for compensation theorem is shown in Figure E5.9(*b*).

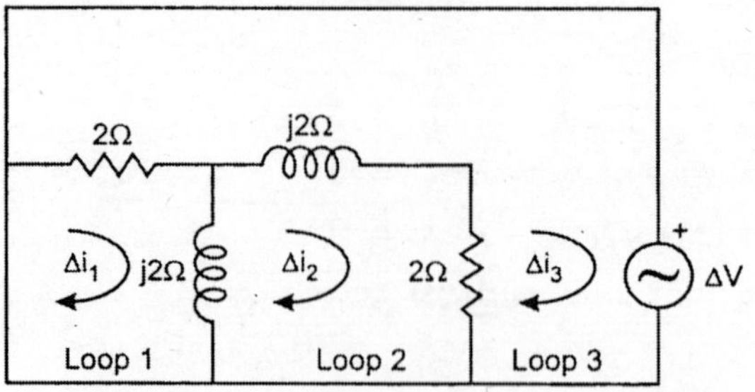

Fig. E5.9(b)

Here $\Delta X_c = -j2 - 0 = -2\ \Omega$

$\Delta V = -2*(-1.25 + j2.17) = 2.5000 - j4.3400$ V

The loop equations are

Loop 1 : $(2 + j2)\Delta i_1 - j2\Delta i_2 - 2\Delta i_3 = 0$ ...(*iv*)

Loop 2 : $(2 - j4)\Delta i_2 - j2\Delta i_1 - (2 + j2)\Delta i_3 = 0$ ...(*v*)

Loop 3 : $(4 + j2)\Delta i_3 - 2\Delta i_1 - (2 + j2)\Delta i_2 = \Delta V$ ...(*vi*)

Solving equations (*iv*), (*v*) and (*vi*) we get the change in loop currents

$\Delta i_1 = 0.54 + j0.14$ A, $\Delta i_2 = 1.15 + j0.64$ A and $\Delta i_3 = 0.83 - j0.33$ A

5.5 THEVENIN'S THEOREM

Statement : Any two terminal linear, active, bilateral a.c. circuit can be replaced by an equivalent circuit consisting of a voltage source (V_{TH}) and a series impedance (Z_{TH}).

Where V_{TH} is called the Thevenin's voltage and Z_{TH} is called the Thevenin's impedance.

SOLVED EXAMPLES

EXAMPLE 5.10. *Find the current flowing through the capacitor for the circuit shown in Figure E5.10, using Thevenin's theorem.*

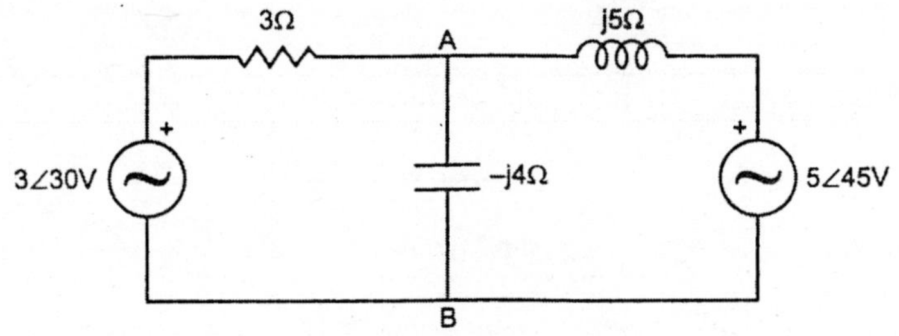

Fig. E5.10

SOLUTION. *To find V_{TH}*

The circuit is shown in Figure E5.10(*a*). The capacitor has been removed from the circuit.

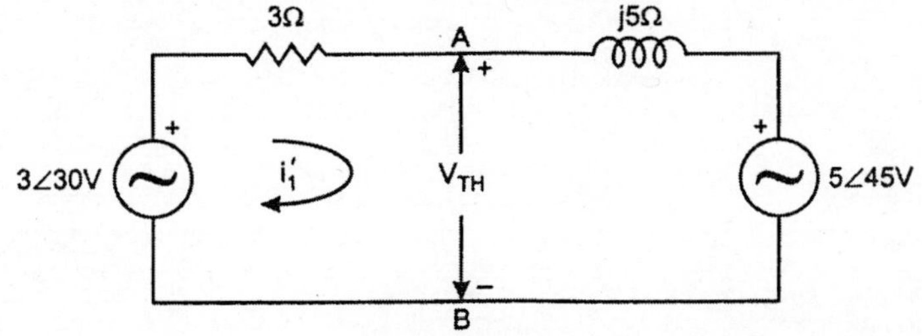

Fig. E5.10(a)

The current $i_1' = \dfrac{3\angle 30 - 5\angle 45}{3 + j5} = -0.38 - j0.04$ A

Then $$V_{TH} = 3\angle 30 - 3i_1' = 3.73 + j1.62 \text{ V}$$

To find Z_{TH}

The circuit to find Z_{TH} is shown in Figure E5.10(*b*). Here all the independent voltage sources has been replaced by their internal impedance (short circuit).

$$Z_{TH} = \frac{3 \times j5}{3 + j5} = 2.21 + j1.32 \ \Omega$$

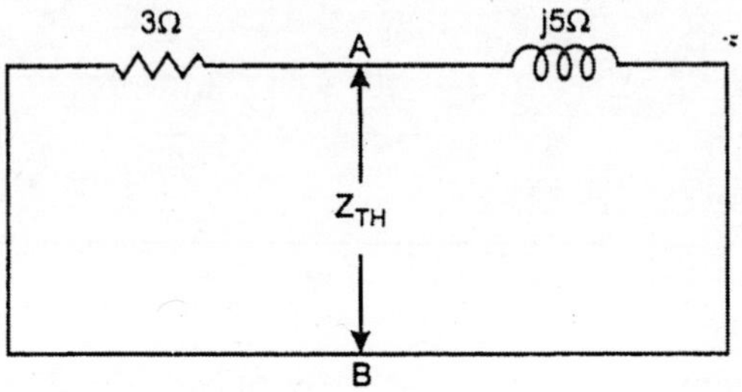

Fig. E5.10(b)

Thus the Thevenin's equivalent circuit is shown in Figure E5.10(*c*).

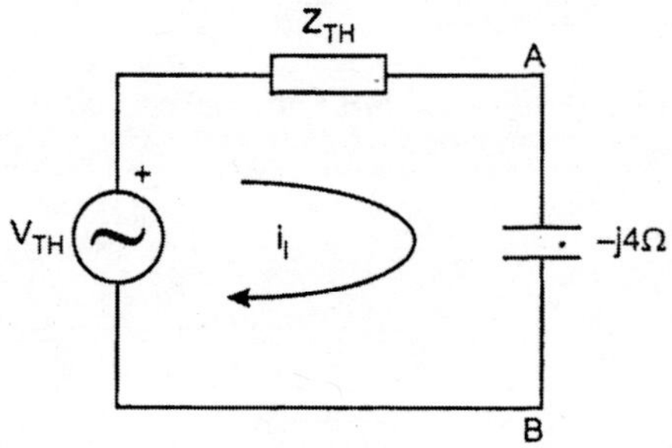

Fig. E5.10(c)

From Figure E5.10(*c*), we get

$$i_1 = \frac{V_{TH}}{Z_{TH} - j4} = 0.32 + j1.12 \text{ A}$$

EXAMPLE 5.11. *Find the current flowing through –j1 Ω for the circuit shown in Figure E5.11, using Thevenin's theorem.*

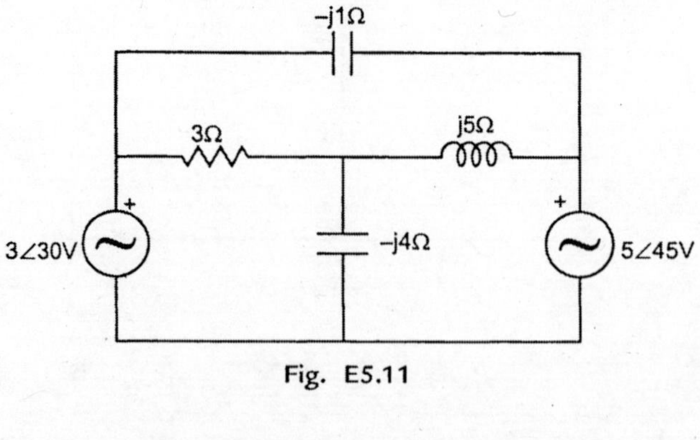

Fig. E5.11

SOLUTION. *To find* V_{TH}

The circuit is shown in Figure E5.11(*a*). The capacitor has been removed from the circuit.

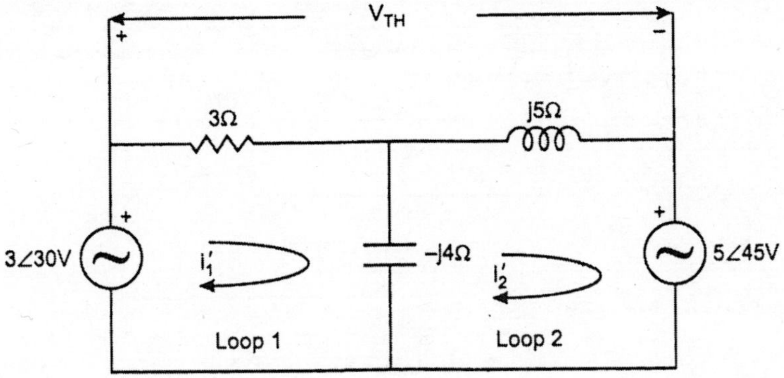

Fig. E5.11(a)

The loop equations are

Loop 1 : $(3 - j4)i_1' + j4i_2' = 3\angle 30$ ...(*i*)

Loop 2 : $j1i_2' + j4i_1' = -5\angle 45$ ...(*ii*)

Solving equations (*i*) and (*ii*) we get

$$i_1' = -0.64 + j0.93 \text{ A}$$

and $$i_2' = -0.96 - j0.19 \text{ A}$$

Then $$V_{TH} = -3i_1' - j5i_2' = 0.97 + j2 \text{ V}$$

To find Z_{TH}

The circuit to find Z_{TH} is shown in Figure E5.11(*b*). Here all the independent voltage sources has been replaced by their internal impedance (short circuit).

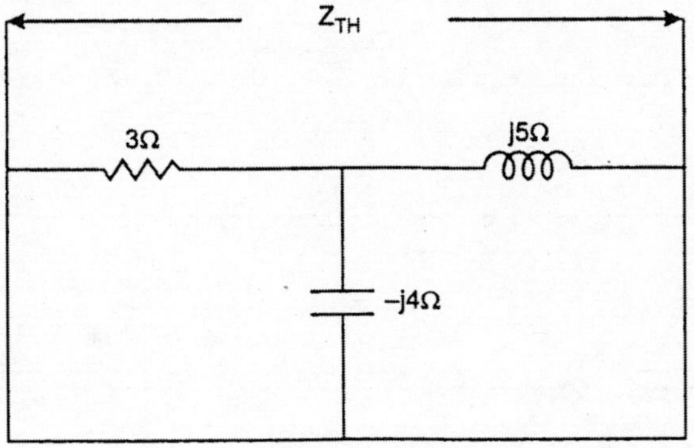

Fig. E5.11(b)

From Figure E5.11(*b*) we get

$$Z_{TH} = 0 \ \Omega$$

Thus, the Thevenin's equivalent circuit is shown in Figure E5.11(c).

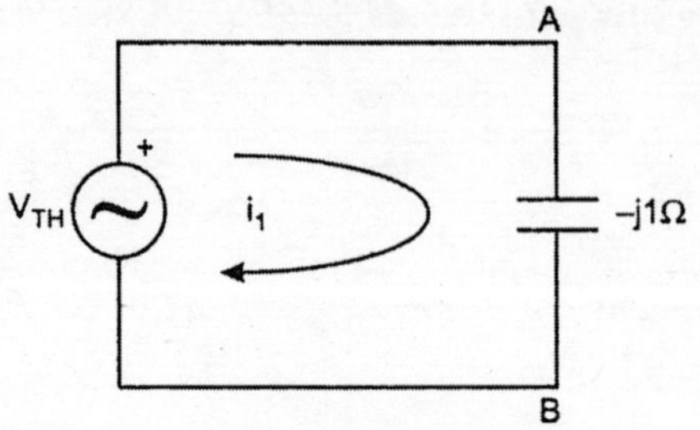

Fig. E5.11(c)

From Figure E5.11(c) we get

$$i_1 = \frac{V_{TH}}{-j1} = -2 + j0.97 \text{ A}$$

Example 5.12. *Find the drop across X_L using Thevenin's theorem, for the circuit shown in Figure E5.12.*

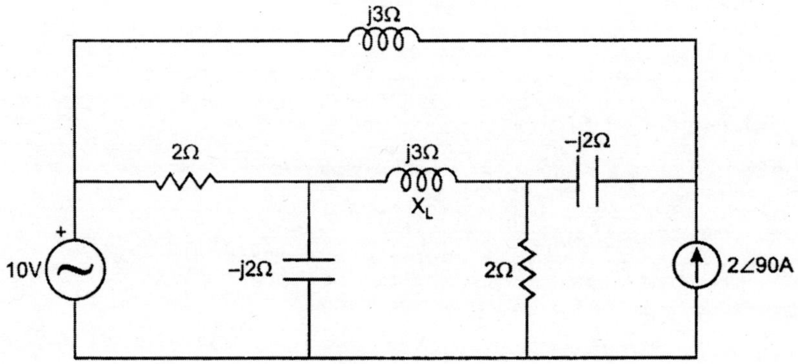

Fig. E5.12

Solution. *To find V_{TH}*

The circuit is shown in Figure E5.12(a). The inductor X_L has been removed from the circuit.

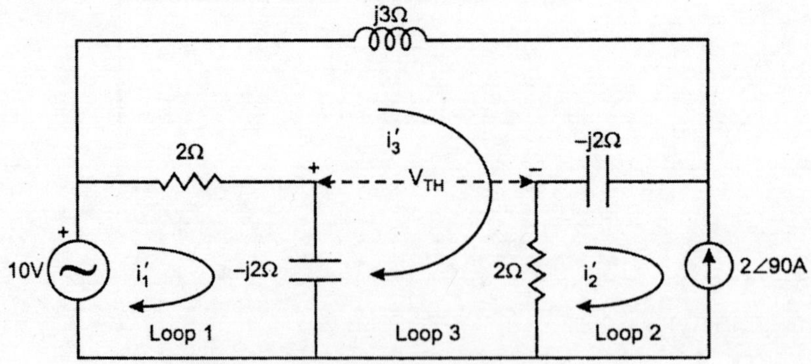

Fig. E5.12(a)

The loop equations are

Loop 1 : $(2 - j2)i_1' - (2 - j2)i_3' = 10$ ...(*i*)

Loop 2 : $i_2' = -j2$ Amp ...(*ii*)

Loop 3 : $(4 - j1)i_3' + j2i_1' - 2i_2' = 0$ ...(*iii*)

Substituting equation (*ii*) in equation (*iii*) we get

$$(4 - j1)i_3' + j2i_1' = -j4 \quad ...(iv)$$

Solving equations (*i*) and (*iv*) we get,

$$i_1' = 3.15 + j0.08 \text{ A} \quad \text{and} \quad i_3' = 0.65 - j2.4 \text{ A}$$

Then $V_{TH} = -j2(i_1' - i_3') - 2(i_2' - i_3') = 6.26 - j5.8$ V

To find Z_{TH}

The circuit to find Z_{TH} is shown in Figure E5.12(*b*). Here the independent voltage source has been replaced by short circuit and independent current source has been replaced by open circuit.

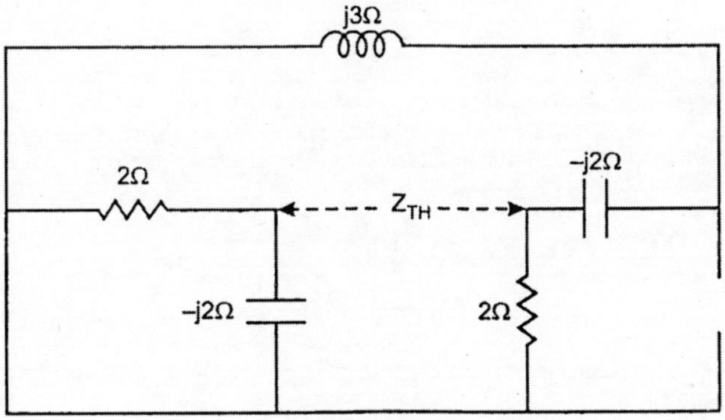

Fig. E5.12(b)

$$Z_{TH} = \frac{2\times(-j2)}{2-j2} + \frac{2\times j1}{2+j1} = 1.4 - j0.2 \text{ A}$$

Thus, the Thevenin's equivalent circuit is shown in Figure E5.12(*c*).

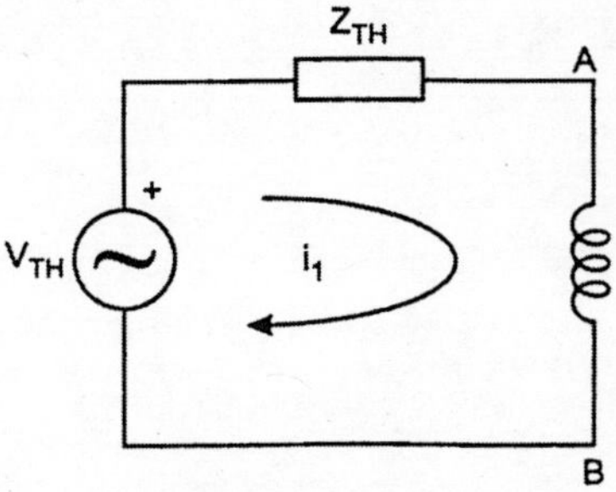

Fig. E5.12(c)

From Figure E5.12(c) we get

$$i_1 = \frac{V_{TH}}{Z_{TH} + j3} = -0.76 - j2.62 \text{ A}$$

Then
$$V_{XL} = j3 \times i_1 = 7.85 - j2.29 \text{ V}$$

EXAMPLE 5.13. *Find the drop across R_L using Thevenin's theorem, for the circuit shown in Figure E5.13.*

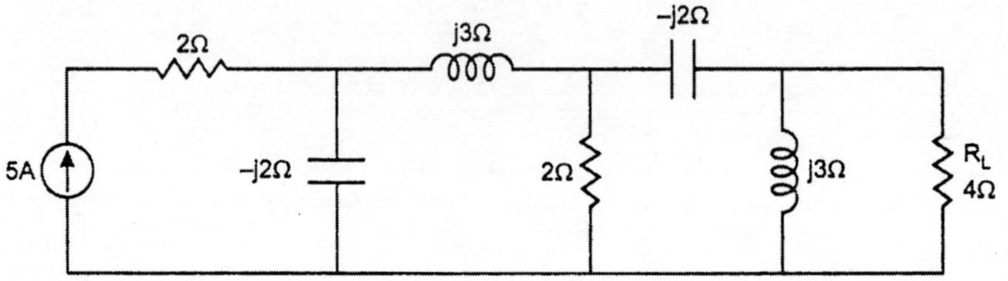

Fig. E5.13

SOLUTION. *To find V_{TH}*

The circuit is shown in Figure E5.13(a). The resistor R_L has been removed from the circuit.

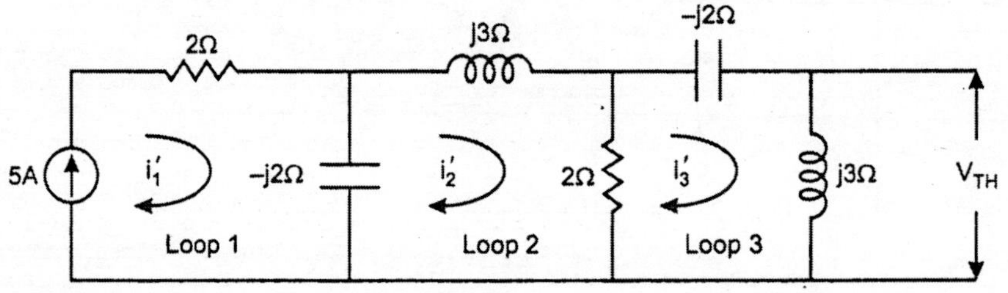

Fig. E5.13(a)

The loop equations are

Loop 1 : $i_1' = 5 \text{ A}$ ...(i)

Loop 2 : $(2 + j1)i_2' + j2i_1' - 2i_3' = 0$ ...(ii)

Substituting equation (i) in equation (ii) we get

$$(2 + j1)i_2' - 2i_3' = -j10 \quad \text{...}(iii)$$

Loop 3 : $(2 + j1)i_3' - 2i_2' = 0$ ...(iv)

Solving equations (iii) and (iv) we get

$$i_3' = -4.71 + j1.17 \text{ A}$$

and
$$V_{TH} = j3 \times i_3' = -3.53 - j14.12 \text{ V}$$

To find Z_{TH}

The circuit to find Z_{TH} is shown in Figure E5.13(*b*). Here the independent current source has been replaced by open circuit.

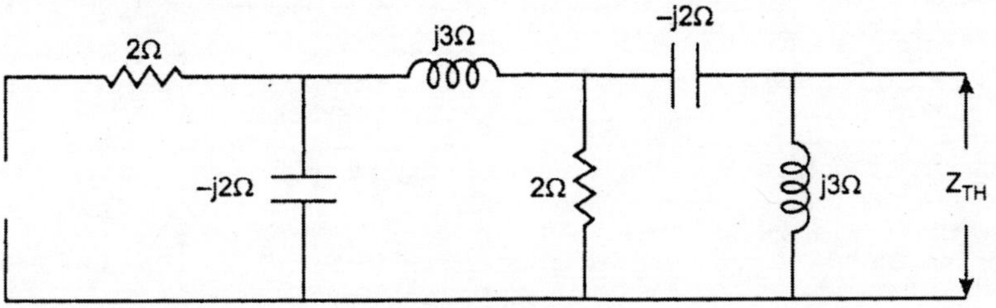

Fig. E5.13(b)

From Figure E5.13(*b*) we get

$$Z_{TH} = 1.06 - j1.76\ \Omega$$

Thus, the Thevenin's equivalent circuit is shown in Figure E5.13(*c*).

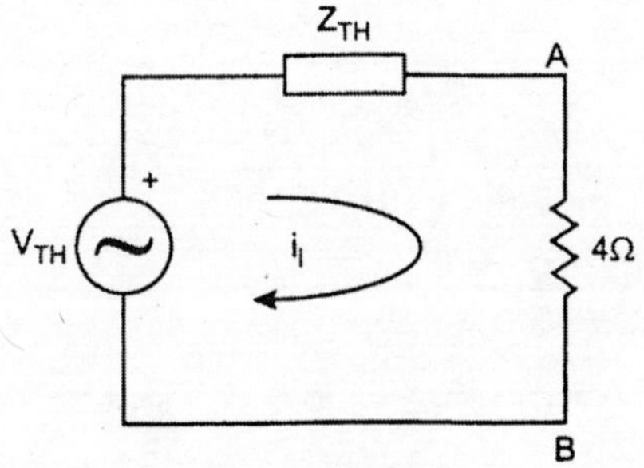

Fig. E5.13(c)

From Figure E5.13(*c*) we get

$$i_1 = \frac{V_{TH}}{Z_{TH}+4} = 0.24 - j2.71 \text{ A}$$

Then $$V_{RL} = 4 \times i_1 = 0.97 - j10.83 \text{ V}$$

EXAMPLE 5.14. *Find the Thevenin's equivalent across A-B, for the circuit shown in Figure E5.14.*

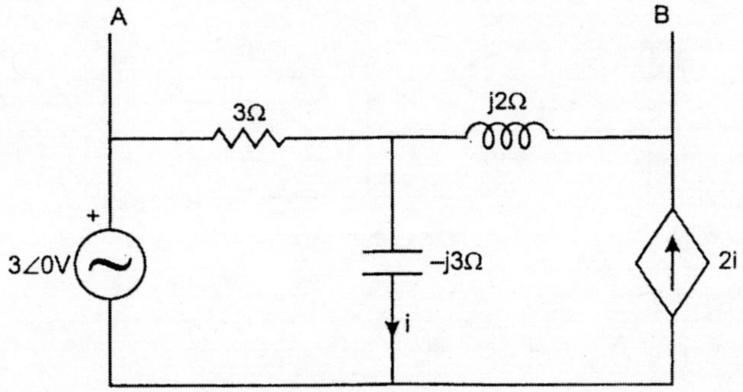

Fig. E5.14

SOLUTION. *To find* V_{TH}

The circuit is redrawn in Figure E5.14(*a*), with loop currents.

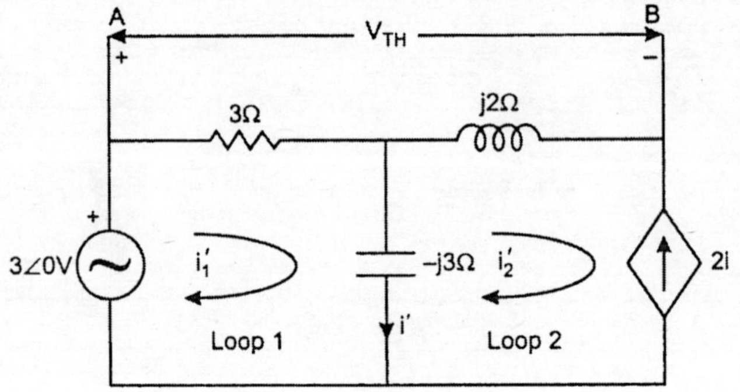

Fig. E5.14(a)

The loop equations are

Loop 1 : $(3 - j3)i_1' + j3i_2' = 3$ ...(*i*)

Loop 2 : $i_2' = -2i'$ ...(*ii*)

Again $i' = i_1' - i_2'$ ...(*iii*)

Substituting equation (*iii*) in equation (*ii*) we get

$$2i_1' - i_2' = 0 \quad ...(iv)$$

Solving equations (*i*) and (*iv*), we get

$$i_1' = 0.5 - j0.5 \text{ A} \quad \text{and} \quad i_2' = 1 - j1 \text{ A}$$

Then from Figure E5.14(*a*), $V_{TH} = -3i_1' - j2i_2' = -3.5 - j0.5$ V

To find i_{sc}

The circuit to find i_{sc} is shown in Figure E5.14(*b*). Here $i_{sc} = i_3''$. The loop equations are

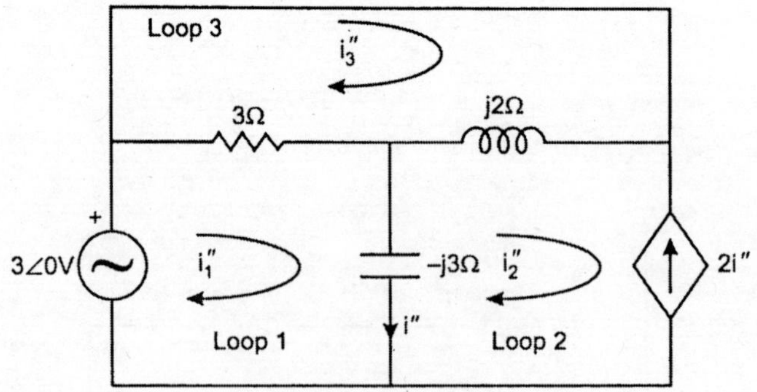

Fig. E5.14(b)

Loop 1 : $(3 - j3)i_1'' + j3i_2'' - 3i_3'' = 3$ ...(*v*)

Loop 2 : $i_2'' = -2i''$ ...(*vi*)

Again $i'' = i_1'' - i_2''$ ...(*vii*)

Substituting equation (*vii*) in equation (*vi*) we get

$$2i_1'' - i_2'' = 0 \qquad ...(viii)$$

Loop 3 : $(3 + j2)i_3'' - 3i_1'' - j2i_2'' = 0$ ...(*ix*)

Solving equations (*v*), (*viii*) and (*ix*), we get

$$i_{sc} = i_3'' = -0.4 - j2.2 \text{ A}$$

Thus $Z_{TH} = \dfrac{V_{TH}}{i_{sc}} = 0.5 - j1.5\ \Omega$

Then, the Thevenin's equivalent circuit is shown in Figure E5.14(*c*).

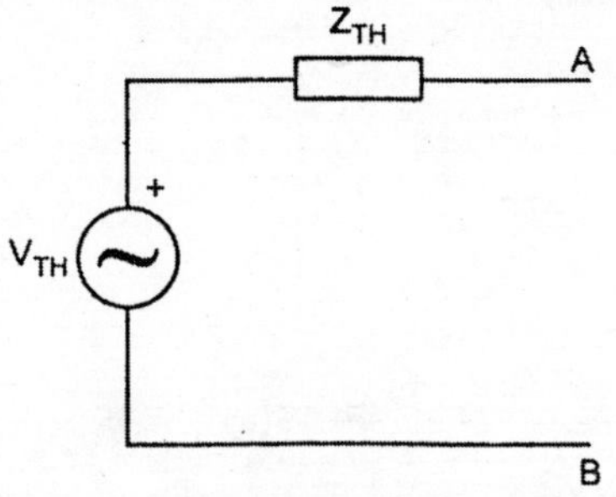

Fig. E5.14(c)

EXAMPLE 5.15. *For the circuit shown in Figure E5.15, find the current through j4 Ω using Thevenin's theorem.*

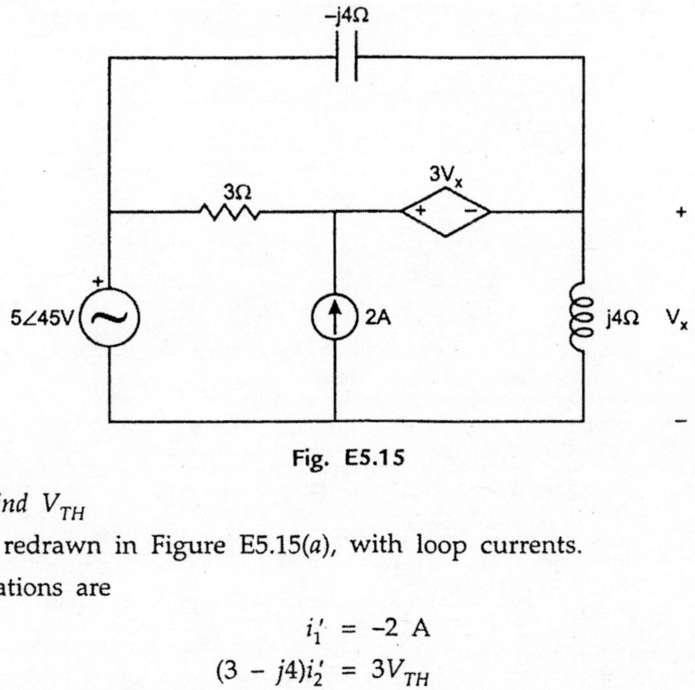

Fig. E5.15

SOLUTION. *To find V_{TH}*

The circuit is redrawn in Figure E5.15(*a*), with loop currents.

The loop equations are

Loop 1 : $i_1' = -2$ A ...(*i*)

Loop 2 : $(3 - j4)i_2' = 3V_{TH}$ ...(*ii*)

Again applying KVL in the loop *ABCDEFA* we get

$$3(i_1' - i_2') + 3V_{TH} + V_{TH} - 5\angle 45 = 0 \qquad ...(iii)$$

Substituting equation (*i*) in equation (*iii*) we get

$$3i_2' + 4V_{TH} = 6 + 5\angle 45 \quad \text{...(iv)}$$

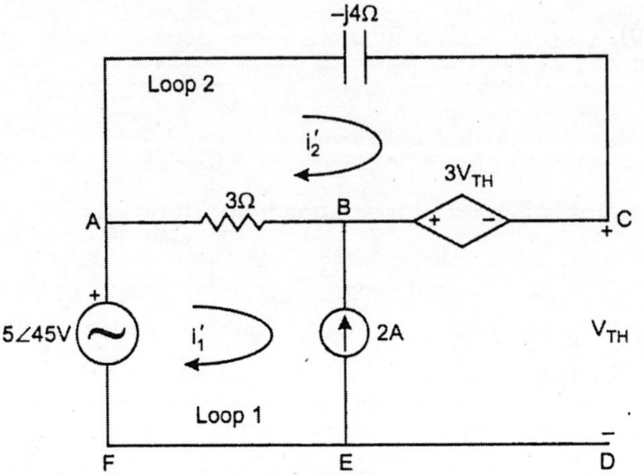

Fig. E5.15(a)

Solving equations (*ii*) and (*iv*) we get

$$V_{TH} = 1.92 + j0.15 \text{ V}$$

To find i_{sc}

The circuit to find i_{sc} is shown in Figure E5.15(*b*). Here $i_{sc} = i_2''$. The loop equations are

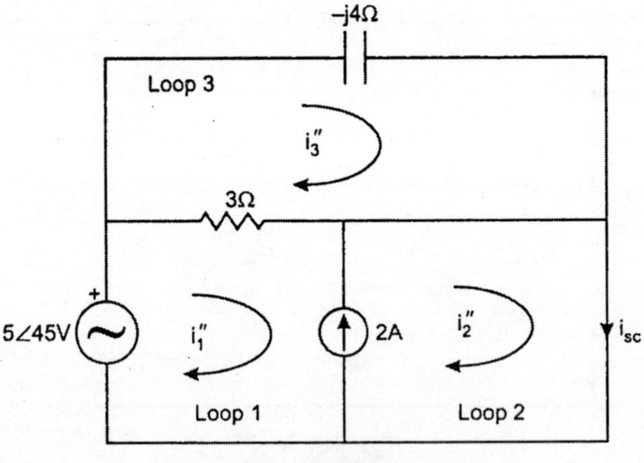

Fig. E5.15(b)

Loop 1 and Loop 2 : $i_2'' - i_1'' = 2$ ...(*v*)

And $3(i_1'' - i_3'') = 5\angle 45$ ...(*vi*)

Loop 3 : $(3 - j4)i_3'' - 3i_1'' = 0$ ...(*vii*)

Solving equations (*v*), (*vi*) and (*vii*), we get

$$i_{sc} = i_2'' = 2.29 + j2.06 \text{ A}$$

Thus

$$Z_{TH} = \frac{V_{TH}}{i_{sc}} = 0.49 - j0.38\ \Omega$$

Then the Thevenin's equivalent circuit is shown in Figure E5.15(*c*).

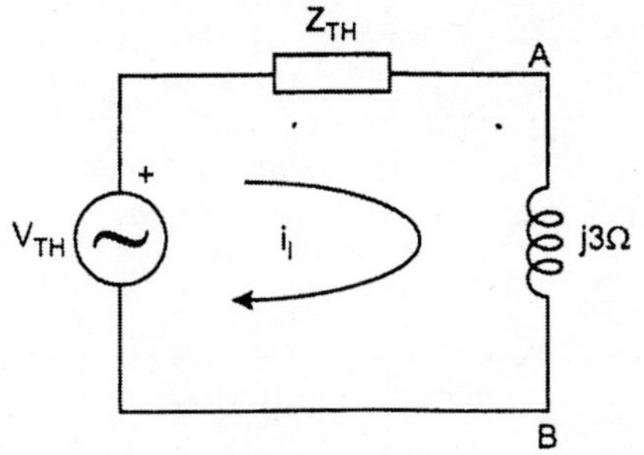

Fig. E5.15(c)

From Figure E5.15(*c*) we get $i_1 = \dfrac{V_{TH}}{Z_{TH} + j4} = 0.11 - j0.52$ A

5.6 NORTON'S THEOREM

Statement : Any two terminal linear, active, bilateral a.c. circuit can be replaced by an equivalent circuit consisting of a current source (I_N) and a parallel impedance (Z_N).

Where I_N is called the Norton's current and Z_N is called the Norton's impedance.

SOLVED EXAMPLES

EXAMPLE 5.16. *Find the Norton's equivalent circuit across A-B, for the circuit shown in Figure E5.16.*

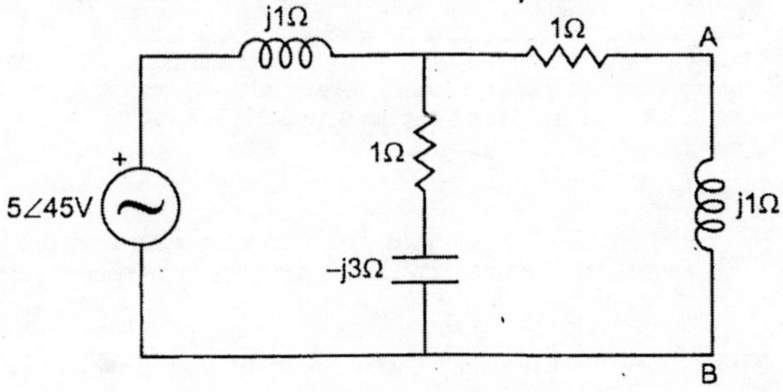

Fig. E5.16

SOLUTION. *To find* i_{sc}

To find i_{sc} the load resistance is removed from the network and it is replaced by a short circuit. We will have to find this short circuit current. The modified circuit is shown in Figure E5.16(*a*), with loop currents.

The loop equations are

Loop 1 : $(1 - j2)i_1' - (1 - j3)i_2' = 5\angle 45$ ...(*i*)

Loop 2 : $(2 - j2)i_2 - (1 - j3)i_1 = 0$ ...(*ii*)

Solving equations (*i*) and (*ii*) we get

$$i_{sc} = i_2' = (2.35 - 1.18j) \text{ A}$$

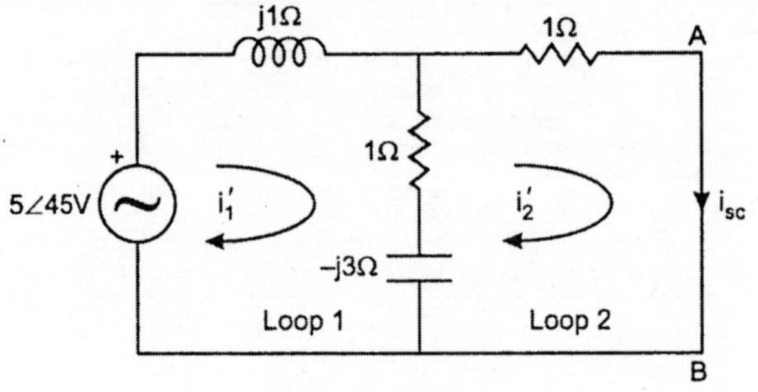

Fig. E5.16(a)

To find Z_n

The circuit to find Z_n is shown in Figure E5.16(*b*), here the independent voltage source has been replaced by short circuit (the internal resistance) and the load has been removed from the circuit.

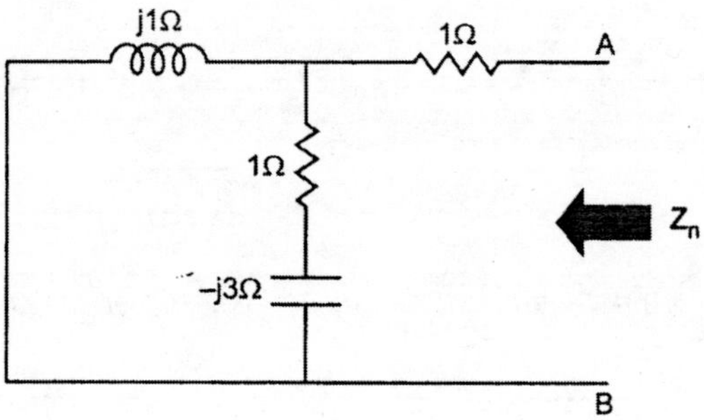

Fig. E5.16(b)

From Figure E5.16(*b*) we get

$$Z_n = \frac{(1-j3)\times(j1)}{1-j2} + 1 = 1.2 + j1.4 \ \Omega$$

Thus, the Norton's equivalent circuit is shown in Figure E5.16(*c*).

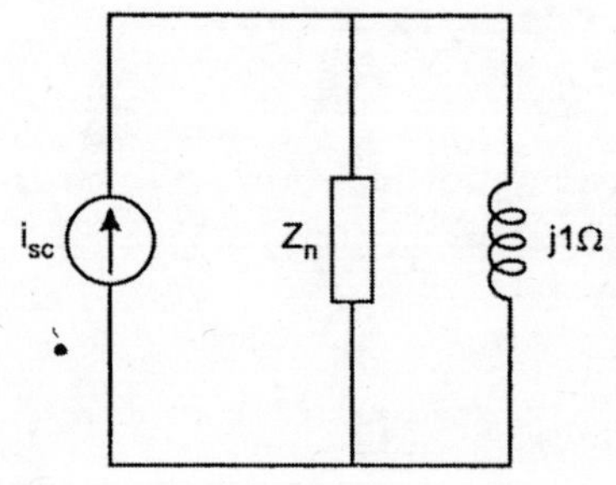

Fig. E5.16(c)

EXAMPLE 5.17. *For the circuit shown in Figure E5.17, find the current through C_1 using Norton's theorem.*

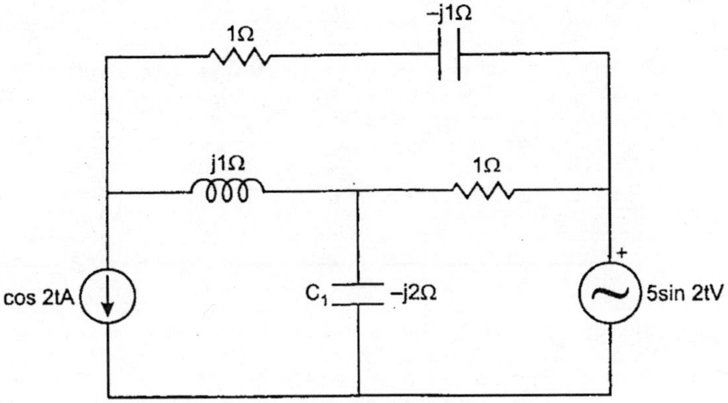

Fig. E5.17

SOLUTION. *To find i_{sc}*

To find i_{sc} the capacitor C_1 is removed from the network and it is replaced by a short circuit. We will have to find this short circuit current. The modified circuit is shown in Figure E5.17(*a*), with loop currents.

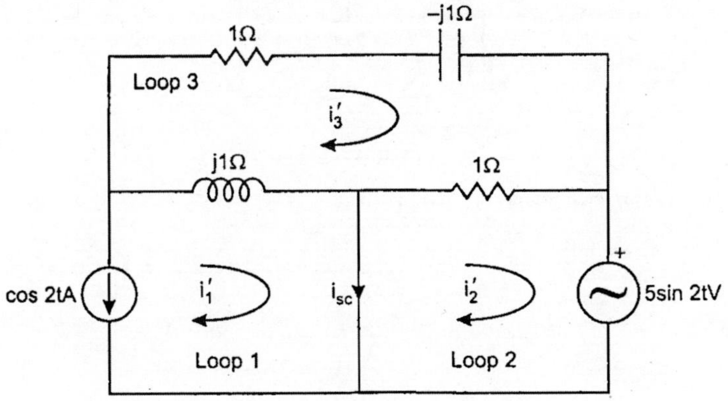

Fig. E5.17(a)

The loop equations are

Loop 1 : $i_1' = -\cos 2t$ ...(*i*)

Loop 2 : $i_2' = 5 \sin 2t$ ...(*ii*)

Form Figure E5.17(*a*) we get

$$i_{sc} = i_1' - i_2' = -(\cos 2t + 5 \sin 2t) \text{ A}$$

To find Z_n

The circuit to find Z_n is shown in Figure E5.17(*b*), here the independent voltage source has been replaced by short circuit (the internal resistance), the independent current source has been replaced by open circuit (the internal resistance) and load has been removed from the circuit.

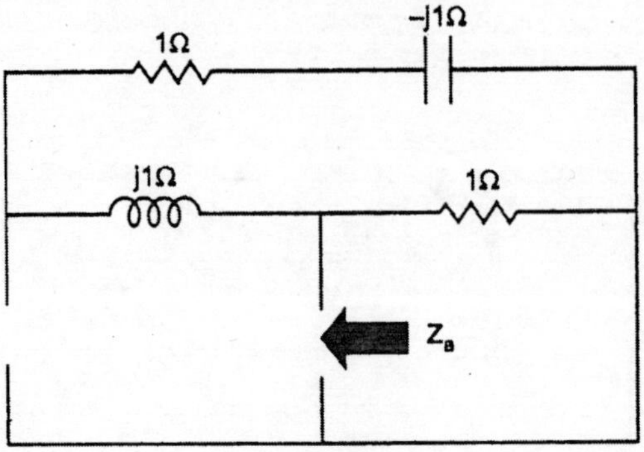

Fig. E5.17(b)

From Figure E5.17(*b*) we get

$$Z_n = \frac{1 \times 1}{2} = 0.5\ \Omega$$

Thus, the Norton's equivalent circuit is shown in Figure E5.17(*c*)

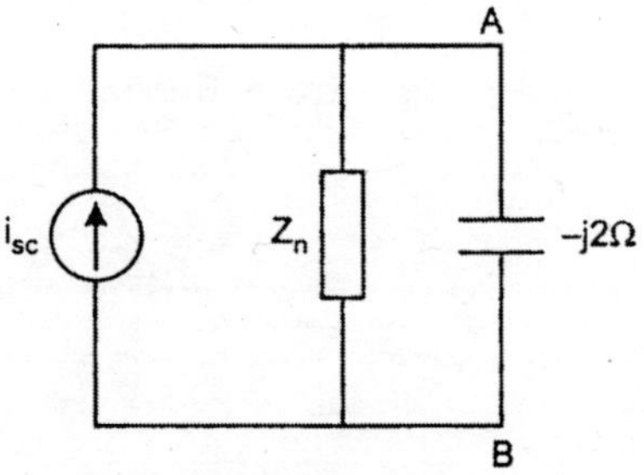

Fig. E5.17(c)

From Figure E5.17(*c*), the current through C_1 is

$$i_{sc} \times \frac{Z_n}{Z_n - j2} = -(\cos 2t + 5 \sin 2t) \times (0.06 + j0.24)\ \text{A}$$

EXAMPLE 5.18. *For the circuit shown in Figure E5.18, find the* V_0 *using Norton's theorem.*

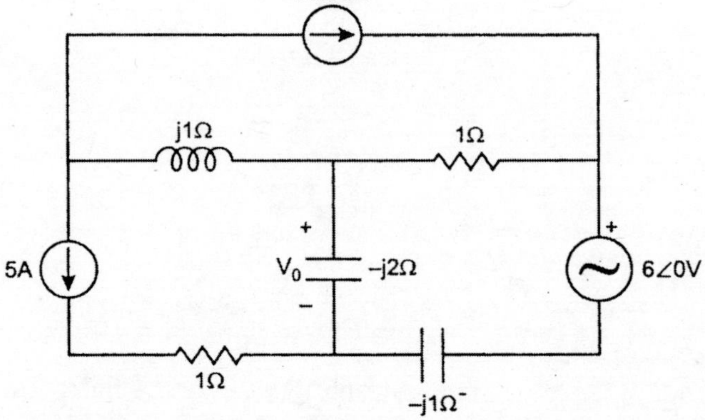

Fig. E5.18

SOLUTION. *To find* i_{sc}

To find i_{sc} the load is removed from the network and it is replaced by a short circuit. We will have to find this short circuit current. The modified circuit is shown in Figure E5.18(*a*), with loop currents.

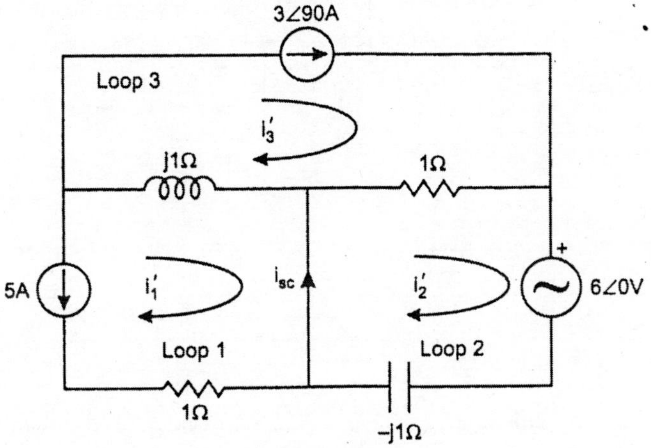

Fig. E5.18(a)

The loop equations are

Loop 1 : $i_1' = 5 \text{ A}$ ...(*i*)

Loop 2 : $i_2' = -\dfrac{6}{1-j1} = (3 + j3) \text{ A}$ ...(*ii*)

Form Figure E5.18(*a*) we get $i_{sc} = i_2' - i_1' = (-2 + j3) \text{ A}$

To find Z_n

The circuit to find Z_n is shown in Figure E5.18(*b*), here the independent voltage source has been replaced by short circuit (the internal resistance), the independent current source has been replaced by open circuit (the internal resistance) and load has been removed from the circuit.

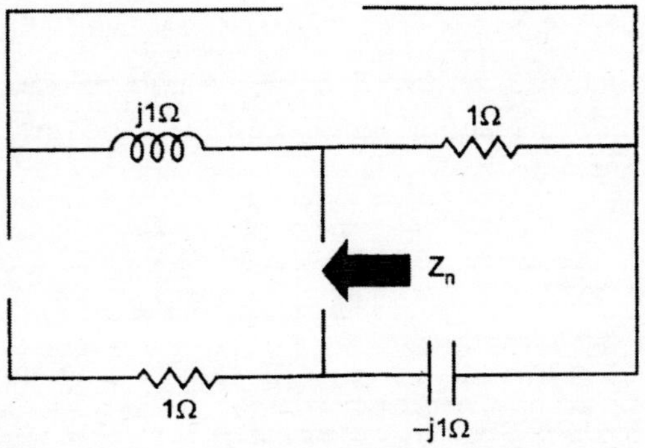

Fig. E5.18(b)

From Figure E5.18(*b*) we get $Z_n = (1 - j1) \ \Omega$

Thus, the Norton's equivalent circuit is shown in Figure E5.18(c)

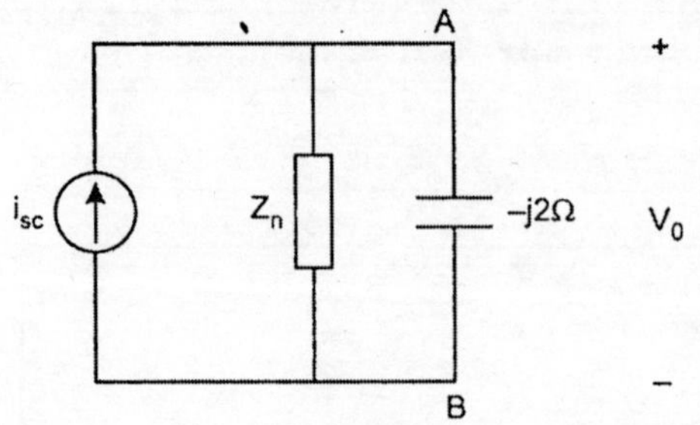

Fig. E5.18(c)

From Figure E5.18(c), the current through $-j2\ \Omega$ is

$$i = i_{sc} \times \frac{Z_n}{Z_n - j2} = -1.4 + j0.8 \text{ A}$$

Thus $$V_0 = -j2 \times i = 1.6 + j2.8 \text{ V}$$

EXAMPLE 5.19. *For the circuit shown in Figure E5.19, find the drop across the inductor using Norton's theorem.*

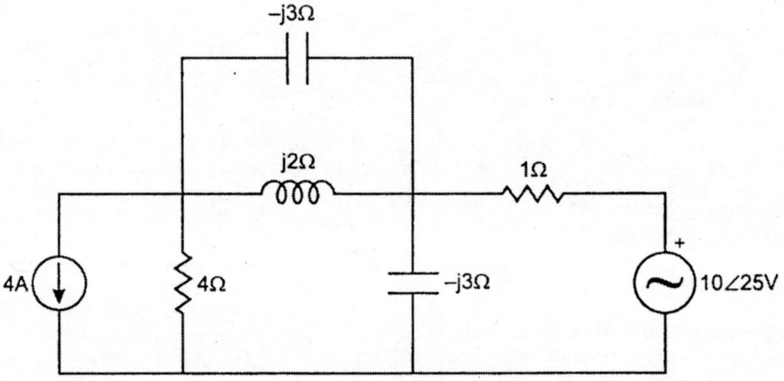

Fig. E5.19

SOLUTION. *To find i_{sc}*

To find i_{sc} the load (inductor) is removed from the network and it is replaced by a short circuit. We will have to find this short circuit current. The modified circuit is shown in Figure E5.19(*a*), with loop currents.

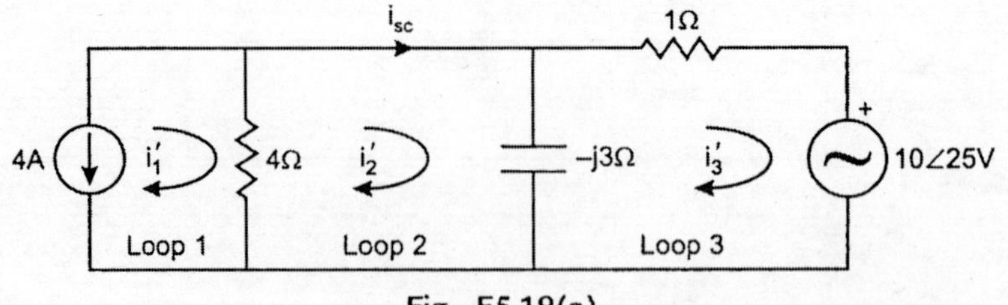

Fig. E5.19(a)

The loop equations are

Loop 1 : $i_1' = -4$ A ...(i)

Loop 2 : $(4 - j3)i_2' - 4i_1' + j3i_3' = 0$ ...(ii)

Substituting equation (i) in equation (ii) we get

$$(4 - j3)i_2' + j3i_3' = -16 \quad ...(iii)$$

Loop 3 : $(1 - j3)i_3' + j3i_2' = -10\angle 25 = 9.06 + j4.23$ ...(iv)

Solving equations (iii) and (iv) we get

$$i_{sc} = i_2' = -1.35 + j0.14 \text{ A}$$

To find Z_n

The circuit to find Z_n is shown in Figure E5.19(b), here the independent voltage source has been replaced by short circuit (the internal resistance), the ndependent current source has been replaced by open circuit (the internal resistance) and load has been removed from the circuit.

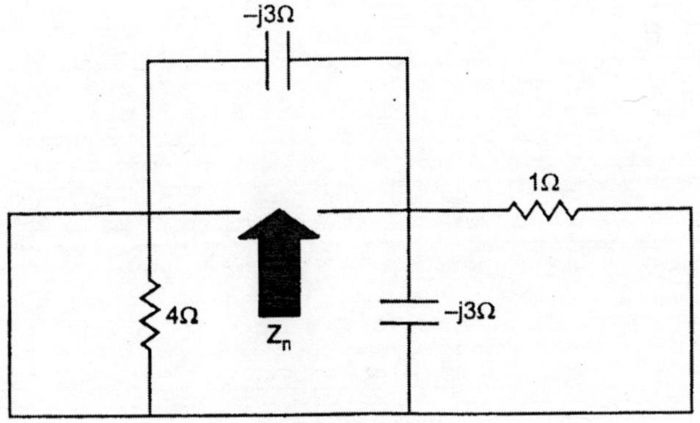

Fig. E5.19(b)

From Figure E5.19(b) we get

$$Z_n = \frac{\left[\frac{1\times(-j3)}{1-j3}+4\right]\times(-j3)}{\left[\frac{1\times(-j3)}{1-j3}+4\right]+(-j3)} = 1.26 - j2.15\ \Omega$$

Thus, the Norton's equivalent circuit is shown in Figure E5.19(c).

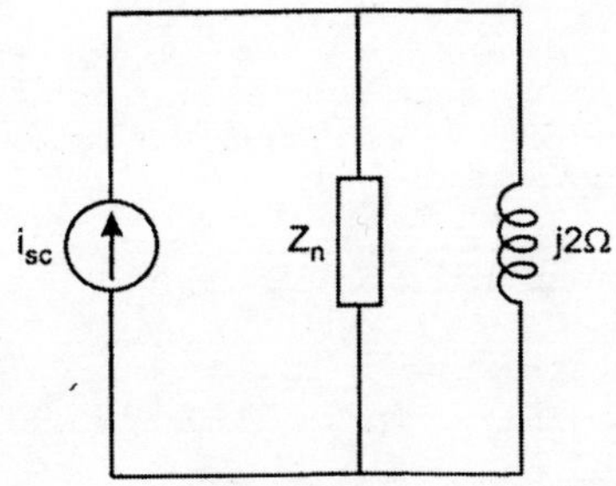

Fig. E5.19(c)

From Figure E5.19(*c*), the current through $j2\ \Omega$ is

$$i = i_{sc} \times \frac{Z_n}{Z_n + j2} = -1.38 + j2.28 \text{ A}$$

Thus $$V_0 = j2 \times i = -4.56 - j2.76 \text{ V}$$

EXAMPLE 5.20. *Find the Norton's equivalent across A-B, for the circuit shown in Figure E5.20.*

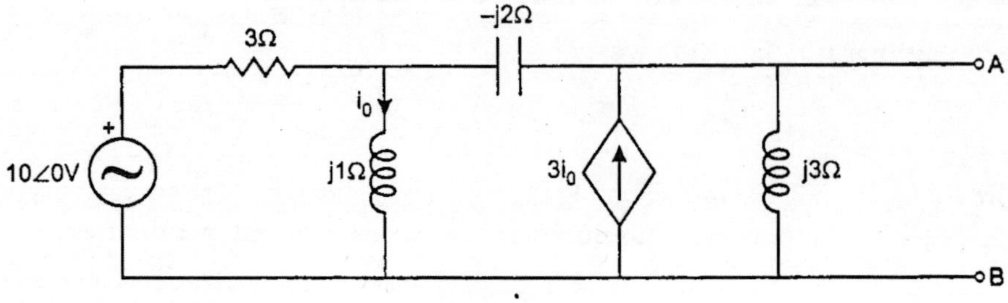

Fig. E5.20

SOLUTION. *To find* V_{TH}

The circuit is redrawn in Figure E5.20(*a*) with loop currents

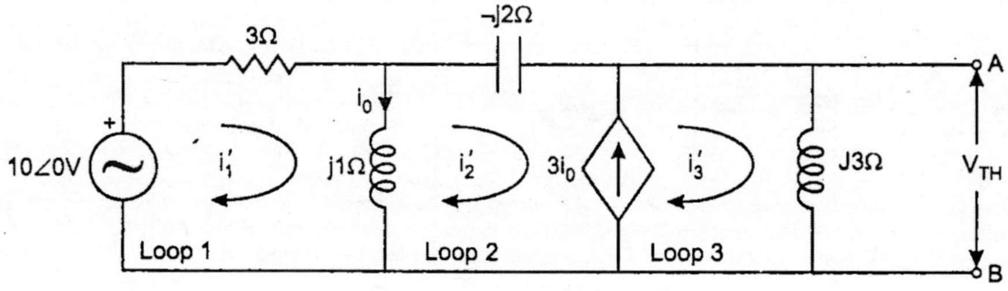

Fig. E5.20(a)

Loop 1 : $(3 + j1)i_1' - j1i_2' = 10$ ...(*i*)

Loop 2 and Loop 3 : $-j1i_2' + j3i_3' - j1i_1' = 0$ ...(*ii*)

$$i_3' - i_2' = 3i_0 \quad \text{...(iii)}$$

and $$i_0 = i_1' - i_2' \quad \text{...(iv)}$$

Substituting equation (*iv*) in equation (*iii*) we get

$$3i_1' - 2i_2' - i_3' = 0 \quad \text{...(v)}$$

Solving equations (*i*), (*ii*) and (*v*) we get

$$i_3' = 2.38 + j0.11 \text{ A}$$

Thus, $$V_{TH} = j3 \times i_3' = (-0.34 + j7.13) \text{ V}$$

To find i_{sc}

The circuit to find i_{sc} is shown in Figure E5.20(*b*).

The loop equations are

Loop 1 : $(3 + j1)i_1'' - j1i_2'' = 10$ ...(*vi*)

Loop 2 and Loop 3 : $-j1i_2'' - j1i_1'' = 0$ ...(*vii*)

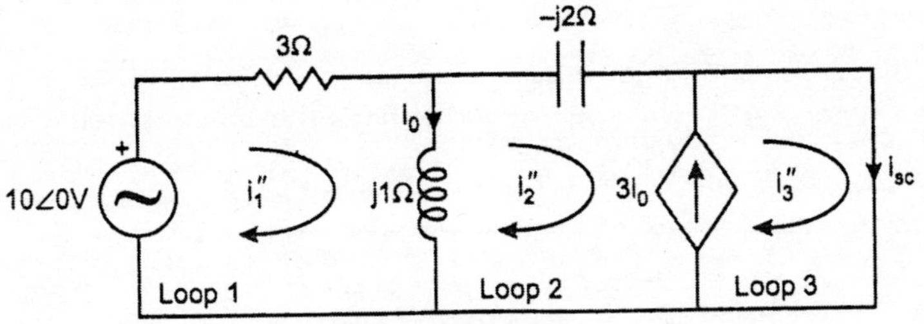

Fig. E5.20(b)

$$i_3'' - i_2'' = 3i_0 \quad ...(viii)$$

and

$$i_0 = i_1'' - i_2'' \quad ...(ix)$$

Substituting equation (*ix*) in equation (*viii*) we get

$$3i_1'' - 2i_2'' - i_3'' = 0 \quad ...(x)$$

Solving equations (*vi*), (*vii*) and (*x*) we get

$$i_3'' = i_{sc} = 11.54 - j7.69 \text{ A}$$

Then

$$Z_n = \frac{V_{TH}}{i_{sc}} = -0.31 + j0.41 \ \Omega$$

Thus, the Norton's equivalent circuit is shown in Figure E5.20(*c*).

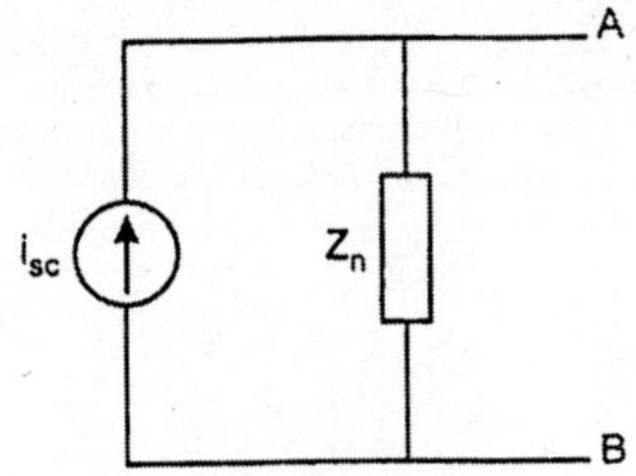

Fig. E5.20(c)

EXAMPLE 5.21. *Find the Norton's equivalent across A-B, for the circuit shown in Figure E5.21.*

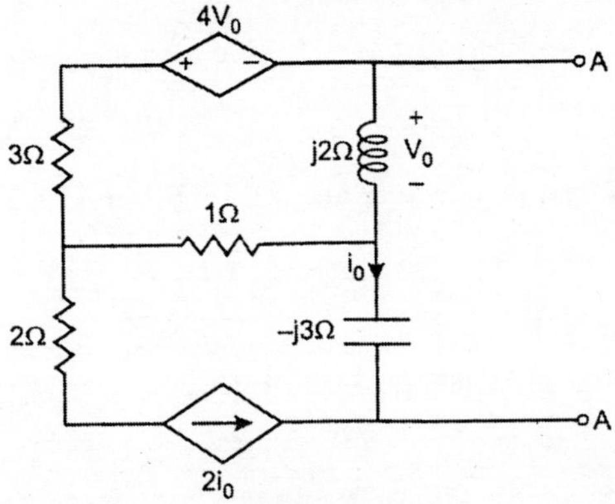

Fig. E5.21

SOLUTION. Since there is no independent source present in the network, it is a dead network so, $i_{sc} = 0$ A

To find Z_n

The circuit to find Z_{sc} is shown in Figure E5.21(*a*). A dummy independent voltage source of 1 V is connected across *A-B*, we will now calculate the current delivered by the source.

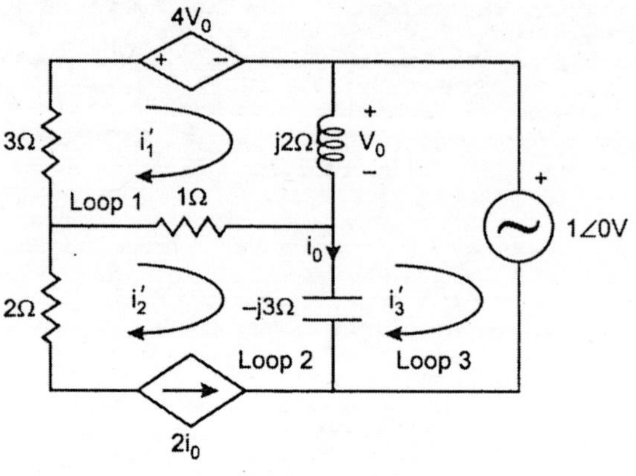

Fig. E5.21(a)

The loop equations are

Loop 1 : $(4 + j2)i_1' - i_2' - j2i_3' = -4V_0$ ...(*i*)

Loop 2 : $i_2' = -2i_0$ ...(*ii*)

Loop 3 : $-j1i_3' + j3i_2' - j2i_1' = -1$ ...(*iii*)

Again $V_0 = j2 \times (i_1' - i_3')$ ...(*iv*)

And $i_0 = i_2' - i_3'$ ...(*v*)

Substituting equation (*iv*) in equation (*i*) we get

$$(4 + j10)i_1' - i_2' - j10i_3' = 0 \qquad \text{...(vi)}$$

Substituting equation (*v*) in equation (*ii*) we get

$$3i_2' - 2i_3' = 0 \qquad \text{...(vii)}$$

Solving equations (*iii*), (*vi*) and (*vii*) we get

The source current $i_3' = -0.62 - j0.83$ A

Thus $Z_n = \dfrac{1}{i_3'} = -0.57 + j0.77\ \Omega$

Thus, the Norton's equivalent circuit is shown in Figure E5.21(*b*).

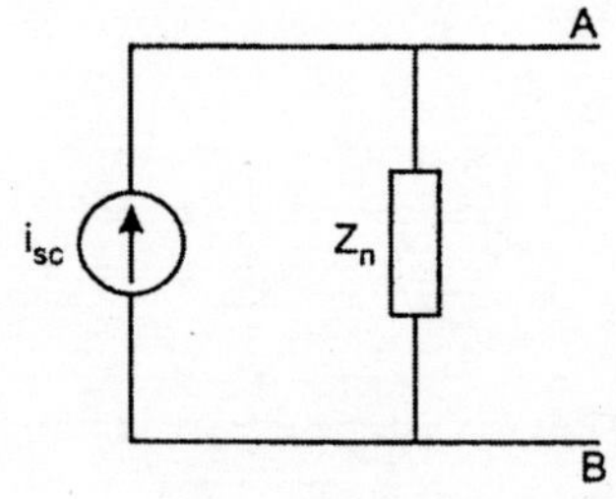

Fig. E5.21(b)

EXAMPLE 5.22. *For the circuit shown in Figure E5.14, of Example 14, find the Norton's equivalent across A-B.*

SOLUTION. The circuit is redrawn as shown in Figure E5.22(*a*).

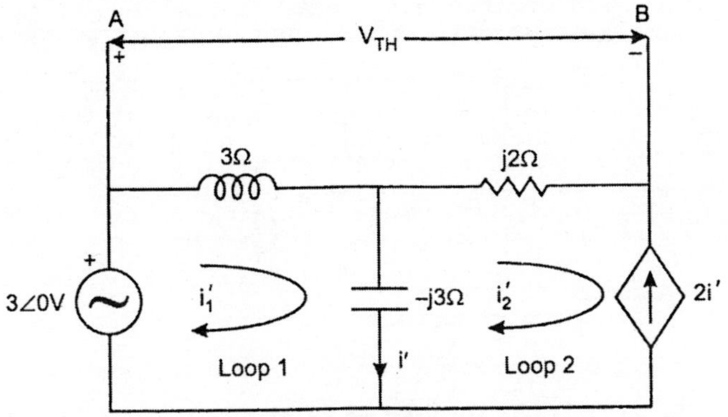

Fig. E5.22(a)

From the solution of Example 14 we get

$$i_{sc} = 0 \text{ A}$$

and
$$Z_{TH} = Z_n = -0.57 + j0.77\ \Omega$$

Thus, the Norton's equivalent across *A-B* is shown in Figure E5.22(*b*).

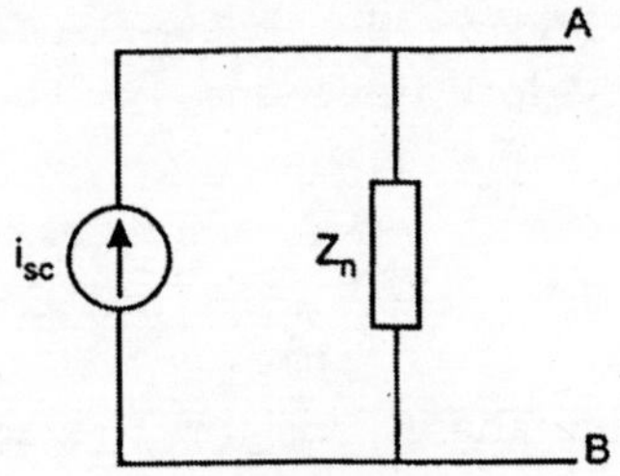

Fig. E5.22(b)

EXAMPLE 5.23. *For the circuit shown in Figure E5.15 of Example 15, find the current through j4 Ω using Norton's theorem.*

SOLUTION. The circuit is redrawn as shown in Figure E5.23(*a*).

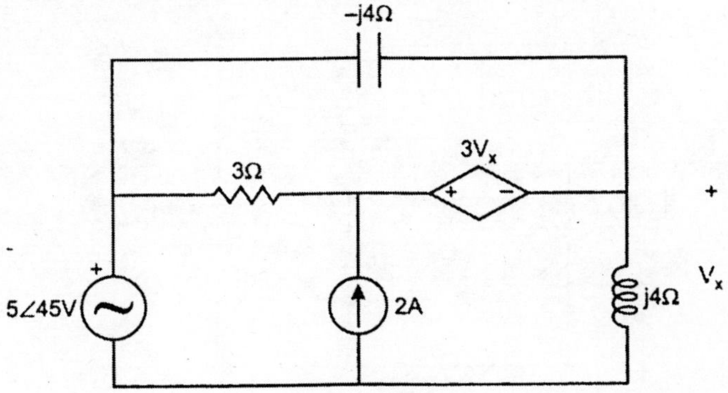

Fig. E5.23(a)

From the solution of Example 15 we get,

$$i_{sc} = 2.29 + j2.06 \text{ A}$$

and
$$Z_{TH} = Z_n = 0.49 - j0.38\ \Omega$$

Thus, the Norton's equivalent across *A-B* is shown in Figure E5.23(*b*)

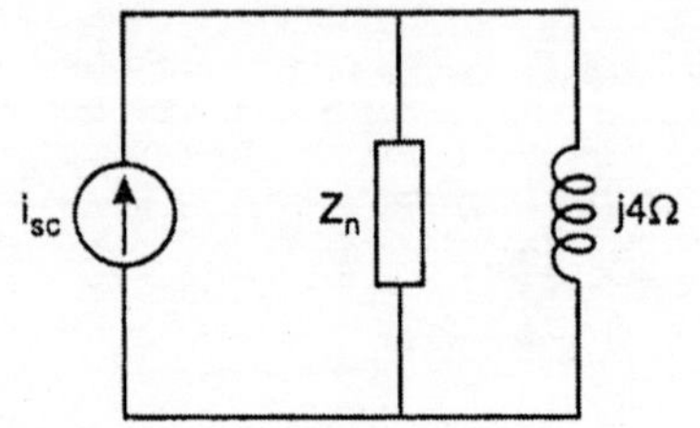

Fig. E5.23(b)

From Figure E5.23(*b*), the current through $j4\ \Omega$ is

$$i = i_{sc} \times \frac{Z_n}{Z_n + j4} = 0.11 - j0.51 \text{ A}$$

5.7 MAXIMUM POWER TRANSFER THEOREM

Statement : A load impedance in a linear, active, bilateral a.c. circuit will receive the maximum power, if and only if the value of the load impedance becomes equal to the complex conjugate of Thevenin's impedance (Z_{TH}). And the maximum power will be

$$P_m = \frac{V_{TH}^2}{4 \times R_{TH}}$$

where V_{TH} is the Thevenin's voltage and R_{TH} is the Thevenin's resistance.

SOLVED EXAMPLES

EXAMPLE 5.24. *Find the value of Z that needs to be connected across A-B, for the circuit shown in Figure E5.24, for maximum power transfer. Also find the power.*

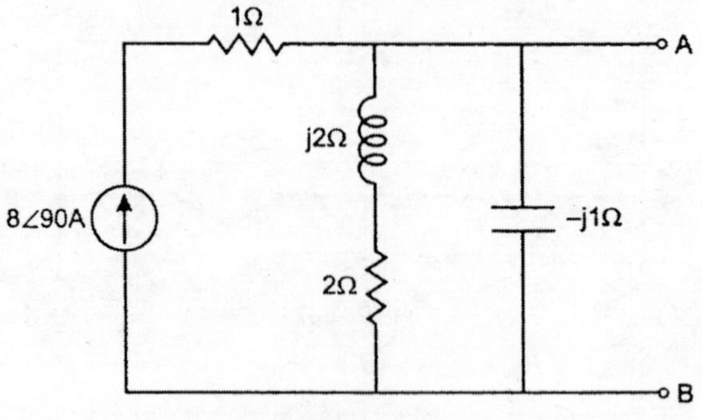

Fig. E5.24

SOLUTION. *To find* V_{TH} -

The circuit is redrawn in Figure E5.24(*a*), with loop currents.

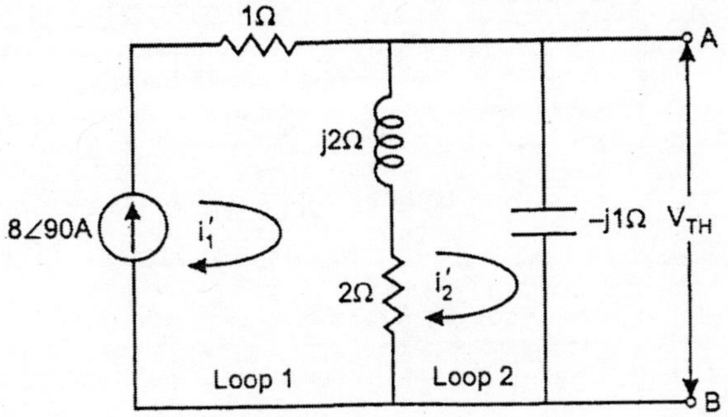

Fig. E5.24(a)

The loop equations are

Loop 1 : $i_1' = j8$ A ...(*i*)

Loop 2 : $(2 + j1)i_2' - (2 + j2)i_1' = 0$ ...(*ii*)

Substituting equation (*i*) in equation (*ii*) we get

$$i_2' = (-3.2 + j9.6) \text{ A}$$

Thus $V_{TH} = (-3.2 + j9.6) \times (-j1) = (9.6 + j3.2) \text{ V} = 10\angle 18.43 \text{ V}$

To find Z_{TH}

The circuit to find Z_{TH} is shown in Figure E5.24(*b*). Here the independent current source has been replaced by open circuit (internal resistance).

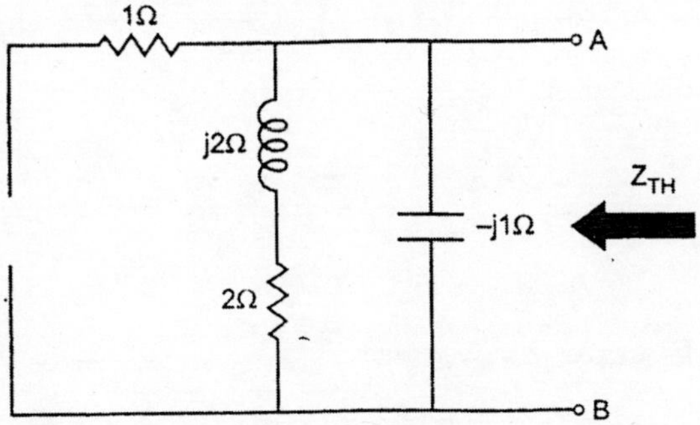

Fig. E5.24(b)

From Figure E5.24(*b*) we get

$$Z_{TH} = \frac{(2+j2)\times(-j1)}{2+j1} = 0.4 - j1.2\ \Omega$$

Thus, the value of Z for maximum power transfer is $= Z_{TH}^* = 0.4 + j1.2\ \Omega$

And the power is $= \dfrac{V_{TH}^2}{4 \times R_{TH}^*} = 62.5$ W

EXAMPLE 5.25. *Find the value of Z, for the circuit shown in Figure E5.25, for maximum power transfer. Also find the power.*

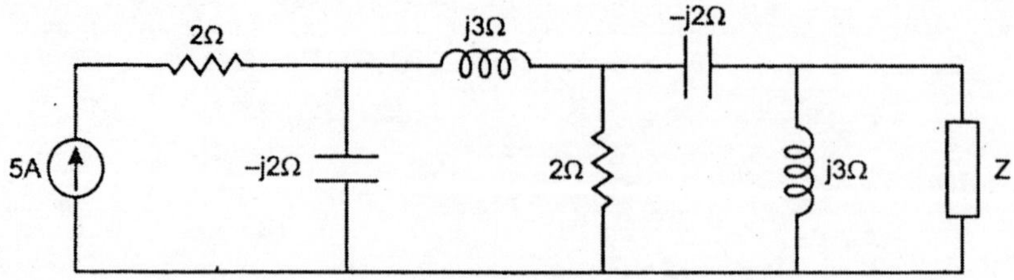

Fig. E5.25

SOLUTION. The circuit is same as of the problem shown in Example 13. From the solution of Example 13, we get

$$V_{TH} = -3.53 - j14.12 \text{ V} = 14.55\angle{-104} \text{ V}$$

and

$$Z_{TH} = 1.06 - j1.76\ \Omega$$

Thus, the value of Z for maximum power transfer is $= Z_{TH}^{*} = 1.06 + j1.76\ \Omega$

And the power is $= \dfrac{V_{TH}^2}{4 \times R_{TH}^{*}} = 50$ W

EXAMPLE 5.26. *Consider the circuit of Example 20, find the value of Z, for maximum power transfer to be connected across A-B and also find the power.*

SOLUTION. The circuit is again redrawn in Figure E5.26.

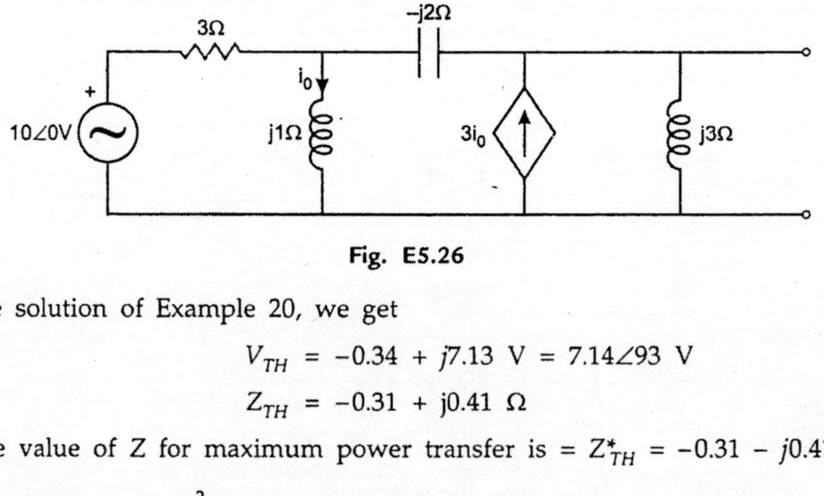

Fig. E5.26

From the solution of Example 20, we get

$$V_{TH} = -0.34 + j7.13 \text{ V} = 7.14\angle 93 \text{ V}$$

and

$$Z_{TH} = -0.31 + j0.41\ \Omega$$

Thus, the value of Z for maximum power transfer is $= Z_{TH}^{*} = -0.31 - j0.41\ \Omega$

And the power is $= \dfrac{V_{TH}^2}{4 \times R_{TH}^{*}} = 41$ W

EXAMPLE 5.27. *Consider the circuit of Example 21, find the value of Z, for maximum power transfer to be connected across A-B and also find the power.*

SOLUTION. The circuit is again redrawn in Figure E5.27

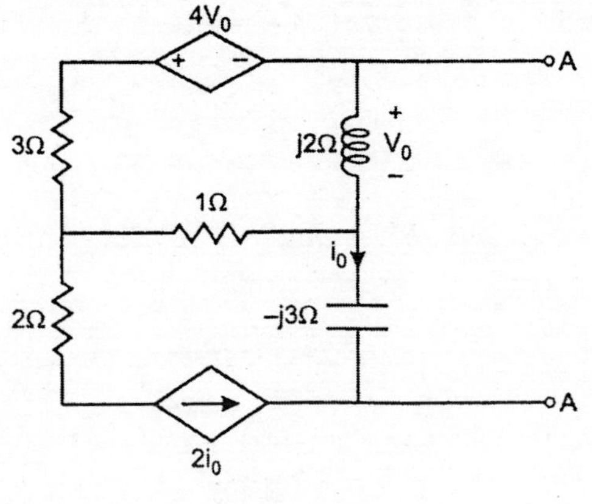

Fig. E5.27

From the solution of Example 21 we get

$$V_{TH} = 0 \text{ V}$$

and

$$Z_{TH} = -0.57 + j0.77\ \Omega$$

Thus, the value of Z for maximum power transfer is $= Z^*_{TH} = -0.57 - j0.77\ \Omega$

And the power is = 0 W

5.8 MILLMAN'S THEOREM

Statement : If a linear active bilateral a.c. circuit consists of n voltage sources (V_1 to V_n) connected in parallel to each other, with their internal impedances (z_1 to z_n) connected in series with the source respectively (Figure 5.1). Then such a circuit can be replaced by a single voltage source E_m with a series impedance z_m (Figure 5.2). Where

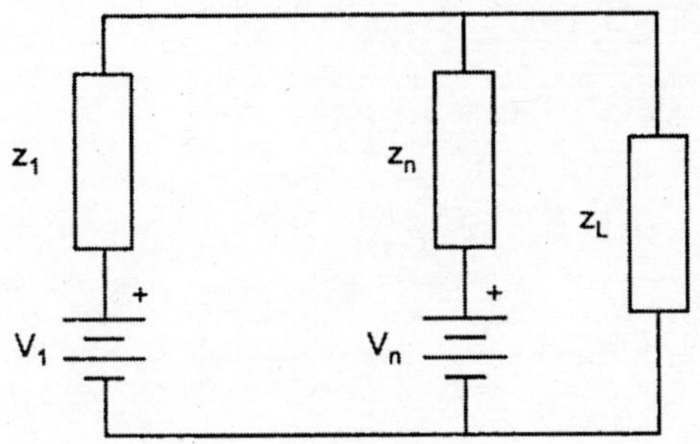

Fig. 5.1

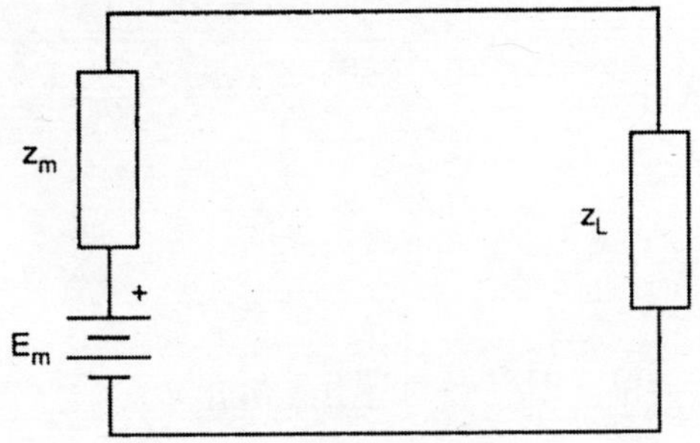

Fig. 5.2

Where

$$E_m = \frac{V_1 \times \frac{1}{R_1} + \dots\dots\dots\dots + V_n \times \frac{1}{R_n}}{\frac{1}{R_1} + \dots\dots\dots\dots + \frac{1}{R_n}}$$

and
$$R_m = \frac{1}{\frac{1}{R_1} + \ldots\ldots\ldots\ldots\ldots + \frac{1}{R_n}}$$

SOLVED EXAMPLES

EXAMPLE 5.28. *Consider the circuit shown in Figure E5.28, find the current through Z_L using Millman's theorem.*

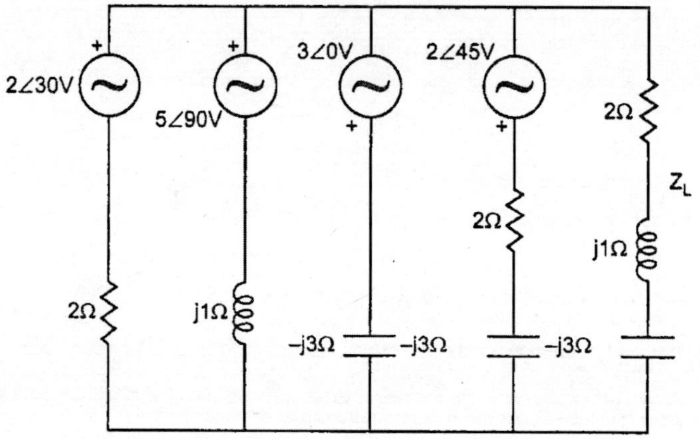

Fig. E5.28

SOLUTION. From Figure E5.28 we get

$$E = \frac{2\angle 30 \times \frac{1}{2} + 5\angle 90 \times \frac{1}{j1} + 3\angle 0 \times \frac{1}{-j3} + 2\angle 45 \times \frac{1}{2-j3}}{\frac{1}{2} + \frac{1}{j1} + \frac{1}{-j3} + \frac{1}{2-j3}}$$

$$= 14.35 + j4.99 \text{ V}$$

$$Z = \frac{1}{\frac{1}{2} + \frac{1}{j1} + \frac{1}{-j3} + \frac{1}{2-j3}} = 1.06 + j0.71\ \Omega$$

Then the current through Z_L is

$$I_L = \frac{E}{Z+Z_L} = 3.39 + j3.06 \text{ A}$$

EXAMPLE 5.29. *Consider the circuit shown in Figure E5.29, find the current through* Z_L *using Millman's theorem.*

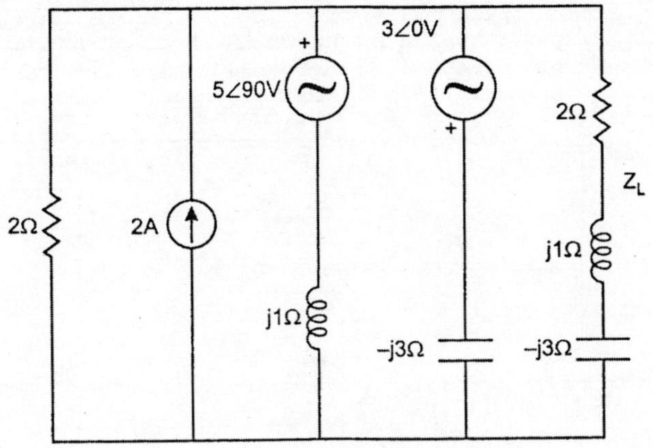

Fig. E5.29

SOLUTION. Converting the current source into voltage source we get the circuit shown in Figure E5.29(*a*).

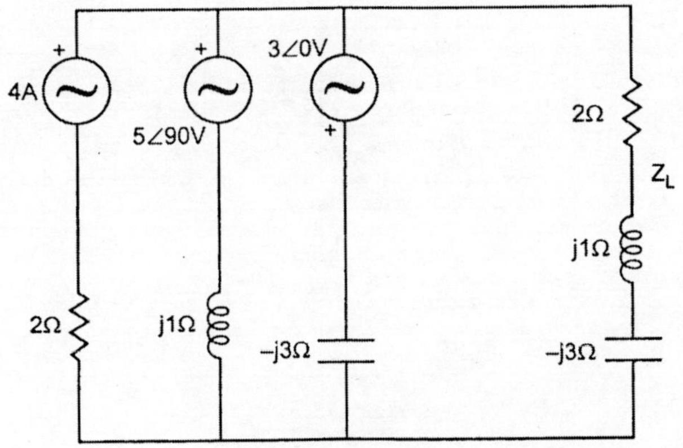

Fig. E5.29(a)

From Figure E5.29(*a*) we get

$$E = \frac{4\times\frac{1}{2}+5\angle 90\times\frac{1}{j1}+3\angle 0\times\frac{1}{-j3}}{\frac{1}{2}+\frac{1}{j1}+\frac{1}{-j3}} = 06.04 + j0.72 \text{ V}$$

$$Z = \frac{1}{\frac{1}{2}+\frac{1}{j1}+\frac{1}{-j3}} = 0.72 + j0.96\ \Omega$$

Then the current through Z_L is

$$I_L = \frac{E}{Z+Z_L} = 1.85 + j0.97 \text{ A}$$

EXERCISES

1. State and explain with a simple example all theorems discussed in this chapter.
2. Find V_x using Superposition theorem for the circuit shown in Figure Q5.2.

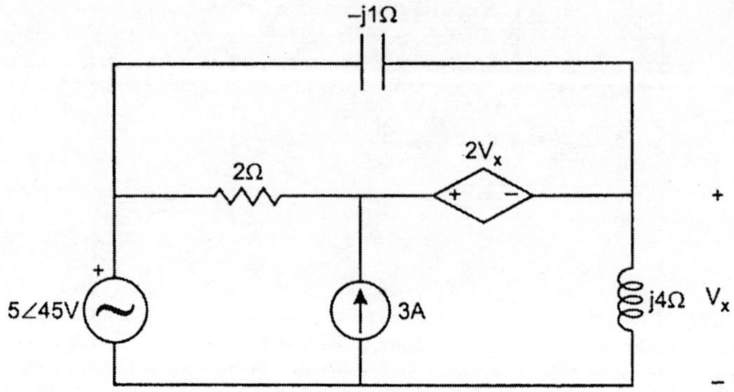

Fig. Q5.2

3. Find the current through $-j3\ \Omega$, using Superposition theorem, for the circuit shown in Figure Q5.3.

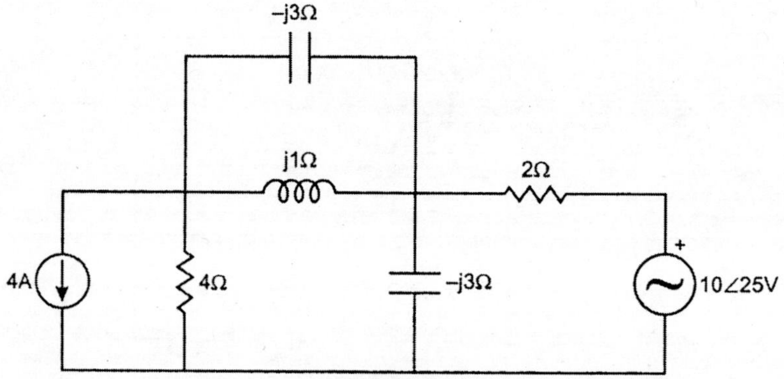

Fig. Q5.3

4. Find the current through $-j6\ \Omega$, using Thevenin's theorem, for the circuit shown in Figure Q5.4. Verify your answer using Norton's theorem.

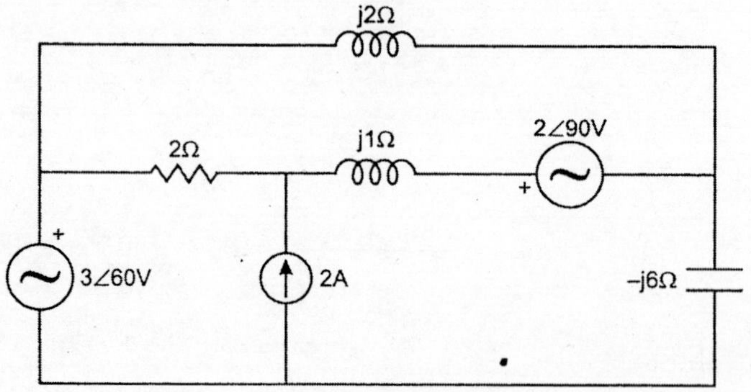

Fig. Q5.4

5. Find the current through $j6\ \Omega$, using Thevenin's theorem, for the circuit shown in Figure Q5.5. Verify your answer using Norton's theorem.

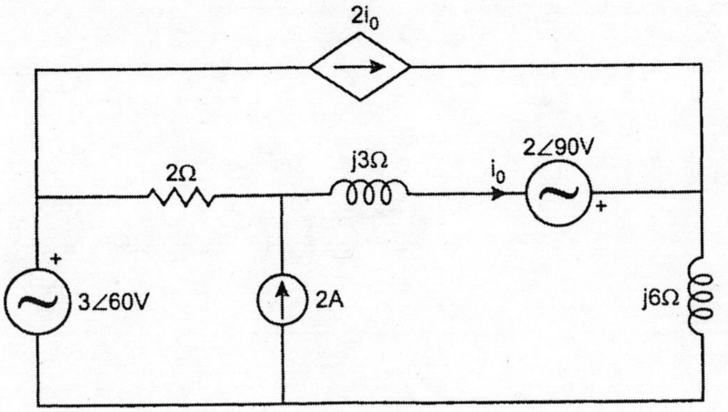

Fig. Q5.5

6. With respect to maximum power transfer theorem prove that when the load impedance is equal to the complex conjugate of Zth, the maximum power transferred is

$$P_m = \frac{V_{TH}^2}{4 \times R_{TH}}$$

and the efficiency is only 50%.

Chapter 6

COUPLED CIRCUITS

6.1 INTRODUCTION

Whatever circuits we studied so for in our previous chapters were all conductively coupled circuits. Here in this chapter we will consider the effect of magnetic coupling between the different coils present in the network. We say the coils are magnetically coupled when the flux produced by one coils affects the response on the other coils and vice versa. The most common example of a magnetically coupled circuit is the transformer.

6.2 CONCEPT OF SELF INDUCED EMF

Let us consider the circuit shown in Figure 6.1. Two coils coil 1 and coil 2 are connected to an alternating voltage source V_1 and V_2 respectively. They produce current i_1 and i_2 in the coil 1 and coil 2 respectively. Let the rms value of flux produced in coil 1 due to the current i_1 be φ_1, and the rms value of flux produced in coil 2 due to the current i_2 be φ_2.

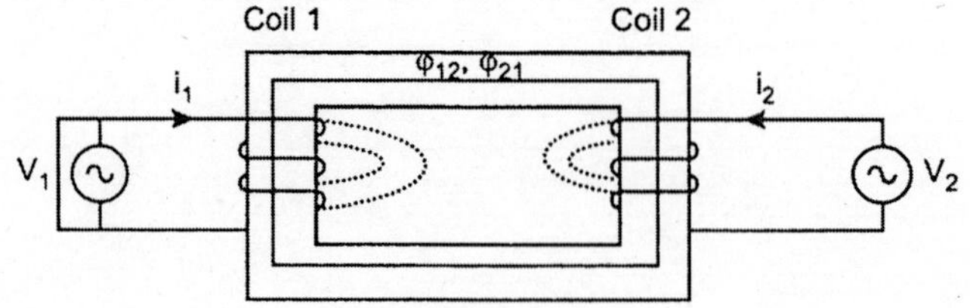

Fig. 6.1

Now if the current through the coils are changed, then the flux produced in the coil will also change and thus from Faraday's law an emf will be induced in the coils. This emf is called the self induced emf and it is proportional to the rate of change of current that is taking place in the circuit.

Thus the self induced emf produced in coil 1 due to the change of current i_1 is given by

$$e_1 = -L_1 \frac{di_1}{dt} \quad ...(6.1)$$

and similarly the self induced emf produced in coil 2 due to the change of current i_2 is given by

$$e_2 = -L_2 \frac{di_2}{dt} \quad ...(6.2)$$

where L_1 and L_2 are the self inductances of coil 1 and coil 2 respectively and is given by

$$L_1 = \frac{N_1\phi_1}{i_1} \quad ...(6.3)$$

and

$$L_2 = \frac{N_2\phi_2}{i_2} \quad ...(6.4)$$

respectively. Where N_1 and N_2 are the number of turns of coil 1 and coil 2 respectively. The unit of self inductance is Henry and it is denoted by H.

6.3 CONCEPT OF MUTUALLY INDUCED EMF

Now, let us assume that some fraction k_1 of the total flux φ_1 produced by the coil 1 link with coil 2 as shown in Figure 6.1. Let us denote that fraction of flux by φ_{12}, then we can write the following relation

$$\Phi_{12} = k_1\varphi_1 \quad ...(6.5)$$

Similarly, let us assume that some fraction k_2 of the total flux φ_2 produced by the coil 2 link with coil 1 as shown in Figure 6.1. Let us denote that fraction of flux by φ_{21}, then we can write the following relation

$$\Phi_{21} = k_2\varphi_2 \quad ...(6.6)$$

The flux Φ_{12} and Φ_{21} will also produce some emf in coil 2 and coil 1 respectively. This emf is called mutually induced emf and is given by,

The mutually induced emf of coil 2 $e_{12} = -M_{12} \dfrac{di_1}{dt}$ ...(6.7)

Similarly the mutually induced emf of coil 1 $e_{21} = -M_{21} \dfrac{di_2}{dt}$ ...(6.8)

where M_{12} and M_{21} are called the mutual inductance of the coil 2 and coil 1 respectively and is given by

$$M_{12} = \frac{N_2\phi_{12}}{i_1} \quad ...(6.9)$$

and

$$M_{21} = \frac{N_1\phi_{21}}{i_2} \quad (6.10)$$

6.4 COEFFICIENT OF COUPLING

From equation (6.5) and equation (6.6) we can write

$$k_1 = \frac{\phi_{12}}{\phi_1} \quad ...(6.11)$$

and

$$k_2 = \frac{\phi_{21}}{\phi_2} \quad ...(6.12)$$

where k_1 and k_2 is termed as coefficient of coupling, and it is defined as the ratio of the fraction of the total flux produced by one coil that links the other coil to the total flux produced by the coil.

Replacing ϕ_{12} of equation (6.11) from equation (6.9) we get

$$k_1 = \frac{M_{12} \times i_1}{N_2 \times \phi_1} \quad ...(6.13)$$

again replacing ϕ_1 of equation (6.13) from equation (6.3) we get

$$k_1 = \frac{M_{12} \times i_1 \times N_1}{N_2 \times L_1 \times i_1} \quad ...(6.14)$$

or

$$k_1 = \frac{M_{12} \times N_1}{N_2 \times L_1} \quad ...(6.15)$$

Similarly substituting equation (6.10) and equation (6.4) in equation (6.12) and simplifying we get

$$k_2 = \frac{M_{21} \times N_2}{N_1 \times L_2} \quad ...(6.16)$$

Now assuming $k_1 = k_2 = k$ and $M_{12} = M_{21} = M$ for equation (6.15) and equation (6.16) and then multiplying the two equations we get

$$k^2 = \frac{M^2}{L_1 L_2} \quad ...(6.17)$$

or

$$M = k\sqrt{L_1 L_2} \quad ...(6.18)$$

6.5 DOT RULE

In a magnetically coupled circuit the dot rule is used to determine the sign of M, that is the mutual inductance.

To explain dot rule let us first consider a simple series magnetically coupled circuit as shown in Figure 6.2. To determine the sign of M the following steps are followed.

(*i*) Place a dot in each coil as shown in Figure 6.2.

(*ii*) If in both the coils current enter the coil through the dot as shown in Figure 6.2(*a*) or if in both the coils current leaves the coil through the dot as shown in Figure 6.2(*b*), then the sign of M is positive.

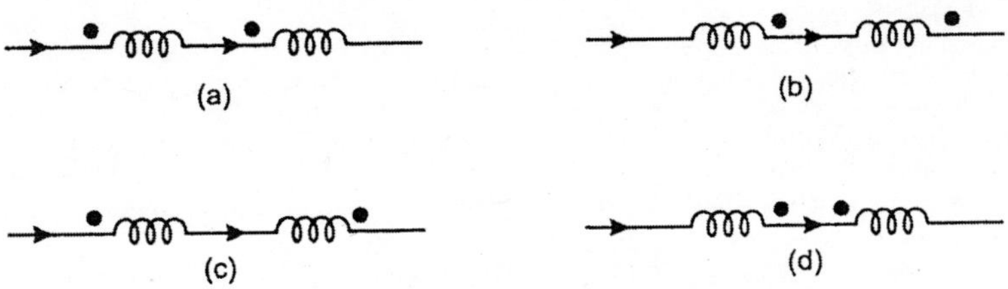

Fig. 6.2

(*iii*) If in one coil current enters through the dot and in the other coil the current leaves through the dot as shown in Figure 6.2(*c*) or vice versa, shown in Figure 6.2(*d*), then the sign of M is negative.

6.6 TUNED CIRCUITS

Such circuits are used for amplification of signals, especially radio frequency signals. Such kind of tuned circuits contains capacitors either in the primary side or in the secondary side or in both the primary and secondary side of a coupled circuit. Depending upon how the capacitors are placed, tuned circuits is classified into,

(*i*) Single Tuned Circuits

(*ii*) Double Tuned Circuits.

6.6.1 Single Tuned Circuit

In such circuits capacitors are present either in the primary side or in the secondary side of the coupled circuit.

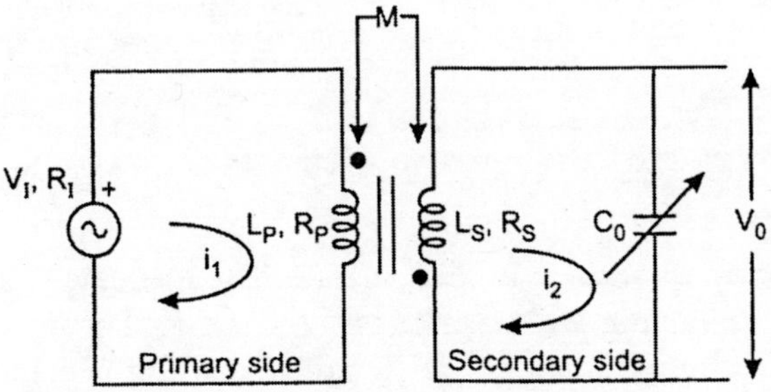

Fig. 6.3

A single tuned circuit is shown in Figure 6.3. The primary side is connected to an alternating voltage source V_I, having an internal resistance of R_I. Let R_P and L_P be the resistance and the inductance of the primary side and let R_S and L_S be the resistance and the inductance of the secondary side respectively. Let M be the mutual inductance between the primary and secondary. Let i_1 and i_2 be the primary and secondary currents respectively. And let C_0 be a variable capacitor connected to the secondary side. The loop equations are given by,

Primary side : $$(R_I + R_P + j\omega L_P)i_1 + j\omega M i_2 = V_I \quad ...(6.19)$$

let $R_{11} = R_I + R_P$, then equation (6.19) becomes

$$(R_{11} + j\omega L_P)i_1 + j\omega M i_2 = V_I \quad ...(6.20)$$

Secondary side :
$$\left(R_S + j\omega L_S - \frac{j}{\omega C_0}\right)i_2 + j\omega M i_1 = 0 \qquad ...(6.21)$$

Let $X_S = \omega L_S - \dfrac{1}{\omega C_0}$, then equation (6.21) becomes

$$(R_S + jX_S)i_2 + j\omega M i_1 = 0 \qquad ...(6.22)$$

Writing equation (6.20) and equation (6.22) in matrix form we get

$$\begin{bmatrix} (R_{11} + j\omega L_P) & j\omega M \\ j\omega M & (R_S + jX_S) \end{bmatrix} \times \begin{bmatrix} i_1 \\ i_2 \end{bmatrix} = \begin{bmatrix} V_I \\ 0 \end{bmatrix} \qquad ...(6.23)$$

Let Z_{11} + $(R_{11} + j\omega L_P)$ and $Z_{22} = (R_S + jX_S)$ then equation (6.23) becomes

$$\begin{bmatrix} Z_{11} & j\omega M \\ j\omega M & Z_{22} \end{bmatrix} \times \begin{bmatrix} i_1 \\ i_2 \end{bmatrix} = \begin{bmatrix} V_I \\ 0 \end{bmatrix} \qquad ...(6.24)$$

Let
$$\Delta = \begin{vmatrix} Z_{11} & j\omega M \\ j\omega M & Z_{22} \end{vmatrix} = Z_{11}Z_{22} + \omega^2 M^2$$

Then
$$i_2 = \frac{\begin{vmatrix} Z_{11} & V_I \\ j\omega M & 0 \end{vmatrix}}{\Delta} = \frac{-j\omega M V_I}{Z_{11}Z_{22} + \omega^2 M^2} \qquad ...(6.25)$$

Then
$$V_0 = i_2 \times \frac{1}{j\omega C_0} = \frac{j\omega M V_I}{Z_{11}Z_{22} + \omega^2 M^2} \times \frac{1}{j\omega C_0} \qquad ...(6.26)$$

Then amplification
$$A = \frac{V_0}{V_I} = \frac{M}{Z_{11}Z_{22} + \omega^2 M^2} \times \frac{1}{C_0} \qquad ...(6.27)$$

6.6.2 Doubled Tuned Circuit

In such circuit capacitors are present in the primary side and also in the secondary side of the coupled circuit.

A doubled tuned circuit is shown in Figure 6.4. The primary side is connected to an alternating voltage source V_I, having an internal resistance of R_I. Let R_P and L_P be the resistance and the inductance of the primary side and let R_S and L_S be the resistance and the inductance of the secondary side respectively.

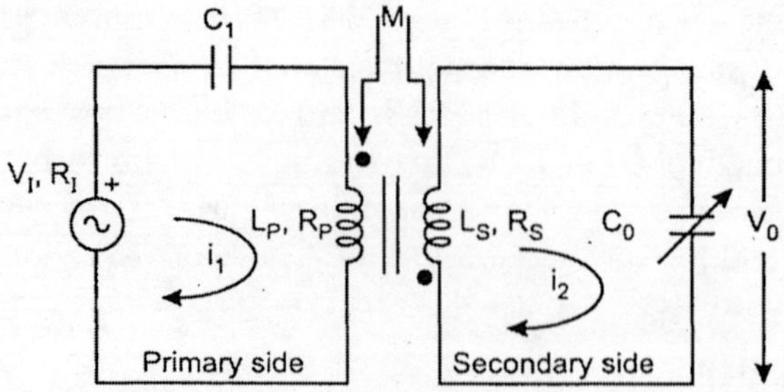

Fig. 6.4

Let M be the mutual inductance between the primary and secondary. Let i_1 and i_2 be the primary and secondary currents respectively. The loop equations are given by,

Primary side : $$\left(R_I + R_P + j\omega L_P - j\frac{1}{\omega C_1}\right)i_1 + j\omega M i_2 = V_I \quad ...(6.28)$$

let $R_{11} = R_I + R_P$, and $X_{11} = \omega L_P - \dfrac{1}{\omega C_1}$ then equation (6.28) becomes

$$(R_{11} + jX_{11}) + j\omega M i_2 = V_I \quad ...(6.29)$$

Secondary side : $$\left(R_S + j\omega L_S - \frac{j}{\omega C_0}\right)i_2 + j\omega M i_1 = 0 \quad ...(6.30)$$

Let $X_S = \omega L_S - \dfrac{1}{\omega C_0}$, then equation (6.30) becomes

$$(R_S + jX_S)i_2 + j\omega M i_1 = 0 \quad ...(6.31)$$

Writing equation (6.29) and equation (6.31) in matrix form we get

$$\begin{bmatrix} R_{11} + jX_{11} & j\omega M \\ j\omega M & R_S + jX_S \end{bmatrix} \times \begin{bmatrix} i_1 \\ i_2 \end{bmatrix} = \begin{bmatrix} V_I \\ 0 \end{bmatrix} \quad ...(6.32)$$

Let $Z_{11} = R_{11} + jX_{11}$ and $Z_{22} = R_S + jX_S$, the equation (6.32) becomes

$$\begin{bmatrix} Z_{11} & j\omega M \\ j\omega M & Z_{22} \end{bmatrix} \times \begin{bmatrix} i_1 \\ i_2 \end{bmatrix} = \begin{bmatrix} V_I \\ 0 \end{bmatrix}$$

Let $$\Delta = \begin{vmatrix} Z_{11} & j\omega M \\ j\omega M & Z_{22} \end{vmatrix} = Z_{11}Z_{22} + \omega^2 M^2$$

Then $$i_2 = \frac{\begin{vmatrix} Z_{11} & V_I \\ j\omega M & 0 \end{vmatrix}}{\Delta} = \frac{-j\omega M V_I}{Z_{11}Z_{22} + \omega^2 M^2} \quad ...(6.33)$$

Then $$V_0 = i_2 \times \frac{1}{j\omega C_0} = \frac{MV_I}{Z_{11}Z_{22} + \omega^2 M^2} \times \frac{1}{C_0} \quad ...(6.34)$$

Then amplification $$A = \frac{V_0}{V_I} = \frac{M}{Z_{11}Z_{22} + \omega^2 M^2} \times \frac{1}{C_0} \quad ...(6.35)$$

SOLVED EXAMPLES

Series Circuits

EXAMPLE 6.1. *For the circuit shown in Figure E6.1, determine the equivalent reactance.*

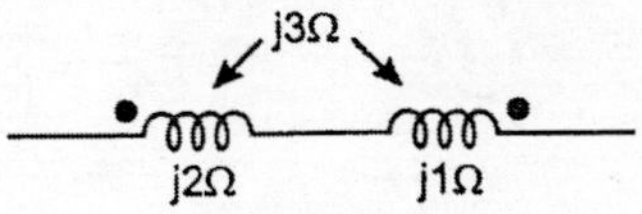

Fig. E6.1

SOLUTION. From the Figure E6.1 we get

$$Z_{eq} = j2 + j1 - j3 = j0\ \Omega$$

EXAMPLE 6.2. *For the circuit shown in Figure E6.2, determine the equivalent reactance.*

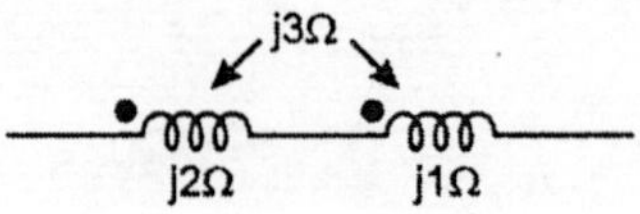

Fig. E6.2

SOLUTION. From Figure E6.2 we get

$$Z_{eq} = j2 + j1 + j3 = j6\ \Omega$$

EXAMPLE 6.3. *For the circuit shown in Figure E6.3, determine the equivalent reactance.*

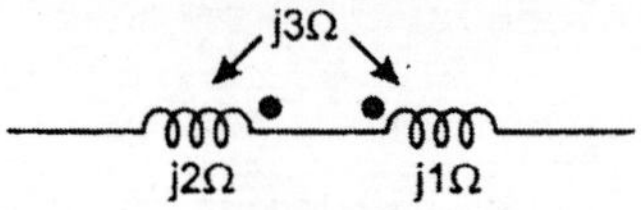

Fig. E6.3

SOLUTION. From Figure E6.3 we get

$$Z_{eq} = j2 + j1 - j3 = j0\ \Omega$$

EXAMPLE 6.4. *For the circuit shown in Figure E6.4, determine the equivalent reactance.*

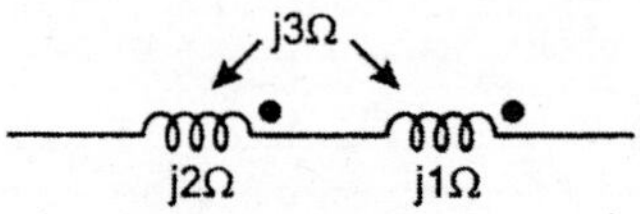

Fig. E6.4

SOLUTION. From Figure E6.4 we get

$$Z_{eq} = j2 + j1 + j3 = j6\ \Omega$$

EXAMPLE 6.5. *For the circuit shown in Figure E6.5, determine the equivalent reactance.*

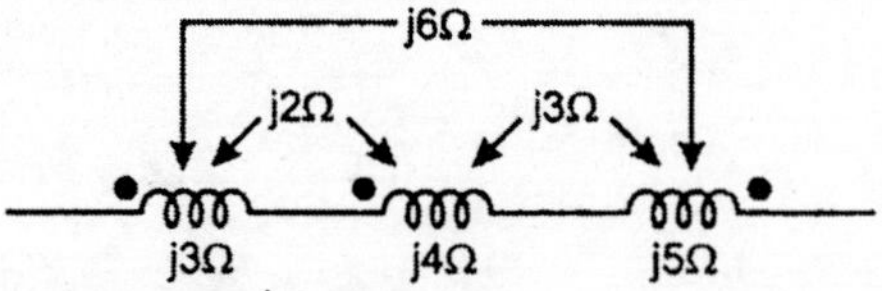

Fig. E6.5

SOLUTION. From Figure E6.5 we get

$$Z_{eq} = j3 + j4 + j5 + j2 - j3 - j6 = j5\ \Omega$$

EXAMPLE 6.6. *For the circuit shown in Figure E6.6, determine the equivalent reactance.*

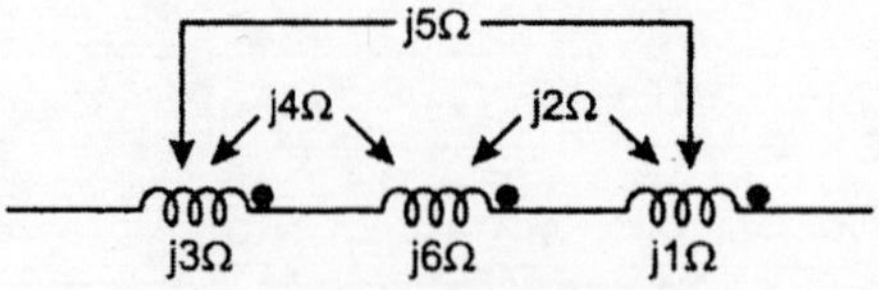

Fig. E6.6

SOLUTION. From Figure E6.6 we get

$$Z_{eq} = j3 + j6 + j1 + j4 + j2 + j5 = j21\ \Omega$$

EXAMPLE 6.7. *For the circuit shown in Figure E6.7, determine the equivalent reactance.*

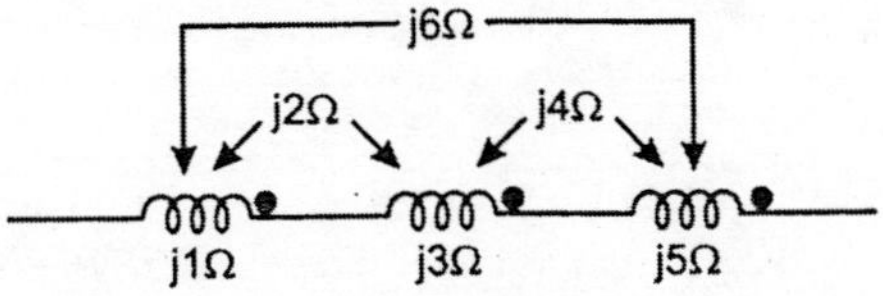

Fig. E6.7

SOLUTION. From Figure E6.7 we get

$$Z_{eq} = j1 + j3 + j5 + j2 + j4 + j6 = j21\ \Omega$$

EXAMPLE 6.8. *For the circuit shown in Figure E6.8, determine the equivalent reactance.*

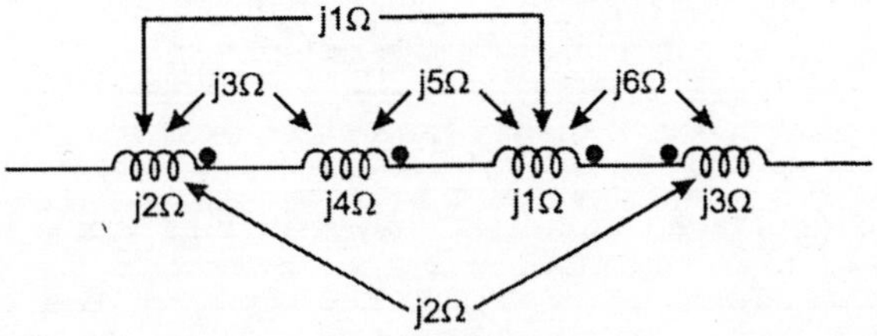

Fig. E6.8

SOLUTION. From Figure E6.8 we get

$$Z_{eq} = j2 + j4 + j1 + j3 + j3 - j5 - j6 + j1 + j2 = j5\ \Omega$$

EXAMPLE 6.9. *For the circuit shown in Figure E6.9, determine the equivalent reactance.*

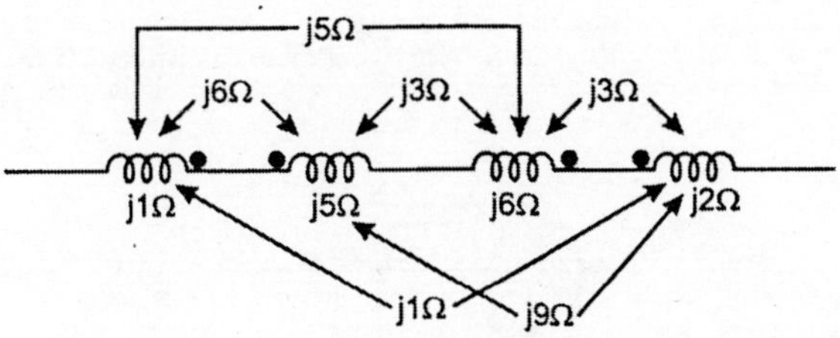

Fig. E6.9

SOLUTION. From Figure E6.9 we get

$$Z_{eq} = j1 + j5 + j6 + j2 - j6 - j3 - j3 + j9 + j5 - j1 = j15\ \Omega$$

EXAMPLE 6.10. *For the circuit shown in Figure E6.10, draw the dotted equivalent circuit and then find the equivalent inductance.*

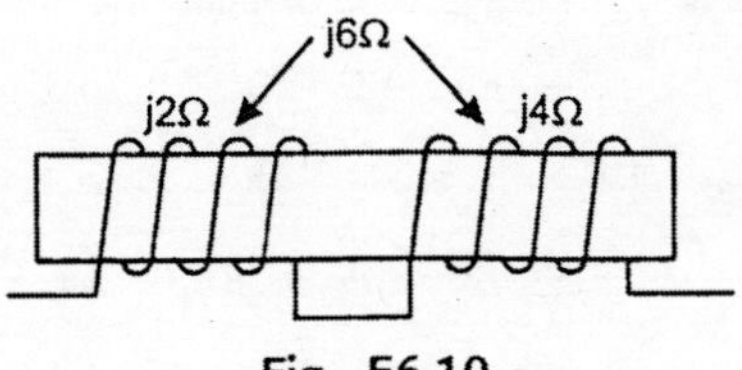

Fig. E6.10

SOLUTION. The dotted equivalent circuit is shown in Figure E6.10(*a*).

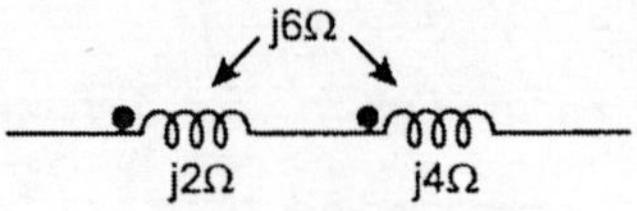

Fig. E6.10(a)

From Figure E6.10(*a*) we get

$$Z_{eq} = j2 + j4 + j6 = j12\ \Omega$$

EXAMPLE 6.11. *For the circuit shown in Figure E6.11, draw the dotted equivalent circuit and then find the equivalent inductance.*

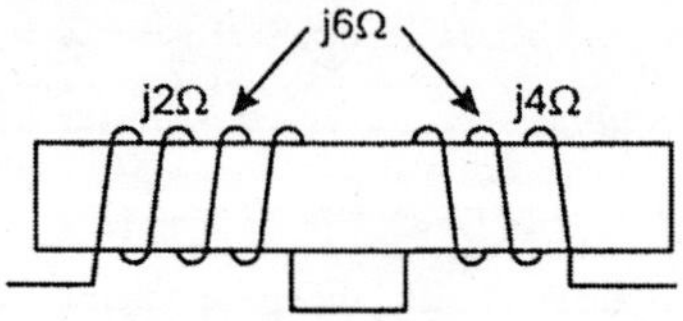

Fig. E6.11

SOLUTION. The dotted equivalent circuit is shown in Figure E6.11(*a*).

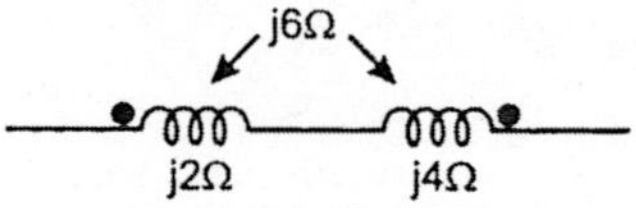

Fig. E6.11(a)

From Figure E6.11(*a*) we get

$$Z_{eq} = j2 + j4 - j6 = j0\ \Omega$$

EXAMPLE 6.12. *For the circuit shown in Figure E6.12, draw the dotted equivalent circuit and then find the equivalent inductance.*

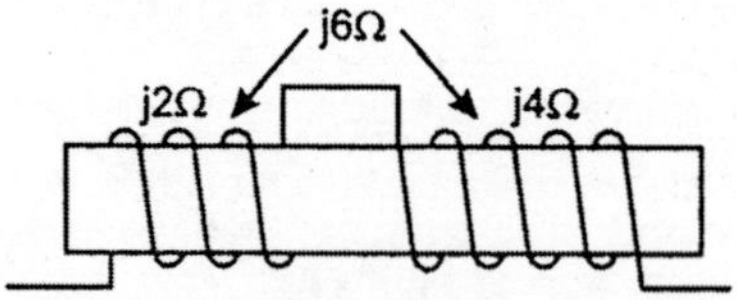

Fig. E6.12

SOLUTION. The dotted equivalent circuit is shown in Figure E6.12(*a*).

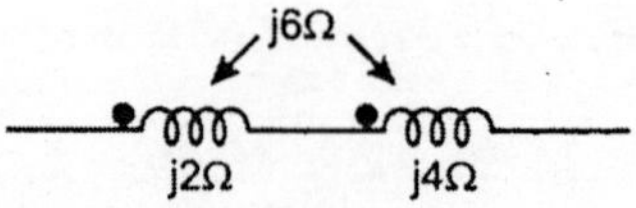

Fig. E6.12(a)

From Figure E6.12(*a*) we get

$$Z_{eq} = j2 + j4 + j6 = j12\ \Omega$$

EXAMPLE 6.13. *For the circuit shown in Figure E6.13, draw the dotted equivalent circuit and then find the equivalent inductance.*

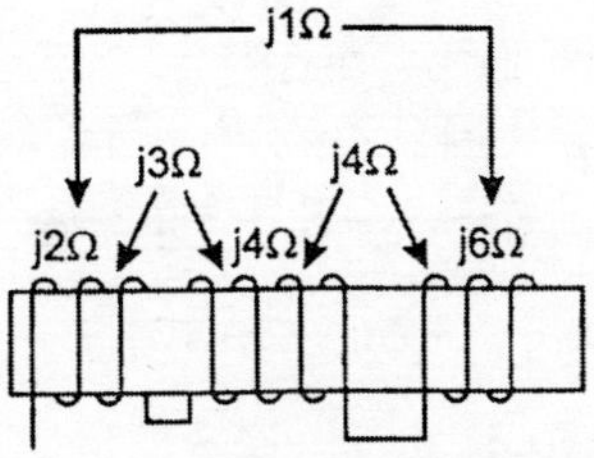

Fig. E6.13

SOLUTION. The dotted equivalent circuit is shown in Figure E6.13(*a*).

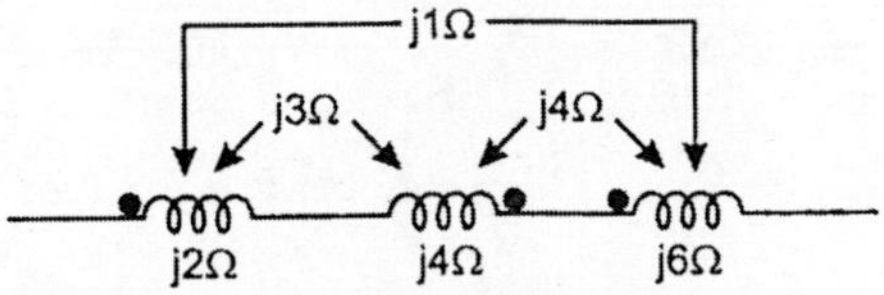

Fig. E6.13(a)

From Figure E6.13(*a*) we get

$$Z_{eq} = j2 + j4 + j6 - j3 - j4 = j4\ \Omega$$

EXAMPLE 6.14. *For the circuit shown in Figure E6.14, draw the dotted equivalent circuit and then find the equivalent inductance.*

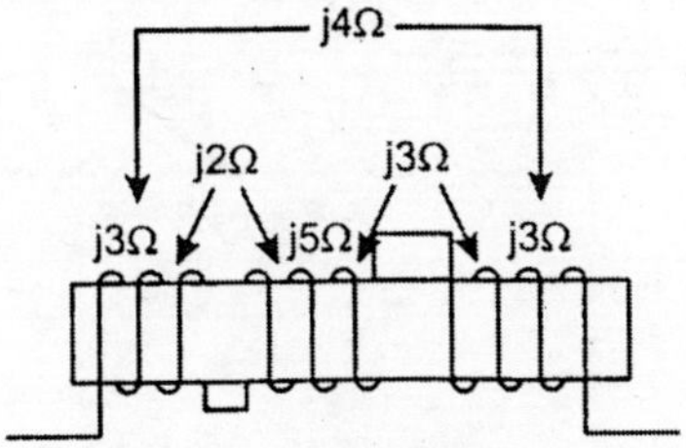

Fig. E6.14

SOLUTION. The dotted equivalent circuit is shown in Figure E6.14(*a*).

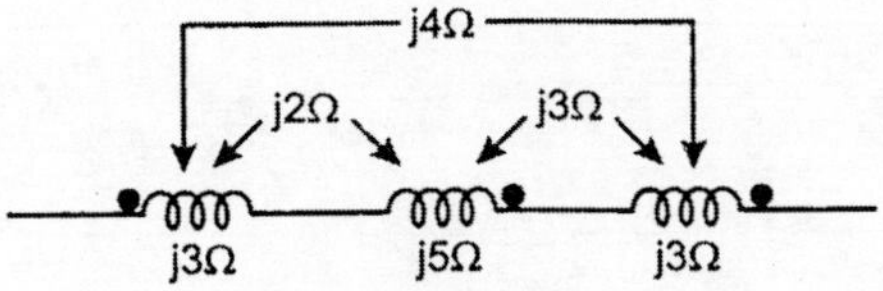

Fig. E6.14(a)

From Figure E6.14(*a*) we get

$$Z_{eq} = j3 + j5 + j3 - j2 + j3 + j4 = j6\ \Omega$$

EXAMPLE 6.15. *For the circuit shown in Figure E6.15, draw the dotted equivalent circuit and then find the equivalent inductance.*

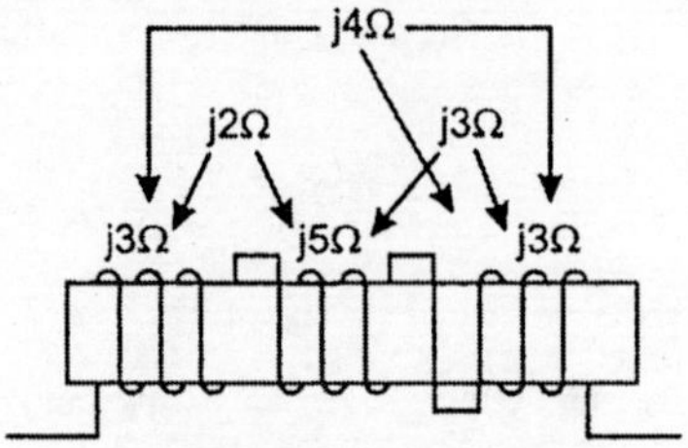

Fig. E6.15

SOLUTION. The dotted equivalent circuit is shown in Figure E6.15(*a*).

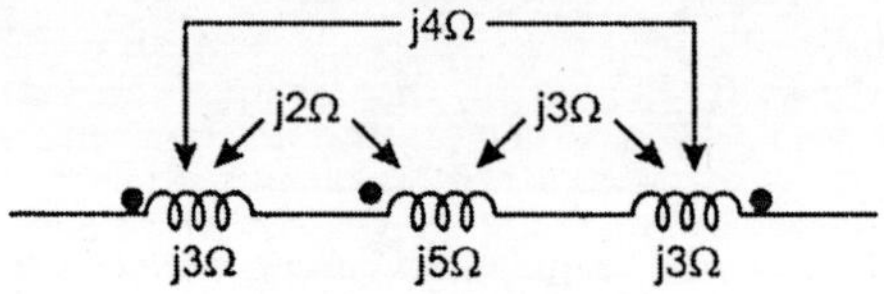

Fig. E6.15(a)

From Figure E6.15(*a*) we get

$$Z_{eq} = j3 + j5 + j3 + j2 - j3 - j4 = j6\ \Omega$$

EXAMPLE 6.16. *For the circuit shown in Figure E6.16, draw the dotted equivalent circuit and then find the equivalent inductance.*

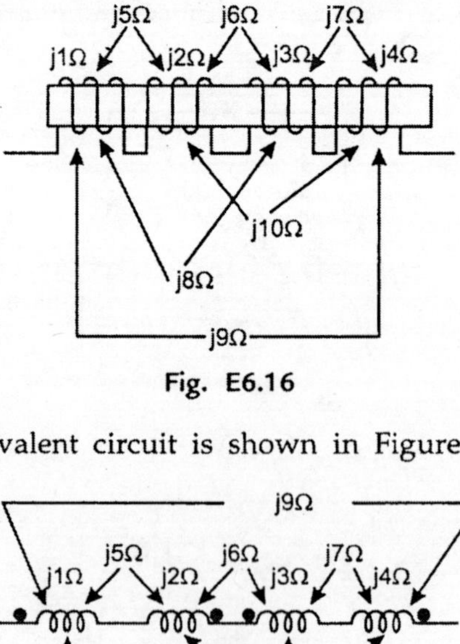

Fig. E6.16

SOLUTION. The dotted equivalent circuit is shown in Figure E6.16(*a*).

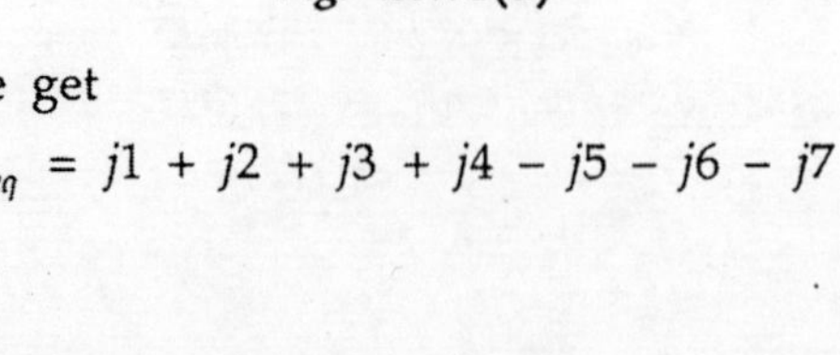

Fig. E6.16(a)

From Figure E6.16(*a*) we get

$$Z_{eq} = j1 + j2 + j3 + j4 - j5 - j6 - j7 + j8 - j9 + j10 = j1\ \Omega$$

EXAMPLE 6.17. *For the circuit shown in Figure E6.17, draw the dotted equivalent circuit and then find the equivalent inductance.*

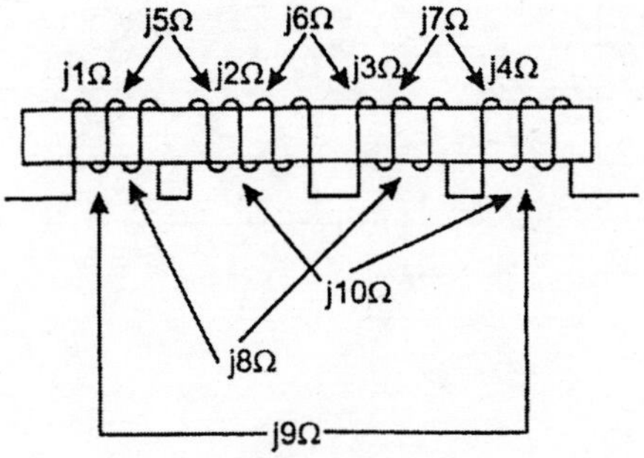

Fig. E6.17

SOLUTION. The dotted equivalent circuit is shown in Figure E6.17(*a*).

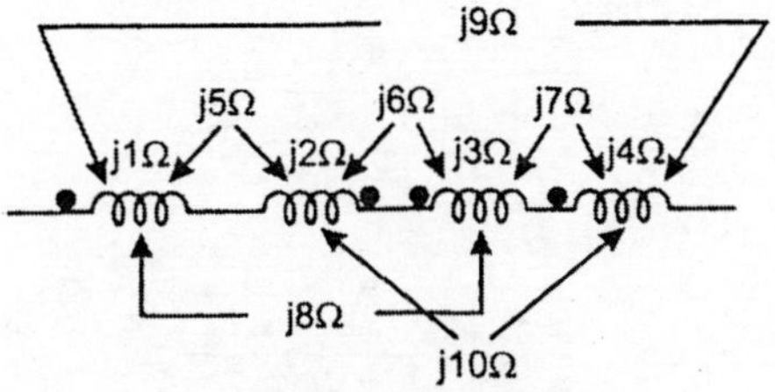

Fig. E6.17(a)

From Figure E6.17(*a*) we get

$$Z_{eq} = j1 + j2 + j3 + j4 - j5 - j6 + j7 + j8 + j9 + j10 = j33\ \Omega.$$

Parallel Circuits

EXAMPLE 6.18. *For the circuit shown in Figure E6.18, find the input impedance and also find the loop currents.*

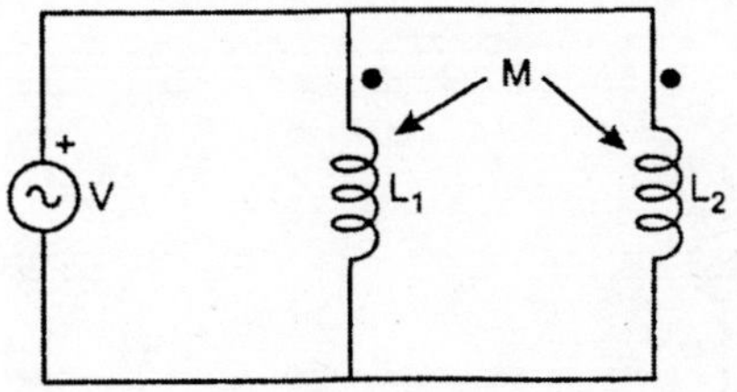

Fig. E6.18

SOLUTION. The loop currents are indicated in Figure E6.18(*a*). Writing the KVL equations we get

Loop 1 : $$j\omega L_1(i_1 - i_2) + j\omega M i_2 = V$$

or $$j\omega L_1 i_1 + j\omega i_2(M - L_1) = V \quad ...(i)$$

Loop 2 : $$j\omega L_2 i_2 + j\omega L_1(i_2 - i_1) + j\omega M i_1 - j\omega M i_2 - j\omega M i_2 = 0$$

or $$j\omega i_2(L_2 + L_1 - 2M) + j\omega i_1(M - L_1) = 0 \quad ...(ii)$$

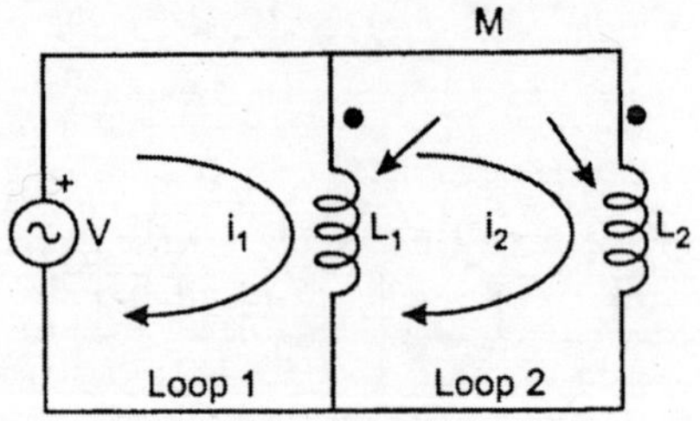

Fig. E6.18(a)

From equation (*ii*) we get

$$i_2 = \frac{L_1 - M}{L_1 + L_2 - 2M} i_1 \quad \text{...(iii)}$$

Substituting equation (*iii*) in equation (*i*) we get

$$j\omega L_1 i_1 + j\omega \frac{(M - L_1)\times(L_1 - M)}{L_1 + L_2 - 2M} i_1 = V$$

or

$$j\omega i_1 \frac{L_1 L_2 - M^2}{L_1 + L_2 - 2M} = V$$

$$\text{the input impedance} = \frac{V}{i_1} = \frac{L_1 L_2 - M^2}{L_1 + L_2 - 2M} j\omega \quad \text{...(iv)}$$

From equation (*iv*) we get $\quad i_1 = \dfrac{V(L_1 + L_2 - 2M)}{L_1 L_2 - M^2} \times \dfrac{1}{j\omega}$ ...(*v*)

Substituting equation (*v*) in equation (*iii*) we get

$$i_2 = \frac{V(L_1 - M)}{(L_1 L_2 - M^2)} \times \frac{1}{j\omega} \quad \text{...(vi)}$$

EXAMPLE 6.19. *For the circuit shown in Figure E6.19, find the input impedance and also find the loop currents.*

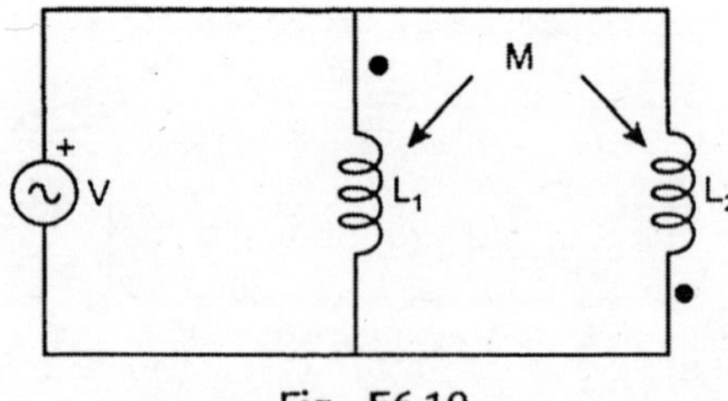

Fig. E6.19

SOLUTION. The loop currents are indicated in Figure E6.19(*a*). Writing the KVL equations we get

Loop 1 : $\quad j\omega L_1(i_1 - i_2) - j\omega M i_2 = V$

or $\quad j\omega L_1 i_1 - j\omega i_2(M + L_1) = V$ ...(*i*)

Loop 2 : $\quad j\omega L_2 i_2 + j\omega L_1(i_2 - i_1) - j\omega M i_1 + j\omega M i_2 + j\omega M i_2 = 0$

or $$j\omega i_2(L_2 + L_1 + 2M) + j\omega i_1(M + L_1) = 0 \quad ...(ii)$$

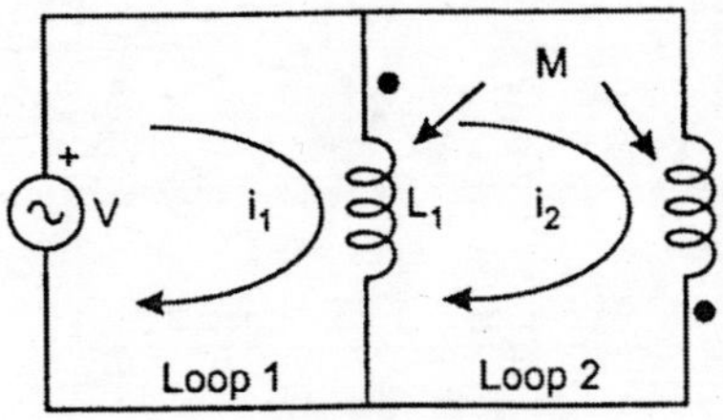

Fig. E6.19(a)

From equation (*ii*) we get

$$i_2 = \frac{L_1 + M}{L_1 + L_2 + 2M} i_1 \quad ...(iii)$$

Substituting equation (*iii*) in equation (*i*) we get

$$j\omega L_1 i_1 - j\omega \frac{(M + L_1) \times (L_1 + M)}{L_1 + L_2 + 2M} i_1 = V$$

or $$j\omega i_1 \frac{L_1 L_2 - M^2}{L_1 + L_2 + 2M} = V$$

$$\text{the input impedance} = \frac{V}{i_1} = \frac{L_1 L_2 - M^2}{L_1 + L_2 + 2M} j\omega \quad ...(iv)$$

From equation (*iv*) we get $$i_1 = \frac{V(L_1 + L_2 + 2M)}{L_1 L_2 - M^2} \times \frac{1}{j\omega} \quad ...(v)$$

Substituting equation (*v*) in equation (*iii*) we get

$$i_2 = \frac{V(L_1 + M)}{(L_1 L_2 - M^2)} \times \frac{1}{j\omega} \quad ...(vi)$$

EXAMPLE 6.20. *For the circuit shown in Figure E6.20, find the input impedance and also find the loop currents.*

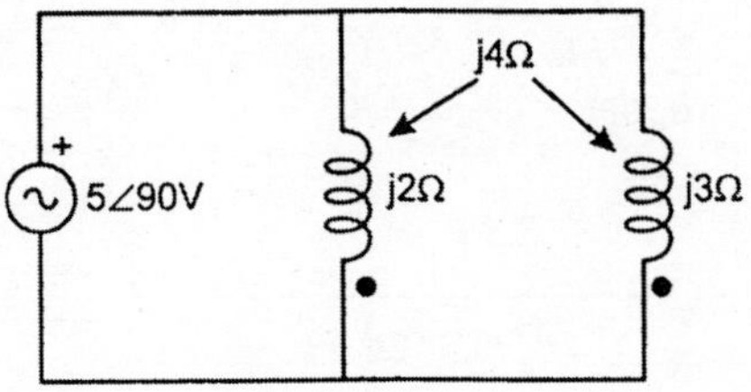

Fig. E6.20

SOLUTION. The loop currents are indicated in Figure E6.20(*a*). Writing the KVL equations we get

Loop 1 : $$j2(i_1 - i_2) + j4i_2 = j5$$

or $$j2i_1 + j2_2 = j5 \quad ...(i)$$

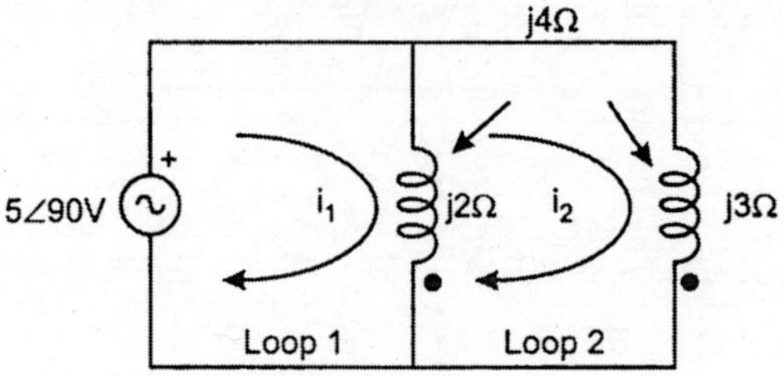

Fig. E6.20(a)

Loop 2 : $j3i_2 + j2(i_2 - i_1) + j4i_1 - j4i_2 - j4i_2 = 0$

or $j2i_1 - j3i_2 = 0$ ...(*ii*)

Writing equations (*i*) and (*ii*) in matrix form we get

$$\begin{bmatrix} j2 & j2 \\ j2 & -j3 \end{bmatrix} \times \begin{bmatrix} i_1 \\ i_2 \end{bmatrix} = \begin{bmatrix} j5 \\ 0 \end{bmatrix}$$

Let $$\Delta = \begin{vmatrix} j2 & j2 \\ j2 & -j3 \end{vmatrix} = 10$$

$$i_1 = \frac{\begin{vmatrix} j5 & j2 \\ 0 & -j3 \end{vmatrix}}{\Delta} = \frac{15}{10} = 1.5 \text{ Amp}$$

$$i_2 = \frac{\begin{vmatrix} j2 & j5 \\ j2 & 0 \end{vmatrix}}{\Delta} = \frac{10}{10} = 1 \text{ Amp}$$

The input impedance is $Z_{in} = \frac{5\angle 90}{i_1} = \frac{j5}{1.5} = j3.33\ \Omega$

EXAMPLE 6.21. *For the circuit shown in Figure E6.21, find the input impedance and also find the loop currents.*

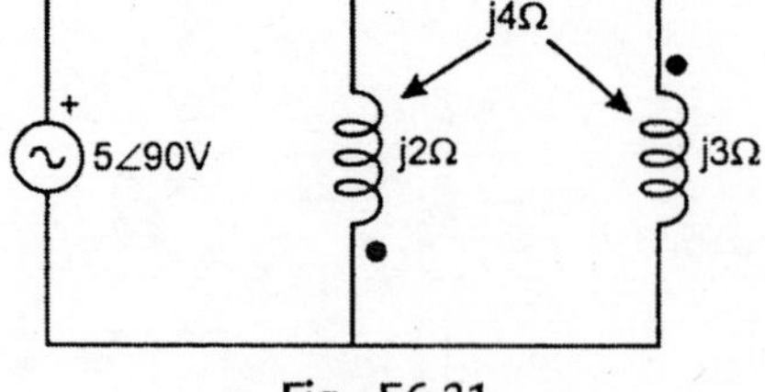

Fig. E6.21

SOLUTION. The loop currents are indicated in Figure E6.21(*a*). Writing the KVL equations we get

Loop 1 : $j2(i_1 - i_2) - j4i_2 = j5$

or $j2i_1 - j6 = j5$ ...(*i*)

Loop 2 : $j3i_2 + j2(i_2 - i_1) - j4i_1 + j4i_2 + j4i_2 = 0$

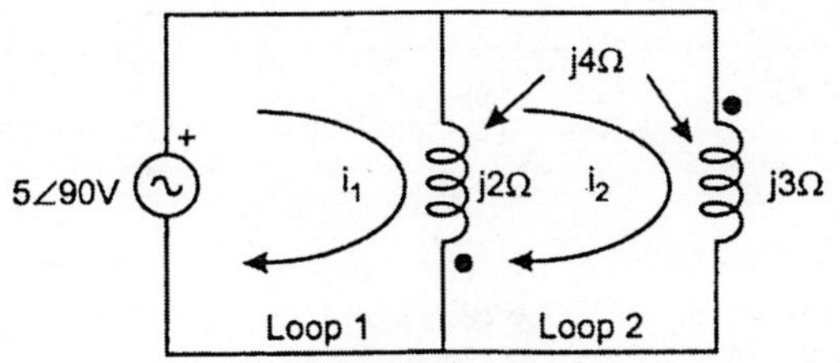

Fig. E6.21(a)

or $$-j6i_1 + j13i_2 = 0 \qquad ...(ii)$$

Writing equations (*i*) and (*ii*) in matrix form we get

$$\begin{bmatrix} j2 & -j6 \\ -j6 & j13 \end{bmatrix} \times \begin{bmatrix} i_1 \\ i_2 \end{bmatrix} = \begin{bmatrix} j5 \\ 0 \end{bmatrix}$$

Let $$\Delta = \begin{vmatrix} j2 & -j6 \\ -j6 & j13 \end{vmatrix} = 10$$

$$i_1 = \frac{\begin{vmatrix} j5 & -j6 \\ 0 & j13 \end{vmatrix}}{\Delta} = \frac{-65}{10} = -6.5 \text{ Amp}$$

$$i_2 = \frac{\begin{vmatrix} j2 & j5 \\ -j6 & 0 \end{vmatrix}}{\Delta} = \frac{-30}{10} = -3 \text{ Amp}$$

The input impedance is $$Z_{in} = \frac{5\angle 90}{i_1} = \frac{j5}{6.5} = j0.769 \text{ Amp}$$

EXAMPLE 6.22. *For the circuit shown in Figure E6.22, find the drop across the two inductor coils.*

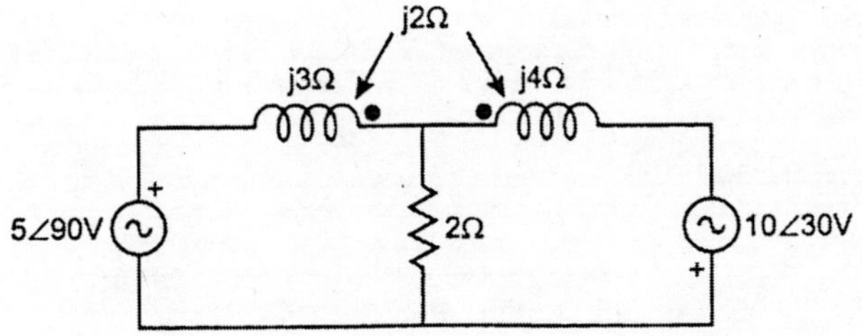

Fig. E6.22

SOLUTION. The loop currents are indicated in Figure E6.22(*a*).

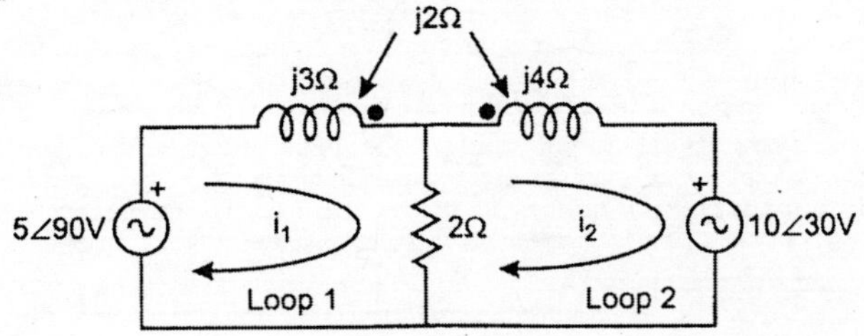

Fig. E6.22(a)

Writing the loop equations we get

Loop 1 : $(2 + j3)i_1 - j2i_2 - 2i_2 = 5\angle 90$

or $(2 + j3)i_1 - (2 + j2)i_2 = j5$ ...(*i*)

Loop 2 : $(2 + j4)i_2 - j2i_1 - 2i_1 = -(8.66 + j5)$

or $(2 + j4)i_2 - (2 + j2)i_1 = -(8.66 + j5)$ ...(*ii*)

Writing equations (*i*) and (*ii*) in matrix form we get

$$\begin{bmatrix} 2+j3 & -(2+j2) \\ -(2+j2) & 2+j4 \end{bmatrix} \times \begin{bmatrix} i_1 \\ i_2 \end{bmatrix} = \begin{bmatrix} j5 \\ -(8.66+j5) \end{bmatrix}$$

Let
$$\Delta = \begin{vmatrix} 2+j3 & -(2+j2) \\ -(2+j2) & 2+j4 \end{vmatrix} = -8 + j6$$

Then
$$i_1 = \frac{\begin{vmatrix} j5 & -(2+j2) \\ -(8.66+j5) & 2+j4 \end{vmatrix}}{\Delta} = \frac{-27.32 - j17.32}{-8+j6} = 1.146 + j3.024 \text{ Amp}$$

$$i_2 = \frac{\begin{vmatrix} 2+j3 & j5 \\ -(2+j2) & -(8.66+j5) \end{vmatrix}}{\Delta} = \frac{-12.32 - j25.98}{-8+j6} = -0.573 + j2.818 \text{ Amp}$$

Then
$$V(j3\ \Omega) = j3i_1 - j2i_2 = -3.436 + j4.584 \text{ V}$$

and
$$V(j4\ \Omega) = j4i_2 - j2i_1 = 12.026 - 2.292j \text{ V}$$

EXAMPLE 6.23. *For the circuit shown in Figure E6.23, find the ration* $\frac{V_1}{V_2}$.

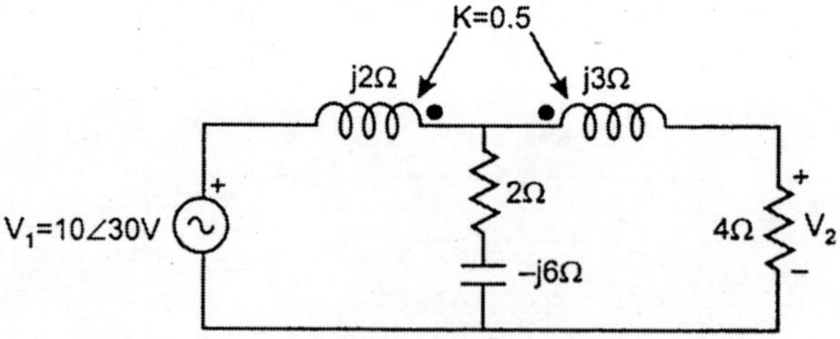

Fig. E6.23

SOLUTION. The loop currents are indicated in Figure E6.23(*a*).

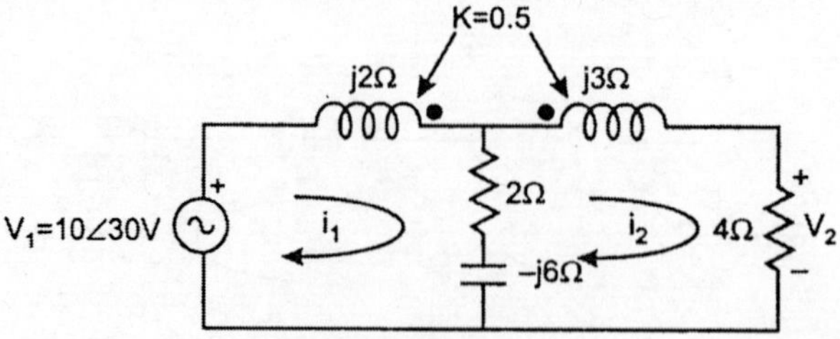

Fig. E6.23(a)

$$x_m = K\sqrt{x_{L1}x_{L2}} = 0.5\sqrt{2\times 3} = 1.22\ \Omega$$

Writing the loop equations we get

Loop 1 : $(2 + j2 - j6)i_1 - (2 - j6)i_2 + j1.22i_2 = 10\angle 30$

or $(2 + j2 - j6)i_1 - (2 - j6)i_2 + j1.22i_2 = (8.66 + j5)$

or $(2 - j4)i_1 + (-2 + j7.22)i_2 = (8.66 + j5)$ ...(*i*)

Loop 2 : $(2 + 4 + j3 - j6)i_2 - (2 - j6)i_1 + j1.22i_1 = 0$

or $(6 - j3)i_2 + (-2 + j7.22)i_1 = 0$ ...(*ii*)

Writing equations (*i*) and (*ii*) in matrix form we get

$$\begin{bmatrix} 2-j4 & -2+j7.22 \\ -2+j7.22 & 6-j3 \end{bmatrix} \times \begin{bmatrix} i_1 \\ i_2 \end{bmatrix} = \begin{bmatrix} 8.66+j5 \\ 0 \end{bmatrix}$$

Let $$\Delta = \begin{vmatrix} 2-j4 & -2+j7.22 \\ -2+j7.22 & 6-j3 \end{vmatrix} = 48.128 - j1.12$$

Then $$i_2 = \frac{\begin{vmatrix} 2-j4 & 8.66+j5 \\ -2+j7.22 & 0 \end{vmatrix}}{\Delta} = \frac{53.42-j52.525}{48.128-j1.12} = 1.135 - j1.065 \text{ Amp}$$

So $V_2 = 4 \times i_2 = 4.54 - j4.26$ V

EXAMPLE 6.24. *For the circuit shown in Figure E6.24, find the drop across all the elements.*

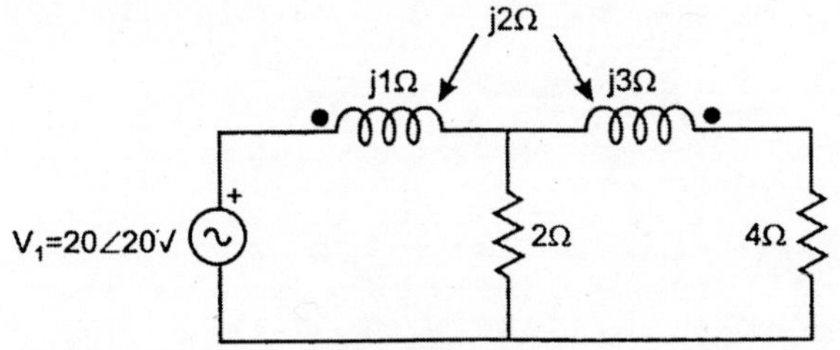

Fig. E6.24

SOLUTION. The loop currents are indicated in Figure E6.24(*a*).

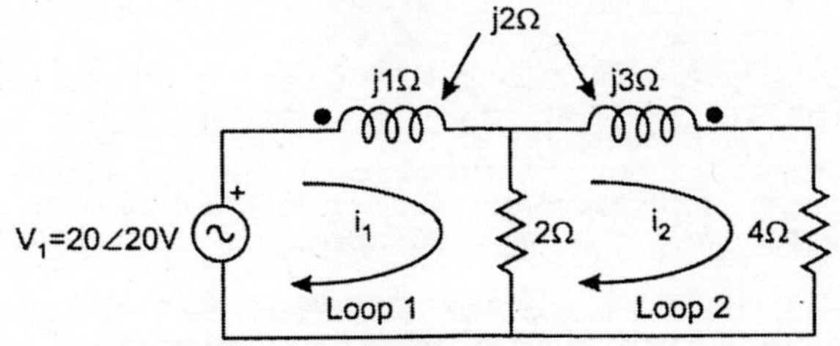

Fig. E6.24(a)

Writing the loop equations we get

Loop 1 : $(2 + j1)i_1 - 2i_2 - j2i_2 = 20\angle 20$

$(2 + j1)i_1 - (2 + j2)i_2 = 20\angle 20$ ...(*i*)

Loop 2 : $(6 + j3)i_2 - 2i_1 - j2i_1 = 0$

$$(6 + j3)i_2 - (2 + j2)i_1 = 0 \quad \text{...(ii)}$$

Writing the equations (*i*) and (*ii*) in matrix form we get

$$\begin{bmatrix} 2+j1 & -2-j2 \\ -2-j2 & 6+j3 \end{bmatrix} \times \begin{bmatrix} i_1 \\ i_2 \end{bmatrix} = \begin{bmatrix} 18.79+j6.84 \\ 0 \end{bmatrix}$$

Let $$\Delta = \begin{vmatrix} 2+j1 & -2-j2 \\ -2-j2 & 6+j3 \end{vmatrix} = 9 + j4$$

Then $$i_1 = \frac{\begin{vmatrix} 18.79+j6.84 & -2-j2 \\ 0 & 6+j3 \end{vmatrix}}{\Delta} = \frac{92.22+j97.41}{9+j4} = 12.57 + j5.24 \text{ Amp}$$

$$i_2 = \frac{\begin{vmatrix} 2+j1 & 18.79+j6.84 \\ -2-j2 & 0 \end{vmatrix}}{\Delta} = \frac{23.9+j51.26}{9+j4} = 4.33 + j3.77 \text{ Amp}$$

then the drop across $j1\ \Omega = j1i_1 - j2i_2 = 2.3 + j3.91$ V

the drop across $j3\ \Omega = j3i_2 - j2i_1 = -0.83 - j12.15$ V

the drop across $4\ \Omega = 4i_2 = 17.32 + j15.08$ V

the drop across $2\ \Omega = 2(i_1 - i_2) = 16.48 - j18.02$ V

EXAMPLE 6.25. *For the circuit shown in Figure E6.25, find the loop currents.*

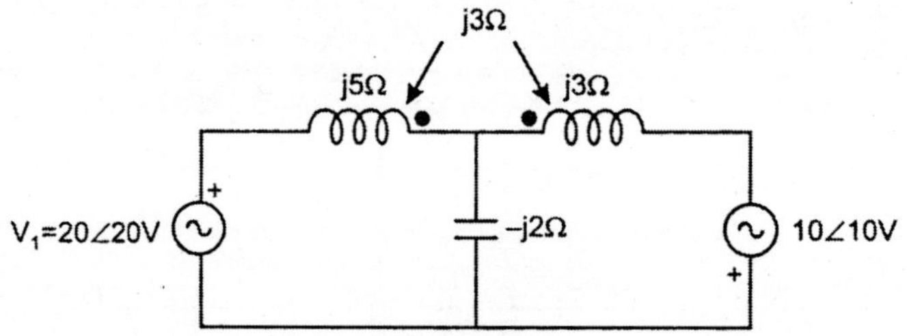

Fig. E6.25

SOLUTION. The loop currents are indicated in Figure E6.25(*a*).

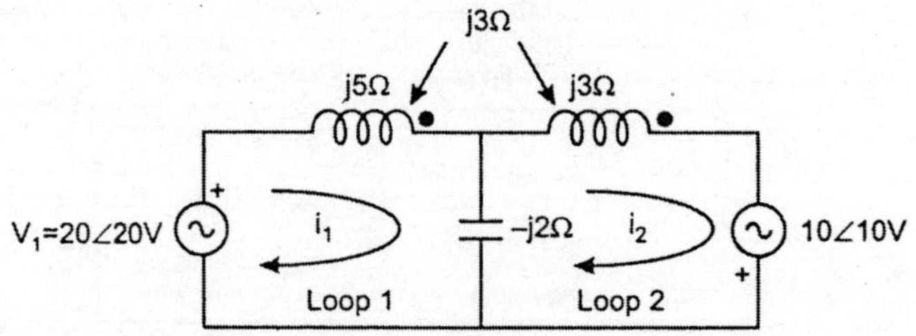

Fig. E6.25(a)

Writing the loop equations we get

Loop 1 : $(j5 - j2)i_1 + j2i_2 + j3i_2 = 20\angle 20$

$$j3i_1 + j5i_2 = 20\angle 20 \quad \text{...(i)}$$

Loop 2 : $(j3 - j2)i_2 + j2i_1 + j3i_1 = 10\angle 10$

$$j1i_2 + j5i_1 = 10\angle 10 \qquad ...(ii)$$

Writing equations (*i*) and (*ii*) in matrix form we get

$$\begin{bmatrix} j3 & j5 \\ j5 & j1 \end{bmatrix} \times \begin{bmatrix} i_1 \\ i_2 \end{bmatrix} = \begin{bmatrix} 18.79 + j6.84 \\ 9.85 + j1.74 \end{bmatrix}$$

Let
$$\Delta = \begin{vmatrix} j3 & j5 \\ j5 & j1 \end{vmatrix} = 22$$

Then
$$i_1 = \frac{\begin{vmatrix} 18.79 + j6.84 & j5 \\ 9.85 + j1.74 & j1 \end{vmatrix}}{\Delta} = \frac{1.86 - j30.46}{22} = 0.08 - j1.38 \text{ Amp}$$

$$i_2 = \frac{\begin{vmatrix} j3 & 18.79 + j6.84 \\ j5 & 9.85 + j1.74 \end{vmatrix}}{\Delta} = \frac{28.98 - j64.4}{22} = 1.32 - j2.93 \text{ Amp}$$

EXAMPLE 6.26. *For the circuit shown in Figure E6.26, find the loop currents, and the drop across j4 Ω.*

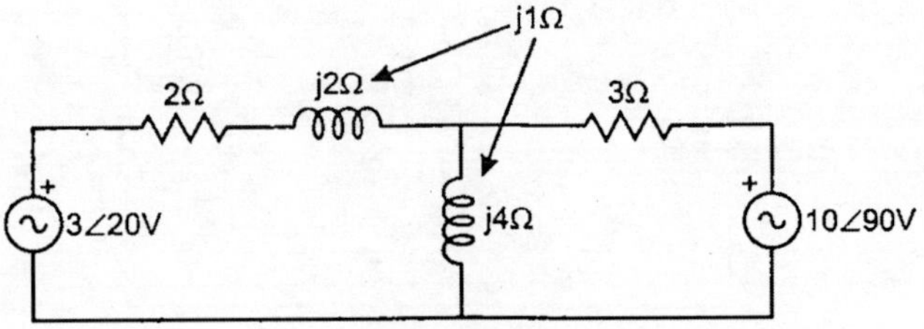

Fig. E6.26

SOLUTION. The loop currents are indicated in Figure E6.26(*a*).

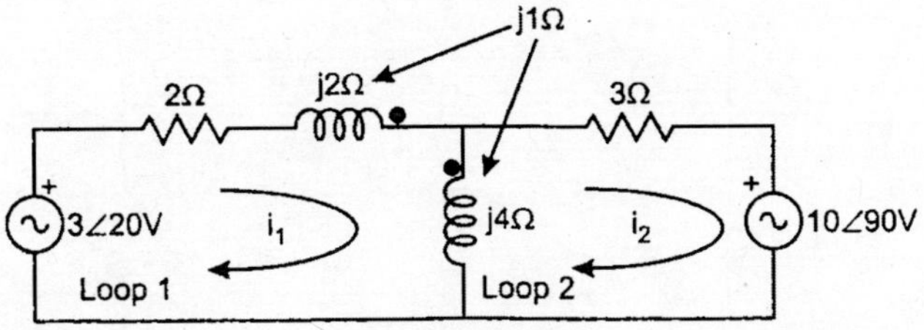

Fig. E6.26(a)

Writing the loop equations we get

Loop 1 : $(2 + j3 + j4)i_1 - j4i_2 - j1i_1 + j1i_2 - j1i_1 = 3\angle 20$

or
$$(2 + j5)i_1 - j3i_2 = (2.82 + j1.03) \qquad ...(i)$$

Loop 2 :
$$(3 + j4)i_2 - j4i_1 + j1i_1 = -j10$$

or
$$(3 + j4)i_2 - j3i_1 = -j10 \qquad ...(ii)$$

Writing equations (*i*) and (*ii*) in the matrix form we get

$$\begin{bmatrix} 2+j5 & -j3 \\ -j3 & 3+j4 \end{bmatrix} \times \begin{bmatrix} i_1 \\ i_2 \end{bmatrix} = \begin{bmatrix} 2.82+j1.03 \\ -j10 \end{bmatrix}$$

Let $$\Delta = \begin{vmatrix} 2+j5 & -j3 \\ -j3 & 3+j4 \end{vmatrix} = -5 + j23$$

Then $$i_1 = \frac{\begin{vmatrix} 2.82+j1.03 & -j3 \\ -j10 & 3+j4 \end{vmatrix}}{\Delta} = \frac{34.34+j14.37}{-5+j23} = 0.27 - j1.56 \text{ Amp}$$

$$i_2 = \frac{\begin{vmatrix} 2+j5 & 2.82+j1.03 \\ -j3 & -j10 \end{vmatrix}}{\Delta} = \frac{46.91-j11.54}{-5+j23} = -0.90 - j1.84 \text{ Amp}$$

and the drop across $j4\ \Omega = j4(i_1 - i_2) - j1i_1 = 0.44 + j4.95$ V

EXAMPLE 6.27. *For the circuit shown in Figure E6.27, find the loop currents.*

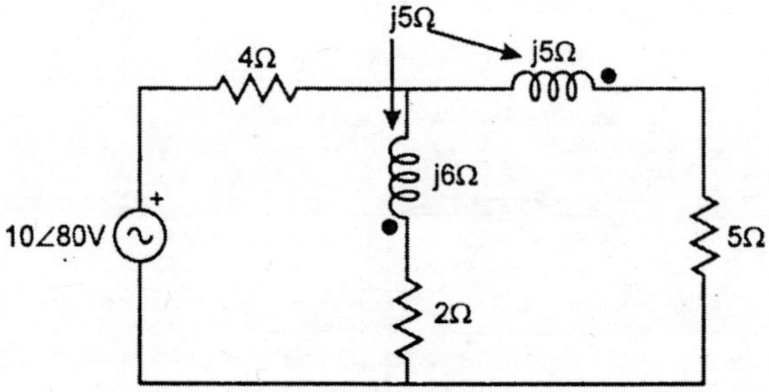

Fig. E6.27

SOLUTION. The loop currents are indicated in Figure E6.27(*a*).

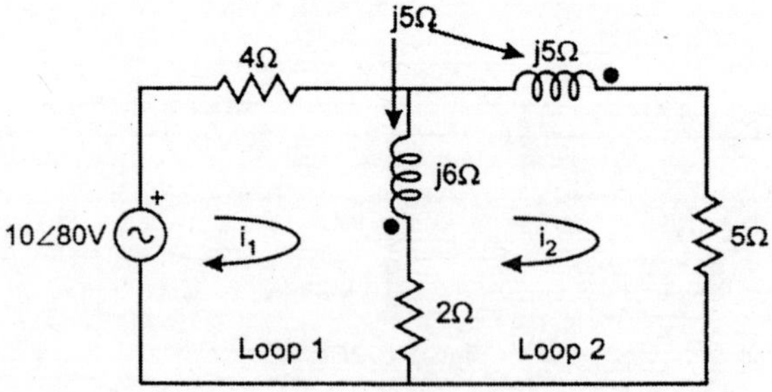

Fig. E6.27(a)

Writing the loop equations we get

Loop 1 : $(6 + j6)i_1 - (2 + j6)i_2 + j5i_2 = 10\angle 80$

or $(6 + j6)i_1 - (2 + j)i_2 = 10\angle 80$ ...(*i*)

Loop 2 : $(7 + j11)i_2 - (2 + j6)i_1 + j5i_1 - j5i_2 - j5i_2 = 0$

or $(7 + j1)i_2 - (2 + j1)i_1 = 0$ ...(*ii*)

Writing equations (*i*) and (*ii*) in matrix form we get

$$\begin{bmatrix} 6+j6 & -2-j1 \\ -2-j1 & 7+j11 \end{bmatrix} \times \begin{bmatrix} i_1 \\ i_2 \end{bmatrix} = \begin{bmatrix} 10\angle 80 \\ 0 \end{bmatrix}$$

Let
$$\Delta = \begin{vmatrix} 6+j6 & -2-j1 \\ -2-j1 & 7+j11 \end{vmatrix} = 27 + j104$$

Then
$$i_1 = \frac{\begin{vmatrix} 10\angle 80 & -2-j1 \\ 0 & 7+j11 \end{vmatrix}}{\Delta} = \frac{-6.96+j31.32}{27+j104} = 0.27 + j0.14 \text{ Amp}$$

and
$$i_2 = \frac{\begin{vmatrix} 6+j6 & 10\angle 80 \\ -2-j1 & 0 \end{vmatrix}}{\Delta} = \frac{1.74+j5.22}{27+j104} = 0.05 - j3.47 \times 10^{-3} \text{ Amp}$$

EXAMPLE 6.28. *For the circuit shown in Figure E6.28, find the loop currents.*

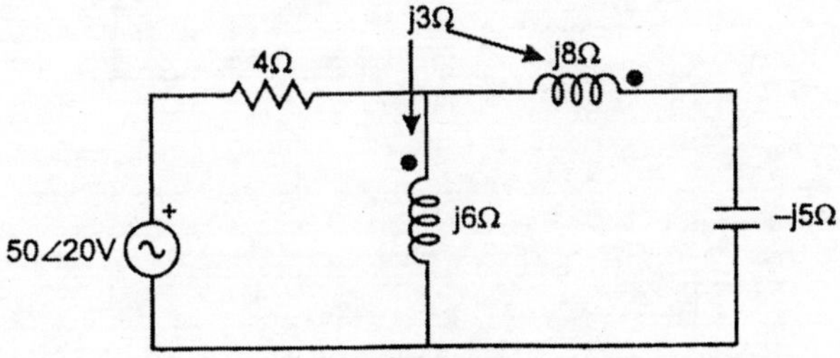

Fig. E6.28

SOLUTION. The loop currents are indicated in Figure E6.28(*a*).

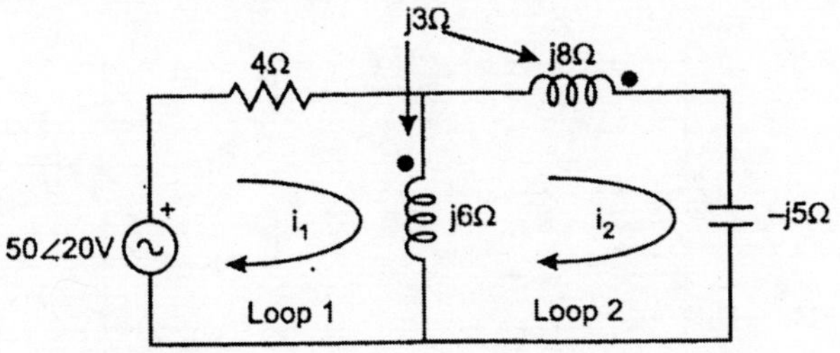

Fig. E6.28(a)

Writing the loop equations we get

Loop 1 : $(4 + j6)i_1 - j6i_2 - j3i_2 = 50\angle 20$

or $(4 + j6)i_1 - j9i_2 = (46.98 + j17.10)$ ...(*i*)

Loop 2 : $(j8 + j6 - j5)i_2 - j6i_1 - j3i_1 + j3i_2 + j3i_2 = 0$

or $j15i_2 - j9i_1 = 0$ ...(*ii*)

Writing equations (*i*) and (*ii*) in matrix form we get

$$\begin{bmatrix} 4+j6 & -j9 \\ -j9 & j15 \end{bmatrix} \times \begin{bmatrix} i_1 \\ i_2 \end{bmatrix} = \begin{bmatrix} 46.98+j17.10 \\ 0 \end{bmatrix}$$

Let $$\Delta = \begin{vmatrix} 4+j6 & -j9 \\ -j9 & j15 \end{vmatrix} = -9 + j60$$

Then $$i_1 = \frac{\begin{vmatrix} 46.98+j17.1 & -j9 \\ 0 & j15 \end{vmatrix}}{\Delta} = \frac{-256.5+j704.7}{-9+j60} = 12.11 + j2.45 \text{ Amp}$$

$$i_2 = \frac{\begin{vmatrix} 4+j6 & 46.98+j17.1 \\ -j9 & 0 \end{vmatrix}}{\Delta} = \frac{-153.9+j422.82}{-9+j60} = 7.27 + j1.47 \text{ Amp}$$

EXAMPLE 6.29. *For the circuit shown in Figure E6.29, find the loop currents, and the drop across j5 Ω.*

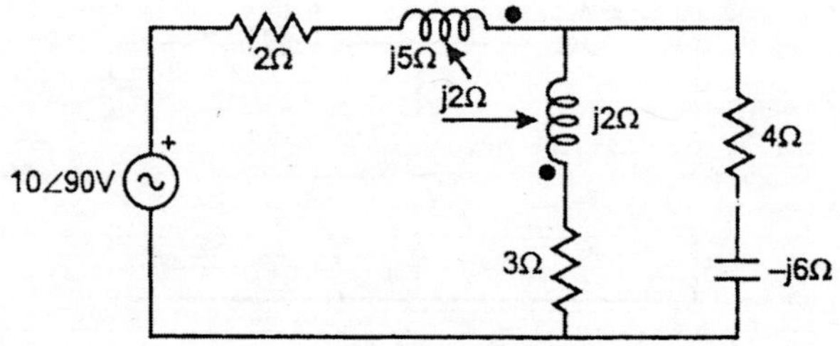

Fig. E6.29

SOLUTION. The loop currents are indicated in Figure E6.29(*a*).

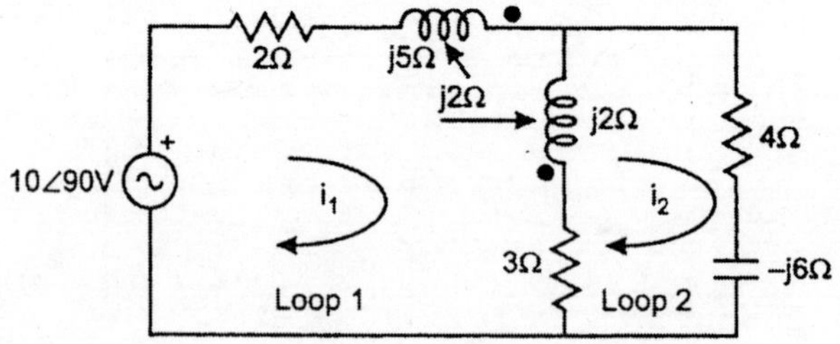

Fig. E6.29(a)

Writing the KVL equation for each loop we get

Loop 1 : $(5 + j7)i_1 - (3 + j2)i_2 - j2i_2 + j2i_1 + j2i_1 = 100j$

or $$(5 + j11)i_1 - (3 + j4)i_2 = j100 \quad ...(i)$$

Loop 2 : $(7 + j2 - j6)i_2 - (3 + j2)i_1 - j2i_1 = 0$

or $$(7 - j4)i_2 - (3 + j4)i_1 = 0 \quad ...(ii)$$

Writing equations (*i*) and (*ii*) in matrix form we get

$$\begin{bmatrix} (5+j11) & -(3+j4) \\ -(3+j4) & (7-j4) \end{bmatrix} \times \begin{bmatrix} i_1 \\ i_2 \end{bmatrix} = \begin{bmatrix} j100 \\ 0 \end{bmatrix}$$

Let
$$\Delta = \begin{vmatrix} (5+j11) & -(3+j4) \\ -(3+j4) & (7-j4) \end{vmatrix} = 86 + j33$$

Then
$$i_1 = \frac{\begin{vmatrix} j100 & -(3+j4) \\ 0 & (7-j4) \end{vmatrix}}{\Delta} = \frac{400+j700}{86+j33} = 6.78 + j5.54 \text{ Amp}$$

$$i_2 = \frac{\begin{vmatrix} (5+j11) & j100 \\ -(3+j4) & 0 \end{vmatrix}}{\Delta} = \frac{-400+j300}{86+j33} = -2.88 + j4.59 \text{ Amp}$$

then the drop across $j5\ \Omega = j5 \times i_1 - j2 \times i_2 + j2 \times i_1 = -29.6 + j53.22$ V

EXAMPLE 6.30. *For the circuit shown in Figure E6.30, find the drop across $-j6\ \Omega$.*

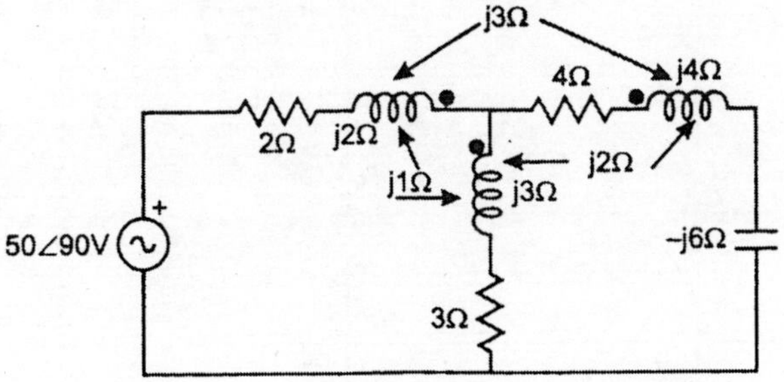

Fig. E6.30

SOLUTION. The loop currents are shown in Figure E6.30(*a*).

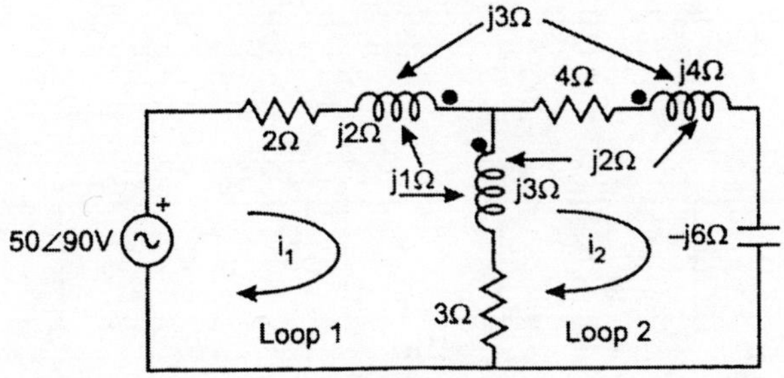

Fig. E6.30(a)

Writing the loop equations we get

Loop 1 : $(2 + 3 + j2 + j3)i_1 - (3 + j3)i_2 - j3i_2 - j1i_1 + j1i_1 = 50\angle 90$

or $$(5 + j4)i_1 - (3 + j5)i_2 = j50 \quad ...(i)$$

Loop 2 : $(4 + 3 + j4 + j3 - j6)i_2 - (3 + j3)i_1 - j3i_1 - j2i_2 + j2i_1 = 0$

or $$(7 - j)i_2 - (3 + j4)i_1 = 0 \quad ...(ii)$$

Writing equations (*i*) and (*ii*) in matrix form we get

$$\begin{bmatrix} (5+j4) & -(3+j5) \\ -(3+j4) & (7-j) \end{bmatrix} \times \begin{bmatrix} i_1 \\ i_2 \end{bmatrix} = \begin{bmatrix} j50 \\ 0 \end{bmatrix}$$

Let
$$\Delta = \begin{vmatrix} (5+j4) & -(3+j5) \\ -(3+j4) & (7-j) \end{vmatrix} = 50 - j4$$

Then
$$i_2 = \frac{\begin{vmatrix} (5+j4) & j50 \\ -(3+j4) & 0 \end{vmatrix}}{\Delta} = \frac{-200+j150}{50-j4} = -4.21 + j2.66 \text{ Amp}$$

The drop across $-j6\ \Omega = i_2 \times (-j6) = 15.96 + j25.26$ V

EXAMPLE 6.31. *For the circuit shown in Figure E6.31, find the loop currents.*

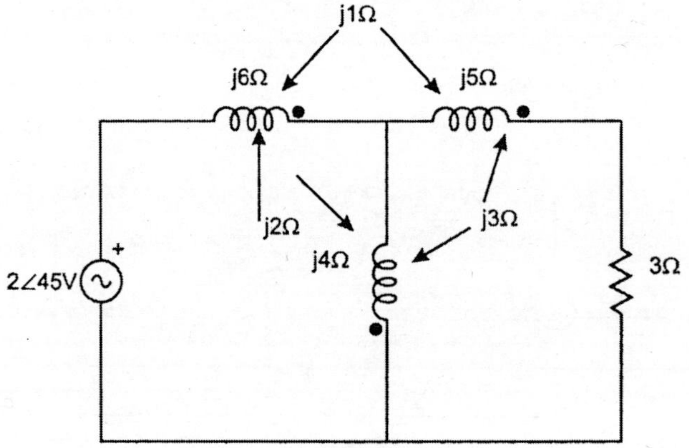

Fig. E6.31

SOLUTION. The loop currents are indicated in Figure E6.31(*a*).

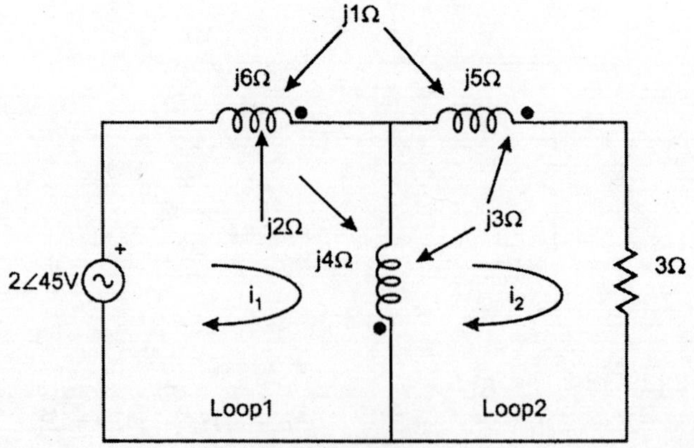

Fig. E6.31(a)

Writing the KVL equation for each loop we get

Loop 1 : $j10i_1 - j4i_2 + j1i_2 + j2i_1 - j2i_2 = 2\angle 45$

$$j12i_1 - j5i_2 = 2\angle 45$$

Loop 2 : $(3 + j9)i_2 - j4i_1 + j1i_1 + j3i_1 - j3i_2 = 0$

$$(3 + j6)i_2 = 0$$

$\Rightarrow$ $i_2 = 0$

therefore $i_1 = \dfrac{2\angle 45}{12\angle 90} = \dfrac{1.41 + j1.41}{j12} = 0.12 - j0.12$ Amp

Transformer Circuits

EXAMPLE 6.32. *For the circuit shown in Figure E6.32, prove the following.*

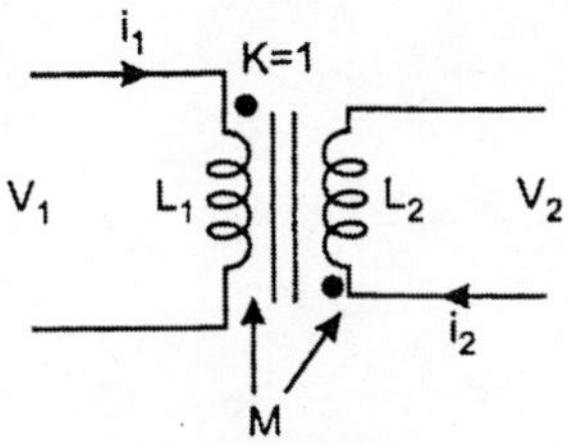

Fig. E6.32

(*i*) $\dfrac{i_1}{i_2} = \sqrt{\dfrac{L_2}{L_1}}$ (*ii*) $\dfrac{N_1}{N_2} = \sqrt{\dfrac{L_1}{L_2}}$

SOLUTION. From Figure E6.32(*a*), writing the loop equations we get

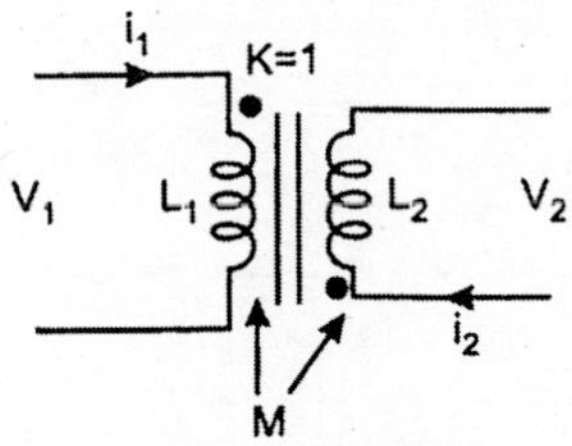

Fig. E6.32(a)

Loop 1 : $j\omega L_1 i_1 + j\omega M i_2 = V_1$ ...(*i*)

Loop 2 : $j\omega L_2 i_2 + j\omega M i_1 = V_2$ ...(*ii*)

Writing equations (*i*) and (*ii*) in matrix form we get

$$\begin{bmatrix} j\omega L_1 & j\omega M \\ j\omega M & j\omega L_2 \end{bmatrix} \times \begin{bmatrix} i_1 \\ i_2 \end{bmatrix} = \begin{bmatrix} V_1 \\ V_2 \end{bmatrix}$$

Let $\Delta = \begin{vmatrix} j\omega L_1 & j\omega M \\ j\omega M & j\omega L_2 \end{vmatrix} = -\omega^2 L_1 L_2 + \omega^2 M^2$

Then $i_1 = \dfrac{\begin{vmatrix} V_1 & j\omega M \\ V_2 & j\omega L_2 \end{vmatrix}}{\Delta} = \dfrac{j\omega V_1 L_2 - j\omega M V_2}{\Delta}$ ...(*iii*)

and
$$i_2 = \frac{\begin{vmatrix} j\omega L_1 & V_2 \\ j\omega M & V_2 \end{vmatrix}}{\Delta} = \frac{j\omega V_2 L_1 - j\omega M V_1}{\Delta} \quad ...(iv)$$

$$\frac{i_1}{i_2} = \frac{j\omega V_1 L_2 - j\omega M V_2}{j\omega V_2 L_1 - j\omega M V_1} = \frac{V_1 L_2 - M V_2}{V_2 L_1 - M V_1}$$

$$= \frac{V_1\sqrt{L_2}\sqrt{L_2} - \sqrt{L_1}\sqrt{L_2}V_2}{V_2\sqrt{L_1}\sqrt{L_1} - \sqrt{L_1}\sqrt{L_2}V_1} \quad [\because M = \sqrt{L_1 L_2}\,]$$

$$= \frac{\sqrt{L_2}(V_1\sqrt{L_2} - \sqrt{L_1}V_2)}{\sqrt{L_1}(\sqrt{L_1}V_2 - \sqrt{L_2}V_1} = \mp\sqrt{\frac{L_2}{L_1}}$$

$$\frac{i_1}{i_2} = \sqrt{\frac{L_2}{L_1}}$$

Again we know that

$$L_1 = \frac{N_1\phi}{i_1} \quad ...(v)$$

$$L_2 = \frac{N_2\phi}{i_2} \quad ...(vi)$$

Dividing equation (*v*) by equation (*vi*) we get

$$\frac{L_1}{L_2} = \frac{N_1\phi}{i_1}\times\frac{i_2}{N_2\phi} = \frac{i_2}{i_1}\times\frac{N_1}{N_2}$$

$$\frac{L_1}{L_2} = \sqrt{\frac{L_1}{L_2}}\times\frac{N_1}{N_2} \Rightarrow \frac{L_1^2 L_2}{L_2^2 L_1} = \frac{N_1^2}{N_2^2} \Rightarrow \sqrt{\frac{L_1}{L_2}} = \frac{N_1}{N_2}$$

EXAMPLE 6.33. *For the circuit shown in Figure E6.33, find V_0*

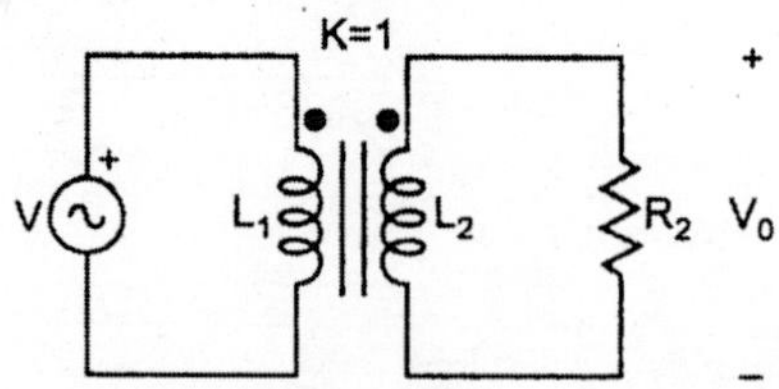

Fig. E6.33

SOLUTION. The loop currents are indicated in Figure E6.33(*a*).

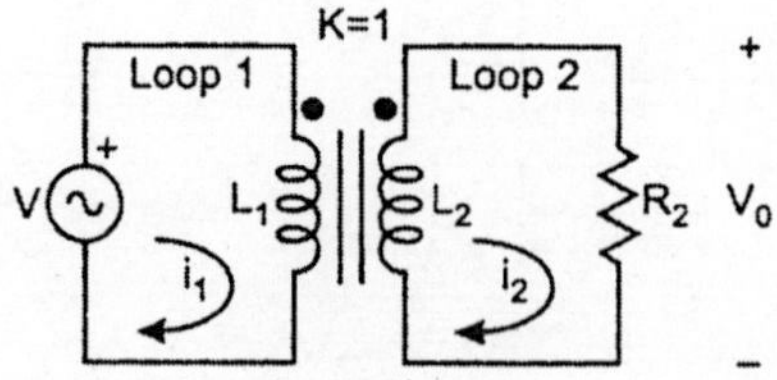

Fig. E6.33(a)

Writing the loop equations we get

Loop 1 : $j\omega L_1 i_1 - j\omega M i_2 = V$ ...(*i*)

Loop 2 : $i_2 R_2 + j\omega L_2 i_2 - j\omega M i_1 = 0$ ...(*ii*)

$$i_2(R_2 + j\omega L_2) - j\omega M i_1 = 0 \quad ...(iii)$$

From equation (*iii*) we get

$$i_1 = \frac{R_2 + j\omega L_2}{j\omega M} i_2 \quad ...(iv)$$

Replacing equation (*iv*) in equation (*i*) we get

$$j\omega L_1\left(\frac{R_2 + j\omega L_2}{j\omega M}\right)i_2 - j\omega M i_2 = V$$

$$i_2\left(\frac{R_2 L_1 + j\omega L_1 L_2}{M} - j\omega M\right) = V$$

$$i_2\left(\frac{R_2 L_1 + j\omega L_1 L_2 - j\omega M^2}{M}\right) = V$$

$$\frac{R_2 L_1 i_2}{M} = V$$

$$i_2 = \frac{VM}{R_2 L_1}$$

$$V_0 = i_2 R_2 = \frac{VM}{L_1} = \frac{V\sqrt{L_1}\sqrt{L_2}}{\sqrt{L_1}\sqrt{L_1}} = \frac{V\sqrt{L_2}}{\sqrt{L_1}}$$

$$V_0 = V\sqrt{\frac{L_2}{L_1}}.$$

Tuned Circuits

EXAMPLE 6.34. *For the circuit shown in Figure E6.34, find V_0.*

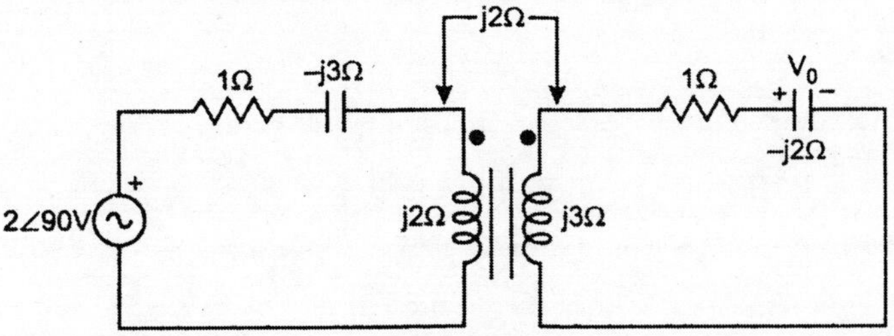

Fig. E6.34

SOLUTION. The Loop currents are shown in Figure E6.34(*a*).

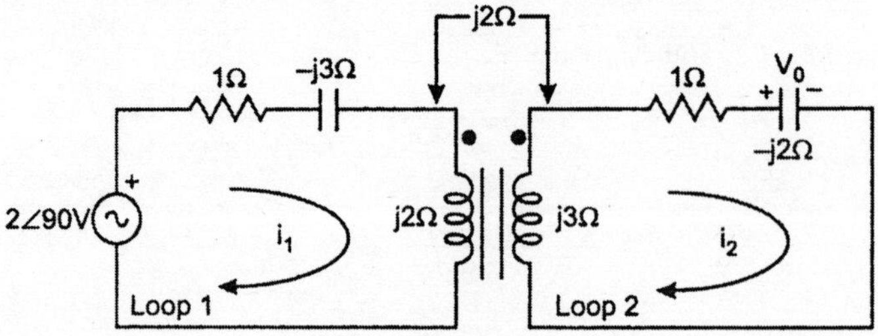

Fig. E6.34(a)

Writing the loop equations we get

Loop 1 : $(1 - j1)i_1 - j2i_2 = 2\angle 90$ ...(*i*)

Loop 2 : $(1 + j1)i_2 - j2i_1 = 0$ ...(*ii*)

Writing the equations (*i*) and (*ii*) in matrix form we get

$$\begin{bmatrix} (1-j1) & -j2 \\ -j2 & (1+j1) \end{bmatrix} \times \begin{bmatrix} i_1 \\ i_2 \end{bmatrix} = \begin{bmatrix} j2 \\ 0 \end{bmatrix}$$

Let $$\Delta = \begin{vmatrix} (1-j1) & -j2 \\ -j2 & (1+j1) \end{vmatrix} = 6$$

$$i_2 = \frac{\begin{vmatrix} (1-j1) & j2 \\ -j2 & 0 \end{vmatrix}}{\Delta} = \frac{-4}{6} = -0.67 \text{ Amp}$$

Therefore $v_0 = i_2 \times (-j2) = j1.34$ V

EXAMPLE 6.35. *For the circuit shown in Figure E6.35, find* $\frac{V}{i_1}$.

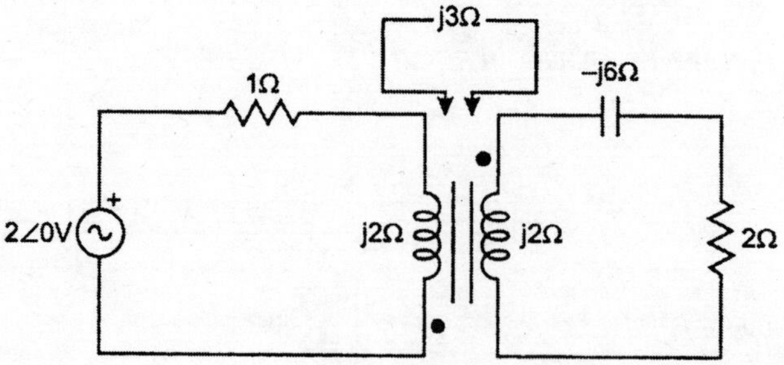

Fig. E6.35

SOLUTION. The loop currents are indicated in Figure E6.35(*a*).

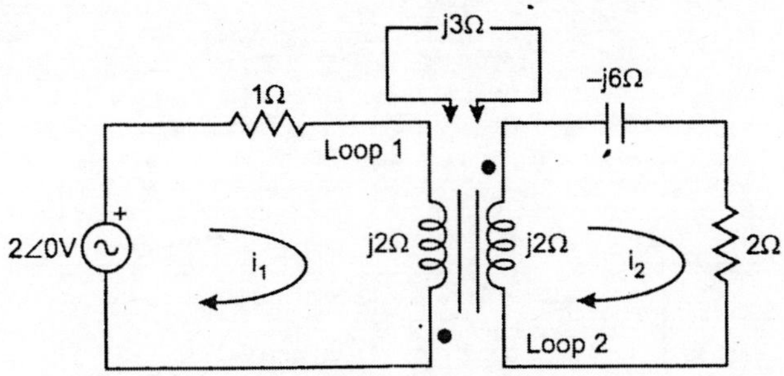

Fig. E6.35(a)

Writing the loop equations we get

Loop 1 : $(1 + j2)i_1 + j3i_2 = 2$ ...(*i*)

Loop 2 : $(2 - j4)i_2 + j3i_1 = 0$ ...(*ii*)

Writing the loop equations in matrix form we get

$$\begin{bmatrix} (1+j2) & j3 \\ j3 & (2-j4) \end{bmatrix} \times \begin{bmatrix} i_1 \\ i_2 \end{bmatrix} = \begin{bmatrix} 2 \\ 0 \end{bmatrix}$$

Let $$\Delta = \begin{vmatrix} (1+j2) & j3 \\ j3 & (2-j4) \end{vmatrix} = 19$$

Then $$i_1 = \frac{\begin{vmatrix} 2 & j3 \\ 0 & (2-j4) \end{vmatrix}}{\Delta} = \frac{4-j8}{19} = 0.21 - j0.42 \text{ Amp}$$

Thus $$\frac{V}{i_1} = \frac{2}{0.21 - j0.42} = 1.9 + j3.81\ \Omega$$

EXAMPLE 6.36. *For the circuit shown in Figure E6.36 draw the dotted equivalent circuit. Then find the drop across 4 Ω resistance.*

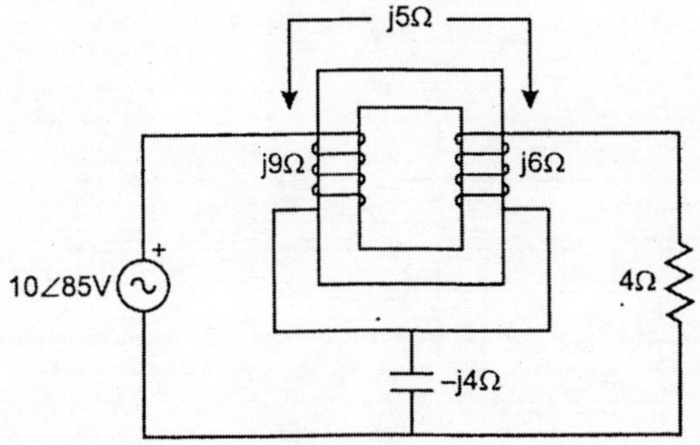

Fig. E6.36

SOLUTION. The dotted equivalent circuit is shown in Figure E6.36(*a*) and also the loop currents are indicated in the Figure E6.36(*a*).

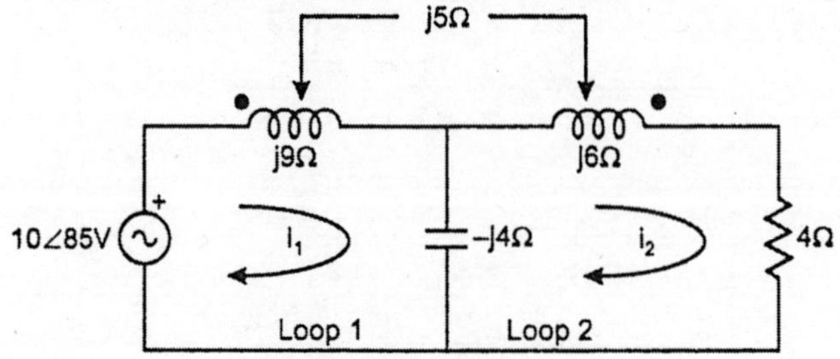

Fig. E6.36(a)

Writing the loop equation we get

Loop 1 : $(j9 - j4)i_1 - j5i_2 = 10\angle 85$ ...(*i*)

or $j5i_1 - j5i_2 = 0.872 + j9.962$ ...(*ii*)

Loop 2 : $(j6 - j4 + 4)i_2 - j5i_1 = 0$ ...(*iii*)

or $(4 + j2)i_2 - j5i_1 = 0$ ...(*iv*)

Writing equations (*ii*) and (*iii*) in matrix form we get

$$\begin{bmatrix} j5 & -j5 \\ -j5 & (4+j2) \end{bmatrix} \times \begin{bmatrix} i_1 \\ i_2 \end{bmatrix} = \begin{bmatrix} 0.872+j9.962 \\ 0 \end{bmatrix}$$

Let $$\Delta = \begin{vmatrix} j5 & -j5 \\ -j5 & (4+j2) \end{vmatrix} = 15 + j20$$

Then $$i_2 = \frac{\begin{vmatrix} j5 & 0.872+j9.962 \\ -j5 & 0 \end{vmatrix}}{\Delta} = \frac{-49.81+j4.36}{15+j20} = -1.06 + j1.69 \text{ Amp}$$

then the drop across $4\ \Omega = 4 \times i_2 = -4.22 + j6.79$ V

EXAMPLE 6.37. *For the circuit shown in Figure E6.37 draw the dotted equivalent circuit. Then find the drop across j2 Ω.*

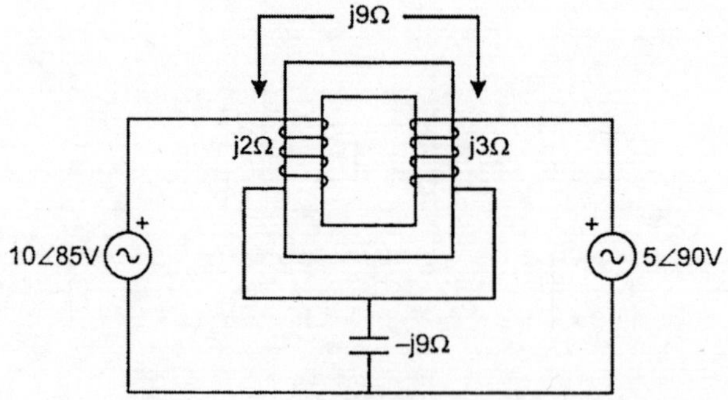

Fig. E6.37

SOLUTION. The dotted equivalent circuit is shown in Figure E6.37(*a*) and also the loop currents are indicated in the Figure E6.37(*a*).

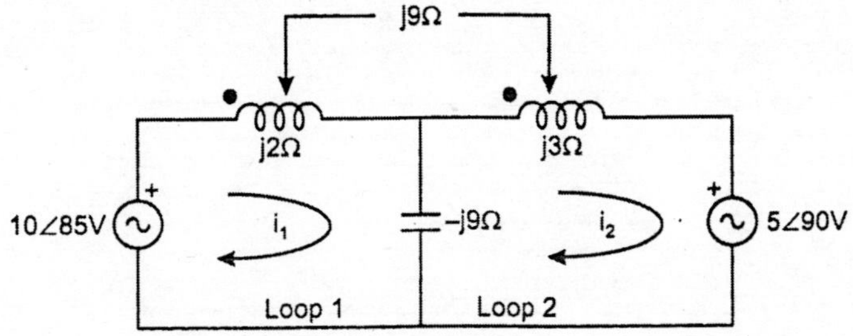

Fig. E6.37(a)

Writing the loop equation we get

Loop 1 : $(j2 - j9)i_1 + j9i_2 = 10\angle 85$ ...(*i*)

or $-j7i_1 + j9i_2 = 0.872 + j9.962$ ...(*ii*)

Loop 2 : $(j3 - j9)i_2 + j9i_1 = 5\angle 90$ ...(*iii*)

or $-j6i_2 + j9i_1 = j5$ ...(*iv*)

Writing equations (*ii*) and (*iv*) in matrix form we get

$$\begin{bmatrix} -j7 & j9 \\ j9 & -j6 \end{bmatrix} \times \begin{bmatrix} i_1 \\ i_2 \end{bmatrix} = \begin{bmatrix} 0.872 + j9.962 \\ j5 \end{bmatrix}$$

Let
$$\Delta = \begin{vmatrix} -j7 & j9 \\ j9 & -j6 \end{vmatrix} = 39$$

$$i_1 = \frac{\begin{vmatrix} 0.872 + j9.962 & j9 \\ j5 & -j6 \end{vmatrix}}{\Delta} = \frac{104.77 - j5.23}{39} = 2.69 - j0.134 \text{ Amp}$$

Then
$$i_2 = \frac{\begin{vmatrix} -j7 & 0.872 + j9.962 \\ j9 & j5 \end{vmatrix}}{\Delta} = \frac{124.66 - j7.84}{39} = 3.19 - j0.20 \text{ Amp}$$

then the drop across $j2\ \Omega = j2i_1 + j9i_2 = 2.07 + j34.09$ V

EXAMPLE 6.38. *For the circuit shown in Figure E6.38 draw the dotted equivalent circuit. Then find the drop across 2 Ω.*

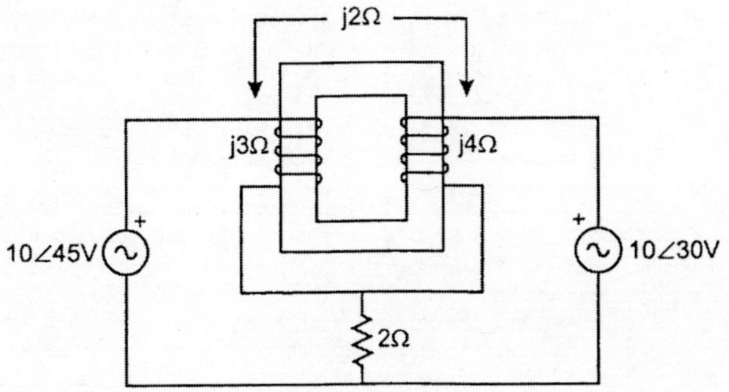

Fig. E6.38

SOLUTION. The dotted equivalent circuit is shown in Figure E6.38(*a*). The loop currents are also indicated in the Figure E6.38(*a*).

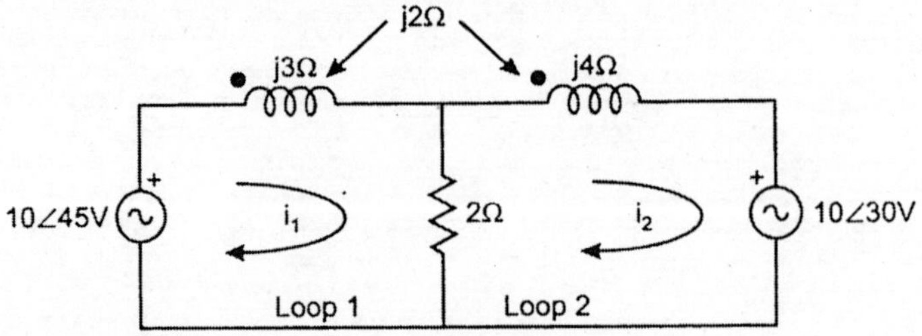

Fig. E6.38(a)

Writing the loop equation we get

Loop 1 : $(2 + j3)i_1 + j2i_2 = 10\angle 45$ ...(*i*)

$(2 + j3)i_1 + j2i_2 = 7.071 + j7.071$ ...(*ii*)

Loop 2 : $(2 + j4)i_2 + j2i_1 = 10\angle 30$ ...(*iii*)

$(2 + j4)i_2 + j2i_1 = 8.66 + j5$ ...(*iv*)

Writing equations (*ii*) and (*iv*) in matrix form we get

$$\begin{bmatrix}(2+j3) & j2 \\ j2 & (2+j4)\end{bmatrix} \times \begin{bmatrix} i_1 \\ i_2 \end{bmatrix} = \begin{bmatrix} 7.071+j7.071 \\ 8.66+j5 \end{bmatrix}$$

Let $$\Delta = \begin{vmatrix}(2+j3) & j2 \\ j2 & (2+j4)\end{vmatrix} = -4 + j14$$

Then $$i_1 = \frac{\begin{vmatrix}7.071+j7.071 & j2 \\ 8.66+j5 & (2+j4)\end{vmatrix}}{\Delta} = \frac{-4.142+j25.1}{-4+j14} = 1.73 - j0.2 \text{ Amp}$$

and $$i_2 = \frac{\begin{vmatrix}(2+j3) & 7.071+j7.071 \\ j2 & 8.66+j5\end{vmatrix}}{\Delta} = \frac{16.46+j21.83}{-4+j14} = 1.13 - j1.49 \text{ Amp}$$

Then the drop across 2 Ω is $= 2(i_1 - i_2) = 1.2 + j2.58$ V

EXAMPLE 6.39. *For the circuit shown in Figure E6.39 find the Norton's equivalent across A-B.*

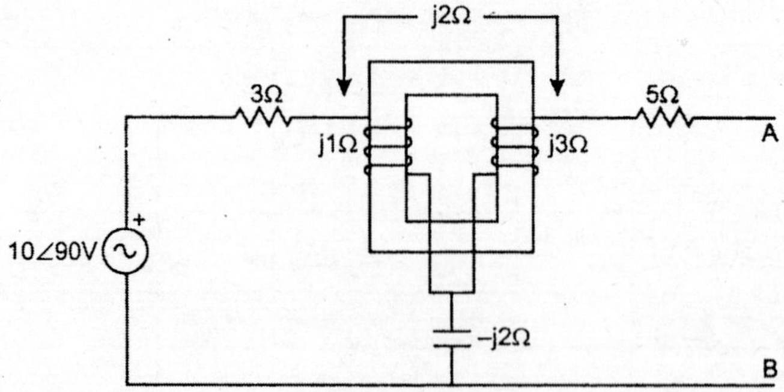

Fig. E6.39

SOLUTION. The dotted equivalent circuit after shorting the terminals *A-B* is shown in Figure E6.39(*a*). The loop currents are shown in the Figure E6.39(*a*).

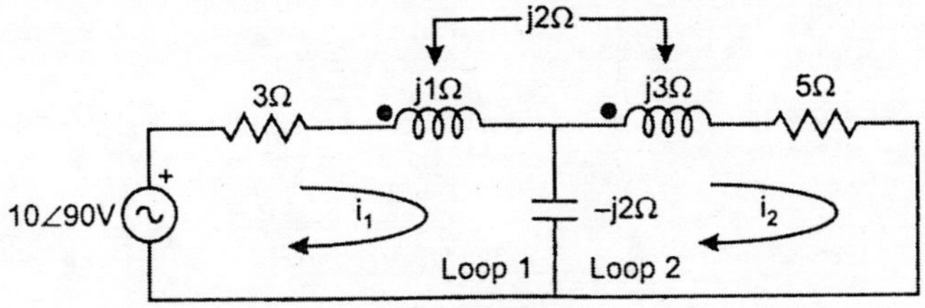

Fig. E6.39(a)

Writing the loop equation we get

Loop 1 : $(3 - j1)i_1 + j2i_2 = 10\angle 90$ ...(*i*)

$(3 - j1)i_1 + j2i_2 = j10$ ...(*ii*)

Loop 2 : $(5 + j1)i_2 + j2i_1 = 0$ ...(*iii*)

Writing equations (*ii*) and (*iii*) in matrix form we get

$$\begin{bmatrix}(3-j1) & j2\\ j2 & (5+j1)\end{bmatrix}\times\begin{bmatrix}i_1\\ i_2\end{bmatrix} = \begin{bmatrix}j10\\ 0\end{bmatrix}$$

Let
$$\Delta = \begin{vmatrix}(3-j1) & j2\\ j2 & (5+j1)\end{vmatrix} = 20 - j2$$

Then the short circuit current is

$$i_2 = \frac{\begin{vmatrix}(3-j1) & j10\\ j2 & 0\end{vmatrix}}{\Delta} = \frac{20}{20-j2} = 0.99 + j0.099 \text{ Amp}$$

To find the open circuit voltage :

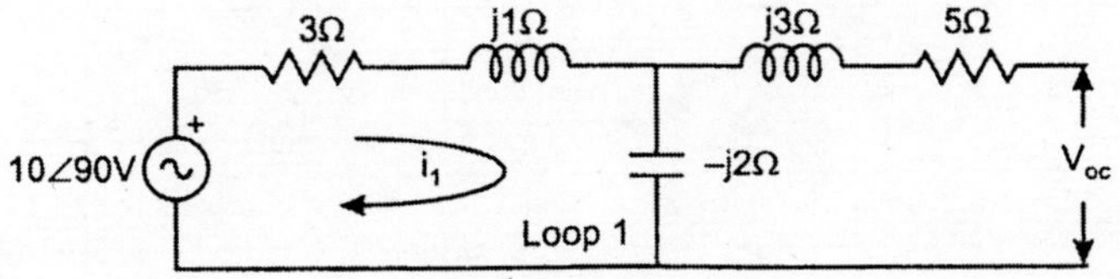

Fig. E6.39(b)

Consider Figure E6.39(*b*) we get

$$i_1 = \frac{10\angle 90}{3-j1} = (-1 + j3) \text{ Amp}$$

Then
$$V_{oc} = i_1 \times (-j2) = (-1 + j3) \times (-j2) = (6 + j2) \text{ V}$$

Then
$$Z_N = \frac{V_{oc}}{i_2} = \frac{(6+j2)}{0.99+j0.099} = 6.2 + j1.4\ \Omega$$

The Norton's equivalent circuit is shown in Figure E6.39(*c*).

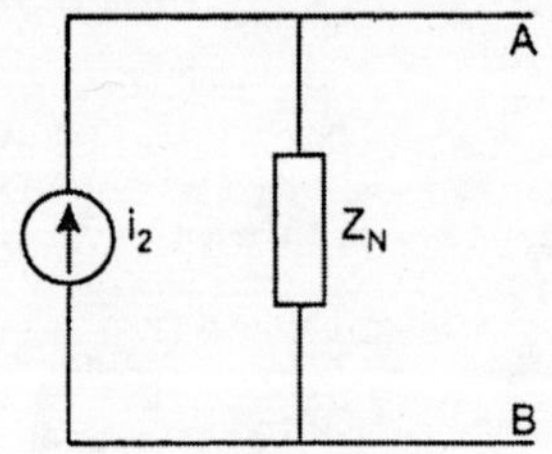

Fig. E6.39(c)

EXAMPLE 6.40. *For the circuit shown in Figure E6.40, find the Norton's equivalent across A-B.*

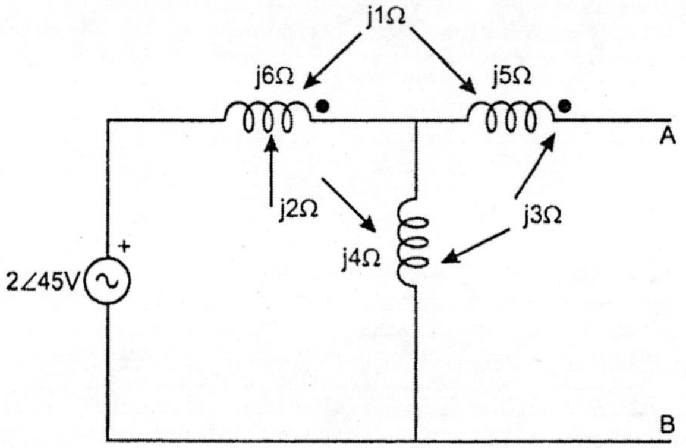

Fig. E6.40

SOLUTION. The loop currents after shorting the terminals *A-B* are shown in Figure E6.40(*a*).

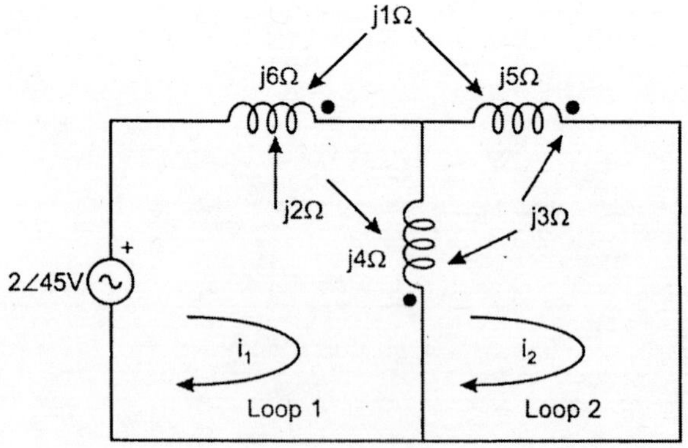

Fig. E6.40(a)

Writing the loop equation we get

Loop 1 : $$j10i_1 - j4i_2 + j1i_2 + j2i_1 - j2i_2 = 2\angle 45 \quad ...(i)$$

$$j12i_1 - j5i_2 = 2\angle 45 \quad ...(ii)$$

Loop 2 : $$j9i_2 - j4i_1 + j1i_1 + j3i_1 - j3i_2 = 0 \quad ...(iii)$$

Again from equation (*iii*) we get $$j6i_2 = 0 \quad ...(iv)$$

Thus the short circuit current $i_2 = 0$ Amp

To find the open circuit voltage :

From Figure E6.40(*b*) we get

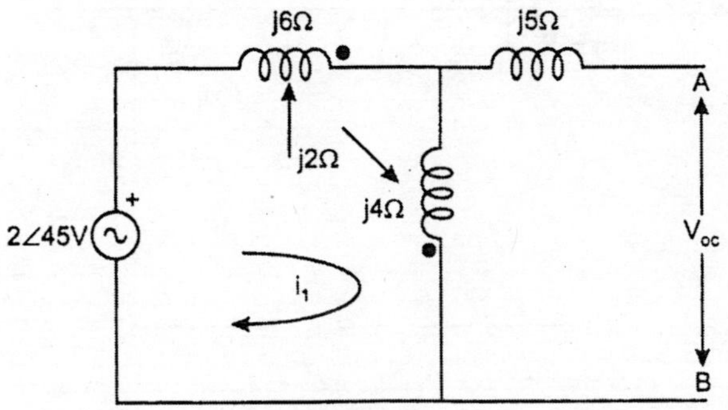

Fig. E6.40(b)

$$i_1 = \frac{2\angle 45}{j14} = \frac{1.414 + j1.414}{j14} = (0.1 - j0.1) \text{ Amp}$$

Then $$V_{oc} = i_1 \times (j4 + j2) = 0.6 + j0.6 \text{ V}$$

Then $$Z_N = \frac{V_{oc}}{i_2} = \frac{0.6 + j0.6}{0} = \infty\ \Omega$$

EXERCISES

1. For the circuit shown in Figure Q6.1, determine the equivalent reactance.

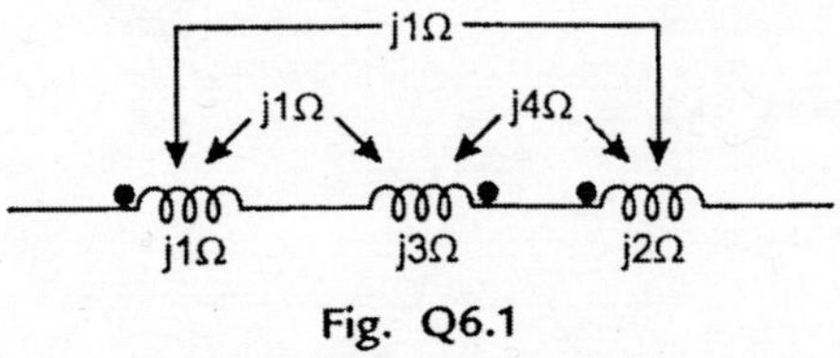

Fig. Q6.1

2. For the circuit shown in Figure Q6.2, determine the equivalent reactance.

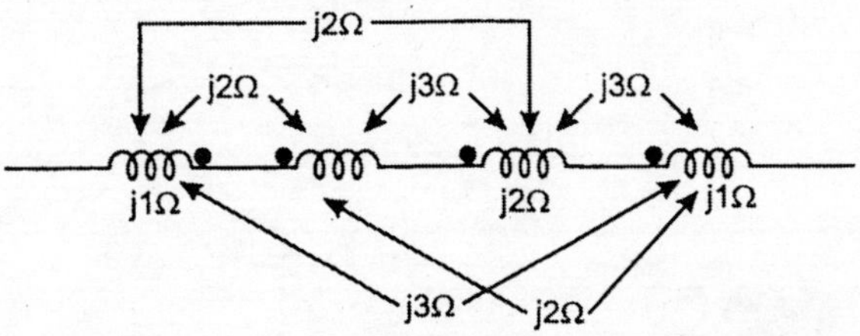

Fig. Q6.2

3. For the circuit shown in Figure Q6.3, draw the dotted equivalent circuit and then find the equivalent inductance.

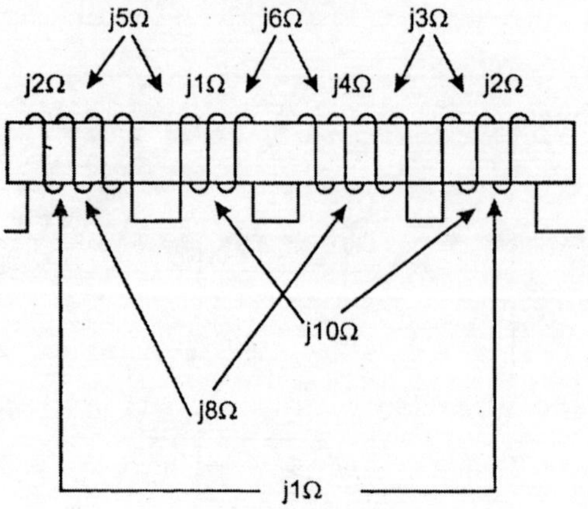

Fig. Q6.3

4. For the circuit shown in Figure Q6.4, find the loop currents, and the drop across $j4\ \Omega$.

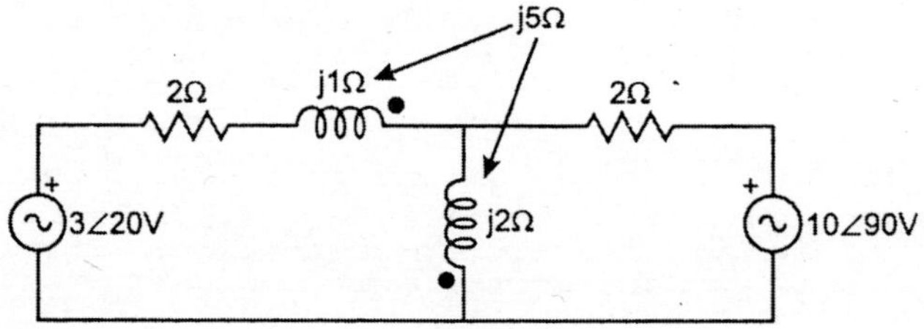

Fig. Q6.4

5. For the circuit shown in Figure Q6.5, find the loop currents, and the drop across $j5\ \Omega$.

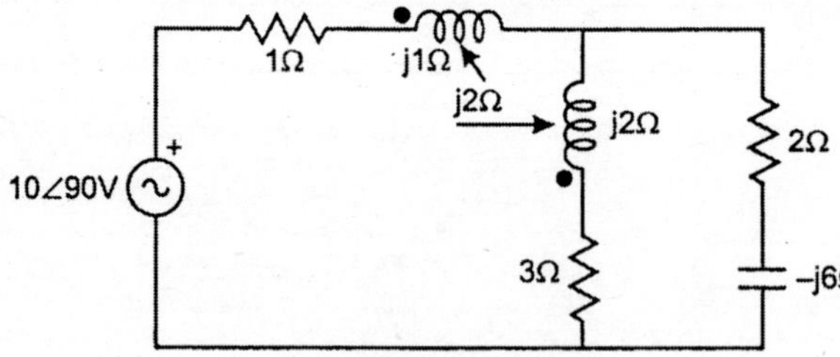

Fig. Q6.5

6. For the circuit shown in Figure Q6.6, find the drop across $-j6\ \Omega$.

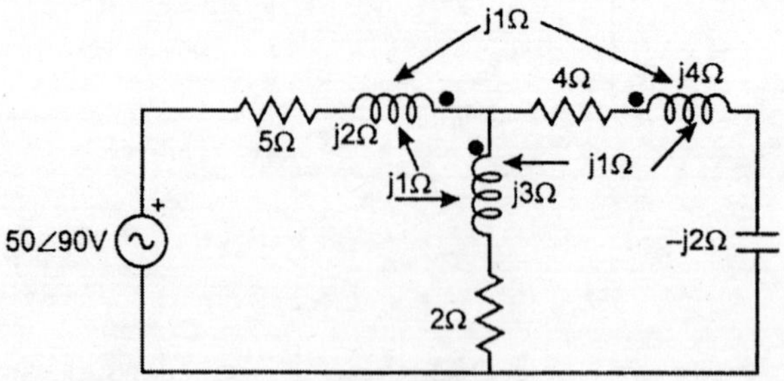

Fig. Q6.6

7. For the circuit shown in Figure Q6.7, find V_0.

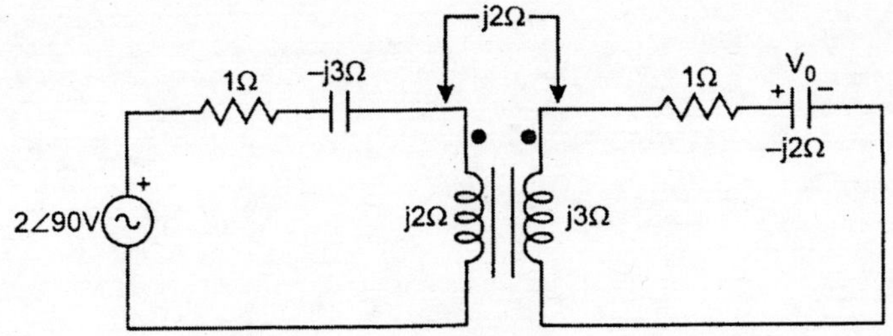

Fig. Q6.7

Chapter 7

RESONANCE

7.1 INTRODUCTION

Resonance is a special case of an RLC circuit where the voltage or the current in the circuit is either maximum or minimum at a particular supply frequency. The phenomenon of resonance can be observed in both series and parallel RLC circuit.

7.2 SERIES RESONANCE IN AC CIRCUIT

Let us consider a series RLC circuit connected to an AC source as shown in Figure 7.1. Since all the three components (R, L and C) are connected in series, same current flows through then as shown in the phasor diagram in Figure 7.2. The resistive drop V_R is in phase with the current I. For the capacitor C, current I leads V_C by 90 degree and for the inductor L, V_L leads the current by 90 degree. The complete phasor is shown in Figure 7.2.

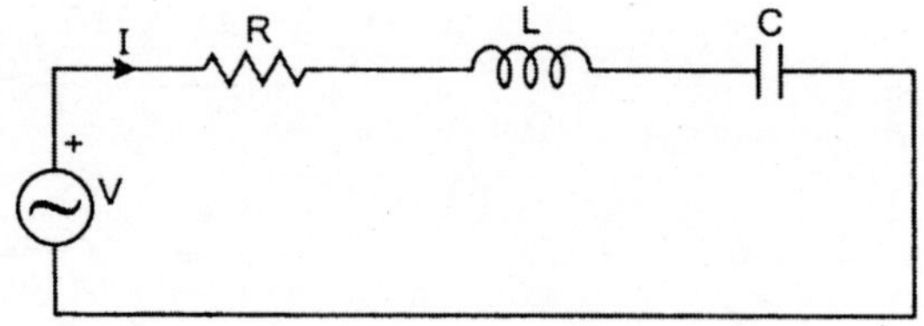

Fig. 7.1

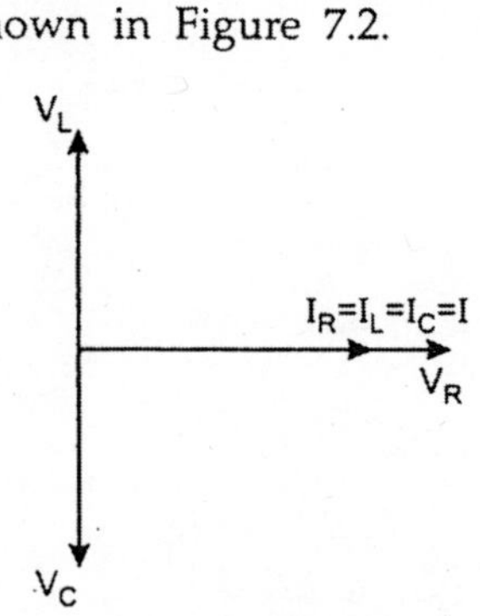

Fig. 7.2

Now let us consider a situation when the drop across the inductor becomes equal to the drop across the capacitor. That is $V_L = V_C$. Since capacitive reactance falls as frequency increases, and inductive reactance falls as frequency decreases, there must be a frequency at which they become equal. The frequency at which they become equal is called the resonant frequency. Thus at resonance $V_L = V_C$ and since they are 180 degree out of phase, their effect will cancel each other and thus the circuit will behave as a purely resistive circuit. That is $Z = R$ and current I through the circuit at resonance is given by,

$$I = \frac{V}{Z} = \frac{V}{R} \quad \text{...(7.1)}$$

Let f_0 be the resonant frequency. That is at this frequency

$$X_L = X_C$$

or

$$2\pi f_0 L = \frac{1}{2\pi f_0 C} \quad \text{...(7.2)}$$

Then from equation (7.2) we get

$$f_0 = \frac{1}{2\pi}\sqrt{\frac{1}{LC}} \text{ Hz}$$

$$\omega_0 = \sqrt{\frac{1}{LC}} \text{ rad/sec} \quad \text{...(7.3)}$$

7.2.1 Properties of a Series RLC Circuit at Resonance

1. The circuit behaves as a purely resistive circuit.
2. The impedance of the circuit is minimum
3. The current through the circuit is maximum
4. Since the circuit behaves like a purely resistive circuit at resonance its power factor is unity.
5. The reactance of the circuit is zero.
6. The power dissipated by the circuit is maximum.
7. the resonance frequency is given by $\omega_0 = \sqrt{\frac{1}{LC}}$ rad/sec.

7.2.2 Variation of R, X_L, X_C, X, Z, I and Power Factor with Frequency

1. *Variation of resistance R with frequency*: The resistance R is independent of frequency. So any variation in frequency will not affect the value of R. Thus if we try to plot R with respect to the frequency, it will be a straight line parallel to x-axis. (Figure 7.3)
2. *Variation of resistance X_L with frequency*: We know $X_L = 2\pi fL$. Thus we can say X_L is directly proportional to the frequency. If frequency increases the value of X_L also increases and vice versa. If we try to plot X_L with respect to the frequency it will be a straight line making an angle of 45 degree with the x-axis. (Figure 7.3)
3. *Variation of resistance X_C with frequency*: We know $X_C = \frac{1}{2\pi fC}$. Thus we can say X_C is inversely proportional to the frequency. If frequency increases the value of X_C also

decreases and vice versa. If we try to plot X_C with respect to the frequency it will be a rectangular hyperbola. X_C is always drawn in the first quadrant because X_C is always regarded as negative. (Figure 7.3)

4. *Variation of net reactance X with frequency*: It is given by $X = X_L \sim X_C$. If it is plotted with respect to frequency it will be a hyperbola. As long as $X_L > X_C$, the graph will be in the second quadrant and when $X_C > X_L$, the graph will be in the first quadrant. At $X_L = X_C$, the graph will cross the ω axis (the point ω_0 in the Figure 7.3) this crossover point gives the resonant frequency.
5. *Variation of net impedance Z with frequency*: The variation of Z with frequency is shown in Figure 7.3. At low frequencies Z is large because $X_C > X_L$. And again at high frequencies Z will have again large value because at high frequencies $X_L > X_C$. But at resonant frequency ω_0 Z will have the minimum value, *i.e.*, $Z = R$, since at resonance $X_L = X_C$, or in other words at resonance $X = X_L \sim X_C = 0$.
6. *Variation of circuit current with frequency*: Circuit current I is given by

$$I = \frac{V}{\sqrt{R^2 + X^2}} = \frac{V}{Z} \qquad ...(7.4)$$

where $Z = \sqrt{R^2 + X^2}$ and $X = X_L \sim X_C$

From equation (7.4), we can see that current I is inversely proportional to impedance Z. Thus when Z is large the current through the circuit is less and vice versa. So at resonance frequency ω_0 the impedance Z is minimum and thus the current will be maximum. The variation of current with frequency is shown in Figure 7.3.

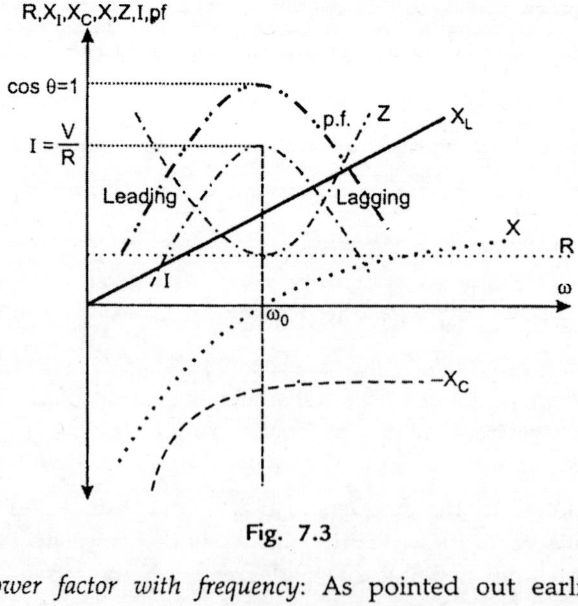

Fig. 7.3

7. *Variation of power factor with frequency*: As pointed out earlier in point 5, at low frequencies Z is large because $X_C > X_L$,and thus the power factor of the circuit will be leading. And again at high frequencies Z will have again large value because at high frequencies $X_L > X_C$, and thus the power factor of the circuit will be lagging. But at resonant frequency ω_0 Z will have the minimum value, *i.e.* $Z = R$, since at resonance

$X_L = X_C$, or in other words at resonance $X = X_L \sim X_C = 0$ and the power factor of the circuit will be unity.

7.2.3 Half Power Frequencies and Bandwidth of a Series RLC Circuit

Let us consider only the current and the impedance graph of Figure 7.3. Variation of only the current and impedance with frequency is redrawn in Figure 7.4.

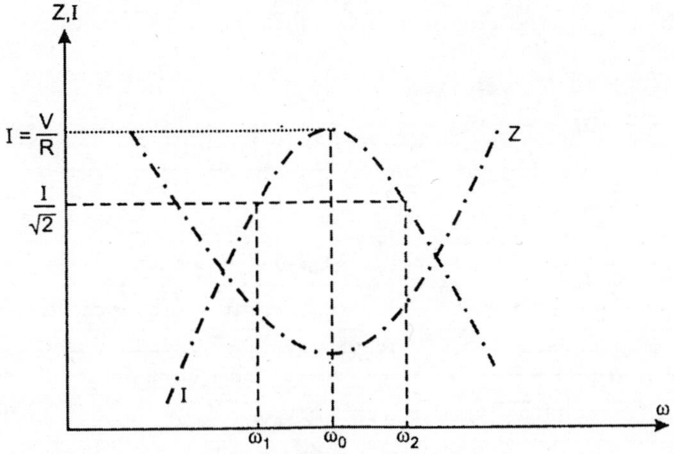

Fig. 7.4

As stated earlier the current through the circuit is maximum at resonance. From Figure 7.4 we can say that at off-resonance for a particular value of current there lie two different values of frequencies. Half power frequencies are two such frequencies at which the current through the circuit falls to $1/\sqrt{2}$ times the maximum value (the current at resonance). In Figure 7.4, ω_1 and ω_2 are the half power frequencies. They are called half power frequencies because at these frequencies the power dissipated by the circuit is half the power dissipated by the circuit at resonance. The band of frequencies that lie between the half power frequencies is called bandwidth.

That is bandwidth $= \omega_2 - \omega_1$ ...(7.5)

7.2.4 To show that the Power Dissipated by the Circuit at Half Power Frequencies is Half the Power Dissipated by the Circuit at Resonance

Let the current through a series RLC circuit at resonance be I_o

The impedance Z of the circuit at resonance is. $Z = R$.

Then the power dissipated by the circuit at resonance is

$$P_0 = I_o^2 R \qquad ...(7.6)$$

At half power frequencies the current through the circuit is $\frac{I_0}{\sqrt{2}}$. Then the power dissipated by the circuit at half power frequencies will be

$$P_1 = P_2 = \left(\frac{I_0}{\sqrt{2}}\right)^2 R = \frac{I_0^2}{2} R \qquad \text{...(7.7)}$$

substituting equation (7.6) in equation (7.7) we get

$$P_1 = P_2 = \frac{P_0}{2} \qquad \text{...(7.8)}$$

Equation (7.8) show that the power dissipated by the circuit at half power frequencies is half the power dissipated by the circuit at resonance.

7.2.5 Expression for Half Power Frequencies in Terms of Circuit Parameters

At half power frequencies, the net reactance of a series RLC resonant circuit is R, where,

$$R = \pm(X_L - X_C) \qquad \text{...(7.9)}$$

$$= \pm\left(\omega L - \frac{1}{\omega C}\right) \qquad \text{...(7.10)}$$

where + sign indicates the condition when $X_L > X_C$, and – sign indicates the condition when $X_C > X_L$, depending on the band of frequencies that lies between the half power frequencies. Let ω_1 and ω_2 be the two half power frequencies. Let at frequency ω_1, $X_C > X_L$ that is the net reactance of the circuit is negative and at frequency ω_2, $X_L > X_C$, that is the net reactance of the circuit is positive. Then at frequencies ω_1 and ω_2 equation (7.10) becomes.

$$\left(\omega_1 L - \frac{1}{\omega_1 C}\right) = -R \qquad \text{...(7.11)}$$

$$\left(\omega_2 L - \frac{1}{\omega_2 C}\right) = R \qquad \text{...(7.12)}$$

From equation (7.11) we get

$$\omega_1^2 LC - 1 - \omega_1 CR = 0$$

or

$$\omega_1 = \frac{-RC \pm \sqrt{(RC)^2 + 4LC}}{2LC} \qquad \text{...(7.13)}$$

Considering only the positive sign of the square root of equation (7.13) we get

$$\omega_1 = \frac{-RC + \sqrt{(RC)^2 + 4LC}}{2LC} \qquad \text{...(7.14)}$$

Similarly solving equation (7.12) we will get

$$\omega_2 = \frac{RC + \sqrt{(RC)^2 + 4LC}}{2LC} \qquad \text{...(7.15)}$$

7.2.6 Expression for Bandwidth in Terms of Circuit Parameters

From equation (7.5) we get

$$\text{Bandwidth} = \omega_2 - \omega_1$$

Substituting equations (7.14) and (7.15) we get

$$\text{Bandwidth} = \frac{2RC}{2LC} = \frac{R}{L} \qquad \text{...(7.16)}$$

7.2.7 Quality Factor

It is also called the Q-factor. It is the voltage magnification in the circuit at resonance. It is defined as the ratio between the voltage drop across the inductor to the applied voltage at resonance. It is also defined as ratio between the voltage drop across the capacitor to the applied voltage at resonance. That is

$$Q = \frac{V_L}{V} = \frac{V_C}{V} \qquad \text{...(7.17)}$$

where,

V_L is the drop across the inductor

V_C is the drop across the inductor and

V is the applied or supply voltage.

7.2.8 Expression for Q-factor in Terms of Circuit Parameters

Considering

$$Q = \frac{V_L}{V} = \frac{I_0 X_L}{I_0 R} = \frac{X_L}{R} = \frac{\omega_o L}{R} \qquad \text{...(7.18)}$$

Substituting equation (7.3) in equation (7.18) we get

$$Q = \frac{1}{\sqrt{LC}} \times \frac{L}{R} = \frac{1}{R}\sqrt{\frac{L}{C}} \qquad \text{...(7.19)}$$

again considering

$$Q = \frac{V_C}{V} = \frac{I_0 X_C}{I_0 R} = \frac{X_C}{R} = \frac{1}{\omega_0 CR} \qquad \text{...(7.20)}$$

Substituting equation (7.3) in equation (7.20) we get

$$Q = \frac{\sqrt{LC}}{CR} = \frac{1}{R}\sqrt{\frac{L}{C}} \qquad \text{...(7.21)}$$

Thus Q-factor in terms of circuit parameter is

$$Q = \frac{1}{R}\sqrt{\frac{L}{C}} \qquad \text{...(7.22)}$$

7.2.9 Selectivity

If ω_0 is the frequency at resonance and BW is the bandwidth of the circuit, then the selectivity is defined as the ratio of ω_0 to bandwidth. That is,

$$\text{Selectivity} = \frac{\omega_0}{BW} = \frac{\omega_0}{\omega_2 - \omega_1} \qquad \text{...(7.23)}$$

where ω_1 and ω_2 are the half power frequencies.

7.2.10 Expression for Selectivity in Terms of Circuit Parameters

Substituting equations (7.3) and (7.16) in equation (7.23) we get

$$\text{Selectivity} = \frac{\omega_0}{BW} = \frac{\omega_0}{\omega_2 - \omega_1} = \frac{\frac{1}{\sqrt{LC}}}{\frac{R}{L}} = \frac{1}{R}\sqrt{\frac{L}{C}} \quad \text{...(7.24)}$$

Equations (7.24) and (7.22) shows that Q-factor and selectivity have the same expression. Thus we can say,

$$\text{Selectivity} = \frac{\omega_0}{\omega_2 - \omega_1} = Q\text{-factor} \quad \text{...(7.25)}$$

7.2.11 Frequency at which the Voltage Across the Inductor (V_L) is Maximum

From Figure 7.1 we get

$$I = \frac{V}{\sqrt{R^2 + (X_L - X_C)^2}}$$

Then
$$V_L = I \times X_L = \frac{V}{\sqrt{R^2 + (X_L - X_C)^2}} \times X_L$$

or
$$V_L^2 = \frac{V^2 \times X_L^2}{R^2 + (X_L - X_c)} = \frac{V^2\omega^2 L^2}{R^2 + \left(\omega L - \frac{1}{\omega C}\right)^2} = \frac{V^2\omega^2 L^2}{R^2 + \left(\frac{\omega^2 LC - 1}{\omega C}\right)^2}$$

Now differentiating V_L^2 with respect to ω we get

$$\frac{dV_L^2}{d\omega} = \frac{\left[R^2\omega^2C^2 + \omega^4L^2C^2 - 2\omega^2LC + 1\right] \times 4\omega^3V^2L^2C^2 - V^2\omega^4L^2C^2(2\omega R^2C^2 + 4\omega^3L^2C^2 - 4\omega LC)}{(R^2\omega^2C^2 + \omega^4L^2C^2 - 2\omega^2LC + 1)^2}$$

The frequency at which V_L is maximum is obtained by equating $\frac{dV_L^2}{d\omega} = 0$

That is

$$\left[R^2\omega^2C^2 + \omega^4L^2C^2 - 2\omega^2LC + 1\right] \times 4\omega^3V^2L^2C^2 - V^2\omega^4L^2C^2(2\omega R^2C^2 + 4\omega^3L^2C^2 - 4\omega LC) = 0$$

or
$$2\omega^2LC - \omega^2C^2R^2 = 2$$

or
$$\omega = \sqrt{\frac{2}{2LC - C^2R^2}} \quad \text{...(7.26)}$$

Equation (7.26) gives the frequency at which the voltage across the inductor is maximum.

7.2.12 Frequency at which the Voltage Across the Capacitor is Maximum

From Figure 7.1 we get

$$I = \frac{V}{\sqrt{R^2 + (X_L - X_C)^2}}$$

Then $$V_C = I \times X_C = \frac{V}{\sqrt{R^2 + (X_L - X_C)^2}} \times X_C$$

$$V_C^2 = \frac{V^2 \times X_C^2}{R^2 + (X_L - X_c)} = \frac{V^2}{R^2 + \left(\omega L - \frac{1}{\omega C}\right)^2} \times \frac{1}{\omega^2 C^2}$$

$$V_C^2 = \frac{V^2}{R^2\omega^2C^2 + \omega^4L^2C^2 - 2\omega^2LC + 1}$$

Now differentiating V_C^2 with respect to ω we get,

$$\frac{dV_C^2}{d\omega} = -\frac{V^2(2R^2\omega C^2 + 4\omega^3L^2C^2 - 4\omega LC)}{(R^2\omega^2C^2 + \omega^4L^2C^2 - 2\omega^2LC + 1)^2}$$

The frequency at which V_C is maximum is obtained by equating

$$\frac{dV_C^2}{d\omega} = 0$$

That is,

$$V^2(2R^2\omega C^2 + 4\omega^3L^2C^2 - 4\omega LC) = 0$$

or $$\omega^2 = \frac{1}{LC} - \frac{R^2}{2L^2}$$

or $$\omega = \sqrt{\frac{1}{LC} - \frac{R^2}{2L^2}} \quad ...(7.27)$$

Equation (7.27) gives the frequency at which the voltage across the capacitor is maximum.

7.3 PARALLEL RESONANCE IN AC CIRCUIT

Let us consider the circuit as shown in Figure 7.5. The corresponding phasor diagram is shown in Figure 7.6.

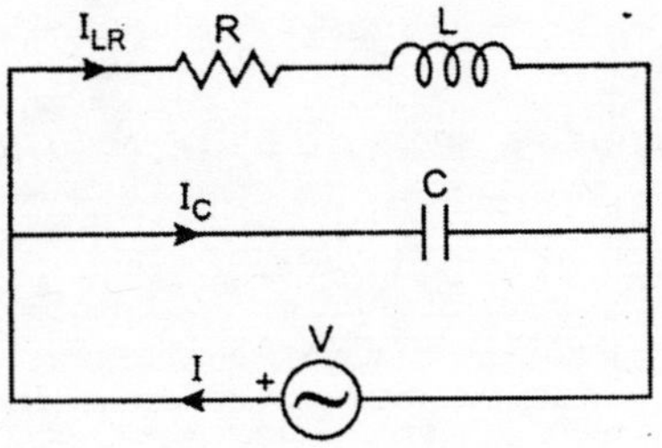

Fig. 7.5

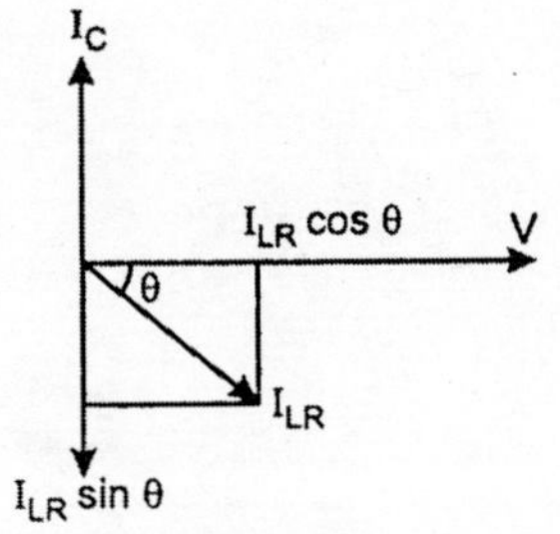

Fig. 7.6

From Figure 7.5 we get $$I_C = \frac{V}{X_C} \quad ...(7.28)$$

and $$I_{LR} = \frac{V}{R+jX_L} = \frac{V}{\sqrt{R^2+X_L^2}} = \frac{V}{Z_{LR}} \qquad ...(7.29)$$

where $Z_{LR} = \sqrt{R^2+X_L^2}$ is the impedance of the coil

The above circuit will be at resonance when the reactive component of the current becomes zero. That is from Figure 7.6, the circuit will be at resonance if,

$$I_C = I_{LR} \sin\theta \qquad ...(7.30)$$

Again we know that

$$\cos\theta = \frac{R}{Z_{LR}}, \text{ then } \sin\theta = \frac{X_L}{Z_{LR}} \qquad ...(7.31)$$

Then substituting equation (7.28), equation (7.29) and equation (7.31) in equation (7.30) we get

$$\frac{V}{X_C} = \frac{V}{Z_{LR}} \times \frac{X_L}{Z_{LR}}$$

or $$Z_{LR}^2 = X_L X_C = \omega_0 L \times \frac{1}{\omega_0 C} = \frac{L}{C} \qquad ...(7.32)$$

where ω_0 is the resonance frequency of the circuit. From equation (7.32), we get

$$Z_{LR} = \sqrt{\frac{L}{C}} \qquad ...(7.33)$$

Substituting the value of Z_{LR} we get

$$\sqrt{R^2+X_L^2} = \sqrt{\frac{L}{C}}$$

$$R^2 + \omega_0^2 L^2 = \frac{L}{C}$$

$$\omega_0 = \sqrt{\frac{1}{LC} - \frac{R^2}{L^2}} \qquad ...(7.34)$$

Equation (7.34) gives the resonant frequency of the circuit

Thus at resonance

$$I = I_{LR} \cos\theta \qquad ...(7.35)$$

That is, if Z_{eq} is the impedance of the circuit as seen from the source terminal, then substituting equations (7.29) and (7.31) in equation (7.35)

$$\frac{V}{Z_{eq}} = \frac{V}{Z_{LR}} \times \frac{R}{Z_{LR}}$$

$$Z_{eq} = \frac{Z_{LR}^2}{R}$$

$$Z_{eq} = \frac{L/C}{R} = \frac{L}{CR} \qquad ...(7.36)$$

Equation (7.36) gives the impedance of the circuit at resonance.

7.3.1 Properties of a Parallel RLC Circuit at Resonance

1. The circuit behaves as a purely resistive circuit.
2. The impedance of the circuit is maximum
3. The current through the circuit is minimum
4. Since the circuit behaves like a purely resistive circuit at resonance its power factor is unity.
5. The susceptance of the circuit is zero.
6. The admittance of the circuit is minimum.

7.3.2 Variation of R, Inductive Suceptance (B_L), Capacitive Suceptance (B_C), Net Suceptance (B), Z, I with Frequency

1. *Variation of resistance R with frequency*: The resistance R is independent of frequency. So any variation in frequency will not affect the value of R. Thus if we try to plot R with respect to the frequency, it will be a straight line parallel to x-axis. (Figure 7.7)
2. *Variation of capacitive suceptance (B_C) with frequency*: We know $B_C = \omega C$. Thus we can say B_C is directly proportional to the frequency. If frequency increases the value of B_C also increases and vice versa. If we try to plot B_C with respect to the frequency it will be a straight line making an angle of 45 degree with the x-axis. (Figure 7.7)
3. *Variation of inductive suceptance (B_L) with frequency*: We know $B_L = -\dfrac{1}{\omega L}$. Thus we can say B_L is inversely proportional to the frequency. If frequency increases the value of B_L also decreases and vice versa. If we try to plot B_L with respect to the frequency it will be a rectangular hyperbola. B_L is always drawn in the first quadrant because B_L is always regarded as negative. (Figure 7.7)
4. *Variation of net suceptance (B) with frequency*: It is given by $B = B_C \sim B_L$. If it is plotted with respect to frequency it will be a hyperbola. As long as $B_C > B_L$, the graph will be in the second quadrant and when $B_L > B_C$, the graph will be in the first quadrant. At $B_L = B_C$, the graph will cross the ω axis (the point ω_0 in the Figure 7.7) this crossover point gives the resonant frequency.
5. *Variation of net impedance Z with frequency*: The variation of Z with frequency is shown in Figure 7.7. At low frequencies Z is small because $B_L > B_C$. And again at high frequencies Z will have again small value because at high frequencies $B_C > B_L$. But at resonant frequency ω_0, Z will have the maximum, *i.e.*, $Z = R$, since at resonance $B_L = B_C$, or in other words at resonance $B = B_L \sim B_C = 0$.
6. *Variation of circuit current with frequency*: Circuit current I is given by

$$I = \frac{V}{\sqrt{R^2 + X^2}} = \frac{V}{Z} \qquad ...(7.37)$$

where $Z = \sqrt{R^2 + X^2}$ and $X = X_L \sim X_C$

From equation (7.37), we can see that current I is inversely proportional to impedance Z. Thus when Z is large the current through the circuit is less and vice versa. So at resonance frequency ω_0 the impedance Z is maximum and thus the current will be minimum. The variation of current with frequency is shown in Figure 7.7.

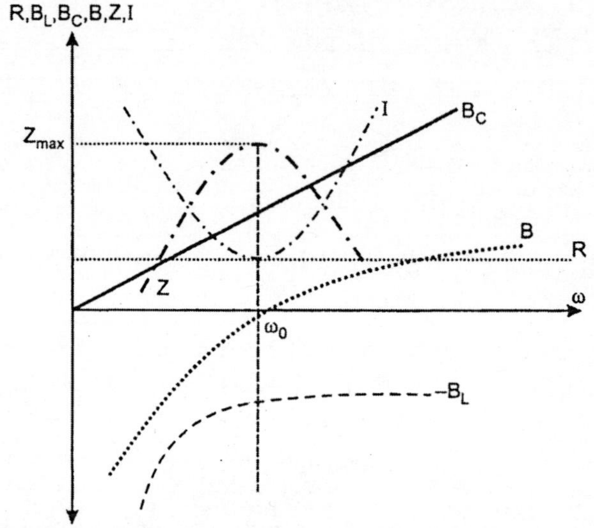

Fig. 7.7

7.3.3 Half Power Frequencies and Bandwidth of a Parallel Circuit

Let us consider only the current and the impedance graph of Figure 7.7. Variation of only the current and impedance with frequency is redrawn in Figure 7.8. As stated earlier the current through the circuit is minimum at resonance.

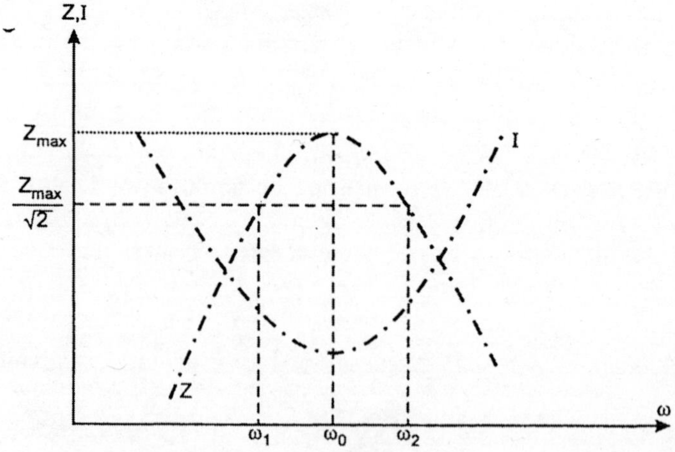

Fig. 7.8

From Figure 7.8 we can say that at off-resonance for a particular value of impedance there lie two different values of frequencies. Half power frequencies are two such frequencies at which the impedance of the circuit falls to $1/\sqrt{2}$ times the maximum value (the impedance at resonance). In Figure 7.8, ω_1 and ω_2 are the half power frequencies. They are called half power frequencies because at these frequencies the power dissipated by the circuit is half the power dissipated by the circuit at resonance. The band of frequencies that lie between the half power frequencies is called bandwidth.

$$\text{That is bandwidth } = \omega_2 - \omega_1 \quad \text{...(7.38)}$$

7.3.4 Quality Factor of a Parallel Circuit

It is also called the Q-factor. It is the current magnification in the circuit at resonance. It is defined as the ratio between the current through the inductor to the supply current at resonance. It is also defined as ratio between the current through capacitor to the supply current at resonance. That is

$$Q = \frac{I_{RL}}{I} = \frac{I_C}{I} \quad \text{...(7.39)}$$

where

I_{RL} is the current through the inductor

I_C is the current through the capacitor and

I is the source current

7.3.5 Expression for Q-factor in Terms of Circuit Parameters

Considering

$$Q = \frac{I_{LR}}{I} = \frac{V/Z_{LR}}{V/Z_{eq}} = \frac{Z_{eq}}{Z_{LR}} = \frac{L/CR}{\omega_0 L} = \frac{1}{\omega_0 LC} \quad \text{...(7.40)}$$

This is same as the Q-factor of series circuit. Thus

$$Q = \frac{1}{R}\sqrt{\frac{L}{C}} \quad \text{...(7.41)}$$

Again considering

$$Q = \frac{I_C}{I} = \frac{V/X_C}{V/Z_{eq}} = \frac{Z_{eq}}{X_C} = \frac{L}{CR} \times \omega_0 C = \frac{\omega_0 L}{R} \quad \text{...(7.42)}$$

This is same as the Q-factor of series circuit. Thus

$$Q = \frac{1}{R}\sqrt{\frac{L}{C}} \quad \text{...(7.43)}$$

SOLVED EXAMPLES

Series Resonance

EXAMPLE 7.1. *In a series RLC circuit, the values of R, L and C are 5 Ω, 4 μH and 1 μF respectively. It is connected to an a.c. source of 100V. Find the following.*

(i) *Resonance frequency.*

(ii) *Current in the circuit at resonance.*

(iii) *Quality factor of the circuit.*

(iv) *Half power frequencies.*

(v) *Bandwidth of the circuit.*

SOLUTION. (i)

$$W_o = \sqrt{\frac{1}{LC}} = \sqrt{\frac{1}{4\times10^{-6}\times1\times10^{-6}}} = 5\times10^5 \text{ rad/sec}$$

(ii)

$$I_0 = \frac{V}{R} = \frac{100}{5} = 20 \text{ A}$$

(iii)

$$Q = \frac{W_o L}{R} = \frac{5\times10^5\times4\times10^{-6}}{5} = 0.4$$

(iv)

$$\omega_1 = -\frac{R}{2L} + \sqrt{\left(\frac{R}{2L}\right)^2 + \frac{1}{LC}}$$

$$= -\frac{5}{2\times4\times10^{-6}} + \sqrt{\left(\frac{5}{2\times4\times10^{-6}}\right)^2 + \frac{1}{4\times10^{-6}\times1\times10^{-6}}}$$

$$= 175391 \text{ rad/sec}$$

and

$$\omega_2 = \frac{R}{2L} + \sqrt{\left(\frac{R}{2L}\right)^2 + \frac{1}{LC}}$$

$$= \frac{5}{2\times4\times10^{-6}} + \sqrt{\left(\frac{5}{2\times4\times10^{-6}}\right)^2 + \frac{1}{4\times10^{-6}\times1\times10^{-6}}}$$

$$= 1425391 \text{ rad/sec}$$

(v)

$$BW = 1425391 - 175391 = 1250000 \text{ rad/sec}$$

EXAMPLE 7.2. *The circuit shown in Figure E7.2 resonates at 1 kHz freq. Find the following:*

(i) *The value of L.*

(ii) *Current at resonance.*

(iii) *Q-Factor.*

(iv) *Half power frequencies.*

(v) *Bandwidth.*

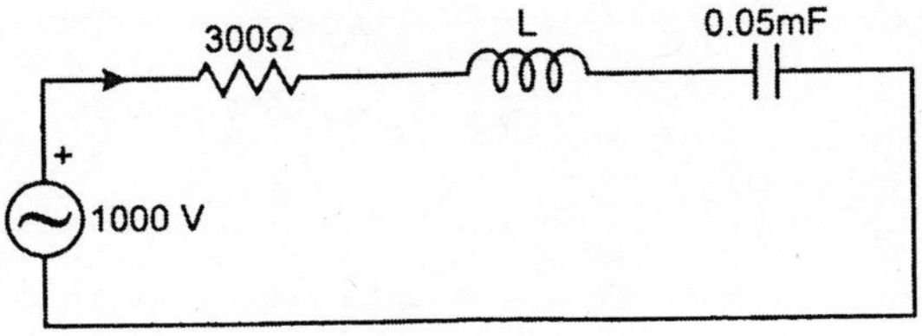

Fig. E7.2

SOLUTION. (*i*)
$$L = \frac{1}{4\pi^2 f_o^2 C} = \frac{1}{4\pi^2 \times (1\times10^3)^2 \times 0.05\times10^{-3}} = 5.066 \times 10^{-4} \text{ H}$$

(*ii*)
$$I_0 = \frac{100}{300} = 0.333 \text{ A}$$

(*iii*)
$$Q = \frac{\omega_0 L}{R} = \frac{2\pi f_0 L}{R} = \frac{2\times\pi\times1\times10^3\times0.5067}{300} = 10.6123$$

(*iv*)
$$\omega_1 = -\frac{R}{2L} + \sqrt{\left(\frac{R}{2L}\right)^2 + \frac{1}{LC}}$$

$$= -\frac{300}{2\times5.066\times10^{-4}} + \sqrt{\left(\frac{300}{2\times5.066\times10^{-4}}\right)^2 + \frac{1}{5.066\times10^{-4}\times0.05\times10^{-3}}}$$

$$= -289.81 \times 10^3 \text{ rad/sec}$$

$$\omega_2 = \frac{R}{2L} + \sqrt{\left(\frac{R}{2L}\right)^2 + \frac{1}{LC}}$$

$$= \frac{300}{2\times5.066\times10^{-4}} + \sqrt{\left(\frac{300}{2\times5.066\times10^{-4}}\right)^2 + \frac{1}{5.066\times10^{-4}\times0.05\times10^{-3}}}$$

$$= 302.38 \times 10^3 \text{ rad/sec}$$

(*v*) Bandwidth $= \omega_2 - \omega_1 = 12.57 \times 10^3$ rad/sec

EXAMPLE 7.3. *In a series RLC circuit, the values of R, L and C are 50 Ω, 5 mH and 0.1 μF respectively. It is connected to an a.c. source of 109 V. Find the following.*

(*i*) *Resonance frequency.*

(*ii*) *Impedance at resonance.*

(*iii*) *Current in the circuit at resonance.*

(*iv*) *Quality factor of the circuit.*

(*v*) *Bandwidth of the circuit.*

(*vi*) *Frequency at which the voltages across the inductor is maximum and also find the voltage across the inductor under the above condition.*

(*vii*) *Frequency at which the voltages across the capacitor is maximum and also find the voltage across the capacitor under the above condition.*

SOLUTION. (*i*) $$W_o = \sqrt{\frac{1}{LC}} = \sqrt{\frac{1}{5\times10^{-3}\times0.1\times10^{-6}}} = 44.72 \times 10^3 \text{ rad/sec}$$

(*ii*) At resonance impedance $Z = R = 50\ \Omega$

(*iii*) $$I_0 = \frac{109}{50} = 2.18 \text{ Amp}$$

(*iv*) $$Q = \frac{W_O L}{R} = \frac{44.72\times10^3\times5\times10^{-3}}{50} = 4.472$$

(*v*) Bandwidth $= \frac{R}{L} = \frac{50}{5\times10^{-3}} = 10000$ rad/sec

(*vii*) $$f_o(l) = \frac{1}{2\pi\sqrt{LC - \frac{C^2R^2}{2}}}$$

$$f_o(l) = \frac{1}{2\pi\sqrt{5\times10^{-3}\times0.1\times10^{-6} - \frac{(0.1\times10^{-6})^2\times50^2}{2}}} = 7208.29 \text{ Hz}$$

$$Z = R + j\left[2\pi f_0(l)L - \frac{1}{2\pi f_0(l)C}\right] = (50 + j5.67)\ \Omega$$

$$I_L = \frac{V}{Z} = \frac{109}{(50+j5.67)} = (2.15 - j0.24) \text{ Amp}$$

$$V_L = I_L \times jX_L = (2.15 - j0.24) \times j226.46 = (54.35 + j486.89) \text{ volt}$$

(*viii*) $$f_o(c) = \frac{1}{2\pi}\sqrt{\frac{1}{LC} - \frac{R^2}{2L^2}}$$

$$f_o(c) = \frac{1}{2\pi}\sqrt{\frac{1}{5\times10^{-3}\times0.1\times10^{-6}} - \frac{50^2}{2\times(5\times10^{-3})^2}} = 7028.09 \text{ Hz}$$

$$Z = R + j\left[2\pi f_0(c)L - \frac{1}{2\pi f_0(c)C}\right] = (50 - j5.67)\ \Omega$$

$$I_C = \frac{V}{Z} = \frac{109}{(50-j5.67)} = (2.15 + j0.244) \text{ Amp}$$

$$V_C = I_C \times (-jX_C) = (2.15 + j0.244) \times (-j226.46) = (55.25 - j486) \text{ volt}$$

EXAMPLE 7.4. *When a series RLC circuit is connected across a 200 volt source, 5 amp current flows through it at resonance. The quality factor and the bandwidth is 62 and 6.283 rad/sec respectively. Find the values of R, L and C and also find the resonance frequency* ω_0.

SOLUTION. $$R = \frac{V}{I} = \frac{200}{5} = 40\ \Omega$$

$$L = \frac{R}{Bandwidth} = \frac{40}{6.283} = 6.366 \text{ H}$$

$$\omega_0 = \frac{QR}{L} = \frac{62 \times 40}{6.366} = 389.57 \text{ rad/sec}$$

$$C = \frac{1}{L\omega_0^2} = \frac{1}{6.366 \times 389.57^2} = 1.035 \text{ μF}$$

EXAMPLE 7.5. *For the circuit shown in Figure E7.5, the quality factor of the circuit is 100 and the resonance frequency is 10 kHz. Find the following:*

(i) the current through the circuit at resonance.

(ii) The values of L and C.

(iii) The drop across each element of the circuit.

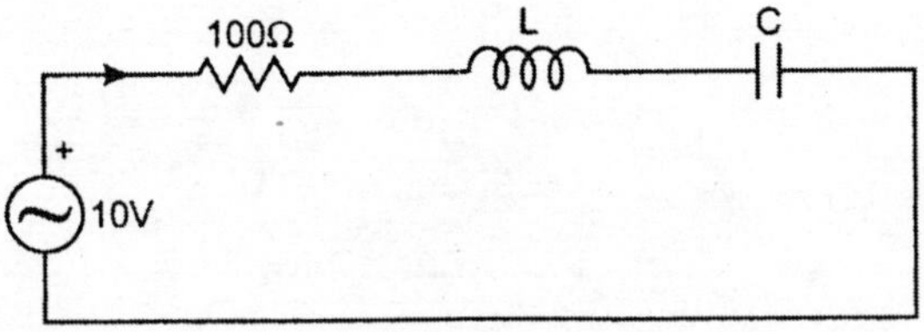

Fig. E7.5

SOLUTION. (*i*) The current through the circuit at resonance is,

$$I_0 = \frac{10}{100} = 0.1 \text{ Amp.}$$

(*ii*)

$$L = \frac{QR}{2\pi f_0} = \frac{100 \times 100}{2\pi \times 10 \times 10^3} = 0.159 \text{ H}$$

$$C = \frac{1}{L \times (2 \times \pi \times f_0)^2} = 1.592 \text{ nF}$$

(*iii*) Drop across

$$L = V_L = I_0 \times X_L = 0.1 \times 2 \times \pi \times 10 \times 10^3 \times 0.159$$

$$= 999.03 \text{ volt}$$

Drop across

$$C = V_C = I_0 \times X_C = \frac{0.1}{2 \times \pi \times 10 \times 10^3 \times 1.593 \times 10^{-9}}$$

$$= 999.08 \text{ volt}$$

EXAMPLE 7.6. *For the circuit shown in Figure E7.6, find an expression for the frequency at resonance.*

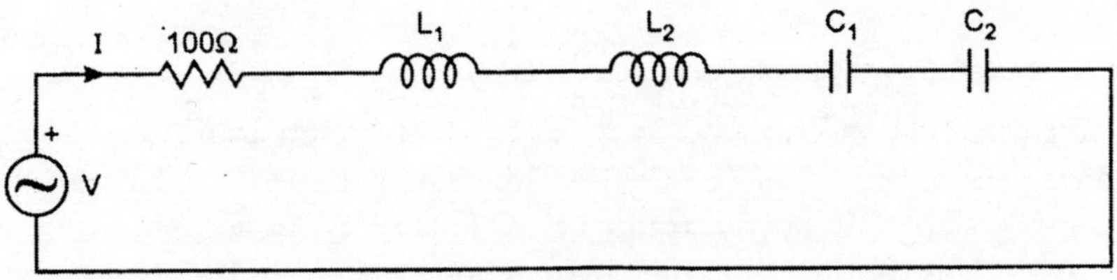

Fig. E7.6

SOLUTION. From the circuit shown in Figure E7.6 we get

$$L_{eq} = L_1 + L_2, \qquad C_{eq} = \frac{C_1C_2}{C_1+C_2}$$

Then

$$\omega_0 = \frac{1}{\sqrt{L_{eq}C_{eq}}} = \frac{1}{\sqrt{\frac{(L_1+L_2)\times C_1C_2}{C_1+C_2}}} = \sqrt{\frac{C_1+C_2}{(L_1+L_2)C_1C_2}}$$

EXAMPLE 7.7. *A capacitor of 30 μF and a resistance of 40 Ω are connected in series with a coil having resistance 5 Ω and inductance L. The circuit resonates at 1.5 kHz frequency. Find the value of L. Also find the current at resonance, Q factor, half power frequencies, and bandwidth.*

SOLUTION. At resonance

$$X_L = X_C \quad i.e., \quad 2\pi f_0 L = \frac{1}{2\pi f_0 C} \qquad \text{...}(i)$$

Then from equation (*i*)

$$L = \frac{1}{4\pi^2 f_0^2 C} = \frac{1}{4\pi^2 \times (1.5\times 10^3)^2 \times 30\times 10^{-6}} = 0.375 \text{ mH}$$

Current at resonance

$$I_0 = \frac{V}{R} = \frac{100}{40+5} = \frac{100}{45} = 2.22 \text{ Amp}$$

$$Q \text{ factor} = \frac{\omega_0 L}{R} = \frac{2\pi f_0 L}{R} = \frac{2\times\pi\times 1.5\times 10^3 \times 0.375\times 10^{-3}}{45} = 0.079$$

Half power frequencies are

$$\omega_1 = -\frac{R}{2L} + \sqrt{\left(\frac{R}{2L}\right)^2 - \frac{1}{LC}}$$

$$= -60 \times 10^3 + [3.6 \times 10^9 - 88.89 \times 10^6] = -745.38 \text{ rad/sec}$$

$$\omega_2 = \frac{R}{2L} + \sqrt{\left(\frac{R}{2L}\right)^2 - \frac{1}{LC}} = 119.25 \times 10^3 \text{ rad/sec}$$

$$\text{Bandwidth} = 119.25 \times 10^3 - 745.38 = 118.5 \times 10^3 \text{ rad/sec}$$

EXAMPLE 7.8. *For a series RLC, prove that the $\omega_2\omega_1 = \omega_0^2$, where ω_1 and ω_2 are the half power frequencies and ω_0 is the resonance frequency.*

SOLUTION. At half power frequencies, the net reactance of a series RLC resonant circuit is R, where,

$$R = \pm(X_L - X_C) \qquad \text{...}(i)$$

$$= \pm\left(\omega L - \frac{1}{\omega C}\right) \qquad \text{...}(ii)$$

where + sign indicates the condition when $X_L > X_C$, and − sign indicates the condition when $X_C > X_L$, depending on the band of frequencies that lies between the half power frequencies. Let ω_1 and ω_2 be the two half power frequencies. Let at frequency ω_1, $X_C > X_L$, that is the net reactance of the circuit is negative and at frequency ω_2, $X_L > X_C$, that is the net reactance of the circuit is positive. Then at frequencies ω_1 and ω_2 equation (*ii*) becomes.

$$\left(\omega_1 L - \frac{1}{\omega_1 C}\right) = -R \qquad ...(iii)$$

$$\left(\omega_2 L - \frac{1}{\omega_2 C}\right) = R \qquad ...(iv)$$

Adding equations (*iii*) and (*iv*) we get

$$L(\omega_1 + \omega_2) - \frac{1}{C}\left(\frac{1}{\omega_1} + \frac{1}{\omega_2}\right) = 0$$

or $$L(\omega_1 + \omega_2) - \frac{1}{C}\left(\frac{\omega_1 + \omega_2}{\omega_1 \omega_2}\right) = 0$$

or $$(\omega_1 + \omega_2)\left[L - \frac{1}{C}\left(\frac{1}{\omega_1 \omega_2}\right)\right] = 0 \qquad ...(v)$$

From equation (*v*) we can say either

$$(\omega_1 + \omega_2) = 0 \quad \text{or} \quad \left[L - \frac{1}{C}\left(\frac{1}{\omega_1 \omega_2}\right)\right] = 0$$

The first case cannot be true because ω_1 and ω_2 are the half power frequencies, so

$$\left[L - \frac{1}{C}\left(\frac{1}{\omega_1 \omega_2}\right)\right] = 0$$

or $$\omega_1 \omega_2 = \frac{1}{LC} \qquad ...(vi)$$

or $$\omega_1 \omega_2 = \omega_0^2$$

since for a series RLC circuit $\omega_0 = \sqrt{\frac{1}{LC}}$, hence proved.

Parallel Resonance

EXAMPLE 7.9. *For the circuit shown in Figure E7.9, find an expression for resonance frequency.*

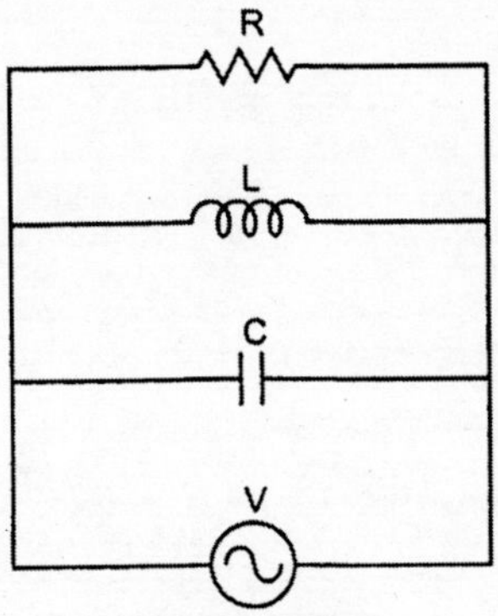

Fig. E7.9

SOLUTION. From Figure E7.9, the total impedance is given by

$$\frac{1}{Z_{eq}} = \frac{1}{R}+\frac{1}{X_L}+\frac{1}{X_C} \qquad ...(i)$$

where $$X_C = j\omega L \qquad ...(ii)$$

and $$X_C = \frac{1}{j\omega C} \qquad ...(iii)$$

$$\frac{1}{Z_{eq}} = \frac{1}{R}+\frac{1}{j\omega L}+j\omega C \qquad ...(iv)$$

or $$\frac{1}{Z_{eq}} = \frac{1}{R}+j\left(\omega C-\frac{1}{\omega L}\right) \qquad ...(v)$$

At resonance the circuit behaves as a purely resistive circuit. In other words the imaginary part of equation (*v*) vanishes. Thus the resonance frequency is obtained by equating the imaginary part to zero. That is

$$\left(\omega_0 C-\frac{1}{\omega_0 L}\right) = 0 \qquad ...(vi)$$

where ω_0 is the resonance frequency. Then from equation (*vi*) we get

$$\omega_0 = \sqrt{\frac{1}{LC}}$$

EXAMPLE 7.10. *For the circuit shown in Figure E7.10, find an expression for resonance frequency.*

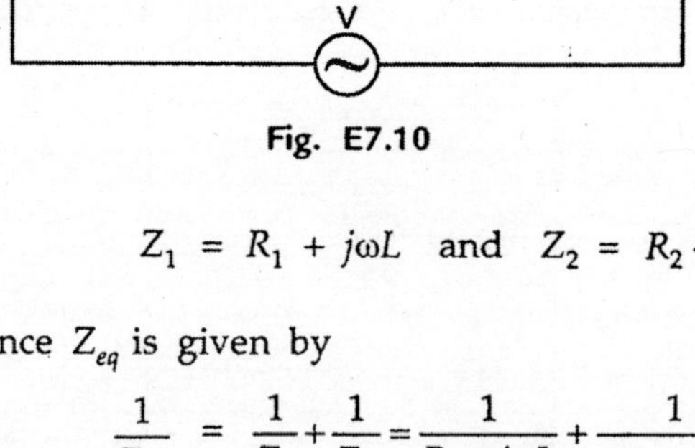

Fig. E7.10

SOLUTION. Let $$Z_1 = R_1 + j\omega L \text{ and } Z_2 = R_2+\frac{1}{j\omega C} = Z_2 = R_2 - j\frac{1}{\omega C}$$

Then the total impedance Z_{eq} is given by

$$\frac{1}{Z_{eq}} = \frac{1}{Z_1}+\frac{1}{Z_2} = \frac{1}{R_1+j\omega L}+\frac{1}{R_2-j\frac{1}{\omega C}}$$

$$= \frac{R_1-j\omega L}{R_1^2+\omega^2L^2}+\frac{\omega C}{R_2\omega C-j} = \frac{R_1-j\omega L}{R_1^2+\omega^2L^2}+\frac{\omega C(R_2\omega C+j)}{R_2^2\omega^2C^2+1}$$

$$= \frac{R_1}{R_1^2+\omega^2L^2}+\frac{R_2\omega^2C^2}{R_2^2\omega^2C^2+1}+j\left(\frac{\omega C}{R_2^2\omega^2C^2}-\frac{\omega L}{R_1^2+\omega^2L^2}\right) \quad ...(i)$$

At resonance the circuit behaves as a purely resistive circuit. In other words the imaginary part of equation (*i*) vanishes. Thus the resonance frequency is obtained by equating the imaginary part to zero. That is

$$\left(\frac{\omega_0 C}{R_2^2\omega_0^2C^2}-\frac{\omega_0 L}{R_1^2+\omega_0^2L^2}\right) = 0 \quad ...(ii)$$

where ω_0 is the resonance frequency. Then from equation (*ii*) we get

$$\omega_0 = \sqrt{\frac{CR_1^2-L}{R_2^2C^2L-CL^2}}$$

EXAMPLE 7.11. *For the circuit shown in Figure E7.11, find an expression for resonance frequency.*

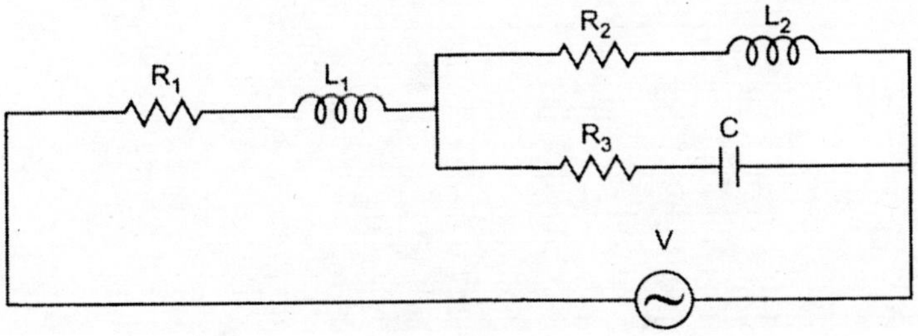

Fig. E7.11

SOLUTION. Let $Z_1 = R_1 + j\omega L_1$ and $Z_2 = R_2 + j\omega L_2$ $Z_3 + R_3 + \frac{1}{j\omega C}$

Then the total impedance Z_{eq} is given by

$$Z_{eq} = Z_1+\frac{Z_2Z_3}{Z_2+Z_3} = Z_1+\frac{(R_2+j\omega L_2)\times\left(R_3-j\frac{1}{\omega C}\right)}{(R_2+j\omega L_2)+\left(R_3-j\frac{1}{\omega C}\right)}$$

$$= Z_1+\frac{R_2R_3+\frac{L_2}{C}+j\left(\omega L_2R_3-\frac{R_2}{\omega C}\right)}{(R_2+R_3)+j\left(\omega L_2-\frac{1}{\omega C}\right)}$$

$$= Z_1+\frac{R_2R_3+\frac{L_2}{C}+j\left(\omega L_2R_3-\frac{R_2}{\omega C}\right)}{(R_2+R_3)+j\left(\omega L_2-\frac{1}{\omega C}\right)}\times\frac{(R_2+R_3)-j\left(\omega L_2-\frac{1}{\omega C}\right)}{(R_2+R_3)-j\left(\omega L_2-\frac{1}{\omega C}\right)}$$

$$= Z_1 + \frac{R_2R_3 + \frac{L_2}{C} + j\left(\omega L_2R_3 - \frac{R_2}{\omega C}\right) \times (R_2 + R_3) - j\left(\omega L_2 - \frac{1}{\omega C}\right)}{(R_2 + R_3)^2 + \left(\omega L_2 - \frac{1}{\omega C}\right)^2}$$

$$= R_1 + j\omega L_1 + \frac{R_2R_3 + \frac{L_2}{C} + j\left(\omega L_2R_3 - \frac{R_2}{\omega C}\right) \times (R_2 + R_3) - j\left(\omega L_2 - \frac{1}{\omega C}\right)}{(R_2 + R_3)^2 + \left(\omega L_2 - \frac{1}{\omega C}\right)^2} \quad ...(i)$$

At resonance the circuit behaves as a purely resistive circuit. In other words the imaginary part of equation (*i*) vanishes. Thus the resonance frequency is obtained by equating the imaginary part to zero. That is

$$j\omega_0 L_1 + \frac{j\left(\omega_0 L_2R_3 - \frac{R_2}{\omega_0 C}\right) \times (R_2 + R_3) - j\left(\omega_0 L_2 - \frac{1}{\omega_0 C}\right)}{(R_2 + R_3)^2 + \left(\omega_0 L_2 - \frac{1}{\omega_0 C}\right)^2} = 0$$

or $$\omega_0 L_1 + \omega_0 L_2 R_3 (R_2 + R_3) - \frac{R_2 \times (R_2 + R_3)}{\omega_0 C} - \left(\omega_0 L_2 - \frac{1}{\omega_0 C}\right) = 0$$

or $$\omega_0^2 L_1 C + \omega_0^2 L_2 R_3 C(R_2 + R_3) - R_2(R_2 + R_3) - \omega_0^2 L_2 C + 1 = 0$$

or $$\omega_0^2 [L_1 C + L_2 R_3 C(R_2 + R_3) - L_2 C] = R_2(R_2 + R_3) - 1$$

or $$\omega_0 = \sqrt{\frac{R_2^2 + R_2R_3 - 1}{L_1C + L_2R_3C(R_2 + R_3) - L_2C}}$$

EXAMPLE 7.12. *For the circuit shown in Figure E7.12, find an expression for resonance frequency.*

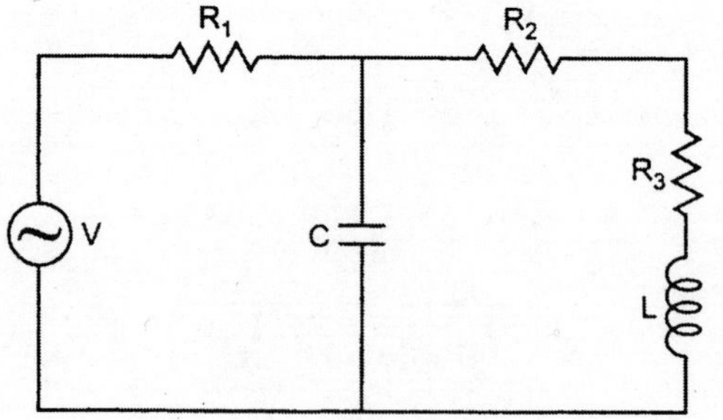

Fig. E7.12

SOLUTION. The equivalent impedance as seen from the source terminal is

$$Z_{eq} = R_1 + \frac{(R_2 + R_3 + j\omega L) \times \frac{1}{j\omega C}}{(R_2 + R_3) + j\omega L + \frac{1}{j\omega C}} = R_1 + \frac{(R_2 + R_3 + j\omega L) \times \left(-j\frac{1}{\omega C}\right)}{(R_2 + R_3) + j\left(\omega L - \frac{1}{\omega C}\right)}$$

$$= R_1 + \frac{\frac{L}{C} - j\left(\frac{R_2+R_3}{\omega C}\right)}{(R_2+R_3)+j\left(\omega L - \frac{1}{\omega C}\right)} \times \frac{(R_2+R_3)-j\left(\omega L - \frac{1}{\omega C}\right)}{(R_2+R_3)-j\left(\omega L - \frac{1}{\omega C}\right)}$$

$$= R_1 + \frac{\left[\frac{L}{C} - j\left(\frac{R_2+R_3}{\omega C}\right)\right] \times \left[(R_2+R_3)-j\left(\omega L - \frac{1}{\omega C}\right)\right]}{(R_2+R_3)^2 + \left(\omega L - \frac{1}{\omega C}\right)^2}$$

$$= R_1 + \frac{\frac{L}{C}(R_2+R_3) - \left(\frac{R_2+R_3}{\omega C}\right) \times \left(\omega L - \frac{1}{\omega C}\right) - j\left(\frac{(R_2+R_3)^2}{\omega C}\right) - j\left(\omega L - \frac{1}{\omega C}\right) \times \frac{L}{C}}{(R_2+R_3)^2 + \left(\omega L - \frac{1}{\omega C}\right)^2}$$

$$= R_1 + \frac{\frac{L}{C}(R_2+R_3) - \left(\frac{R_2+R_3}{\omega C}\right) \times \left(\omega L - \frac{1}{\omega C}\right)}{(R_2+R_3)^2 + \left(\omega L - \frac{1}{\omega C}\right)^2} - \frac{j\left(\frac{(R_2+R_3)^2}{\omega C}\right) + j\left(\omega L - \frac{1}{\omega C}\right) \times \frac{L}{C}}{(R_2+R_3)^2 + \left(\omega L - \frac{1}{\omega C}\right)^2} \quad ...(i)$$

At resonance the circuit behaves as a purely resistive circuit. In other words the imaginary part of equation (*i*) vanishes. Thus the resonance frequency is obtained by equating the imaginary part to zero. That is

$$\frac{j\left(\frac{(R_2+R_3)^2}{\omega_0 C}\right) + j\left(\omega_0 L - \frac{1}{\omega_0 C}\right) \times \frac{L}{C}}{(R_2+R_3)^2 + \left(\omega_0 L - \frac{1}{\omega_0 C}\right)^2} = 0 \quad ...(ii)$$

$$\left(\frac{(R_2+R_3)^2}{\omega_0 C}\right) + \left(\omega_0 L - \frac{1}{\omega_0 C}\right) \times \frac{L}{C} = 0 \quad ...(iii)$$

Solving equation (*iii*) we get

$$\omega_0 = \sqrt{\frac{L - C(R_2+R_3)^2}{L^2 C}}$$

EXERCISES

1. A series RLC circuit with $R = 20\ \Omega$, $L = 5$ mH and $C = 0.2 \times 10^{-6}$ F, is connected across a 100 V A.C. supply. Find the following
   (*a*) Resonance frequency.
   (*b*) Current in the circuit at resonance.
   (*c*) Quality factor of the circuit.

(*d*) Half power frequencies.

(*e*) Bandwidth of the circuit.

2. In a series RLC circuit, the values of *R*, *L* and *C* are 30 Ω, 6 mH and 0.2 μF respectively. It is connected to an a.c. source of 100 V. Find the following.
   (*i*) Resonance frequency.
   (*ii*) Impedance at resonance.
   (*iii*) Current in the circuit at resonance.
   (*iv*) Quality factor of the circuit.
   (*v*) Bandwidth of the circuit.
   (*vi*) Frequency at which the voltages across the inductor is maximum and also find the voltage across the inductor under the above condition.
   (*vii*) Frequency at which the voltages across the capacitor is maximum and also find the voltage across the capacitor under the above condition.

3. For the circuit shown in Figure Q7.3, find an expression for resonance frequency.

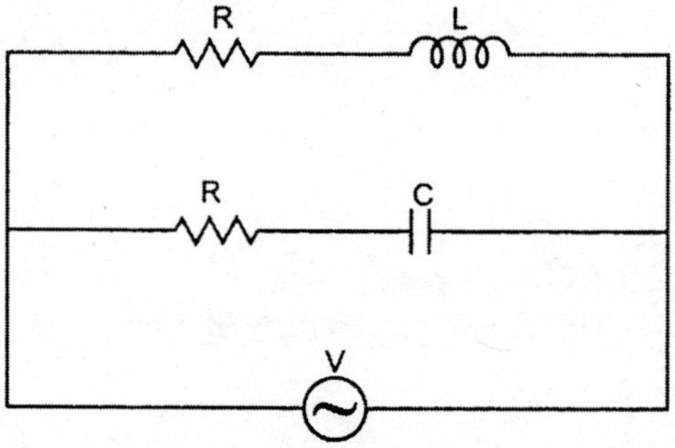

Fig. Q7.3

4. For the circuit shown in Figure Q7.4, find an expression for resonance frequency.

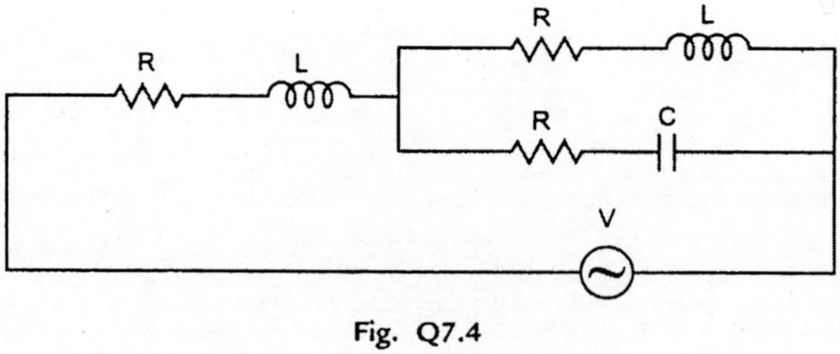

Fig. Q7.4

Chapter 8

LAPLACE TRANSFORMS

8.1 INTRODUCTION

Laplace transform provides a method for representing and analyzing linear systems using algebraic methods. Laplace transforms converts integral and differential equations (calculus problems) into an algebraic equation problem. The solution of algebraic problem is then feel backward through what is called an "Inverse laplace transform" and the solution of the differential equation is obtained.

CALCULUS ⟶ LAPLACE ⟶ ALGEBRA ⟶ INVERSE LAPLACE ⟶ CALCULUS

8.2 DEFINITION

The laplace transform of a given function $f(t)$, defined for $t > 0$, is denoted by $L[f(t)]$ and is defined as

$$L[f(t)] = \int_0^{\infty} e^{-st} f(t)\, dt = F(s) \qquad \text{...(8.1)}$$

where S is an arbitrary positive number. S is called the (complex) frequency variable with units $\sec^{-1}$. t is called the time variable and its unit is sec. and St is unitless.

Note that although S is an arbitrary positive number, it is occasionally necessary to assume that it is large enough to avoid difficulties in the calculation.

8.3 LAPLACE TRANSFORM OF ELEMENTARY FUNCTION

(1) $f(f) = e^t$

$$F(s) = L[f(t)] = \int_0^\infty e^t e^{-st}\, dt = \int_0^\infty e^{(1-s)t} dt = \frac{1}{1-s} e^{(1-s)t}\Big|_0^\infty = \frac{1}{s-1}$$

(2) $f(t) = 1$ for $t \geq 0$. It is called unit step function.

The unit step function is also represented as follows

$$u(t) = \begin{cases} 1 \text{ for } t \geq 0 \\ 0 \text{ for } t < 0 \end{cases}$$

$$F(s) = L[f(t)] = \int_0^\infty e^{-st} dt = -\frac{1}{s} e^{-st}\Big|_0^\infty = \frac{1}{s}$$

(3) $f(t) = \cos \omega t$

It can also be written as

$$f(t) = \frac{1}{2} e^{j\omega t} + \frac{1}{2} e^{-j\omega t}$$

Then,

$$F(s) = L[f(t)] = \int_0^\infty e^{-st} \left(\frac{1}{2} e^{j\omega t} + \frac{1}{2} e^{-j\omega t} \right) dt$$

$$= \frac{1}{2} \int_0^\infty e^{(-s+j\omega)t} dt + \frac{1}{2} \int_0^\infty e^{-(s+j\omega)t} dt$$

$$= \frac{1}{2} \frac{1}{s - j\omega} + \frac{1}{2} \frac{1}{s + j\omega}$$

$$= \frac{s}{s^2 + \omega^2}$$

(4) $f(t) = \sin \omega t$

$$f(t) = \frac{1}{2j} e^{j\omega t} - \frac{1}{2j} e^{-j\omega t}$$

Then,

$$F(s) = L[f(t)] = \int_0^\infty e^{-st} \left(\frac{1}{2j} e^{j\omega t} - \frac{1}{2j} e^{-j\omega t} \right) dt$$

$$= \frac{1}{2j} \int_0^\infty e^{(-s+j\omega)t} dt - \frac{1}{2j} \int_0^\infty e^{-(s+j\omega)t} dt$$

$$= \frac{1}{2j}\frac{1}{s-j\omega} - \frac{1}{2j}\frac{1}{s+j\omega}$$

$$= \frac{\omega}{s^2+\omega^2}$$

(5) $f(t) = t$

$$F(s) = L[f(t)] = \int_0^\infty t \times e^{-st} dt$$

Using integration by parts we get,

$$F(s) = t^n\left[-\frac{te^{-st}}{s}\right]_0^\infty + \frac{1}{s}\int_0^\infty e^{-st} dt = \frac{1}{s^2}$$

(6) $f(t) = t^2$

$$F(s) = L[f(t)] = \int_0^\infty t^2 \times e^{-st} dt$$

Using integration by parts we get,

$$F(s) = t^n\left[-\frac{t^2e^{-st}}{s}\right]_0^\infty + \frac{2}{s}\int_0^\infty te^{-st} dt = \frac{2}{s^3}$$

(7) $f(t) = t^n$ for $n \geq 1$

$$F(s) = L[f(t)] = \int_0^\infty t^n \, e^{-st} dt$$

Using integration by parts we get,

$$t^n\left[\frac{-e^{-st}}{s}\right]_0^\infty + \frac{n}{s}\int_0^\infty t^{n-1}e^{-st} dt = \frac{n}{s}L[t^{n-1}]$$

By applying the formula recursively we get,

$$F(s) = \frac{n!}{s^{n+1}}$$

(8) $f(t) = \delta(t)$ impulse function

$$F(s) = L[f(t)] = L[\delta(t)] = \int_0^\infty \delta(t) \, e^{-st} dt = e^{-st}\Big|_{t=0} = 1$$

(9) $f(t) = \delta^k(t)$

$$F(s) = L[f(t)] = L[\delta^k(t)] = \int_0^\infty \delta^k(t) \, e^{-st} dt$$

$$= (-1)^k \frac{d^k}{dt^k} e^{-st}\Big|_{t=0} = s^k e^{-st}\Big|_{t=0} = s^k$$

(10) $f(t) = e^{-at}$

$$F(s) = L[e^{-at}] = \int_0^\infty e^{-st}e^{-at}dt = \int_0^\infty e^{-(s+a)t}dt = \left[\frac{e^{-(s+a)t}}{-(s+a)}\right]_0^\infty = \frac{1}{s+a}$$

Similarly if $f(t) = e^{bt}$ then

$$F(s) = L[e^{bt}] = \frac{1}{s-b}$$

Here it is assumed that $s > b$ so that $e^{-(s-b)t}$ becoming a decay exponential.

(11) $f(t) = \sinh at$

$$\sinh at = \frac{e^{at} - e^{-at}}{2}$$

Then,
$$L[\sinh at] = L\left[\frac{e^{at} - e^{-at}}{2}\right] = \frac{1}{2}\left[\frac{1}{s-a} - \frac{1}{s+a}\right] = \frac{a}{s^2 - a^2}$$

(12) $f(t) = \cosh at$

$$\cosh at = \frac{e^{at} + e^{-at}}{2}$$

Then,
$$L[\cosh at] = L\left[\frac{e^{at} + e^{-at}}{2}\right] = \frac{1}{2}\left[\frac{1}{s-a} + \frac{1}{s+a}\right] = \frac{s}{s^2 - a^2}$$

Table 8.1

Laplace Transforms of Some Commonly Used Functions

Function [F(t)]	*Laplace Transformations [F(s)]*
$\delta(t)$ (Unit Impulse)	1
$\delta(t - T)$	e^{-Ts}
$u(t)$ (Unit step)	$\frac{1}{s}$
$-u(-t)$	$\frac{1}{s}$
$u(t - T)$	$\frac{1}{s}e^{-Ts}$
A (Step)	$\frac{A}{s}$

t (Ramp)	$\frac{1}{s^2}$
$\frac{t^{n-1}}{(n-1)!}u(t)$	$\frac{1}{s^n}$
$-\frac{t^{n-1}}{(n-1)!}u(-t)$	$\frac{1}{s^n}$
$e^{-\alpha t}u(t)$	$\frac{1}{s+\alpha}$
$-e^{-\alpha t}u(-t)$	$\frac{1}{s+\alpha}$
te^{-at}	$\frac{1}{(s+a)^2}$
t^2e^{-at}	$\frac{2!}{(s+a)^3}$
$-\frac{t^{n-1}}{(n-1)!}e^{-\alpha t}u(-t)$	$\frac{1}{(s+\alpha)^n}$
$\delta(t - T)$	e^{-sT}
$\cos \omega t \ u(t)$	$\frac{s}{s^2+\omega^2}$
$\sin \omega t \ u(t)$	$\frac{\omega}{s^2+\omega^2}$
$\sin (\omega t + \theta)$	$\frac{s \sin\theta + \omega\cos\theta}{s^2+\omega^2}$
$e^{-at} \sin \omega t \ u(t)$	$\frac{\omega}{(s+a)^2+\omega^2}$
t^2	$\frac{2}{s^3}$
t^n for $n > 0$	$\frac{n!}{s^{n+1}}$
$\frac{e^{-at}-e^{-bt}}{b-a}$	$\frac{1}{(s+a)(s+b)}$

8.4 LAPLACE TRANSFORM RULES

(i) **Linearity :** Laplace transform is linear, *i.e.*

$$L[f_1(t) + f_2(t)] = L[f_1(t)] = L[f_2(t)] = F_1(s) + F_2(s)$$

and $$L[k\, f(t)] = kL[f(t)] = kF(s)$$

where k is a constant. The above example also shows that homogeneity and superposition holds for laplace transforms

Example : $$L[3\delta(t) - 2e^t] = 3L[\delta(t)] - 2L[e^t] = 3 - \frac{2}{s-1} = \frac{3s-5}{s-1}$$

(ii) **One-to-one property :** The laplace transform is one-to-one. If $L[f(t)] = L[g(t)]$ then $f(t) = g(t)$.

(iii) **Laplace transform of a derivatives :** If $f(t)$ is continuous at $t = 0$ then,

$$L[f'(t)] = sL[f(t)] - f(0) = sF(s) - f(0)$$

Proof :

$$L[f'(t)] = \int_0^\infty e^{-st} f'(t)\, dt = [e^{-st} f(t)]_0^\infty + s\int_0^\infty e^{-st} f(t)\, dt$$

Using integration by parts we get,

$$L[f'(t) = sL[f(t)] - f(0) = sF(s)] - f(0)$$

Alternatively,

$$L\left[\frac{df}{dt}\right] = sF(s) - f(0)$$

where $f(0)$ is called the initial condition. That is the value of $f(t)$ at $t = 0$.

Similarly,

$$L[f''(t)] = sL[f'(t)] - f'(0) = s[sF(s) - f(0)] - f'(0)$$
$$= s^2F(s) - sf(0) - f'(0)$$

Alternatively,

$$L\left[\frac{d^2 f}{dt^2}\right] = s^2F(s) - sf(0) - f'(0)$$

In general,

$$L\left[\frac{d^n f}{dt^n}\right] = snF(s) - s^{n-1} f(0) - s^{n-2}f'(0) - \ldots - f^{n-1}(0).$$

(iv) **Time scaling property :** Let us define a function $g(t) = f(at)$. Where $a > 0$ is a constant value. If $F(s)$ is the laplace transform of $f(t)$, then

$$L[g(t)] = L[f(at)] = \int_0^\infty f(at)\, e^{-st} dt = \frac{1}{a}\int_0^\infty f(x) e^{-(s/a)x} dx \qquad \ldots(8.2)$$

where $x = at$ and $\dfrac{dx}{dt} = a$

Comparing the integral of equation (8.2) with equation (8.1), we find that s has been replaced by s/a and t has been replaced by x. Thus

$$L[f(at)] = \frac{1}{a}F\left(\frac{s}{a}\right)$$

Example : If $L[e^{t}] = \dfrac{1}{s-a}$ then $L[e^{-at}] = \dfrac{1}{a}\dfrac{1}{\left(\frac{s}{a}\right)-a}$

Example : If $L[u(t)\sin\omega t] = \dfrac{\omega}{\omega^2+s^2}$ then $L[u(t)\sin 2\omega t] = \dfrac{1}{2}\times\dfrac{\omega}{\left(\frac{s}{2}\right)^2+\omega^2}$

(*v*) **Time delay :** This property is also called the time shift property. Let us consider a function as shown below.

$$g(t) = \begin{cases} 0 & t > T \\ f(t-T) & t \geq T \end{cases}$$

Then $$L[g(t) = G(s) = e^{-sT}F(s)$$

Proof :

$$G(s) = \int_0^\infty e^{-sT}\, g(t)\,dt = \int_T^\infty e^{-sT} f(t-T)\,dt$$

Let $$x = t - T \quad \text{then} \quad dx = dt$$

$$G(s) = \int_0^\infty f(x)e^{-s(x+T)}dx = e^{-Ts}\int_0^\infty f(x)e^{-sx}dx = e^{-Ts}F(s)$$

(*vi*) **Derivation of derivative formula :**

$$G(s) = \int_0^\infty f'(t)e^{-st}dt$$

$$= e^{-st}f(t)\Big|_0^\infty - \int_0^\infty f(t)-(se^{-st})dt$$

$$= \lim_{t\to\infty} f(t)e^{-st} - f(0) + sF(s)$$

$$= sF(s) - f(0)$$

(*vii*) **Frequency shift (First shift theorem) :** If $f(t)$ is a function of t and $F(s)$ is its laplace transform, then

$$L[u(t)e^{-at}f(t)] = \int_0^\infty e^{-at}f(t)e^{-st}dt$$

$$= \int_0^\infty f(t)e^{-(s+a)t}\,dt$$

$$= F(s + a)$$

(viii) Time integration : If $f(t)$ is a function of t and $F(s)$ is its laplace transform, then

$$= L\left[\int_0^t f(t)\,dt\right] = \int_0^\infty \left[\int_0^t f(x)\,dx\right] e^{-st}\,dt$$

Using integration by parts we get,

$$u = \int_0^t f(x)dx \qquad du = f(t)\,dt$$

$$dv = e^{-st}\,dt \qquad v = -\frac{1}{s}e^{-st}$$

$$L\left[\int_0^t f(t)\,dt\right] = \left[\int_0^t f(x)\,dx\right] \times \left(-\frac{1}{s}e^{-st}\right)\Bigg|_0^\infty - \int_0^\infty -\frac{1}{s}e^{-st} f(t)\,dt$$

$$= \frac{1}{s}F(s)$$

(ix) Periodic function : if $f(t)$ is a periodic function with period $T > 0$, then,

$$L[f(t)] = \frac{\int_0^T f(t)e^{-st}dt}{1-e^{-sT}}$$

Proof : Since $f(t)$ is a periodic function with period T, we have $f(t + T) = f(t)$

$$L[f(t)] = \int_0^\infty f(t)e^{-st}dt = \int_0^T f(t)e^{-st}dt + \int_T^\infty f(t)e^{-st}dt \qquad ...(8.3)$$

Considering only $\int_T^\infty f(t)e^{-st}dt$

$$t = u + T \text{ so that } dt = du$$

$$\int_T^\infty f(t)e^{-st}dt = \int_0^\infty f(u+T)e^{-s(u+T)}\,du$$

$$= e^{-sT}\int_0^\infty f(u)e^{-su}du = e^{-sT}\int_0^\infty f(t)e^{-st}du$$

$$= e^{-sT}L[f(t)]$$

From equation (8.3)

$$L[f(t)](1 - e^{-sT}) = \int_0^T f(t)e^{-dt}\, dt$$

$$L[f(t)] = \frac{1}{1-e^{-sT}}\int_0^T f(t)e^{-st}\, dt$$

(x) Error function : The error function is defined by,

$$\text{erf}(t) = \frac{2}{\sqrt{\pi}}\int_0^t e^{-\xi^2}\, d\xi$$

whereas the complementary error function erfc(t) is given by,

$$\text{erfc}(t) = 1 - \text{ref}(t)$$

Then

$$L[\text{erf}(\sqrt{t})] = \frac{2}{\sqrt{\pi}}\int_0^\infty e^{-st}\int_0^{\sqrt{t}} e^{-\xi^2}\, d\xi\, dt$$

$$= \frac{2}{\sqrt{\pi}}\int_0^\infty e^{-\xi^2}\int_{\xi^2}^\infty e^{-st}\, dt\, d\xi$$

on changing the order of integration

$$= \frac{2}{s\sqrt{\pi}}\int_0^\infty e^{-(1+s)\xi^2}\, d\xi$$

on evaluating the inner integral

$$= \frac{1}{s\sqrt{1+s}}$$

(xi) Frequency differentiation : If $f(t)$ is a function of t and $F(s)$ is its laplace transform, then

$$F(s) = \int_0^\infty f(t)e^{-st}\, dt$$

$$\frac{dF(s)}{ds} = \int_0^\infty f(t)(-te^{-st})\, dt$$

$$= \int_0^\infty (-tf(t))e^{-st}\, dt$$

$$= L[-t\, f(t)]$$

Repeated application of this equation gives

$$L[t^n f(t)] = (-1)^n \frac{d^n F(s)}{ds^n}$$

Example : $$L[te^{-at}u(t)] = -\frac{d}{ds}\left(\frac{1}{s+a}\right) = \frac{1}{(s+a)^2}$$

(*xii*) Multiplication by *t* : Let $g(t)$ be a function defined by $g(t) = tf(t)$, then $G(s) = -F'(s)$

$$F(s) = \int_0^\infty f(t)e^{-st}\,dt$$

differentiating with respect to s we get,

$$F'(s) = \int_0^\infty (-t)e^{-st} f(t)\,dt$$
$$= -F'(s)$$

Initial Value Theorem

If $f(t)$ and $f'(t)$ are laplace transformable and if $L[f(t)] = F(s)$, then

$$\lim_{s\to\infty} sF(s) = \lim_{t\to 0} f(t)$$

Proof :

We know that, $$L[f'(t)] = sL[f(t)] - f(0)$$
$$= sF(s) = sL[f(t)]$$
$$= L[f'(t)] + f(0)$$
$$= \int_0^\infty e^{-st} f'(t)\,dt + f(0)$$

Taking limit as $s \to \infty$ on both sides we have,

$$\lim_{s\to\infty} sF(s) = \lim_{s\to\infty} \int_0^\infty e^{-st} f'(t)\,dt + f(0)$$
$$= \int_0^\infty \left[\lim_{s\to\infty} e^{-st} f'(t)\right] dt + f(0)$$
$$= f(0)$$
$$= \lim_{t\to 0} f(t)$$

Final Value Theorem

If $f(t)$ and $f'(t)$ are laplace transformable and if $L[f(t)] = F(s)$, then

$$\lim_{s\to 0} sF(s) = \lim_{t\to\infty} f(t)$$

Proof :

We know that,
$$L[f'(t)] = sL[f(t)] - f(0)$$
$$sF(s) = sL[f(t)]$$
$$= L[f'(t) + f(0)$$
$$= \int_0^\infty e^{-st} f'(t)dt + f(0)$$

Taking limit as $s \to 0$ on both sides we have,

$$\lim_{s\to 0} sF(s) = \lim_{s\to 0} \int_0^\infty e^{-st} f'(t)dt + f(0)$$
$$= \int_0^\infty \left[\lim_{s\to 0} e^{-st} f'(t)\right] dt + f(0)$$
$$= \int_0^\infty f'(t)dt + f(0)$$
$$= \lim_{t\to\infty} [f(t)]_0^t + f(0)$$
$$= \lim_{t\to\infty} f(t) - f(0) + f(0)$$
$$= \lim_{t\to\infty} f(t)$$

SOLVED EXAMPLES

EXAMPLE 8.1. *Find the laplace transform of* $f(t) = 5e^{-6t} + t^2 + \sin 4t - 5 \cos 2t$

SOLUTION.
$$L[f(t)] = L[5e^{-6t} + t^2 + \sin 4t - 5 \cos 2t]$$
$$= \frac{5}{s+6} + \frac{2!}{s^3} + \frac{4}{s^2+16} - 5\frac{s}{s^2+4}$$

EXAMPLE 8.2. *Find the laplace transform of* $f(t) = t^5 e^{-6t}$

SOLUTION.
$$L[f(t)] = L[t^5 e^{-6t}]$$

Then
$$L[t^5] = \frac{5!}{s^6}$$

So
$$L[t^5 e^{-6t}] = \frac{5!}{(s+6)^6}$$

EXAMPLE 8.3. *Find the laplace transform of* $f(t) = 6 \cos 4t + 2 \sin 10t$

SOLUTION.
$$L[f(t)] = L[6 \cos 4t + 2 \sin 10t]$$
$$= \frac{6s}{s+16} + \frac{20}{s^2+100}$$

EXAMPLE 8.4. *Find the laplace transform of* $f(t) = e^{-4t}\sin 3t$

SOLUTION. $$L[f(t)] = L[e^{-4t}\sin 3t]$$

Then $$L[\sin 3t] = \frac{3}{s^2+9}$$

So $$L[e^{-4t}\sin 3t] = \frac{3}{(s+4)^2+9}$$

EXAMPLE 8.5. *Find the laplace transform of* $f(t) = t^4e^{4t}$

SOLUTION. $$L[f(t)] = L[t^4e^{4t}]$$

Then $$L[t^4] = \frac{4!}{s^5}$$

So $$L[t^4e^{4t}] = \frac{4!}{(s-4)^5}$$

EXAMPLE 8.6. *Find the laplace transform of* $f(t) = t\sin t$

SOLUTION. $$L[f(t)] = L[t\sin t]$$

Then $$L[\sin t] = \frac{1}{s^2+1}$$

So $$L[t\sin t] = -\frac{d}{ds}\left(\frac{1}{(s^2+1)}\right) = \frac{2s}{(s^2+1)^2}$$

EXAMPLE 8.7. *Find the laplace transform of* $f(t) = t^2\cos t$

SOLUTION. Then $$L[\cos t] = \frac{s}{s^2+1}$$

So $$L[t^2\cos t] = \frac{d^2}{ds^2}\left(\frac{s}{(s^2+1)}\right) = \frac{2s^3-6s}{(s^2+1)^3}$$

EXAMPLE 8.8. *Find the laplace transform of* $f(t) = e^t(\cos t + \sin t)$

SOLUTION. $$L[f(t)] = L[e^t(\cos t + \sin t)]$$

Then $$L[\cos t + \sin t] = \frac{s}{s^2+1} + \frac{1}{s^2+1}$$

So $$L[e^t(\cos t + \sin t)] = \frac{s-1}{(s-1)^2+1} + \frac{1}{(s-1)^2+1} = \frac{s}{(s-1)^2+1}$$

EXAMPLE 8.9. *Find the laplace transform of* $f(t) = \cos t \,.\, \sin t$

SOLUTION. $$\cos t = \frac{e^t+e^{-t}}{2}$$

Then $$L(\cos t \,.\, \sin t) = \frac{1}{2}[L(t^t \sin t) + L(e^{-t}\sin t)]$$

$$= \frac{1}{2}\left[\frac{1}{(s-1)^2+1} + \frac{1}{(s+1)^2+1}\right]$$

EXAMPLE 8.10. *Find the laplace transform of* $f(t) = te^{2t}\cos 5t$

SOLUTION. $$L(\cos 5t) = \frac{s}{s^2+25}$$

Then $$L(e^{2t}\cos 5t) = \frac{s-2}{(s-2)^2+25}$$

Then $$L(te^{2t}\cos 5t) = -\frac{d}{ds}\left[\frac{s-2}{(s-2)^2+25}\right] = \frac{(s-2)^2-25}{[(s-2)^2+25]^2}$$

EXAMPLE 8.11. *Find the laplace transform of the signal given in Figure E8.11.*

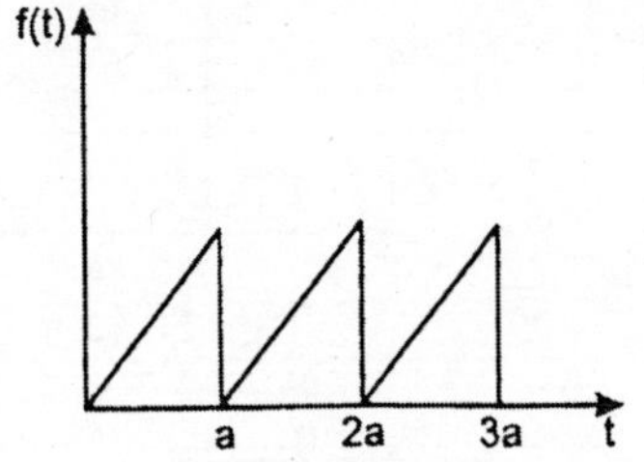

Fig. E8.11

SOLUTION. $$L[f(t)] = \frac{1}{1-e^{-as}}\int_0^a te^{-st}\,dt$$

$$= \frac{1}{1-e^{-as}}\left[\frac{e^{-st}}{s^2}(-st-1)\right]_0^a$$

$$= \frac{1-(1+as)e^{-as}}{s^2(1-e^{-as})} = \frac{(1+as)-(1+as)e^{-as}-as}{s^2(1-e^{-as})}$$

$$= \frac{(1+as)(1-e^{-as})-as}{s^2(1-e^{-as})} = \frac{1+as}{s^2} - \frac{a}{s(1-e^{-as})}$$

EXAMPLE 8.12. *Find the laplace transform of the signal given in Figure E8.12.*

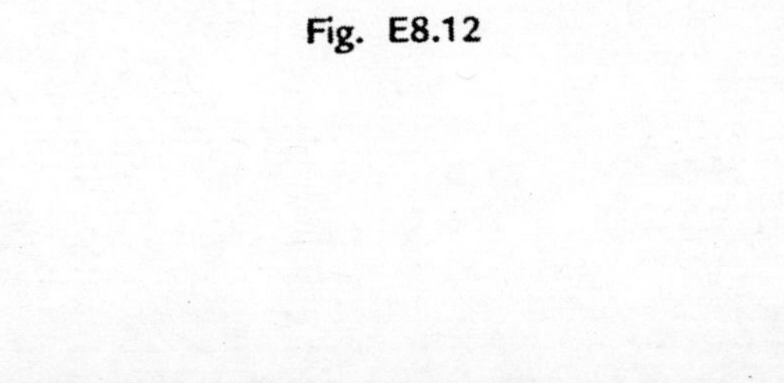

Fig. E8.12

SOLUTION.

$$f(t) = \begin{matrix} 0 & t<0 \\ 1 & t>0 \end{matrix}$$

$$L[f(t)] = \int_0^\infty f(t)e^{-st}dt = \int_0^\infty e^{-st}dt = \left.\frac{e^{-st}}{-s}\right|_0^\infty = \frac{1}{s}$$

EXAMPLE 8.13. *Find the laplace transform of the signal given in Figure E8.13.*

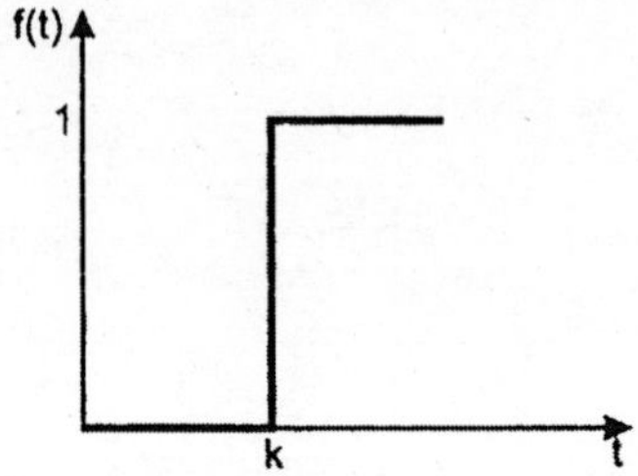

Fig. E8.13

SOLUTION.

$$f(t-k) = \begin{matrix} 0 & t<k \\ 1 & t>k \end{matrix}$$

$$L[f(t-k)] = \int_0^\infty f(t-k)e^{-st}dt = \int_k^\infty e^{-st}dt = \left.\frac{e^{-st}}{-s}\right|_k^\infty = \frac{e^{-ks}}{s}$$

EXAMPLE 8.14. *Find the laplace transform of the signal given in Figure E8.14.*

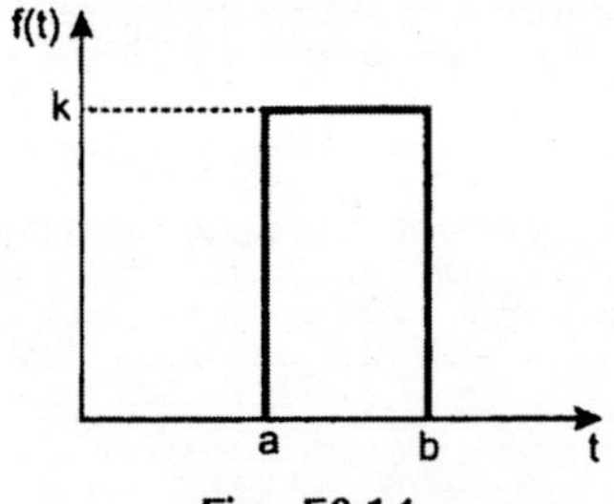

Fig. E8.14

SOLUTION.

$$f(t) = \begin{matrix} 0 & t<a \\ k & a<t<b \\ 0 & t>b \end{matrix}$$

$$L[f(t)] = \int_a^b ke^{-st}dt = \frac{k}{s}\left[e^{-as} - e^{-bs}\right]$$

8.5 INVERSE LAPLACE TRANSFORM

To return to the time domain from s-domain inverse laplace transforms are used. To find the inverse laplace transforms the following notation is used

$$f(t) = L^{-1}[f(s)]$$

SOLVED EXAMPLES

EXAMPLE 8.15. *Find the inverse laplace transform of* $F(s) = \frac{6}{s} - \frac{1}{s-5} - \frac{3}{s-2}$.

SOLUTION. Let
$$F(s) = \frac{6}{s} - \frac{1}{s-5} - \frac{3}{s-2} = 6\frac{1}{s} - \frac{1}{s-5} - 3\frac{1}{s-2}$$

Then
$$L^{-1}[F(s)] = f(t) = 6L^{-1}\left[\frac{1}{s}\right] - L^{-1}\left[\frac{1}{s-5}\right] - 3L^{-1}\left[\frac{1}{s-2}\right]$$

Thus
$$f(t) = 6 - e^{5t} - 3e^{2t}$$

EXAMPLE 8.16. *Find the inverse laplace transform of* $F(s) = \frac{19}{s+2} - \frac{8}{3s-5} + \frac{1}{s^5}$.

SOLUTION. Let
$$F(s) = \frac{19}{s+2} - \frac{8}{3\left(s-\frac{5}{3}\right)} + \frac{1}{s^{4+1}} = \frac{19}{s+2} - \frac{8}{3\left(s-\frac{5}{3}\right)} + \frac{1}{4!}\frac{4!}{s^{4+1}}$$

Then
$$L^{-1}[F(s)] = f(t) = 19e^{-2t} - \frac{8}{3}e^{\frac{5t}{3}} + \frac{1}{4!}t^4$$

EXAMPLE 8.17. *Find the inverse laplace transform of* $F(s) = \frac{8}{3s^2+12} + \frac{3}{s^2-49}$.

SOLUTION. Let
$$F(s) = \frac{1}{3}\frac{8}{s^2+4} + \frac{3}{s^2-49} = \frac{1}{3}\frac{4\times 2}{s^2+2^2} + \frac{3}{7}\frac{7}{s^2-7^2}$$

Then
$$L^{-1}[F(s)] = f(t) = \frac{4}{3}\sin 2t + \frac{3}{7}\sinh 7t$$

EXAMPLE 8.18. *Find the inverse laplace transform of* $F(s) = \frac{s+7}{s^2-3s-10}$.

SOLUTION. Let
$$F(s) = \frac{s+7}{s^2-3s-10} = \frac{s+7}{(s+2)(s-5)}$$

Using partial fraction let

$$\frac{s+7}{(s+2)(s-5)} = \frac{A}{s+2} + \frac{B}{s-5} \quad \text{...}(i)$$

Then solving the partial fraction of equation (i) we get

$$A = -0.71 \quad \text{and} \quad B = 1.71$$

Thus $$F(s) = -\frac{0.71}{s+2}+\frac{1.71}{s-5}$$

Then $$L^{-1}[F(s)] = f(t) = -0.71e^{-2t} + 1.71e^{5t}$$

EXAMPLE 8.19. *Find the inverse laplace transform of* $F(s) = \frac{3s-2}{2s^2-6s-2}$.

SOLUTION. $$F(s) = \frac{3s-2}{2s^2-6s-2} = \frac{3s-2}{(s-2.62)(s-0.38)}$$

Let $$\frac{3s-2}{(s-2.62)(s-0.38)} = \frac{A}{s-2.62}+\frac{B}{s-0.38} \quad ...(i)$$

Solving equation (*i*), we get

$$A = 2.62 \text{ and } B = 0.38$$

Then $$F(s) = \frac{2.62}{s-2.62}+\frac{0.38}{s-0.38}$$

Thus $$L^{-1}[F(s)] = f(t) = 2.62e^{2.62t} + 0.38e^{0.38t}$$

EXAMPLE 8.20. *Find the inverse laplace transform of* $F(s) = \frac{5}{s^3+5s^2+8s+6}$.

SOLUTION. Let $$F(s) = \frac{5}{s^3+5s^2+8s+6} = \frac{5}{(s+3)(s^2+2s+2)} = \frac{5}{(s+3)(s+1+j1)(s+1-j1)}$$

Let $$F(s) = \frac{5}{(s+3)(s+1+j1)(s+1-j1)} = \frac{A}{s+3}+\frac{B}{s+1+j1}+\frac{C}{s+1-j1} \quad ...(i)$$

Solving the equation (1) using partial fraction we get

$$A = 1 \quad B = -1 \quad \text{and} \quad C = 1$$

$$F(s) = \frac{1}{s+3}-\frac{1}{s+1+j1}+\frac{1}{s+1-j1}$$

Then $$L^{-1}[F(s)] = f(t) = e^{-3t} - e^{-(1+j1)t} + e^{-(1-j1)t}$$

EXAMPLE 8.21. *Find the inverse laplace transform of* $F(s) = \frac{15s^2-15s-11}{(s+1)(s-2)^3}$.

SOLUTION. Let $$F(s) = \frac{15s^2-15s-11}{(s+1)(s-2)^3} = \frac{A}{s+1}+\frac{B}{(s-2)^3}+\frac{C}{(s-2)^2}+\frac{D}{s-2} \quad ...(i)$$

Solving equation (*i*) using partial fraction we get

$$A = -\frac{1}{3} \quad B = -7 \quad C = 4 \quad D = \frac{1}{3}$$

$$F(s) = -\frac{1}{3}\frac{1}{s+1}-\frac{7}{(s-2)^3}+\frac{4}{(s-2)^2}+\frac{1}{3}\frac{1}{s-2}$$

Then $$L^{-1}[F(s)] = f(t) = \frac{1}{3}e^{-t}-\frac{7}{2}t^2e^{2t}+4te^{2t}+\frac{1}{3}e^{2t}$$

EXAMPLE 8.22. *Find the inverse laplace transform of* $F(s) = \dfrac{s^2+2s+3}{(s^2+2s+2)(s^2+2s+5)}$.

SOLUTION.
$$F(s) = \frac{s^2+2s+3}{(s^2+2s+2)(s^2+2s+5)} = \frac{As+B}{s^2+2s+2} + \frac{Cs+D}{s^2+2s+5}$$

Solving the partial fraction we get

$$F(s) = \frac{3}{(s+1)^2+1} + \frac{2}{3} \times \frac{1}{(s+1)^2+4}$$

Then
$$L^{-1}[F(s)] = f(t) = \frac{1}{3}e^{-t}\sin t + \frac{1}{3}e^{-t}\sin 2t$$

8.6 CONVOLUTION

Let $f(t)$ and $g(t)$ be the two signals. Then the convolution of these two signal is denoted by $h = f * g$, where

$$h(t) = \int_0^t f(\tau)g(t-\tau)d\tau \qquad \text{...(8.4)}$$

Equation (8.4), can also be written as,

$$h(t) = \int_0^t f(t-\tau)g(\tau)\,d\tau \qquad \text{...(8.5)}$$

From equation (8.4) and equation (8.5), we can say,

$$f * g = g * f$$

Laplace transform method turns convolution into simple multiplication. That is,

$$H(s) = L[f * g] = F(s)G(s)$$

Proof :

$$H(s) = \int_{t=0}^{\infty} e^{-st}\left(\int_{\tau=0}^{t} f(\tau)g(t-\tau)\,d\tau\right)dt$$

$$= \int_{t=0}^{\infty}\int_{\tau=0}^{t} e^{-st} f(\tau)g(t-\tau)\,d\tau\,dt$$

Changing the order of integration

$$= \int_{\tau=0}^{\infty}\int_{t=\tau}^{t} e^{-st} f(\tau)g(t-\tau)\,dt\,d\tau \qquad \text{...(8.6)}$$

Let $t_1 = t - \tau$, then $dt_1 = dt$. Equation (8.6) becomes,

$$H(s) = \int_{\tau=0}^{\infty}\int_{t_1=0}^{\infty} e^{-s(t_1+\tau)} f(\tau) g(t_1)\, dt_1\, d\tau$$

$$= \left(\int_{\tau=0}^{\infty} e^{-s\tau} f(\tau) d\tau\right)\left(\int_{t_1=0}^{\infty} e^{-st_1} g(t_1) dt_1\right)$$

$$= F(s)G(s)$$

SOLVED EXAMPLES

EXAMPLE 8.23. *Find the inverse laplace transform of* $F(s) = \dfrac{s}{(s^2+4^2)^2}$.

SOLUTION. We know that

$$L^{-1}\left[\frac{s}{s^2+4^2}\right] = \cos 4t \quad \text{and} \quad L^{-1}\left[\frac{1}{s^2+4^2}\right] = \frac{\sin 4t}{4}$$

From convolution theorem

$$L^{-1}\left[\frac{s}{s^2+4^2}\cdot\frac{1}{s^2+4^2}\right] = \int_0^t \cos 4\tau \cdot \frac{\sin 4(t-\tau)}{4}\, d\tau$$

$$= \frac{1}{4}\int_0^t \cos 4\tau(\sin 4t \cos 4\tau - \cos 4t \sin 4\tau)\, d\tau$$

$$= \frac{t \sin 4t}{8}$$

EXAMPLE 8.24. *Find the inverse laplace transform of* $F(s) = \dfrac{s}{(s^2+4^2)^2}$.

SOLUTION. We know that

$$L^{-1}\left[\frac{1}{s^2}\right] = t \quad \text{and} \quad L^{-1}\left[\frac{1}{(s+1)^2}\right] = te^{-t}$$

From convolution theorem

$$L^{-1}\left[\frac{1}{s^2}\cdot\frac{1}{(s+1)^2}\right] = \int_0^t \tau\, e^{-\xi}(t-\tau)\, d\tau$$

$$= te^{-t} + 2e^{-t} + t - 2$$

EXERCISES

1. Find the inverse Laplace transform of the following:

(i) $\dfrac{s+5}{s(s^2+4s+3)}$ (ii) $\dfrac{s}{(s+1)(s^2-4)(s+2)(s-4)}$

(iii) $\dfrac{4(s+5)}{s(s^2+4s+3)}$ (iv) $\dfrac{2s+14}{(s+1)(s^2-4)(s+2)(s+4)}$

2. Using Laplace transform solve the following : (assume initial conditions to be zero)

(i) $\dfrac{d^2x}{dy^2}+3\dfrac{dx}{dy}+x=5u(y)$ (ii) $\dfrac{d^2x}{dy^2}+3\dfrac{dx}{dy}+3x+\delta(y)=0$

(iii) $\dfrac{d^2x}{dy^2}+5\dfrac{dx}{dy}+x=\cos y$

3. Find $f(t)$, using convolution when

(i) $F(s) = \dfrac{1}{(s+1)(s+3)}$ (ii) $F(s) = \dfrac{4}{(s^2+2s+5)^2}$

4. Find the final value of the function whose Laplace transform is $F(s) = \dfrac{s+6}{s(s+3)}$

5. Find the initial value of $f(t)$ whose Laplace transform is $F(s) = \dfrac{s+1}{(s+2)(s+3)}$

Chapter **9**

TRANSIENT ANALYSIS (WITH LAPLACE TRANSFORMATIONS)

9.1 INTRODUCTION

The transients are produced in the circuit, when a network or circuit changes its state. For example, if a circuit is switched ON from its OFF state or vice versa, transients are produced in the circuit. Transients can also appear in a circuit if the excitation is suddenly changed.

9.2 ANALYSIS OF RL CIRCUIT

9.2.1 Analysis of RL Circuit Excited with Step Input

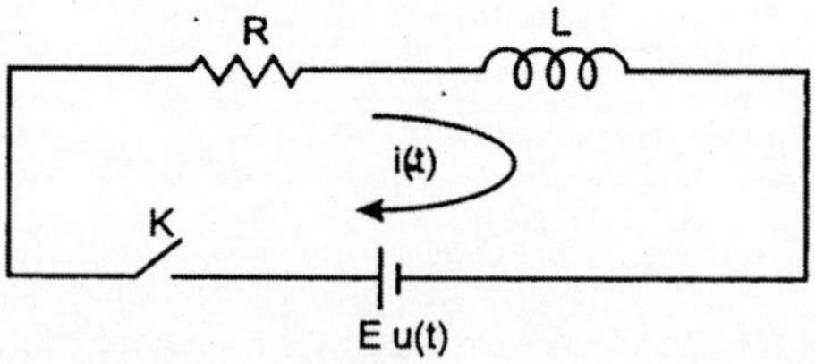

Fig. 9.1

Consider the circuit shown in Figure 9.1. Let at time $t = 0$ the switch K is closed, then the loop equation is given by,

$$Ri(t) + L\frac{di(t)}{dt} = E\ u(t) \quad \text{...(9.1)}$$

Taking laplace transformation on both sides of equation (9.1) we get

$$RI(s) + L[sI(s) - i(0\_)] = \frac{E}{s} \qquad ...(9.2)$$

where $i(0\_)$ denotes current through the circuit of Figure 9.1 before the switch is closed. It is zero (since before the closing of the switch K the circuit was an open circuit.). Then equation (9.2) can be written as

$$I(s)\,[R + Ls] = \frac{E}{s} \qquad ...(9.3)$$

or

$$I(s) = \frac{E}{s(R+Ls)} = \frac{E/L}{s\left(\frac{R}{L}+s\right)} \qquad ...(9.4)$$

Let

$$\frac{E/L}{s\left(\frac{R}{L}+s\right)} = \frac{A}{s}+\frac{B}{s+\frac{R}{L}} \qquad ...(9.5)$$

where A and B are constants.

To *find the values of constants A and B* :

Form equation (9.5) we get

$$A\left(s+\frac{R}{L}\right)+Bs = \frac{E}{L} \qquad ...(9.6)$$

Then equating the coefficient of s we get

$$A + B = 0 \qquad ...(9.7)$$

And equating the constants we get

$$A\frac{R}{L} = \frac{E}{L}$$

or

$$A = \frac{E}{R}$$

Substituting the value of the constant A in equation (9.7) we get

$$B = -\frac{E}{R}$$

Substituting the value of A and B in equation (9.5) we get

$$I(s) = \frac{E}{R}\left[\frac{1}{s}-\frac{1}{s+\frac{R}{L}}\right] \qquad ...(9.8)$$

Taking the inverse laplace transformation of equation (9.8) we get

$$i(t) = \frac{E}{R}\left[1-e^{-\left(R/L\right)t}\right] \text{ amp}$$

9.2.2 Analysis of RL Circuit Excited with Impulse Input

Consider the circuit shown in Figure 9.2. Let at time $t = 0$ the switch K is c sed, then the loop equation is given by,

$$Ri(t) + L\frac{di(t)}{dt} = E\delta(t) \qquad ...(9.9)$$

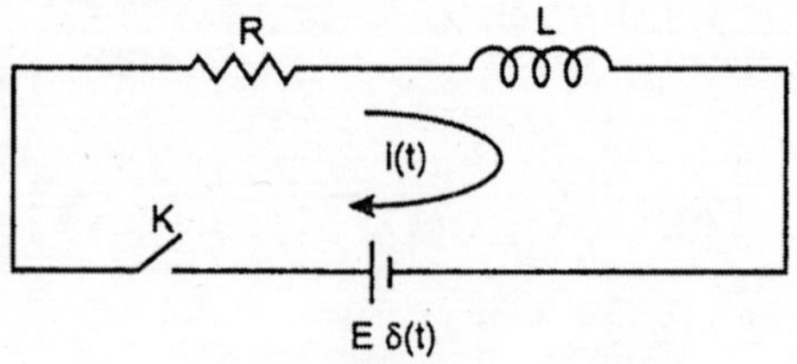

Fig. 9.2

Taking laplace transformation on both sides of equation (9.9) we get

$$RI(s) + L[sI(s) - i(s) - i(0\_)] = E \qquad ...(9.10)$$

where denotes current through the circuit of Figure 9.2 before the switch is closed. It is zero (since before the closing of the switch K the circuit was an open circuit.). Then equation (9.10) can be written as

$$I(s)\ [R + Ls] = E \qquad ...(9.11)$$

or

$$I(s) = \frac{E}{(R+Ls)} = \frac{E/L}{\left(\frac{R}{L}+s\right)} \qquad ...(9.12)$$

Taking the inverse laplace transformation of equation (9.12) we get,

$$i(t) = \frac{E}{L}e^{-\left(R/L\right)t} \text{ amp}$$

9.2.3 Analysis of RL Circuit Excited with Ramp Input

Consider the circuit shown in Figure 9.3. Let at time $t = 0$ the switch K is closed, then the loop equation is given by,

$$Ri(t) + L\frac{di(t)}{dt} = Er(t) \qquad ...(9.13)$$

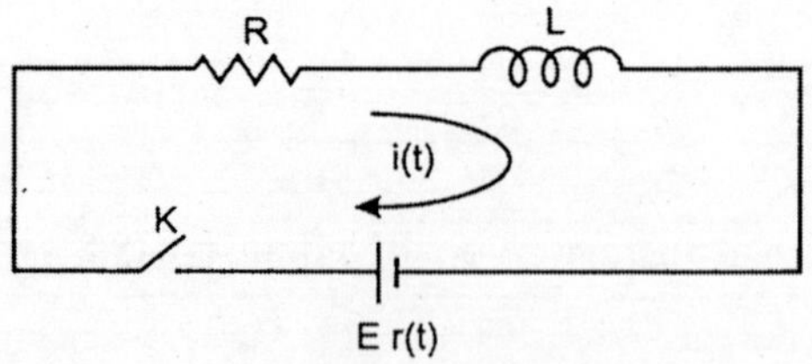

Fig. 9.3

Taking laplace transformation on both sides of equation (9.13) we get

$$RI(s) + L[sI(s) - i(0-)] = \frac{E}{s^2} \quad ...(9.14)$$

where $i(0\_)$ denotes current through the circuit of Figure 9.3 before the switch is closed. It is zero (since before the closing of the switch K the circuit was an open circuit.). Then equation (9.14) can be written as

$$I(s)\ [R + Ls] = \frac{E}{s^2} \quad ...(9.15)$$

or

$$I(s) = \frac{E}{s^2(R+Ls)} = \frac{E/L}{s^2\left(\frac{R}{L}+s\right)} \quad ...(9.16)$$

Let

$$\frac{E/L}{s^2\left(\frac{R}{L}+s\right)} = \frac{A}{s}+\frac{B}{s^2}+\frac{C}{s+\frac{R}{L}} \quad ...(9.17)$$

where A, B and C are constants.

To *find the values of constants A, B and C* :

$$A = \frac{d}{ds}\left\{s^2\frac{E/L}{s^2\left(\frac{R}{L}+s\right)}\right\}\Bigg|_{s=0} = \frac{RL-EL}{R^2}$$

$$B = s^2\frac{E/L}{s^2\left(\frac{R}{L}+s\right)}\Bigg|_{s=0} = \frac{E}{R}$$

$$C = \left(s+\frac{R}{L}\right)\frac{E/L}{s^2\left(\frac{R}{L}+s\right)}\Bigg|_{s=-\frac{R}{L}} = \frac{EL}{R^2}$$

Substituting the value of A, B and C in equation (9.17) we get

$$I(s) = \frac{E/L}{s^2\left(\frac{R}{L}+s\right)} = \frac{RL-EL}{R^2}\times\frac{1}{s}+\frac{E}{R}\times\frac{1}{s^2}+\frac{EL}{R^2}\frac{1}{s+\frac{R}{L}} \quad ...(9.18)$$

Taking the inverse laplace transformation of equation (9.18) we get

$$i(t) = \frac{RL-EL}{R^2}\times u(t)+\frac{E}{R}\times t+\frac{EL}{R^2}e^{-Rt/L} \text{ amp}$$

9.2.4 Analysis of RL Circuit Excited with Pulse Input

Consider the circuit shown in Figure 9.4. Let at time $t = 0$ the switch K is closed, then the loop equation is given by,

$$Ri(t) + L\frac{di(t)}{dt} = E(t) \qquad ...(9.19)$$

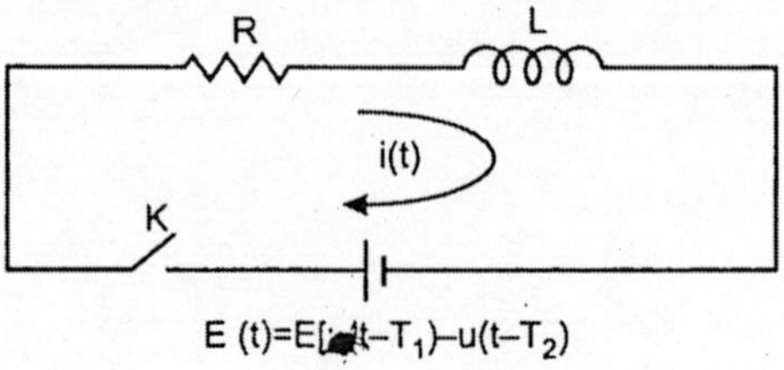

Fig. 9.4

Taking laplace transformation on both sides of equation (9.19) we get,

$$RI(s) + L[sI(s) - i(0\_)] = \frac{E}{s}(e^{-sT_1} - e^{-sT_2}) \qquad ...(9.20)$$

where $i(0\_)$ denotes current through the circuit of Figure 9.4 before the switch is closed. It is zero (since before the closing of the switch K the circuit was an open circuit.). Then equation (9.20) can be written as

$$I(s) = [R + Ls] = \frac{E}{s}(e^{-sT_1} - e^{-sT_2}) \qquad ...(9.21)$$

or

$$I(s) = \frac{E}{s(R+Ls)}(e^{-sT_1} - e^{-sT_2}) = \frac{E/L}{s\left(\frac{R}{L}+s\right)}(e^{-sT_1} - e^{-sT_2}) \qquad ...(9.22)$$

or

$$I(s) = \frac{E}{R}\left[\frac{e^{-sT_1}}{s} - \frac{e^{-sT_2}}{s} - \frac{e^{-sT_1}}{s+R/L} + \frac{e^{-eT_2}}{s+R/L}\right] \qquad ...(9.23)$$

Taking the inverse laplace transformation of equation (9.23) we get

$$i(t) = \frac{E}{R}\left[u(t-T_1) - u(t-T_2) - e^{-\frac{R(t-T_1)}{L}}u(t-T_1) + e^{-\frac{R(t-T_1)}{L}}u(t-T_2)\right] \text{ amp}$$

9.2.5 Analysis of RL Circuit Excited with Sinusoidal Input

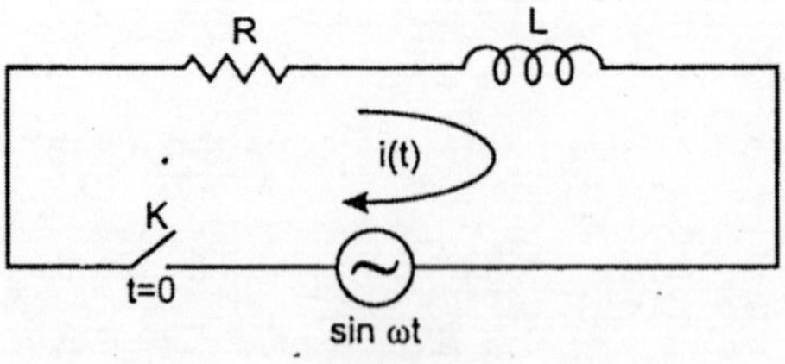

Fig. 9.5

Consider the circuit shown in Figure 9.5. Let at time $t = 0$ the switch K is closed, then the loop equation is given by,

$$RI(t) + L = \frac{di(t)}{dt} = \sin \omega t \qquad ...(9.24)$$

Taking l. place transformation on both sides of equation (9.24) we get

$$RI(s) + L[sI(s) - i(0\_)] = \frac{\omega}{s^2 + \omega^2} \qquad ...(9.25)$$

where $i(0\_)$ denotes current through the circuit of Figure 9.5 before the switch is closed. It is zero (since before the closing of the switch K the circuit was an open circuit.). Then equation (9.25) can be written as

$$I(s)\ [R + Ls] = \frac{\omega}{s^2 + \omega^2} \qquad ...(9.26)$$

or

$$I(s) = \frac{\omega}{(s^2 + \omega^2)(R + Ls)} = \frac{\omega/L}{(s + j\omega)(s - j\omega)\left(s + \frac{R}{L}\right)}$$

Let

$$\frac{\omega/L}{(s + j\omega)(s - j)(s + \frac{R}{L})} = \frac{A}{s + j\omega} + \frac{B}{s - j\omega} + \frac{C}{s + \frac{R}{L}} \qquad ...(9.27)$$

where A, B and C are constants.

To find the values of constants A, B and C :

$$A = \left.\frac{\frac{\omega}{L}(s + j\omega)}{(s + j\omega)(s - j\omega)(s + \frac{R}{L})}\right|_{s=-j\omega} = \left.\frac{\frac{\omega}{L}}{(s - j\omega)(s + \frac{R}{L})}\right|_{s=-j\omega}$$

$$= \frac{\omega}{L}\left[\frac{-2\omega^2 L + j2R}{4L^2\omega^4 + 4R^2}\right]$$

$$B = A^* = \frac{\omega}{L}\left[\frac{-2\omega^2 L + j2R}{4L^2\omega^4 + 4R^2}\right]$$

$$C = \left.\frac{\frac{\omega}{L}\left(s + \frac{R}{L}\right)}{(s + j\omega)(s - j\omega)\left(s + \frac{R}{L}\right)}\right|_{s=-\frac{R}{L}} = \left.\frac{\frac{\omega}{L}}{(s + j\omega)(s - j\omega)}\right|_{s=-\frac{R}{L}}$$

$$= \frac{\omega L}{R^2 + \omega^2 L^2}$$

Then equation (9.26) becomes,

$$I(s) = \frac{\omega}{L}\left[\frac{-2\omega^2 L + j2R}{4L^2\omega^4 + 4R^2}\right]\frac{1}{s + j\omega} + \frac{\omega}{L}\left[\frac{-2\omega^2 L - j2R}{4L^2\omega^4 + 4R^2}\right]$$

$$\frac{1}{s-j\omega}+\frac{\omega L}{R^2+\omega^2L^2}\frac{1}{s+\frac{R}{L}} \qquad ...(9.28)$$

Taking the inverse laplace transformation of equation (9.28) we get

$$i(t) = \frac{\omega}{L}\left[\frac{-2\omega^2L+j2R}{4L^2\omega^4+4R^2}\right]e^{-j\omega t}+\frac{\omega}{L}\left[\frac{-2\omega^2L-2jR}{4L^2\omega^4+4R^2}\right]e^{j\omega t}+\frac{\omega L}{R^2+\omega^2L^2}e^{-\frac{Rt}{L}} \text{ amp}$$

SOLVED EXAMPLES

EXAMPLE 9.1. *For the circuit shown in Figure E9.1, the switch k was closed at time t = 0. Find the complete expression for the current after the switch is closed. Assume all the initial conditions to be zero.*

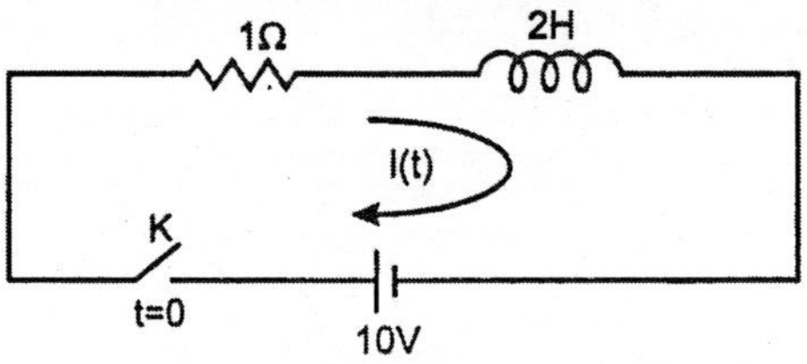

Fig. E9.1

SOLUTION. The loop equation is given by,

$$1i(t) + 2\frac{di(t)}{dt} = 10 \qquad ...(i)$$

Taking laplace transformation on both sides of equation (*i*) we get,

$$1I(s) + 2[sI(s) - i(0\_)] = \frac{10}{s} \qquad ...(ii)$$

where $i(0\_)$ denotes current through the circuit of Figure E9.1 before the switch is closed. $i(0\_)$ = 0 (as indicated in the question). Then equation (*ii*) becomes

$$I(s)\,[1 + 2s] = \frac{10}{s} \qquad ...(iii)$$

or

$$I(s) = \frac{10}{s(1+2s)} = \frac{10/2}{s\left(\frac{1}{2}+s\right)} \qquad ...(iv)$$

Let

$$\frac{10/2}{s\left(\frac{1}{2}+s\right)} = \frac{A}{s}+\frac{B}{s+\frac{1}{2}} \qquad ...(v)$$

where A and B are constants.

To find the values of constants A and B :

Form equation (*v*) we get

$$A\left(s+\frac{1}{2}\right)+Bs = \frac{10}{2} \qquad ...(vi)$$

Then equating the coefficient of s we get

$$A + B = 0 \qquad \text{...(vii)}$$

And equating the constants we get

$$A\frac{1}{2} = \frac{10}{2}$$

or

$$A = 10$$

Substituting the value of the constant A in equation (*vii*) we get

$$B = -10$$

Substituting the value of A and B in equation (*v*) we get

$$I(s) = \frac{10}{1}\left[\frac{1}{s} - \frac{1}{s+\frac{1}{2}}\right] \qquad \text{...(viii)}$$

Taking the inverse laplace transformation of equation (*viii*) we get

$$i(t) = \frac{10}{1}\left[1 - e^{-\left(\frac{1}{2}\right)t}\right] = 10\left[1 - e^{-0.5t}\right] \text{ amp}$$

EXAMPLE 9.2. *For the circuit shown in Figure E9.2, the switch k was closed at time t = 0, connecting the circuit to an impulse source of 10 δ(t). Find the complete expression for the current after the switch is closed. Assume the initial current through the inductor is i(0) = 2A (the direction of i(0) is indicated in the Figure E9.2)*

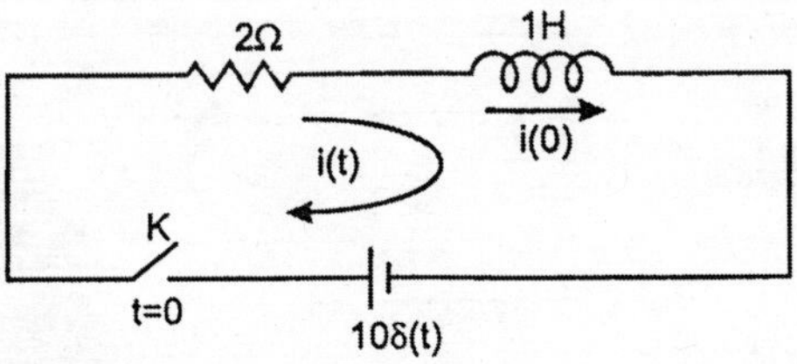

Fig. E9.2

SOLUTION. The loop equation is given by,

$$2i(t) + 1\frac{di(t)}{dt} = 10\delta(t) \qquad \text{...(i)}$$

Taking laplace transformation on both sides of equation (*i*) we get

$$2I(s) + 1[sI(s) - i(0)] = 10 \qquad \text{...(ii)}$$

where $i(0)$ denotes current through the circuit of Figure E9.2 before the switch is closed. $i(0) = 2A$ (as indicated in the question). Then equation (*ii*) becomes

$$2I(s) + 1[sI(s) - 2] = 10 \qquad \text{...(iii)}$$

or

$$I(s) = \frac{10+2}{(2+s)} = \frac{12}{(2+s)} \qquad \text{...(iv)}$$

Taking the inverse laplace transformation of equation (*iv*) we get

$$i(t) = 12e^{-2t} \text{ amp}$$

EXAMPLE 9.3. *For the circuit shown in Figure E9.3, the switch k was closed at time t = 0, connecting the circuit to a source of 10u(t). Find the complete expression for the current after the switch is closed. Assume the initial current through the inductor is i(0) = 2A (the direction of i(0) is indicated in Figure E9.3).*

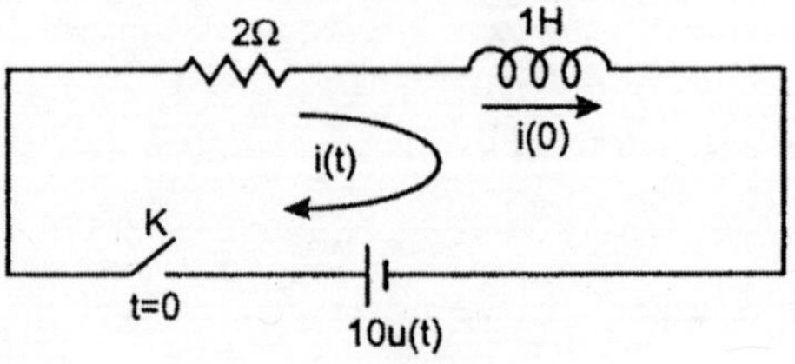

Fig. E9.3

SOLUTION. The loop equation is given by,

$$2i(t) + 1\frac{di(t)}{dt} = 10u(t) \qquad ...(i)$$

Taking laplace transformation on both sides of equation (*i*) we get

$$2I(s) + 1[sI(s) - i(0) = \frac{10}{s} \qquad ...(ii)$$

where $i(0)$ denotes current through the circuit of Figure E9.3 before the switch is closed.

$i(0) = 2A$ (as indicated in the question). Then equation (*ii*) becomes

$$2I(s) + 1[sI(s) - 2] = \frac{10}{s} \qquad ...(iii)$$

or

$$I(s)\ [2 + s] = \frac{10}{s} + 2$$

or

$$I(s) = \frac{10+2s}{s(2+s)} \qquad ...(iv)$$

Let

$$\frac{10+2s}{s(2+s)} = \frac{A}{s} + \frac{B}{s+2} \qquad ...(v)$$

where A and B are constants.

To find the values of constants A and B :

Form equation (*v*) we get

$$A(s + 2) + Bs = 10 + 2s \qquad ...(vi)$$

Then equating the coefficient of s we get

$$A + B = 2 \qquad ...(vii)$$

And equating the constants we get

$$2A = 10$$

or

$$A = 5$$

Substituting the value of the constant A in equation (*vii*) we get

$$B = -3$$

Substituting the value of A and B in equation (*v*) we get

$$I(s) = \left[\frac{5}{s} - \frac{3}{s+2}\right] \quad ...(viii)$$

Taking the inverse laplace transformation of equation (*viii*) we get

$$i(t) = 5 - 3e^{-2t} \text{ amp}$$

EXAMPLE 9.4. *For the circuit shown in Figure E9.4, the switch k was closed at time t = 0, connecting the circuit to a source of 10u(t). Find the complete expression for the current after the switch is closed. Assume the initial current through the inductor is i(0) = 2A (the direction of i(0) is indicated in Figure E9.4).*

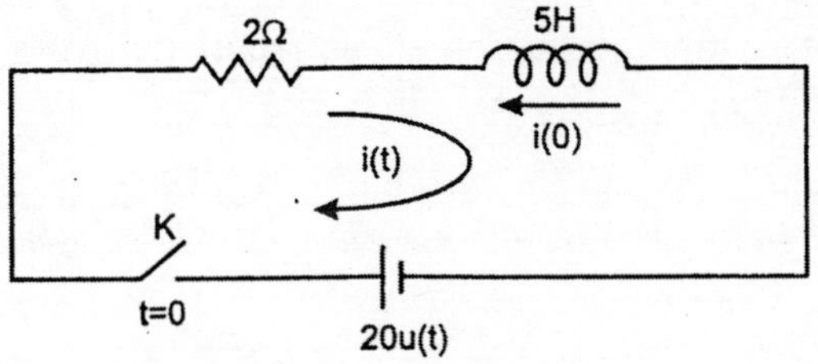

Fig. E9.4

SOLUTION. The loop equation is given by,

$$2i(t) + 5\frac{di(t)}{dt} = 20u(t) \quad ...(i)$$

Taking laplace transformation on both sides of (*i*) we get

$$2I(s) + 5[sI(s) - i(0)] = \frac{20}{s} \quad ...(ii)$$

where $i(0)$ denotes current through the circuit of Figure E9.4 before the switch is closed.

$i(0) = -2A$. Then equation (*ii*) becomes

$$2I(s) + 5[sI(s) + 2] = \frac{20}{s} \quad ...(iii)$$

or

$$I(s)\,[2 + 5s] = \frac{20}{s} - 2$$

or

$$I(s) = \frac{20-2s}{s(2+5s)} = \frac{\frac{20-2s}{5}}{s\left(\frac{2}{5}+s\right)} = \frac{4-0.4s}{s(0.4+s)} \quad ...(iv)$$

Let

$$\frac{4-0.4s}{s(0.4+s)} = \frac{A}{s} + \frac{B}{s+0.4} \quad ...(v)$$

where A and B are constants.

To find the values of constants A and B :

Form equation (*v*) we get

$$A(s + 0.4) + Bs = 4 - 0.4s \quad ...(vi)$$

Then equating the coefficient of s we get

$$A + B = -0.4 \quad ...(vii)$$

And equating the constants we get

$$0.4\,A = 4$$

or $$A = 10$$

Substituting the value of the constant A in equation (*vii*) we get

$$B = -10.4$$

Substituting the value of A and B in equation (*v*) we get

$$I(s) = \left[\frac{10}{s} - \frac{10.4}{s+0.4}\right] \quad ...(viii)$$

Taking the inverse laplace transformation of equation (*viii*) we get

$$i(t) = 10 - 10.4\,e^{-0.4t} \text{ amp}$$

EXAMPLE 9.5. *For the circuit shown in Figure E9.5, the switch k was closed at time t = 0. Find the complete expression for the current after the switch is closed. Assume all the initial conditions to be zero.*

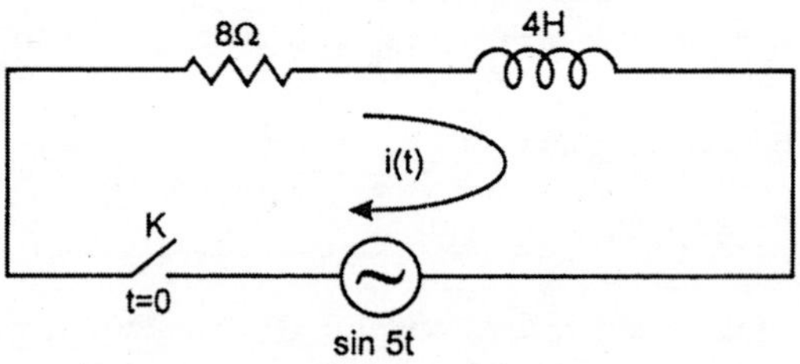

Fig. E9.5

SOLUTION. The loop equation is given by,

$$8i(t) + 4\frac{di(t)}{dt} = \sin 5t \quad ...(i)$$

Taking laplace transformation on both sides of equation (*i*) we get

$$8I(s) + 4[sI(s) - i(0\_)] = \frac{5}{s^2+5^2} \quad (ii)$$

where $i(0\_)$ denotes current through the circuit of Figure E9.5 before the switch is closed. $i(0\_) = 0$ (as indicated in the question). Then equation (*ii*) becomes

$$I(s)\,[8 + 4s] = \frac{5}{s^2+25} \quad ...(iii)$$

or $$I(s) = \frac{5}{(s^2+25)(8+4s)} = \frac{5/4}{(s+j5)(s-j5)\left(s+\frac{8}{4}\right)}$$

$$= \frac{1.25}{(s+j5)(s-j5)(s+2)} \quad ...(iv)$$

Let $$\frac{1.25}{(s+j5)\,(s-j5)\,(s+2)} = \frac{A}{s+j5} + \frac{B}{s-j5} + \frac{C}{s+2} \quad ...(v)$$

where A, B and C are constants.

To find the values of constants A, B and C :

$$A = \left.\frac{1.25(s+j5)}{(s+j5)(s-j5)(s+2)}\right|_{s=-j5} = \left.\frac{1.25}{(s-j5)(s+2)}\right|_{s=-j5} = \frac{-62.5+j25}{2900}$$

$$B = A^* = \frac{-62.5-j25}{2900}$$

$$C = \left.\frac{1.25(s+2)}{(s+j5)(s-j5)(s+2)}\right|_{s=-2} = \left.\frac{1.25}{(s-j5)(s+j5)}\right|_{s=-2}$$

$$= \left.\frac{1.25}{s^2+25}\right|_{s=-2} = \frac{1.25}{29} = 0.043$$

Then equation (v) becomes,

$$I(s) = \frac{1.25}{(s+j5)(s-j5)(s+2)} = \frac{-62.5+j25}{2900}$$

$$\frac{1}{s+j5} + \frac{-62.5-j25}{2900}\frac{1}{s-j5} + \frac{0.043}{s+2} \qquad ...(vi)$$

Taking the inverse laplace transformation of equation (vi) we get

$$I(s) = \frac{-62.5+j25}{2900}e^{-j5t} + \frac{-62.5-j25}{2900}e^{j5t} + 0.043e^{-2t} \text{ amp}$$

EXAMPLE 9.6. *Consider the circuit shown in Figure E9.6. Initially the switch S_1 was closed and the switch S_2 was open for a long period of time. At time $t = 0$ the switch S_1 was opened and the switch S_2 was closed. Find an expression for the current $i(t)$. Also find the drop across each circuit elements.*

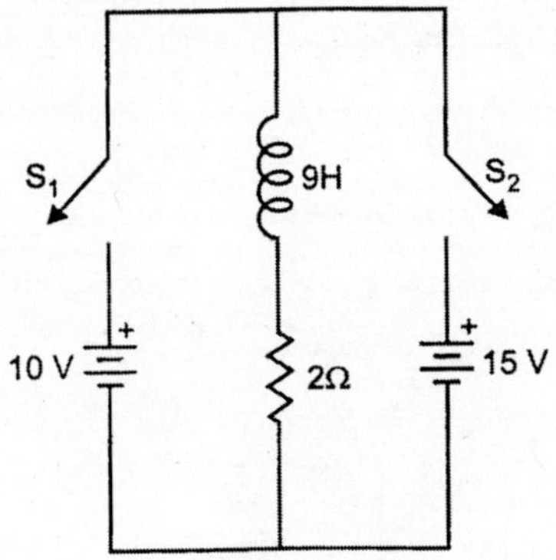

Fig. E9.6

SOLUTION. When the switch S_1 was closed and the switch S_2 was open for a long period of time, the current through the circuit is

$$i_1(t) = \frac{10}{2} = 5 \text{ amp}$$

At time $t = 0$, when the switch S_1 was opened and the switch S_2 was closed, the expression for current is

$$2i(t) + 9\frac{di(t)}{dt} = 15 \qquad ...(i)$$

Taking laplace transformation on both sides of equation (*i*) we get

$$2I(s) + 9[sI(s) - i(0)] = \frac{15}{s} \quad ...(ii)$$

where $i(0)$ denotes current through the circuit of Figure E9.6 before the switch is closed. $i(0) = i_1(t) = 5A$ (as indicated in the question). Then equation (*ii*) becomes

$$2I(s) + 9[sI(s) - 5] = \frac{15}{s} \quad ...(iii)$$

or

$$I(s)\,[2 + 9s] = \frac{15}{s} + 45$$

or

$$I(s) = \frac{15+45s}{s(2+9s)} = \frac{1.67+5s}{s\left(s+\frac{2}{9}\right)} \quad ...(iv)$$

Let

$$\frac{1.67+5s}{s\left(s+\frac{2}{9}\right)} = \frac{A}{s} + \frac{B}{s+\frac{2}{9}} \quad ...(v)$$

where A and B are constants.

To find the values of constants A and B :

Form equation (*v*) we get

$$A\left(s+\frac{2}{9}\right) + Bs = 1.67 + .5s \quad ...(vi)$$

Then equating the coefficient of s we get

$$A + B = 5 \quad ...(vii)$$

And equating the constants we get

$$\frac{2}{9}A = 1.67$$

or

$$A = 7.51$$

Substituting the value of the constant A in equation (*vii*) we get

$$B = -2.51$$

Substituting the value of A and B in equation (*v*) we get

$$I(s) = \left[\frac{7.51}{s} - \frac{2.51}{s+\frac{2}{9}}\right] \quad ...(viii)$$

Taking the inverse laplace transformation of equation (*viii*) we get

$$i(t) = 7.51 - 2.51\,e^{-\frac{2t}{9}} \text{ amp}$$

$$e_R = R \times i(t) = 15.02 - 5.02\,e^{-\frac{2t}{9}} \text{ amp}$$

$$e_L = V - e_R = 0.02 + 5.02\,e^{-\frac{2t}{9}} \text{ amp}$$

EXAMPLE 9.7. *Consider the circuit shown in Figure E9.7. Initially the switch S_1 was closed and the switch S_2 was open for 1 sec. Then the switch S_1 was opened and the switch S_2 was closed. Find an expression for the current $i(t)$.*

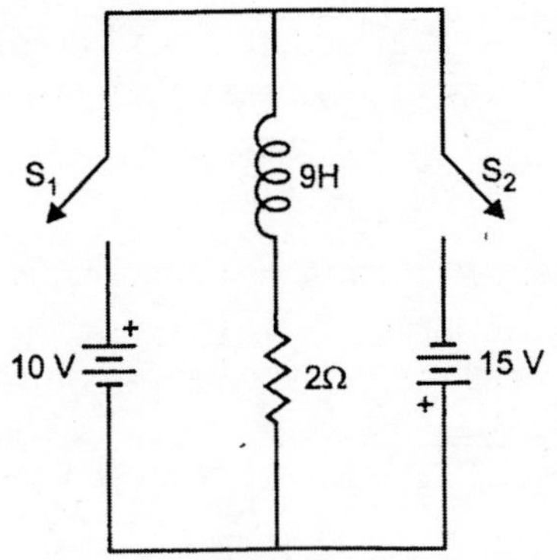

Fig. E9.7

SOLUTION. When the switch S_1 was closed and the switch S_2 was open the switch S_1 was closed and the switch S_2 was open, the expression for current is

$$2i(t) + 9\frac{di(t)}{dt} = 10 \qquad ...(i)$$

Taking laplace transformation on both sides of equation (*i*) we get

$$2I(s) + 9[sI(s) - i(0)] = \frac{10}{s} \qquad ...(ii)$$

where $i(0)$ denotes current through the circuit of Figure E9.7 before the switch is closed. $i(0) = 0$. Then equation (*ii*) becomes

or $$I(s)\,[2 + 9s] = \frac{10}{s} \qquad ...(iii)$$

or $$I(s) = \frac{10}{s(2+9s)} = \frac{1.11}{s\left(s+\frac{2}{9}\right)} \qquad ...(iv)$$

Let $$\frac{1.11}{s\left(s+\frac{2}{9}\right)} = \frac{A}{s} + \frac{B}{s+\frac{2}{9}} \qquad ...(v)$$

where A and B are constants.

To find the values of constants A and B :

Form equation (*v*) we get

$$A\left(s+\frac{2}{9}\right) + Bs = 1.11 \qquad ...(vi)$$

Then equating the coefficient of s we get

$$A + B = 0 \qquad ...(vii)$$

And equating the constants we get

$$\frac{2}{9}A = 1.11$$

or
$$A = 4.99$$

Substituting the value of the constant A in equation (*vii*) we get

$$B = -4.99$$

Substituting the value of A and B in equation (*v*) we get

$$I(s) = \left[\frac{4.99}{s} - \frac{4.99}{s+\frac{2}{9}}\right] \qquad \text{...(viii)}$$

Taking the inverse laplace transformation of equation (*viii*) we get

$$i(t) = 4.99 - 4.99\,e^{-\frac{2t}{9}} \text{ amp} \qquad \text{...(ix)}$$

At $t = 1$ sec equation (*ix*) becomes

$$i(t) = 0.99 \text{ amp}$$

After time $t = 1$ sec, when the switch S_1 was opened and the switch S_2 was closed, the expression for current is

$$2i(t) + 9\frac{d\,i(t)}{dt} = 15 \qquad \text{...(x)}$$

Taking laplace transformation on both sides of equation (*x*) we get

$$2I(s) + 9[sI(s) - i(0)] = \frac{15}{s} \qquad \text{...(xi)}$$

where $i(0)$ denotes current through the circuit of Figure E9.7 before the switch is closed. $i(0) = i_1(t) = -0.99$ A. Then equation (*xi*) becomes

$$2I(s) + 9[sI(s) + 0.99] = \frac{15}{s} \qquad \text{...(xii)}$$

or
$$I(s)\,[2 + 9s] = \frac{15}{s} - 8.91$$

or
$$I(s) = \frac{15-8.91s}{s(2+9s)} = \frac{1.67-0.99s}{s\left(s+\frac{2}{9}\right)} \qquad \text{...(xiii)}$$

Let
$$\frac{1.67-0.99s}{s\left(s+\frac{2}{9}\right)} = \frac{A}{s} + \frac{B}{s+\frac{2}{9}} \qquad \text{...(xiv)}$$

where A and B are constants.

To find the values of constants A and B :

Form equation (*xiv*) we get

$$A\left(s+\frac{2}{9}\right) + Bs = 1.67 - 0.99s \qquad \text{...(xv)}$$

Then equating the coefficient of s we get

$$A + B = 0.99 \qquad \text{...(xvi)}$$

And equating the constants we get

$$\frac{2}{9}A = 1.67$$

or $$A = 7.52$$

Substituting the value of the constant A in equation (*xvi*) we get

$$B = -6.53$$

Substituting the value of A and B in equation (*xiv*) we get

$$I(s) = \left[\frac{7.53}{s} - \frac{6.53}{s+\frac{2}{9}}\right] \qquad ...(xvii)$$

Taking the inverse laplace transformation of equation (*xvii*) we get

$$i(t) = 7.53 - 6.53\,e^{-\frac{2t}{9}} \text{ amp}$$

EXAMPLE 9.8. *Consider the circuit shown in Figure E9.8. Initially the switch S_1 was closed and the switch S_2 was open for 1 sec. Then the switch S_1 was opened and the switch S_2 was closed. Find an expression for the current i(t).*

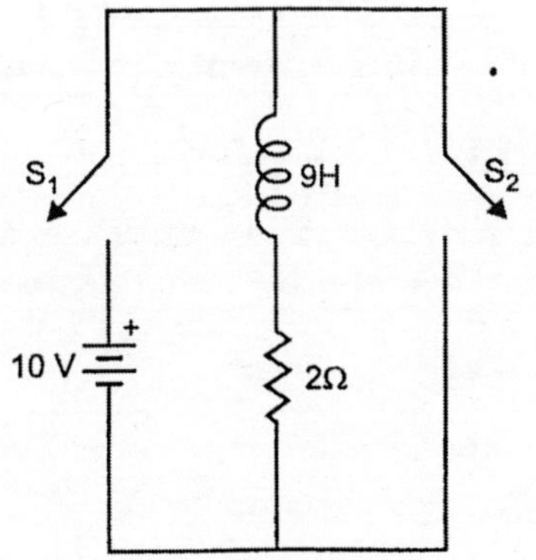

Fig. E9.8

SOLUTION. When the switch S_1 was closed and the switch S_2 was open the switch S_1 was closed and the switch S_2 was open, the expression for current is

$$2i(t) + 9\frac{di(t)}{dt} = 10 \qquad ...(i)$$

Taking laplace transformation on both sides of equation (*i*) we get

$$2I(s) + 9[sI(s) - i(0)] = \frac{10}{s} \qquad ...(ii)$$

where $i(0)$ denotes current through the circuit of Figure E9.8 before the switch is closed. $i(0) = 0$. Then equation (*ii*) becomes

or $$I(s)[2 + 9s] = \frac{10}{s} \qquad ...(iii)$$

or $$I(s) = \frac{10}{s(2+9s)} = \frac{1.11}{s\left(s+\frac{2}{9}\right)} \qquad ...(iv)$$

Let
$$\frac{1.11}{s\left(s+\frac{2}{9}\right)} = \frac{A}{s}+\frac{B}{s+\frac{2}{9}} \qquad ...(v)$$

where A and B are constants.

To find the values of constants A and B :

Form equation (v) we get

$$A\left(s+\frac{2}{9}\right) + Bs = 1.11 \qquad ...(vi)$$

Then equating the coefficient of s we get

$$A + B = 0 \qquad ...(vii)$$

And equating the constants we get

$$\frac{2}{9}A = 1.11$$

or
$$A = 4.99$$

Substituting the value of the constant A in equation (vii) we get

$$B = -\ 4.99$$

Substituting the value of A and B in equation (v) we get

$$I(s) = \left[\frac{4.99}{s} - \frac{4.99}{s+\frac{2}{9}}\right] \qquad ...(viii)$$

Taking the inverse laplace transformation of equation $(viii)$ we get

$$i(t) = 4.99 - 4.99\,e^{-\frac{2t}{9}} \text{ amp} \qquad ...(ix)$$

At $t = 1$ sec equation (ix) becomes

$$i(t) = 0.99 \text{ amp}$$

After time $t = 1$ sec, when the switch S_1 was opened and the switch S_2 was closed, the expression for current is

$$2i(t) + 9\frac{di(t)}{dt} = 0 \qquad ...(x)$$

Taking laplace transformation on both sides of equation (x) we get

$$2I(s) + 9[sI(s) - i(0)] = 0 \qquad ...(xi)$$

where $i(0)$ denotes current through the circuit of Figure E9.8 before the switch is closed. $i(0) = i_1(t) = -\ 0.99$ A. Then equation (xi) becomes

$$2I(s) + 9[sI(s) + 0.99] = 0 \qquad ...(xii)$$

or
$$I(s)\ [2 + 9s] = -\ 8.91$$

or
$$I(s) = \frac{8.91}{(2+9s)} = \frac{0.99}{\left(s+\frac{2}{9}\right)} \qquad ...(xiii)$$

Taking inverse laplace transformation of equation $(xiii)$ we get

EXAMPLE 9.9. *Consider the circuit shown in Figure E9.9. Initially the switch S was open. At time $t = 0$ the switch S is closed. Find an expression for the transient current.*

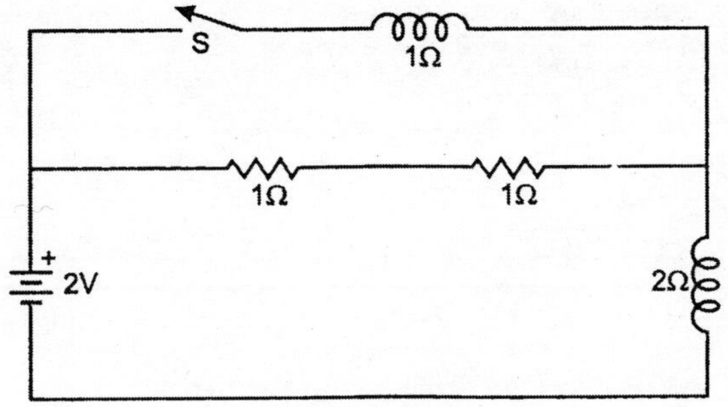

Fig. E9.9

SOLUTION. When the switch was open,

$$i(0) = \frac{2}{1+1} = 1 \text{ amp}$$

When the switch is closed at time $t = 0$, we get a circuit as shown in Figure E9.9(*a*). Writing the loop equations we get

Loop 1 : $$2i_1 - 2i_2 + 2\frac{di_1}{dt} = 2 \quad ...(i)$$

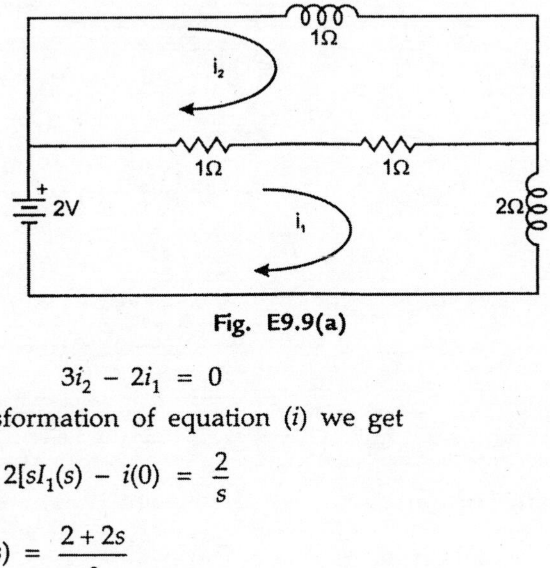

Fig. E9.9(a)

Loop 2 : $$3i_2 - 2i_1 = 0 \quad ...(ii)$$

Taking laplace transformation of equation (*i*) we get

$$2I_1(s) - 2I_2(s) + 2[sI_1(s) - i(0)) = \frac{2}{s}$$

$$I_1(s)\ [2 + 2s] - 2I_2(s) = \frac{2+2s}{s} \quad ...(iii)$$

Taking laplace transformation of equation (*ii*) we get

$$3I_2(s) - 2I_1(s) = 0 \quad ...(iv)$$

Solving equations (*iii*) and (*iv*) we get

$$I_1(s) = \frac{3}{s} - \frac{2}{s+0.333} \quad \text{and}$$

$$I_2(s) = \frac{2}{s} - \frac{4}{3} \times \frac{1}{s+0.333}$$

Taking inverse laplace transformation we get

$$i_1(t) = 3 - 2e^{-0.333t} \text{ amp}$$

$$i_2(t) = 2 - 1.333e^{-0.333t} \text{ amp}$$

9.3 ANALYSIS OF RC CIRCUIT

9.3.1 Analysis of RC Circuit Excited with Step Input

Consider the circuit shown in Figure 9.6. Let at time $t = 0$ the switch K is closed, then the loop equation is given by,

$$Ri(t) + \frac{1}{C}\int_{-\infty}^{\infty} i(t)\, dt = E\, u(t) \qquad ...(9.29)$$

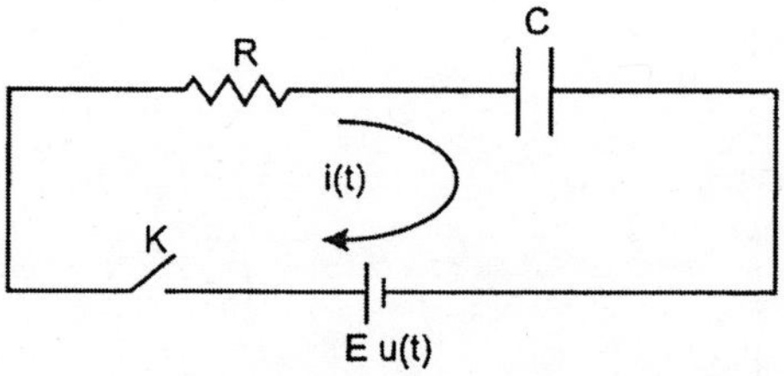

Fig. 9.6

$$Ri(t) + \frac{1}{C}\int_{-\infty}^{0} i(t)\, dt + \frac{1}{C}\int_{-\infty}^{\infty} i(t)\, dt = E\, u(t) \qquad ...(9.30)$$

Let, $e_0 = \frac{1}{C}\int_{-\infty}^{\infty} i(t)\, dt$ is the voltage stored on the capacitor before the switch is closed.

Then equation (9.30) becomes,

$$Ri(t) + e_0 + \frac{1}{C}\int_{-\infty}^{\infty} i(t)\, dt = E\, u(t) \qquad ...(9.31)$$

Taking laplace transformation on both sides of equation (9.31) we get

$$RI(s) + \frac{e_0}{s} + \frac{1}{Cs}I(s) = \frac{E}{s} \qquad ...(9.32)$$

Let us assume $e(0)$. Then equation (9.32) can be written as

$$I(s)\left[\frac{1}{Cs} + R\right] = \frac{E}{s}$$

or

$$I(s)\left[\frac{1+CRs}{Cs}\right] = \frac{E}{s}$$

or
$$I(s) = \frac{EC}{1+RCs}$$

or
$$I(s) = \frac{EC/RC}{\frac{1}{RC}+s} = \frac{E}{R}\left[\frac{1}{s+\frac{1}{RC}}\right] \quad ...(9.33)$$

Taking the inverse laplace transformation of equation (9.33) we get

$$i(t) = \frac{E}{R}\left[e^{-\left(1/RC\right)t}\right] \text{ amp} \quad ...(9.34)$$

9.3.2 Analysis of RC Circuit Excited with Impulse Input

Consider the circuit shown in Figure 9.7. Let at time $t = 0$ the switch K is closed, then the loop equation is given by,

$$Ri(t) + \frac{1}{C}\int_{-\infty}^{\infty} i(t)\,dt = E\,\delta(t) \quad ...(9.35)$$

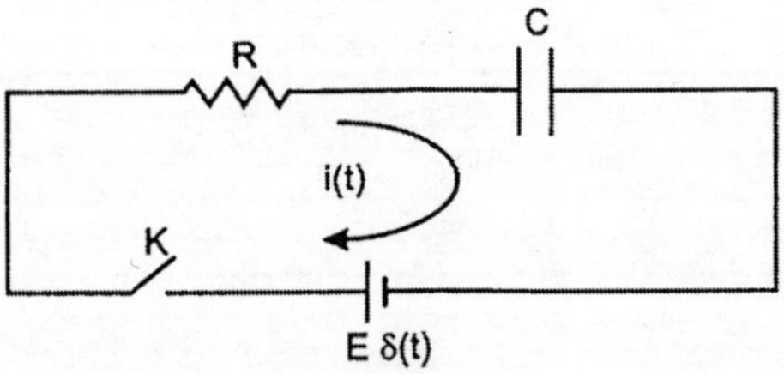

Fig. 9.7

$$Ri(t) + \frac{1}{C}\int_{-\infty}^{0} i(t)\,dt + \frac{1}{C}\int_{0}^{\infty} i(t)\,dt = E\,\delta(t) \quad ...(9.36)$$

Let, $e_0 = \frac{1}{C}\int_{-\infty}^{0} i(t)\,dt$ is the voltage stored on the capacitor before the switch is closed. Then equation (9.36) becomes,

$$Ri(t) + e_0 + \frac{1}{C}\int_{0}^{\infty} i(t)\,dt = E\,\delta(t) \quad ...(9.37)$$

Taking laplace transformation on both sides of equation (9.37) we get

$$RI(s) + \frac{e_0}{s} + \frac{1}{Cs}\, I(s) = E \quad ...(9.38)$$

Let us assume $e(0) = 0$. Then equation (9.38) can be written as

$$I(s)\left[\frac{1}{Cs}+R\right] = E$$

or
$$I(s)\left[\frac{1+CRs}{Cs}\right] = E$$

or
$$I(s) = \frac{ECs}{1+RCs}$$

or
$$I(s) = \frac{ECs/RC}{\frac{1}{RC}+s} = \frac{E}{R}\left[\frac{s}{s+\frac{1}{RC}}\right] \quad ...(9.39)$$

Taking the inverse laplace transformation of equation (10.39) we get,

$$i(t) = \frac{E}{R}\left[te^{\frac{-t}{RC}}\right] \text{ amp} \quad ...(9.40)$$

9.3.3 Analysis of RC Circuit Excited with Ramp Input

Consider the circuit shown in Figure 9.8. Let at time $t = 0$ the switch K is closed, then the loop equation is given by,

$$Ri(t) + \frac{1}{C}\int_{-\infty}^{\infty} i(t)\, dt = E\, r(t)$$

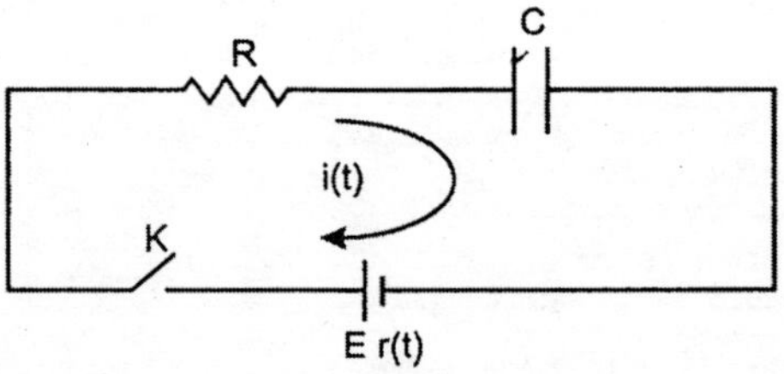

Fig. 9.8

$$Ri(t) + \frac{1}{C}\int_{-\infty}^{0} i(t)\, dt + \frac{1}{C}\int_{0}^{\infty} i(t)\, dt = E\, r(t) \quad ...(9.41)$$

Let, $e_0 = \frac{1}{C}\int_{-\infty}^{0} i(t)\, dt$ is the voltage stored on the capacitor before the switch is closed. Then equation (9.41) becomes,

$$Ri(t) + e_0 + \frac{1}{C}\int_{0}^{\infty} i(t)\, dt = Er(t) \quad ...(9.42)$$

Taking laplace transformation on both sides of equation (9.42) we get

$$RI(s) + \frac{e_0}{s} + \frac{1}{Cs}I(s) = \frac{E}{s^2} \quad ...(9.43)$$

Let us assume $e(0) = 0$. Then equation (9.43) can be written as

$$I(s)\left[\frac{1}{Cs}+R\right] = \frac{E}{s^2}$$

or
$$I(s)\left[\frac{1+CRs}{Cs}\right] = \frac{E}{s^2}$$

or
$$I(s) = \frac{ECs}{s^2(1+RCs)}$$

or
$$I(s) = \frac{EC/RC}{s\left(\frac{1}{RC}+s\right)} = \frac{E}{R}\left[\frac{1}{s\left(s+\frac{1}{RC}\right)}\right]$$

Let
$$\left[\frac{1}{s\left(s+\frac{1}{RC}\right)}\right] = \frac{A}{s}+\frac{B}{\left(s+\frac{1}{RC}\right)} \quad ...(9.44)$$

where A and B are constants.

To find the values of constants A and B :

Form equation (9.44) we get

$$A\left(s+\frac{1}{RC}\right) + BS = 1 \quad ...(9.45)$$

Then equating the coefficient of s we get

$$A + B = 0 \quad ...(9.46)$$

and equating the constants we get

$$\frac{A}{RC} = 1$$

or
$$A = RC$$

Substituting the value of the constant A in equation (9.46) we get

$$B = -RC$$

Substituting the value of A and B in equation (9.44) we get

$$I(s) = \frac{E}{R}\left[\frac{RC}{s}-\frac{RC}{s+\frac{1}{RC}}\right]$$

$$I(s) = EC\left[\frac{1}{s}-\frac{1}{s+\frac{1}{RC}}\right] \quad ...(9.47)$$

Taking the inverse laplace transformation of equation (9.47) we get

$$I(t) = EC\left[1-e^{-(1/RC)t}\right] \text{ amp}$$

9.3.4 Analysis of RC Circuit Excited with Pulse Input

Consider the circuit shown in Figure 9.9. Let at time $t = 0$ the switch K is closed, then the loop equation is given by,

$$Ri(t) + \frac{1}{C}\int_{-\infty}^{\infty} i(t)\, dt = E(t) \quad ...(9.48)$$

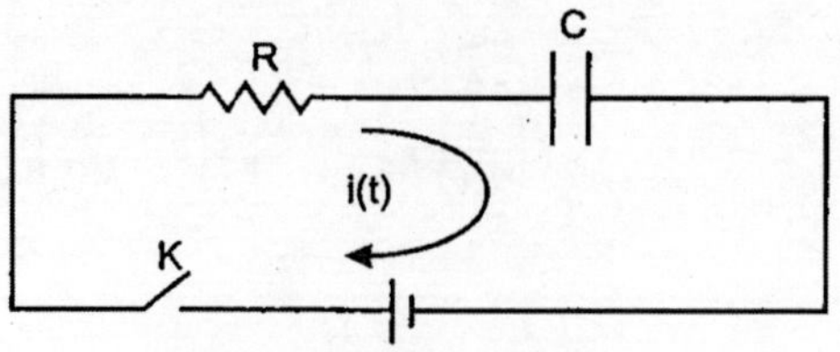

E (t) = E [u (t–T₁) –u(t–T₂)]

Fig. 9.9

$$Ri(t) + \frac{1}{C}\int_{-\infty}^{0} i(t)\ dt + \frac{1}{C}\int_{0}^{\infty} i(t)\, dt = E(t) \qquad \text{...(9.49)}$$

Let, $e_0 = \frac{1}{C}\int_{-\infty}^{0} i(t)\ dt$ is the voltage stored on the capacitor before the switch is closed. Then equation (9.49) becomes,

$$Ri(t) + e_0 + \frac{1}{C}\int_{0}^{\infty} i(t)\ dt = E(t) \qquad \text{...(9.50)}$$

Taking laplace transformation on both sides of equation (9.50) we get

$$RI(s) + \frac{e_0}{s} + \frac{1}{Cs}I(s) = \frac{E}{s}(e^{-sT_1} - e^{-sT_2}) \qquad \text{...(9.51)}$$

Let us assume $e(0) = 0$. Then equation (9.51) can be written as

$$I(s)\left[\frac{1}{Cs} + R\right] = \frac{E}{s}(e^{-sT_1} - e^{-sT_2})$$

or
$$I(s)\left[\frac{1 + CRs}{Cs}\right] = \frac{E}{s}(e^{-sT_1} - e^{-sT_2})$$

or
$$I(s) = \frac{EC}{(1 + RCs)}(e^{-sT_1} - e^{-sT_2})$$

or
$$I(s) = \frac{EC/RC}{\left(\frac{1}{RC} + s\right)}(e^{-sT_1} - e^{-sT_2})$$

$$= \frac{E}{R}\left[\frac{1}{\left(s + \frac{1}{RC}\right)}(e^{-sT_1} - e^{-sT_2})\right]$$

$$= \frac{E}{R}\left[\frac{e^{-sT_1}}{\left(s + \frac{1}{RC}\right)} - \frac{e^{-sT_2}}{\left(s + \frac{1}{RC}\right)}\right] \qquad \text{...(9.52)}$$

Taking the inverse laplace transformation of equation (9.52) we get

$$i(t) = \frac{E}{R}\left[e^{-\frac{t-T_1}{RC}}u(t - T_1) - e^{-\frac{t-T_2}{RC}}u(t - T_2)\right] \text{ amp} \qquad \text{...(9.53)}$$

9.3.5 Analysis of RC Circuit Excited with Sinusoidal Input

Consider the circuit shown in Figure 9.10. Let at time $t = 0$ the switch K is closed, then the loop equation is given by,

$$Ri(t) + \frac{1}{C}\int_{-\infty}^{\infty} i(t)\ dt = E(t) \qquad ...(9.54)$$

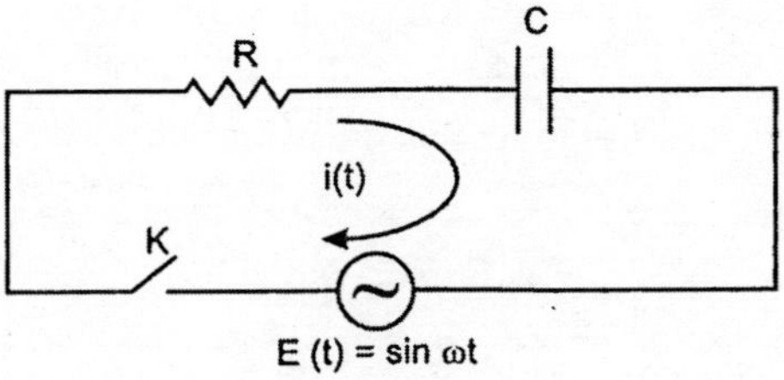

Fig. 9.10

$$Ri(t) + \frac{1}{C}\int_{-\infty}^{0} i(t)\ dt + \frac{1}{C}\int_{0}^{\infty} i(t)\ dt = E(t) \qquad ...(9.55)$$

Let, $e_0 = \frac{1}{C}\int_{-\infty}^{0} i(t)\ dt$ is the voltage stored on the capacitor before the switch is closed.

Then equation (9.55) becomes,

$$Ri(t) + e_0 + \frac{1}{C}\int_{-\infty}^{\infty} i(t)\ dt = E(t) \qquad ...(9.56)$$

Taking laplace transformation on both sides of equation (9.56) we get

$$RI(s) + \frac{e_0}{s} + \frac{1}{Cs}I(s) = \frac{\omega}{\omega^2 + s^2} \qquad ...(9.57)$$

Let us assume $e(0)$. Then equation (9.57) can be written as

$$I(s)\left[\frac{1}{Cs} + R\right] = \frac{\omega}{\omega^2 + s^2}$$

or

$$I(s)\left[\frac{1 + CRs}{Cs}\right] = \frac{\omega}{\omega^2 + s^2}$$

or

$$I(s) = \frac{\omega Cs}{(1 + RCs) \times (\omega^2 + s^2)}$$

or

$$I(s) = \frac{\frac{\omega s}{R}}{\left(s + \frac{1}{RC}\right) \times (s \times j\omega) \times (s - j\omega)} \qquad ...(9.58)$$

Let

$$\frac{\frac{\omega s}{R}}{\left(s + \frac{1}{RC}\right) \times (s \times j\omega) \times (s - j\omega)} = \frac{A}{\left(s + \frac{1}{RC}\right)} + \frac{B}{s + j\omega} + \frac{C}{s - j\omega}$$

where A, B and C are constants.

To find the values of constants A, B and C :

$$A = \left.\frac{\frac{\omega s}{R}\times\left(s+\frac{1}{RC}\right)}{\left(s+\frac{1}{RC}\right)\times(s+j\omega)\times(s-j\omega)}\right|_{s=-\frac{1}{RC}} = -\frac{\omega C}{1+R^2C^2\omega^2}$$

$$B = \left.\frac{\frac{\omega s}{R}\times(s+j\omega)}{\left(s+\frac{1}{RC}\right)\times(s+j\omega)\times(s-j\omega)}\right|_{s=-j\omega} = \frac{\omega}{2R}\left[\frac{RC+j\omega}{R^2C^2+\omega^2}\right]$$

$$C = B^* = \frac{\omega}{2R}\left[\frac{RC-j\omega}{R^2C^2+\omega^2}\right]$$

Substituting the values of A, B and C in equation (9.58) we get

$$I(s) = \frac{\frac{\omega s}{R}}{\left(s+\frac{1}{RC}\right)\times(s+j\omega)\times(s-j\omega)}$$

$$= -\frac{\omega C}{1+R^2C^2\omega^2}\frac{1}{\left(s+\frac{1}{RC}\right)}+\frac{\omega}{2R}\left[\frac{RC+j\omega}{R^2C^2+\omega^2}\right]$$

$$\frac{1}{s+j\omega}+\frac{\omega}{2R}\left[\frac{RC+j\omega}{R^2C^2+\omega^2}\right]\frac{1}{s-j\omega} \qquad ...(9.59)$$

Taking the inverse Laplace Transformation of equation (9.59) we get

$$i(t) = -\frac{\omega C}{1+R^2C^2\omega^2}e^{-\frac{t}{RC}}+\frac{\omega}{2R}\left[\frac{RC+j\omega}{R^2C^2+\omega^2}\right]e^{-j\omega}$$

$$+\frac{\omega}{2R}\left[\frac{RC+j\omega}{R^2C^2+\omega^2}\right]e^{+j\omega}A$$

SOLVED EXAMPLES

EXAMPLE 9.10. *For the circuit shown in Figure E9.10, the switch k was closed at time t = 0. Find the complete expression for the current after the switch is closed. Assume all the initial conditions to be zero.*

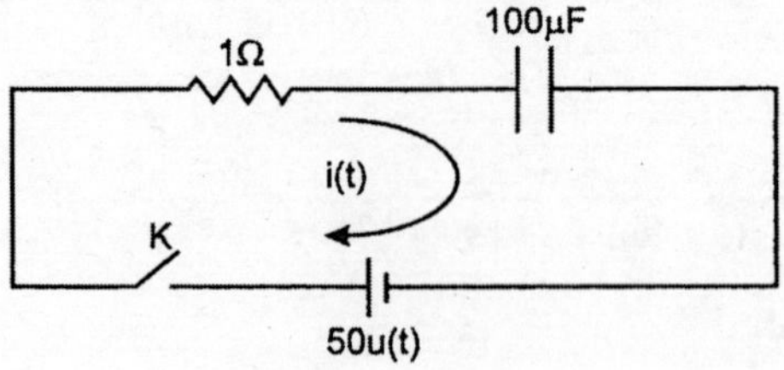

Fig. E9.10

SOLUTION. Consider the circuit shown in Figure E9.10. Let at time $t = 0$ the switch K is closed, then the loop equation is given by,

$$1\, i(t) + \frac{1}{100\times 10^{-6}} \int_{-\infty}^{\infty} i(t)dt = 50u(t) \qquad ...(i)$$

$$1\, i(t) + \frac{1}{100\times 10^{-6}} \int_{-\infty}^{\infty} i(t)dt + \frac{1}{100\times 10^{-6}} \int_{0}^{\infty} i(t)dt = 50u(t) \qquad ...(ii)$$

Let, $e_0 = \frac{1}{100\times 10^{-6}} \int_{-\infty}^{0} i(t)\, dt$ is the voltage stored on the capacitor before the switch is closed. Then equation (*ii*) becomes,

$$1\, i(t) + e_0 + \frac{1}{100\times 10^{-6}} \int_{0}^{\infty} i(t)dt = 50u(t) \qquad ...(iii)$$

Let us assume $e(0)$. Then equation (*iii*) can be written as

$$1\, i(t) + e_0 + \frac{1}{100\times 10^{-6}} \int_{0}^{\infty} i(t)dt = 50u(t)$$

Taking laplace transformation on both sides of equation (*iii*) we get

$$1I(s) + \frac{1}{100\times 10^{-6}s} I(s) = \frac{50}{s} \qquad ...(iv)$$

or

$$I(s)\left[\frac{1}{100\times 10^{-6}s} + 1\right] = \frac{50}{s}$$

or

$$I(s)\left[\frac{1+100\times 10^{-6}\times 1s}{100\times 10^{-6}s}\right] = \frac{50}{s}$$

or

$$I(s) = \frac{50+100\times 10^{-6}}{1+1\times 100\times 10^{-6}s}$$

or

$$I(s) = \frac{50\times 100\times 10^{-6} / 1\times 100\times 10^{-6}}{\frac{1}{1\times 100\times 10^{-6}} + s} = 50\left[\frac{1}{s+\frac{1}{1\times 100\times 10^{-6}}}\right] \qquad ...(v)$$

Taking the inverse laplace transformation of equation (*v*) we get

$$I(t) = 50\left[e^{-\left(1/1\times 100\times 10^{-6}\right)t}\right] \text{ amp} = 50\ [e^{-10000}] \text{ amp} \qquad ...(vi)$$

EXAMPLE 9.11. *For the circuit shown in Figure E9.11, the switch k was closed at time t = 0, connecting the circuit to an impulse source of 10δ(t). Find the complete expression for the current after the switch is closed. Assume* $e_0 = 2V$ *(the polarity is as indicated in the Figure E9.11)*

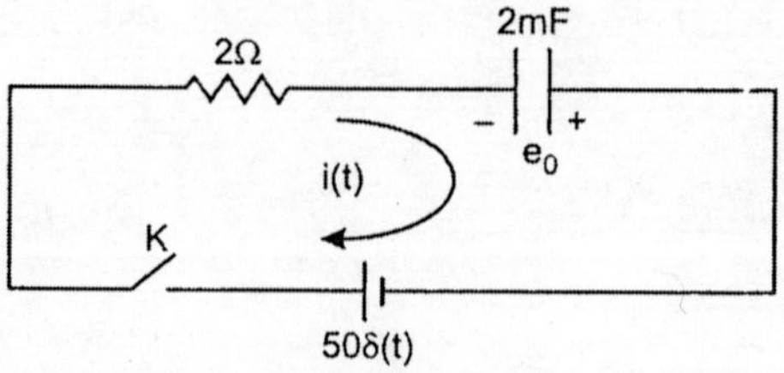

Fig. E9.11

SOLUTION. Consider the circuit shown in Figure E9.11. Let at time $t = 0$ the switch K is closed, then the loop equation is given by,

$$Ri(t) + \frac{1}{C}\int_{-\infty}^{\infty} i(t)\ dt = E\delta(t) \quad ...(i)$$

$$Ri(t) + \frac{1}{C}\int_{-\infty}^{0} i(t)\ dt + \frac{1}{C}\int_{0}^{\infty} i(t)\ dt = E\delta(t) \quad ...(ii)$$

Let, $e_0 = \frac{1}{C}\int_{-\infty}^{0} i(t)\ dt$ is the voltage stored on the capacitor before the switch is closed.

Then equation (*ii*) becomes,

$$Ri(t) - e_0 + \frac{1}{C}\int_{0}^{\infty} i(t)\ dt = E\delta(t) \quad ...(iii)$$

Taking laplace transformation on both sides of equation (*iii*) we get

$$RI(s) - \frac{e_0}{s} + \frac{1}{Cs}\ I(s) = E \quad ...(iv)$$

$$I(s)\left[\frac{1}{Cs} + R\right] = E + \frac{e_0}{s}$$

or $$I(s)\left[\frac{1+CRs}{Cs}\right] = \left(E + \frac{e_0}{s}\right)$$

or $$I(s) = \left(E + \frac{e_0}{s}\right) \times \left(\frac{Cs}{1+CRs}\right)$$

or $$I(s) = \frac{ECs}{1+CRs} + \frac{Ce_0}{1+CRs} \quad ...(v)$$

$$I(s) = \frac{E}{R}\left[\frac{s}{s+\frac{1}{CR}}\right] + \frac{e_0}{R}\left[\frac{1}{s+\frac{1}{CR}}\right] \quad ...(vi)$$

Taking the inverse laplace transformation of equation (*vi*) we get

EXAMPLE 9.12. *For the circuit shown in Figure E9.12, the switch k was closed at time t = 0, connecting the circuit to a ramp input of 10r(t). Find the complete expression for the current after the switch is closed. Assume e_0 = 2V (the polarity is as indicated in the Figure E9.12)*

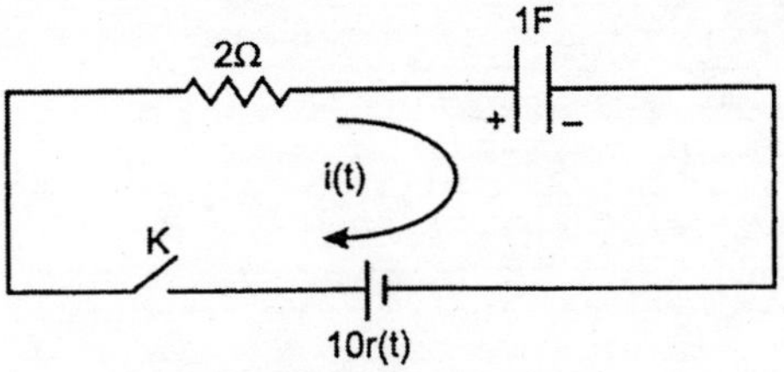

Fig. E9.12

SOLUTION. Consider the circuit shown in Figure E9.12. Let at time $t = 0$ the switch K is closed, then the loop equation is given by,

$$Ri(t) + \frac{1}{C}\int_{-\infty}^{\infty} i(t)\, dt = Er(t) \qquad ...(i)$$

$$Ri(t) + \frac{1}{C}\int_{-\infty}^{0} i(t)\, dt + \frac{1}{C}\int_{0}^{\infty} i(t)\, dt = Er(t) \qquad ...(ii)$$

Let, $e_0 = \frac{1}{C}\int_{-\infty}^{0} i(t)\, dt$ is the voltage stored on the capacitor before the switch is closed.

Then equation (*ii*) becomes,

$$Ri(t) + e_0 + \frac{1}{C}\int_{0}^{\infty} i(t)\, dt = Er(t) \qquad ...(iii)$$

Taking laplace transformation on both sides of equation (*iii*) we get

$$RI(s) + \frac{e_0}{s} + \frac{1}{Cs}I(s) = \frac{E}{s^2} \qquad ...(iv)$$

or
$$I(s)\left[\frac{1}{Cs} + R\right] = \frac{E}{s^2} - \frac{e_0}{s}$$

or
$$I(s)\left[\frac{1+CRs}{Cs}\right] = \frac{E}{s^2} - \frac{e_0}{s}$$

or
$$I(s) = \frac{ECs}{s^2(1+RCs)} - \frac{Ce_0}{1+CRs}$$

or
$$I(s) = \frac{EC/RC}{s\left(\frac{1}{RC}+s\right)} = \frac{E}{R}\left[\frac{1}{s\left(s+\frac{1}{RC}\right)}\right] - \frac{e_0/R}{s+\frac{1}{CR}} \qquad ...(v)$$

Let
$$\left[\frac{1}{s\left(s+\frac{1}{RC}\right)}\right] = \frac{A}{s} + \frac{B}{\left(s+\frac{1}{RC}\right)} \qquad ...(vi)$$

where A and B are constants.

To find the values of constants A and B :

From equation (*vi*) we get

$$A\left(s+\frac{1}{RC}\right) + Bs = 1 \qquad ...(vii)$$

Then equating the coefficient of s we get

$$A + B = 0 \qquad ...(viii)$$

and equating the constants we get

$$\frac{A}{RC} = 1$$

or

$$A = RC$$

Substituting the value of the constant A in equation (*viii*) we get

$$B = -RC$$

Substituting the value of A and B in equation (*v*) we get

$$I(s) = \frac{E}{R}\left[\frac{RC}{s} - \frac{RC}{s+\frac{1}{RC}}\right] - \frac{e_0/R}{s+\frac{1}{CR}}$$

$$I(s) = EC\left[\frac{1}{s} - \frac{1}{s+\frac{1}{RC}}\right] - \frac{e_0/R}{s+\frac{1}{CR}} \qquad ...(ix)$$

Taking the inverse laplace transformation of equation (*ix*) we get

$$i(t) = EC\left[1 - e^{-(1/RC)t}\right] - \frac{e_0}{R} e^{-\frac{t}{CR}} \text{ amp} \qquad ...(x)$$

Substituting the values of R, L, C and E in equation (*x*), we get

$$i(t) = 10\delta(t)\ [1 - e^{-0.5t}] - e^{-0.5t} \text{ amp}$$

EXAMPLE 9.13. *For the circuit shown in Figure E9.13, the switch k was closed at time t = 0. Find the complete expression for the current after the switch is closed.*

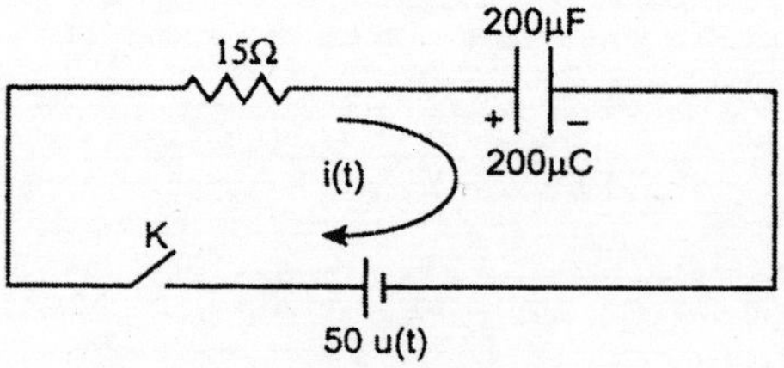

Fig. E9.13

SOLUTION. From the circuit shown in Figure E9.13, the KVL equation is,

$$Ri(t) + \frac{1}{C}\int_{-\infty}^{\infty} i(t)\, dt = Eu(t) \qquad ...(i)$$

$$Ri(t) + \frac{1}{C}\int_{-\infty}^{0} i(t)\, dt + \frac{1}{C}\int_{0}^{\infty} i(t)\, dt = Er(t) \qquad ...(ii)$$

where $$\frac{1}{C}\int_{-\infty}^{0} i(t)\ dt = e_0 \text{ and is given by}$$

$$e_0 = \frac{Q}{C} = \frac{200\times10^{-6}}{200\times10^{-6}} = 1V$$

$$Ri(t) + e_0 + \frac{1}{C}\int_{0}^{\infty} i(t)\ dt = Eu(t) \qquad ...(iii)$$

Taking laplace transformation on both sides of equation (*iii*) we get

$$RI(s) + \frac{e_0}{s} + \frac{1}{Cs}I(s) = \frac{E}{s} \qquad ...(iv)$$

$$I(s)\left[\frac{1}{Cs}+R\right] = \frac{E}{s}-\frac{e_0}{s}$$

or $$I(s)\left[\frac{1+CRs}{Cs}\right] = \frac{E-e_0}{s}$$

or $$I(s) = \frac{(E-e_0)C}{(1+RCs)} = \frac{(E-e_0)/R}{\frac{1}{RC}+s}$$

Taking inverse laplace transform we get

$$i(t) = \frac{E-e_0}{R}e^{-\frac{t}{RC}} \text{ amp} \qquad ...(v)$$

Substituting the values of R, L, C and E in equation (v) we get

$$i(t) = 3.27e^{-333.33t} \text{ amp}$$

EXAMPLE 9.14. *For the circuit shown in Figure E9.14, the switch k was closed at time t = 0. Find the complete expression for the current after the switch is closed. Assume a frequency of 50Hz.*

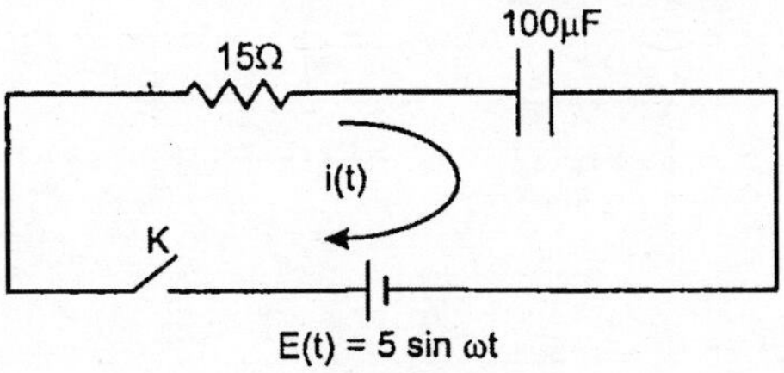

Fig. E9.14

SOLUTION. Consider the circuit shown in Figure E9.14. Let at time t =0 the switch K is closed, then the loop equation is given by,

$$Ri(t) + \frac{1}{C}\int_{-\infty}^{\infty} i(t)\ dt = E(t) \qquad ...(i)$$

$$Ri(t) + \frac{1}{C}\int_{-\infty}^{0} i(t)\ dt + \frac{1}{C}\int_{0}^{\infty} i(t)\,dt = E(t) \qquad ...(ii)$$

Let, $e_0 = \frac{1}{C}\int_{-\infty}^{0} i(t)\, dt$ is the voltage stored on the capacitor before the switch is closed.

Then equation (*ii*) becomes,

$$Ri(t) + e_0 + \frac{1}{C}\int_{0}^{\infty} i(t)\, dt = E(t) \qquad ...(iii)$$

Taking laplace transformation on both sides of equation (*iii*) we get

$$RI(s)+\frac{e_0}{s}+\frac{1}{Cs}I(s) = \frac{\omega}{\omega^2+s^2} \qquad ...(iv)$$

Let us assume $e(0) = 0$. Then equation (*iv*) can be written as

$$I(s)\left[\frac{1}{Cs}+R\right] = \frac{\omega}{\omega^2+s^2}$$

or
$$I(s)\left[\frac{1+CRs}{Cs}\right] = \frac{\omega}{\omega^2+s^2}$$

or
$$I(s) = \frac{\omega Cs}{(1+RCs)\times(\omega^2+s^2)}$$

or
$$I(s) = \frac{\frac{\omega s}{R}}{\left(s+\frac{1}{RC}\right)\times(s+j\omega)\times(s-j\omega)} \qquad ...(v)$$

Let,
$$\frac{\frac{\omega s}{R}}{\left(s+\frac{1}{RC}\right)\times(s+j\omega)\times(s-j\omega)} = \frac{A}{\left(s+\frac{1}{RC}\right)}+\frac{B}{s+j\omega}+\frac{C}{s-j\omega}$$

where A, B and C are constants.

To find the values of constants A, B and C :

$$A = \left.\frac{\frac{\omega s}{R}\times\left(s+\frac{1}{RC}\right)}{\left(s+\frac{1}{RC}\right)\times(s+j\omega)\times(s-j\omega)}\right|_{s=-\frac{1}{RC}} = -\frac{\omega C}{1+R^2C^2\omega^2}$$

$$B = \left.\frac{\frac{\omega s}{R}\times(s+j\omega)}{\left(s+\frac{1}{RC}\right)\times(s+j\omega)\times(s-j\omega)}\right|_{s=-j\omega} = \frac{\omega}{2R}\left[\frac{RC+j\omega}{R^2C^2+\omega^2}\right]$$

$$C = B^* = C=B^* = \frac{\omega}{2R}\left[\frac{RC-j\omega}{R^2C^2+\omega^2}\right]$$

Substituting the values of A, B and C in equation (*v*) we get

$$I(s) = \frac{\frac{\omega s}{R}}{\left(s+\frac{1}{RC}\right)\times(s+j\omega)\times(s-j\omega)}$$

$$= \frac{\omega C}{1+R^2C^2\omega^2}\frac{1}{\left(s+\frac{1}{RC}\right)} + \frac{\omega}{2R}\left[\frac{RC+j\omega}{R^2C^2+\omega^2}\right]$$

$$\frac{1}{s+j\omega} + \frac{\omega}{2R}\left[\frac{RC+j\omega}{R^2C^2+\omega^2}\right]\frac{1}{s-j\omega} \qquad ...(vi)$$

Taking the inverse Laplace Transformation of equation (*vi*) we get

$$i(t) = -0.25e^{-666.66t} + j0.03e^{-j314.2} + j0.03e^{+j314.2} \text{ A}$$

EXAMPLE 9.15. *For the circuit shown in Figure E9.15, the switch S_1 was closed and the switch S_2 was open for a long period of time. Then at time $t = 0$ the switch S_1 was opened and the switch S_2 was closed. Find the expression for transient current.*

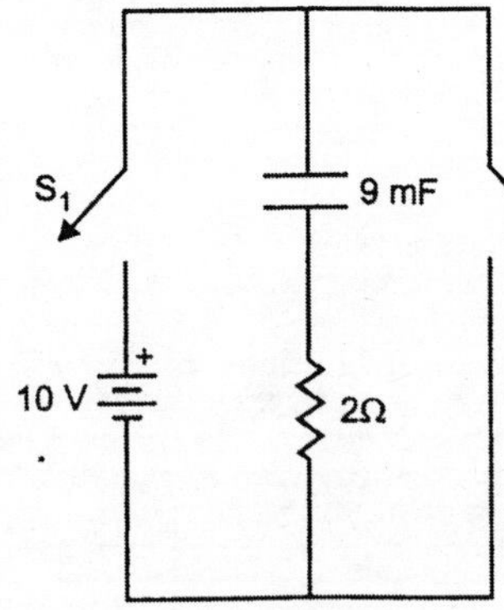

Fig. E9.15

SOLUTION. When the switch S_1 was closed for a long period of time, the circuit becomes as shown in Figure E9.15(*a*), where e_0 = 10V.

When the switch S_1 is opened and the switch S_2 is closed the circuit becomes as shown in Figure E9.15(*b*).

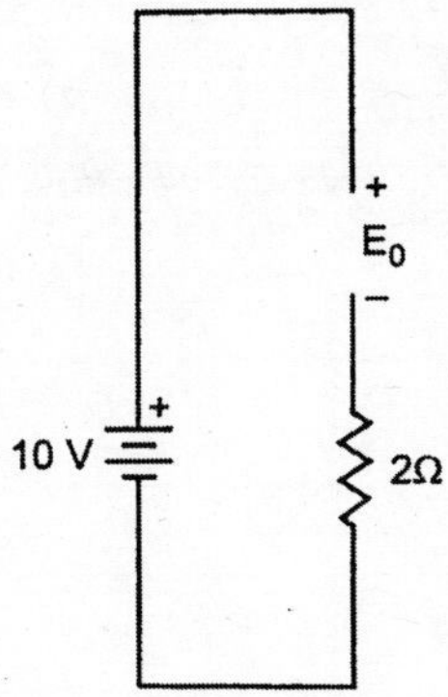

Fig. E9.15(a)

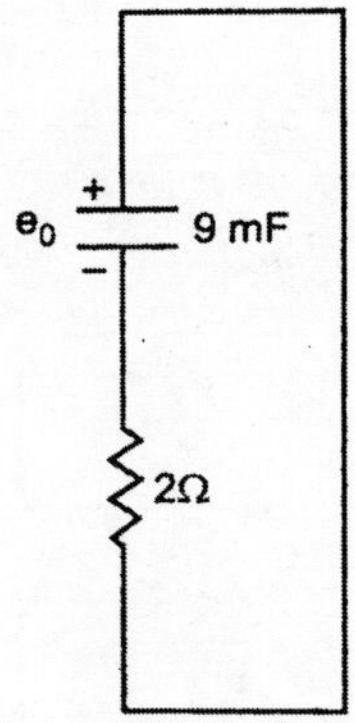

Fig. E9.15(b)

Writing the KVL equation we get

$$Ri(t) + \frac{1}{C}\int_{-\infty}^{\infty} i(t)\, dt = 0 \qquad ...(i)$$

$$Ri(t) + \frac{1}{C}\int_{-\infty}^{0} i(t)\ dt + \frac{1}{C}\int_{0}^{\infty} i(t)\ dt = 0 \qquad ...(ii)$$

Let, $e_0 = \frac{1}{C}\int_{-\infty}^{0} i(t)\ dt$ is the voltage stored on the capacitor before the switch S_2 is closed.

Then equation (*ii*) becomes,

$$Ri(t) - e_0 + \frac{1}{C}\int_{0}^{\infty} i(t)\ dt = 0 \qquad ...(iii)$$

Taking laplace transformation on both sides of equation (*iii*) we get

$$RI(s) + \frac{1}{Cs}I(s) = \frac{e_0}{s}$$

or
$$I(s)\left[\frac{1}{Cs} + R\right] = \frac{e_0}{s}$$

or
$$I(s)\left[\frac{1 + CRs}{Cs}\right] = \frac{e_0}{s}$$

or
$$I(s) = \frac{e_0 C}{1 + RCs}$$

or
$$I(s) = \frac{e_0 C / RC}{\frac{1}{RC} + s} = \frac{e_0}{R}\left[\frac{1}{s + \frac{1}{RC}}\right] \qquad ...(iv)$$

Taking the inverse laplace transformation of equation (*iv*) we get

$$i(t) = \frac{e_0}{R}\left[e^{-(1/RC)t}\right] \text{ amp} \qquad ...(v)$$

Substituting the values of e_0, R, and C we get

$$i(t) = \frac{10}{2}\left[e^{-\left(\frac{1}{2 \times 9 \times 10^{-3}}\right)t}\right] \text{ amp} = 5e^{55.56t} \text{ amp}$$

EXAMPLE 9.16. *For the circuit shown in Figure E9.16, the switch S_1 was closed and the switch S_2 was open for 1ms. Then at time t = 0 the switch S_1 was opened and the switch S_2 was closed. Find the expression for transient current.*

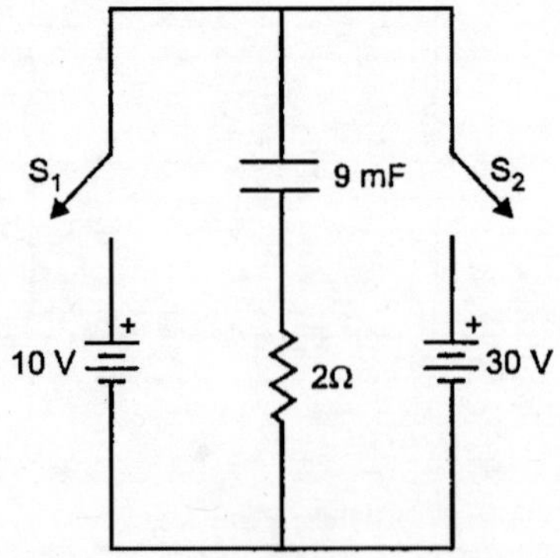

Fig. E9.16

SOLUTION. When the switch S_1 was closed,the circuit becomes as shown in Figure E9.16(*a*).

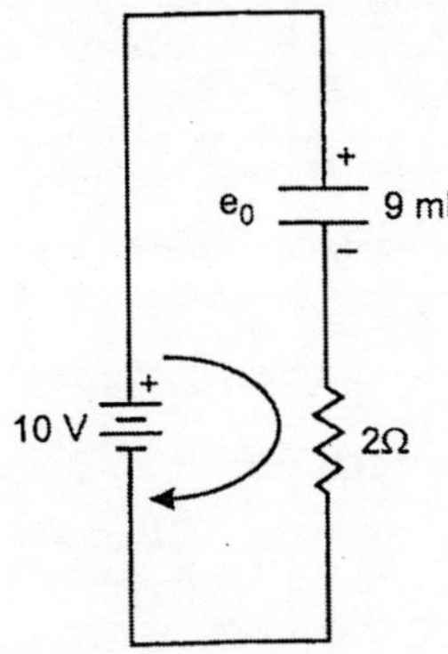

Fig. E9.16(a)

Writing the KVL equation we get,

$$Ri(t) + \frac{1}{C}\int_{-\infty}^{\infty} i(t)\ dt = Eu(t) \quad ...(i)$$

$$Ri(t) + \frac{1}{C}\int_{-\infty}^{0} i(t)\ dt + \frac{1}{C}\int_{0}^{\infty} i(t)\, dt = Eu(t) \quad ...(ii)$$

Let, $e_0 = \frac{1}{C}\int_{-\infty}^{0} i(t)\ dt$ is the voltage stored on the capacitor before the switch is closed.

Then equation (*ii*) becomes,

$$Ri(t) + \frac{1}{C}\int_{0}^{\infty} i(t)\ dt = Eu(t) \quad ...(iii)$$

Taking laplace transformation on both sides of equation (*iii*) we get

$$RI(s) + \frac{1}{Cs}I(s) = \frac{E}{s} \quad ...(iv)$$

or $$I(s)\left[\frac{1}{Cs}+R\right] = \frac{E}{s}$$

or $$I(s)\left[\frac{1+CRs}{Cs}\right] = \frac{E}{s}$$

or $$I(s) = \frac{EC}{1+RCs}$$

or $$I(s) = \frac{EC/RC}{\frac{1}{RC}+s} = \frac{E}{R}\left[\frac{1}{s+\frac{1}{RC}}\right] \quad ...(v)$$

Taking the inverse laplace transformation of equation (*v*) we get

$$i(t) = \frac{E}{R}\left[e^{-\left(1/RC\right)t}\right] \text{ amp} \quad ...(vi)$$

Substituting the values $E = 10V$, $R = 2\Omega$ and $C = 9mF$ we get

$$i(t) = \frac{10}{2}\left[e^{-\left(\frac{1}{2\times 9\times 10^{-3}}\right)t}\right] \text{ amp} = 5e^{-55.55t} \text{ amp}$$

$$V_R = i(t) \times R = 10e^{-55.55t} \text{ V}$$

$$V_C = E_0 = 10 - V_R = 10 \times (1 - e^{-55.55t}) \text{ V}$$

Substituting $t = 1$ mS we get

$$V_C = E_0 = 0.54 \text{ V}$$

When the switch S_2 is closed we get the circuit as shown in Figure E9.16(*b*), where $E_0 = V_C$.

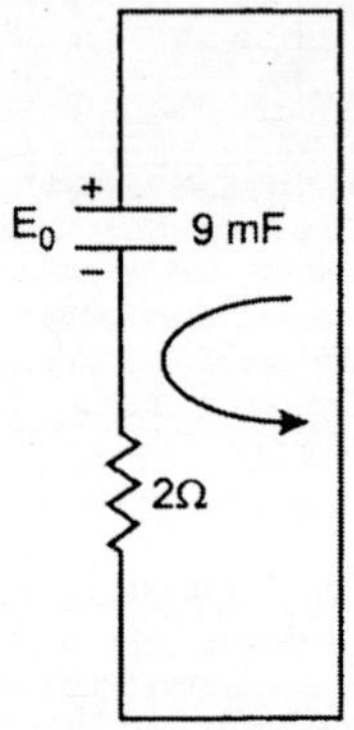

Fig. E9.16(b)

Writing the KVL equation we get

$$Ri(t) + \frac{1}{C}\int_{-\infty}^{\infty} i(t)\, dt = 0 \qquad \text{...(vii)}$$

$$Ri(t) + \frac{1}{C}\int_{-\infty}^{0} i(t)\, dt + \frac{1}{C}\int_{0}^{\infty} i(t)\, dt = 0 \qquad \text{...(viii)}$$

Let, $E_0 = \frac{1}{C}\int_{-\infty}^{0} i(t)\, dt$ is the voltage stored on the capacitor before the switch S_2 is closed.

Then equation (*viii*) becomes,

$$Ri(t) - E_0 + \frac{1}{C}\int_{0}^{\infty} i(t)\, dt = 0 \qquad \text{...(ix)}$$

Taking laplace transformation on both sides of equation (*ix*) we get

$$RI(s) + \frac{1}{Cs}I(s) = \frac{E_0}{s}$$

or

$$I(s)\left[\frac{1}{Cs} + R\right] = \frac{E_0}{s}$$

or

$$I(s)\left[\frac{1+CRs}{Cs}\right] = \frac{E_0}{s}$$

or
$$I(s) = \frac{E_0 C}{1 + RCs}$$

or
$$I(s) = \frac{E_0 C / RC}{\frac{1}{RC} + s} = \frac{E_0}{R}\left[\frac{1}{s + \frac{1}{RC}}\right] \quad ...(x)$$

Taking the inverse laplace transformation of equation (x) we get

$$i(t) = \frac{E_0}{R}\left[e^{-(1/RC)t}\right] \text{ amp} \quad ...(xi)$$

Substituting the values of E_0, R, and C we get

$$i(t) = \frac{0.54}{2}\left[e^{-\left(1/2 \times 9 \times 10^{-3}\right)t}\right] \text{amp}$$
$$= 0.27e^{-55.55t} \text{ amp}$$

EXAMPLE 9.17. *For the circuit shown in Figure E9.16, the switch S_1 was closed and the switch S_2 was open for 1ms. Then at time t = 0 the switch S_1 was opened and the switch S_2 was closed. Find the expression for transient current.*

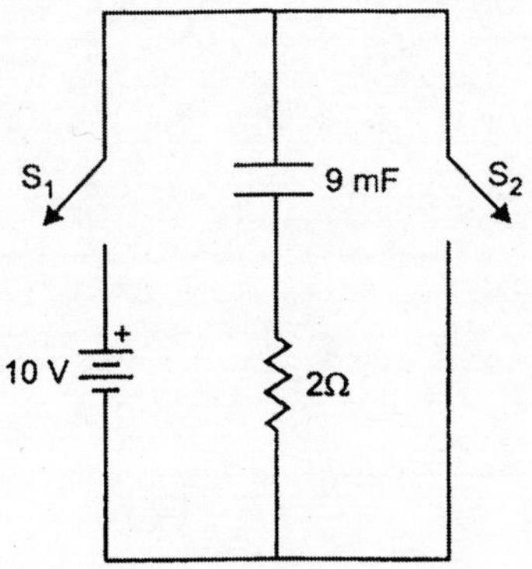

Fig. E9.17

SOLUTION. When the switch S_1 was closed, the circuit becomes as shown in Figure E9.17(a).

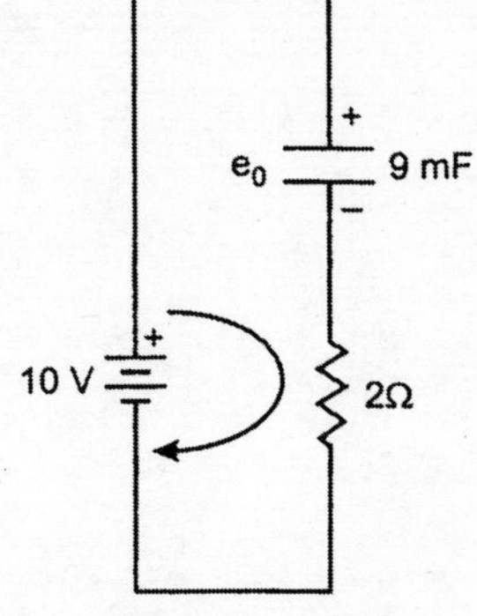

Fig. E9.17(a)

Writing the KVL equation we get

$$Ri(t) + \frac{1}{C}\int_{-\infty}^{\infty} i(t)\,dt = Eu(t) \qquad ...(i)$$

$$Ri(t) + \frac{1}{C}\int_{-\infty}^{0} i(t)\,dt + \frac{1}{C}\int_{0}^{\infty} i(t)\,dt = Eu(t) \qquad ...(ii)$$

Let, $e_0 = \frac{1}{C}\int_{-\infty}^{0} i(t)\,dt$ is the voltage stored on the capacitor before the switch is closed. $e_0 = 0$ V. Then equation (*ii*) becomes,

$$Ri(t) + \frac{1}{C}\int_{-\infty}^{0} i(t)\,dt = Eu(t) \qquad ...(iii)$$

Taking laplace transformation on both sides of equation (*iii*) we get

$$RI(s) + \frac{1}{Cs}I(s) = \frac{E}{s} \qquad ...(iv)$$

or $$I(s)\left[\frac{1}{Cs} + R\right] = \frac{E}{s}$$

or $$I(s)\left[\frac{1 + CRs}{Cs}\right] = \frac{E}{s}$$

or $$I(s) = \frac{EC}{1 + RCs}$$

or $$I(s) = \frac{EC/RC}{\frac{1}{RC} + s} = \frac{E}{R}\left[\frac{1}{s + \frac{1}{RC}}\right] \qquad ...(v)$$

Taking the inverse laplace transformation of equation (*v*) we get

$$i(t) = \frac{E}{R}\left[e^{-\left(1/RC\right)t}\right] \text{ amp} \qquad ...(vi)$$

Substituting the values $E = 10$ V, $R = 2\ \Omega$ and $C = 9$ mF we get

$$i(t) = \frac{10}{2}\left[e^{-\left(1/2 \times 9 \times 10^3\right)t}\right] \text{ amp} = 5e^{-55.55t} \text{ amp}$$

$$V_R = i(t) \times R = 10e^{-55.55t} \text{ V}$$

$$V_C = E_0 = 10 - V_R = 10 \times (1 - e^{-55.55t}) \text{ V}$$

Substituting $t = 1$mS we get

$$V_C = e_0 = 0.54 \text{ V}$$

When the switch S_2 is closed we get the circuit as shown in Figure E9.17(*b*).

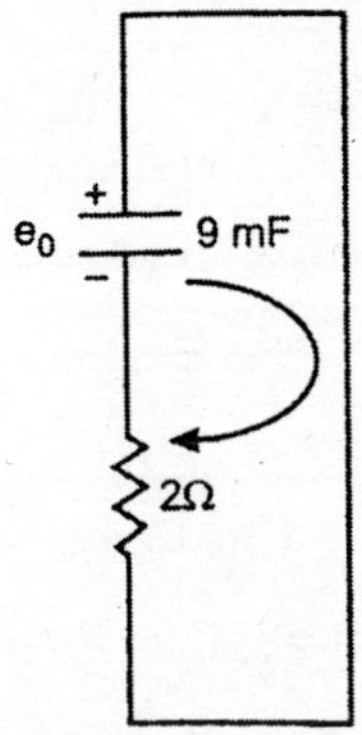

Fig. E9.17(b)

Writing the KVL equation we get

$$Ri(t) + \frac{1}{C}\int_{-\infty}^{\infty} i(t)\ dt = Eu(t) \qquad ...(vii)$$

$$Ri(t) + \frac{1}{C}\int_{-\infty}^{0} i(t)\ dt + \frac{1}{C}\int_{0}^{\infty} i(t)\ dt = Eu(t) \qquad ...(viii)$$

where $$\frac{1}{C}\int_{-\infty}^{0} i(t)\ dt = e_0 = 0.54V$$

$$Ri(t) + e_0 + \frac{1}{C}\int_{0}^{\infty} i(t)\ dt = Eu(t) \qquad ...(ix)$$

Taking laplace transformation on both sides of equation (*ix*) we get

$$RI(s) + \frac{e_0}{s} + \frac{1}{Cs}I(s) = \frac{E}{s} \qquad ...(x)$$

or $$I(s)\left[\frac{1}{Cs}+R\right] = \frac{E}{s} - \frac{e_0}{s}$$

or $$I(s)\left[\frac{1+CRs}{Cs}\right] = \frac{E-e_0}{s}$$

or $$I(s) = \frac{(E-e_0)C}{(1+RCs)} = \frac{(E-e_0)/R}{\frac{1}{RC}+s}$$

Taking inverse laplace transform we get

$$i(t) = \frac{E-e_0}{R}e^{-\frac{t}{RC}} \text{ amp}$$

Substituting the values $E = 30$ V, $e_0 = 0.54$ V, $R = 2\ \Omega$ and $C = 9$ mF we get

$$i(t) = \frac{30-0.54}{2}e^{-\frac{t}{2\times 9\times 10^{-3}}} = 14.73e^{-55.55t} \text{ amp}$$

EXAMPLE 9.18. *For the circuit shown in Figure E9.18, find the loop currents assuming all initial conditions to be zero.*

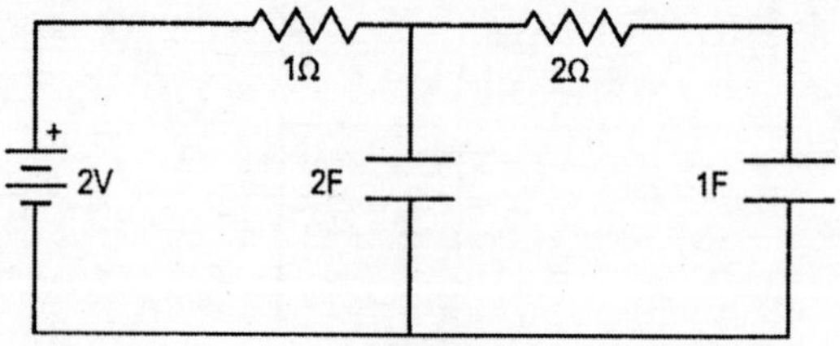

Fig. E9.18

SOLUTION. The loop currents are indicated in Figure E9.18(*a*).

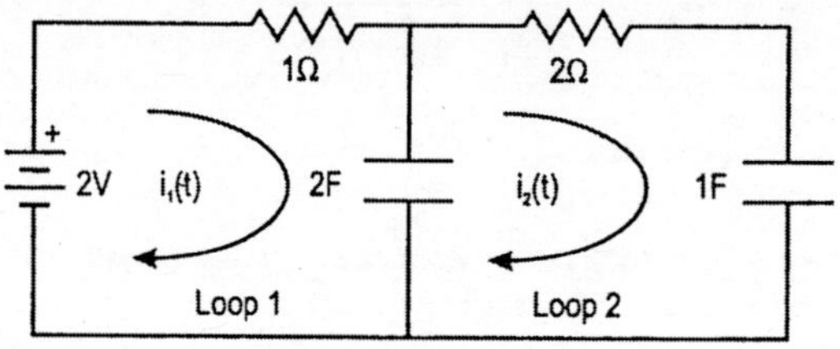

Fig. E9.18(a)

Writing the loop equations we get

Loop 1 : $$1i_1(t) + \frac{1}{2}\int_{-\infty}^{+\infty} i_1(t) - \frac{1}{2}\int_{-\infty}^{+\infty} i_2(t) = 2$$

or $$1i_1(t) + \frac{1}{2}\int_{-\infty}^{0} i_1(t) + \frac{1}{2}\int_{0}^{\infty} i_1(t) - \frac{1}{2}\int_{-\infty}^{0} i_2(t) - \frac{1}{2}\int_{0}^{\infty} i_2(t) = 2$$

or $$1i_1(t) + \frac{1}{2}\int_{0}^{\infty} i_1(t) - \frac{1}{2}\int_{0}^{\infty} i_2(t) = 2 \qquad \left[\text{Since } \frac{1}{2}\int_{-\infty}^{0} i_1(t) = \frac{1}{2}\int_{-\infty}^{0} i_2(t) = 0\right]$$

Taking laplace transformation on both sides we get

$$1I_1(s) + \frac{1}{2s}I_1(s) - \frac{1}{2s}I_2(s) = \frac{2}{s}$$

or $$1I_1(s)\left[1 + \frac{1}{2s}\right] - \frac{1}{2s}I_2(s) = \frac{2}{s} \qquad \text{...(i)}$$

Loop 2 : $$2i_2(t) + 1\int_{-\infty}^{+\infty} i_2(t) + \frac{1}{2}\int_{-\infty}^{+\infty} i_2(t) - \frac{1}{2}\int_{-\infty}^{+\infty} i_1(t) = 0$$

or $$2i_2(t) + 1\int_{-\infty}^{0} i_2(t) + 1\int_{0}^{\infty} i_2(t) + \frac{1}{2}\int_{-\infty}^{0} i_2(t)$$

$$+ \frac{1}{2}\int_{0}^{\infty} i_2(t) - \frac{1}{2}\int_{-\infty}^{0} i_1(t) - \frac{1}{2}\int_{0}^{\infty} i_1(t) = 0$$

or $$2i_2(t) + 1\int_0^{\infty} i_2(t) + \frac{1}{2}\int_0^{\infty} i_2(t) - \frac{1}{2}\int_0^{\infty} i_1(t) = 0$$

$$\left[\text{Since } 1\int_{-\infty}^{0} i_2(t) = \frac{1}{2}\int_{-\infty}^{0} i_2(t) = \frac{1}{2}\int_{-\infty}^{0} i_1(t) = 0\right]$$

Taking laplace transformation on both sides we get

$$2I_2(s) + \frac{1}{s}I_2(s) + \frac{1}{2s}I_2(s) - \frac{1}{2s}I_1(s) = 0$$

or $$I_2(s)\left[2 + \frac{1}{s} + \frac{1}{2s}\right] - \frac{1}{2s}I_1(s) = 0 \qquad \text{...(ii)}$$

Writing equations (*i*) and (*ii*) in matrix form we get

$$\begin{bmatrix} 1+\frac{1}{2s} & -\frac{1}{2s} \\ -\frac{1}{2s} & 2+\frac{1}{s}+\frac{1}{2s} \end{bmatrix} \times \begin{bmatrix} I_1(s) \\ I_2(s) \end{bmatrix} = \begin{bmatrix} \frac{2}{s} \\ 0 \end{bmatrix}$$

Let $$\Delta = \begin{vmatrix} 1+\frac{1}{2s} & -\frac{1}{2s} \\ -\frac{1}{2s} & 2+\frac{1}{s}+\frac{1}{2s} \end{vmatrix} = \frac{8s^2+10s+2}{4s^2}$$

Then $$I_1(s) \frac{\begin{vmatrix} \frac{2}{s} & -\frac{1}{2s} \\ 0 & 2+\frac{1}{s}+\frac{1}{2s} \end{vmatrix}}{\Delta} = \frac{16s+12}{8s^2+10s+2} = \frac{16s+12}{(s+1)(8s+2)} \qquad \text{...(iii)}$$

Let $$\frac{16s+12}{(s+1)(8s+2)} = \frac{A}{s+1} + \frac{B}{8s+2} \qquad \text{...(iv)}$$

Solving equation (*iv*) we get

$$A = \frac{4}{6} \text{ and } B = \frac{32}{3}$$

Then equation (*iii*) can be written as

$$I_1(s) = \frac{16s+12}{(s+1)(8s+2)} = \frac{4}{6}\frac{1}{s+1} + \frac{32}{3}\frac{1}{8s+2} = \frac{4}{6}\frac{1}{s+1} + \frac{32}{3}\frac{1/8}{s+2/8}$$

Taking inverse laplace transformation we get

$$i_1(t) = \frac{4}{6}e^{-t} + \frac{32}{24}e^{-\frac{2t}{8}} \text{ amp}$$

Again $$I_2(s) \frac{\begin{vmatrix} 1+\frac{1}{2s} & \frac{2}{s} \\ -\frac{1}{2s} & 0 \end{vmatrix}}{\Delta} = \frac{4}{(s+1)(8s+2)} \qquad \text{...(v)}$$

Let $$\frac{4}{(s+1)(8s+2)} = \frac{A}{s+1}+\frac{B}{8s+2} \quad ...(vi)$$

Solving equation (*vi*) we get

$$A = -\frac{2}{3} \text{ and } B = \frac{16}{3}$$

Then equation (*v*) becomes,

$$I_2(s) = \frac{4}{(s+1)(8s+2)} = -\frac{2}{3}\times\frac{1}{s+1}+\frac{16}{3}\times\frac{1}{8s+2} = -\frac{2}{3}\frac{1}{s+1}+\frac{2}{3}\frac{1}{s+\frac{2}{8}}$$

Taking inverse laplace transformation we get

$$i_2(t) = -\frac{2}{3}e^{-t}+\frac{2}{3}e^{-\frac{2t}{8}} \text{ amp}$$

9.4 ANALYSIS OF RLC CIRCUIT

9.4.1 Analysis of RLC Circuit Excited with Step Input

Consider the circuit shown in Figure 9.11. Let at time $t = 0$ the switch K is closed, then the loop equation is given by,

$$R\,i(t) + \frac{1}{C}\int i(t)\,dt + L\frac{di}{dt} = Eu(t) \quad ...(9.60)$$

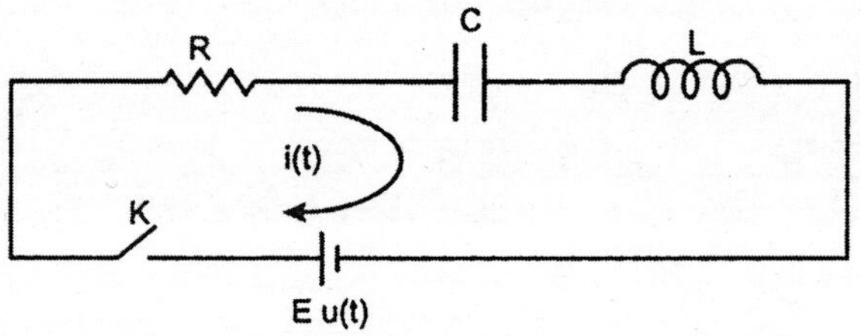

Fig. 9.11

Taking laplace transformation on both sides of equation (9.60) we get

$$RI(s) + L[sI(s) - i(0\_)] + \frac{1}{C}\left[\frac{I(s)}{s}+\frac{e(0\_)}{s}\right] = \frac{E}{s} \quad ...(9.61)$$

Assuming all the initial conditions to be zero the equation (9.61) reduces to,

$$RI(s) + LsI(s) + \frac{1}{C}\left[\frac{I(s)}{s}\right] = \frac{E}{s}$$

or $$I(s)\left[R + Ls + \frac{1}{Cs}\right] = \frac{E}{s}$$

or $$I(s)\left[\frac{RCs + LCs^2 + 1}{Cs}\right] = \frac{E}{s}$$

or $$I(s) = \frac{EC}{RCs + LCs^2 + 1}$$

or
$$I(s) = \frac{E/L}{s^2 + \frac{R}{L}s + \frac{1}{LC}} \quad ...(9.62)$$

Let the roots of the quadratic equation $s^2 + \frac{R}{L}s + \frac{1}{LC}$ be x_1 and x_2 respectively. Then equation (9.62) becomes,

$$I(s) = \frac{E/L}{(s+x_1)(s+x_2)} \quad ...(9.63)$$

Case I : If x_1 and x_2 are real and equal

Let $x_1 = x_2 = x$, then equation (9.63) reduces to

$$I(s) = \frac{E/L}{(s+x)^2} \quad ...(9.64)$$

Taking the inverse laplace transformation of equation (9.64) we get

$$i(t) = \frac{E}{L}\, te^{-xt} \text{ amp}$$

Case II : If x_1 and x_2 are real and not equal.

Then equation (9.63) can be written as

$$I(s) = \frac{E/L}{(s+x_1)(s+x_2)} = \frac{A}{s+x_1} + \frac{B}{s+x_2} \quad ...(9.65)$$

where A and B are constants.

To find the values of constants A and B :

From equation (9.65) we get

$$A(s + x_2) + B(s + x_1) = \frac{E}{L}$$

or
$$As + Bs + Ax_2 + Bx_1 = \frac{E}{L}$$

Equating the coefficient of s we get

$$A + B = 0 \quad ...(9.66)$$

Equating the constants we get

$$Ax_2 + Bx_1 = \frac{E}{L} \quad ...(9.67)$$

Solving equations (9.66) and (9.67) we get

$$A = \frac{E}{L}\left(\frac{1}{x_2 - x_1}\right) \quad (9.68)$$

and
$$B = \frac{E}{L}\left(\frac{1}{x_1 - x_2}\right) \quad ...(9.69)$$

Then substituting the values of equations (9.68) and (9.69) in equation (9.65) we get

$$I(s) = \frac{E}{L}\left[\frac{1}{(x_2 - x_1)(s + x_1)} + \frac{1}{(x_1 - x_2)(s + x_2)}\right] \quad ...(9.70)$$

Taking the inverse laplace transformation of equation (9.70) we get

$$i(t) = \frac{E}{L}\left[\frac{1}{(x_2 - x_1)}e^{-x_1 t} + \frac{1}{x_1 - x_2}e^{-x_2 t}\right] \text{ amp}$$

Case III : If x_1 and x_2 are imaginary.

Let $x_1 = A + jB$ and $x_2 = A - jB$

Then equation (9.63) can be written as

$$I(s) = \frac{E/L}{(s + A + jB)(s + A - jB)} = \frac{K_1}{s + A + jB} + \frac{K_2}{s + A - jB} \quad ...(9.71)$$

where K_1 and K_2 are constants.

To find the values of constants K_1 and K_2 :

From equation (9.71) we get

$$K_1(s + A - jB) + K_2(s + A + jB) = \frac{E}{L}$$

or

$$s(K_1 + K_2) + K_1(A - jB) + K_2(A + jB) = \frac{E}{L}$$

Equating the coefficient of s we get

$$K_1 + K_2 = 0 \quad ...(9.72)$$

Equating the constants we get

$$K_1(A - jB) + K_2(A + jB) = \frac{E}{L} \quad ...(9.73)$$

Solving equations (9.72) and (9.73) we get

$$K_1 = -\frac{E}{L}\left(\frac{1}{2jB}\right) \quad ...(9.74)$$

and

$$K_2 = \frac{E}{L}\left(\frac{1}{2jB}\right) \quad ...(9.75)$$

Then substituting the values of equations (9.74) and (9.75) in equation (9.24) we get

$$I(s) = \frac{E}{L}\left[-\frac{1}{2jB}\frac{1}{s + A + jB} + \frac{1}{2jB}\frac{1}{s + A - jB}\right] \quad (9.76)$$

Taking the inverse laplace transformation of equation (9.70) we get

$$I(t) = \frac{E}{L}\left[-\frac{1}{2jB}e^{-(A + jB)t} + \frac{1}{2jB}e^{-(A - jB)t}\right] \text{ amp}$$

9.4.2 Analysis of RLC Circuit Excited with Impulse Input

Consider the circuit shown in Figure 9.12. Let at time $t = 0$ the switch K is closed, then the loop equation is given by,

$$Ri(t) + \frac{1}{C}\int i(t)\,dt + L\frac{di}{dt} = E\delta(t) \qquad ....(9.77)$$

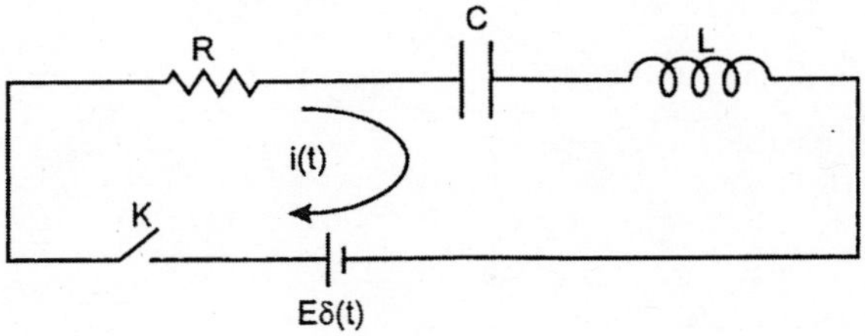

Fig. 9.12

Taking laplace transformation on both sides of equation (9.77) we get

$$RI(s) + L[sI(s) - i(0\_)] + \frac{1}{C}\left[\frac{I(s)}{s} + \frac{e(0\_)}{s}\right] = E \qquad ...(9.78)$$

Assuming all the initial conditions to be zero the equation (9.78) reduces to,

$$RI(s) + LsI(s) + \frac{1}{C}\left[\frac{I(s)}{s}\right] = E$$

or $$I(s)\left[R + Ls + \frac{1}{Cs}\right] = E$$

or $$I(s)\left[\frac{RCs + LCs^2 + 1}{Cs}\right] = E$$

or $$I(s) = \frac{ECs}{RCs + LCs^2 + 1}$$

or $$I(s) = \frac{E}{L} \times \frac{s}{s^2 + \frac{R}{L}s + \frac{1}{LC}} \qquad ...(9.79)$$

Let the roots of the quadratic equation $s^2 + \frac{R}{L}s + \frac{1}{LC}$ be x_1 and x_2 respectively. Then equation (9.79) becomes,

$$I(s) = \frac{E}{L} \times \frac{s}{(s - x_1)(s - x_2)} \qquad ...(9.80)$$

Case I : If x_1 and x_2 are real and equal

Let $x_1 = x_2 = x$, then equation (9.80) reduces to

$$I(s) = \frac{E}{L}\frac{s}{(s - x)^2} \qquad ...(9.81)$$

Taking the inverse laplace transformation of equation (9.81) we get

$$i(t) = \frac{E}{L} te^{-xt} \text{ amp}$$

Case II : If x_1 and x_2 are real and not equal.

Let $$I(s) = \frac{s}{(s-x_1)(s-x_2)} = \frac{A}{s-x_1} + \frac{B}{s-x_2} \qquad ...(9.82)$$

where A and B are constants.

To find the values of constants A and B :

From equation (9.82) we get

$$A(s - x_2) + B(s - x_1) = s$$

or $$As + Bs - Ax_2 - Bx_1 = s$$

Equating the coefficient of s we get

$$A + B = 1 \qquad ...(9.83)$$

Equating the constants we get

$$Ax_2 + Bx_1 = 0 \qquad ...(9.84)$$

Solving equations (9.83) and (9.84) we get

$$A = \left(\frac{x_1}{x_2 - x_1}\right) \qquad ...(9.85)$$

and $$B = \left(\frac{x_2}{x_1 - x_2}\right) \qquad ...(9.86)$$

Then substituting the values of equations (9.85) and (9.86) in equation (9.82) we get

$$I(s) = \frac{E}{L}\left[-\frac{x_1}{(x_2 - x_1)(s - x_1)} + \frac{x_2}{(x_2 - x_1)(s - x_2)}\right] \qquad ...(9.87)$$

Taking the inverse laplace transformation of equation (9.87) we get

$$i(t) = \frac{E}{L}\left[-\frac{x_1}{(x_2 - x_1)} e^{x_1 t} + \frac{x_2}{x_2 - x_1} e^{x_2 t}\right] \text{ amp}$$

Case III : If x_1 and x_2 are imaginary.

Let $x_1 = \alpha + j\beta$ and $x_2 = \alpha - j\beta$

Let, $$I(s) = \frac{s}{(s-\alpha-j\beta)(s-\alpha+j\beta)} = \frac{A}{s-\alpha-j\beta} + \frac{B}{s-\alpha+j\beta} \qquad ...(9.88)$$

Solving equation (9.88) we get

$$A = \frac{\alpha + j\beta}{2j\beta} \quad \text{and} \quad B = \frac{-\alpha + j\beta}{2j\beta}$$

$$I(s) = \frac{E}{L}\frac{s}{(s-\alpha-j\beta)(s-\alpha+j\beta)} = \frac{E}{L} \times \frac{\alpha + j\beta}{2j\beta}$$

$$\times \frac{1}{s-\alpha-j\beta} + \frac{E}{L} \times \frac{j\beta - \alpha}{2j\beta} \times \frac{1}{s-\alpha+j\beta}$$

Taking inverse laplace transformation we get

$$i(t) = \frac{E}{L}\times\frac{\alpha+j\beta}{2j\beta}\times e^{(\alpha+j\beta)t} + \frac{E}{L}\times\frac{j\beta-\alpha}{2j\beta}\times e^{(\alpha-j\beta)t} \text{ amp}$$

9.4.3 Analysis of RLC Circuit Excited with Ramp Input

Consider the circuit shown in Figure 9.13. Let at time $t = 0$ the switch K is closed, then the loop equation is given by,

$$Ri(t) + \frac{1}{C}\int i(t)\,dt + L\frac{di}{dt} = Eu(t) \qquad \text{...(9.89)}$$

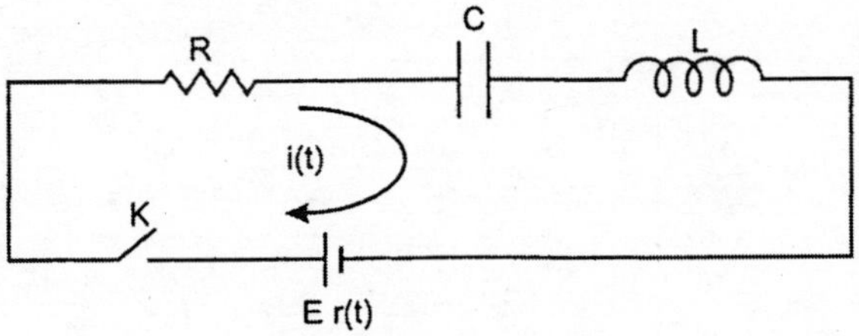

Fig. 9.13

Taking laplace transformation on both sides of equation (9.89) we get

$$RI(s) + L[sI(s) - i(0\_)] + \frac{1}{C}\left[\frac{I(s)}{s}+\frac{e(0\_)}{s}\right] = \frac{E}{s^2} \qquad \text{...(9.90)}$$

Assuming all the initial conditions to be zero equation (9.90) reduces to,

$$RI(s) + LsI(s) + \frac{1}{C}\left[\frac{I(s)}{s}\right] = \frac{E}{s^2}$$

or

$$I(s)\left[R + Ls + \frac{1}{Cs}\right] = \frac{E}{s^2}$$

or

$$I(s)\left[\frac{RCs + LCs^2 + 1}{Cs}\right] = \frac{E}{s^2}$$

or

$$I(s) = \frac{EC}{s\times(RCs + LCs^2 + 1)}$$

or

$$I(s) = \frac{E/L}{s\times\left(s^2 + \frac{R}{L}s + \frac{1}{LC}\right)} \qquad \text{...(9.91)}$$

Let the roots of the quadratic equation $s^2 + \frac{R}{L}s + \frac{1}{LC}$ be x_1 and x_2 respectively.

Case I : Roots are real and unequal

Equation (9.91) becomes

$$I(s) = \frac{E/L}{s\times(s-x_1)(s-x_2)} \qquad \text{...(9.92)}$$

Let $$\frac{1}{s\times(s-x_1)(s-x_2)} = \frac{A}{s}+\frac{B}{s-x_1}+\frac{C}{s-x_2} \qquad ...(9.93)$$

Solving equation (9.93) we get

$$A = \frac{1}{x_1x_2}$$

$$B = -\frac{1}{x_1(x_2-x_1)}$$

$$C = \frac{1}{x_2(x_2-x_1)}$$

Then equation (9.92) becomes,

$$I(s) = \frac{E}{L}\left[\frac{1}{x_1x_2}\times\frac{1}{s}-\frac{1}{x_1(x_2-x_1)}\frac{1}{s-x_1}+\frac{1}{x_2(x_2-x_1)}\times\frac{1}{s-x_2}\right]$$

Taking inverse laplace transformation we get

$$i(t) = \frac{E}{L}\left[\frac{1}{x_1x_2}u(t)-\frac{1}{x_1(x_2-x_1)}e^{x_1t}+\frac{1}{x_2(x_2-x_1)}e^{x_2t}\right] \text{ amp}$$

Case II : Roots are real and equal

Let $$x_1 = x_2 = x$$

Equation (9.91) becomes

$$I(s) = \frac{E/L}{s\times(s-x)^2}$$

Case III : If x_1 are x_2 imaginary.

Let $$x_1 = \alpha + j\beta \text{ and } x_2 = \alpha - j\beta$$

Equation (9.91) becomes

$$I(s) = \frac{E/L}{s\times(s-\alpha-j\beta)\times(s-\alpha+j\beta)} \qquad ..(9.94)$$

Let $$\frac{1}{s\times(s-x_1)(s-x_2)} = \frac{A}{s}+\frac{B}{s-\alpha-j\beta}+\frac{C}{s-\alpha+j\beta} \qquad ...(9.95)$$

Solving equation (9.95) we get

$$A = \frac{1}{\alpha^2+\beta^2}$$

$$B = \frac{1}{2\beta^2+2j\alpha\beta}$$

$$C = \frac{1}{2\beta^2+2j\alpha\beta}$$

Then equation (9.92) becomes,

$$I(s) = \frac{E}{L}\left[\frac{1}{\alpha^2+\beta^2}\times\frac{1}{s}+\frac{1}{2\beta^2+2j\alpha\beta}\times\frac{1}{s-x_1}-\frac{1}{2\beta^2+2j\alpha\beta}\times\frac{1}{s-x_2}\right]$$

Taking inverse laplace transformation we get

$$i(t) = \frac{E}{L}\left[\frac{1}{\alpha^2+\beta^2}u(t)+\frac{1}{2\beta^2+2j\alpha j}e^{x_1 t}-\frac{1}{2\beta^2+2j\alpha j}e^{x_2 t}\right] \text{ amp}$$

9.4.4 Analysis of RLC Circuit Excited with Pulse Input

Consider the circuit shown in Figure 9.14. Let at time $t = 0$ the switch K is closed, then the loop equation is given by,

$$Ri(t) + \frac{1}{C}\int i(t)\ dt + L\frac{di}{dt} = E(t) \quad \text{...(9.96)}$$

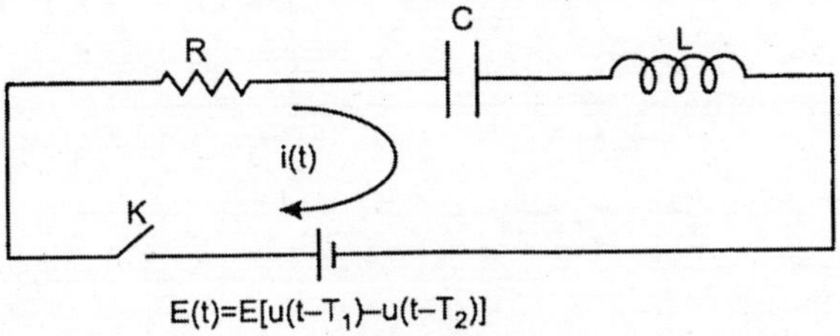

Fig. 9.14

Taking laplace transformation on both sides of equation (9.96) we get

$$RI(s) + L[sI(s) - i(0\_)] + \frac{1}{C}\left[\frac{I(s)}{s}+\frac{e(0\_)}{s}\right] = \frac{E}{s}\left[e^{-sT_1}-e^{-sT_2}\right] \quad \text{...(9.97)}$$

Assuming all the initial conditions to be zero the equation (9.97) reduces to,

$$RI(s) + L[sI(s) + \frac{1}{C}\left[\frac{I(s)}{s}\right] = \frac{E}{s}\left[e^{-sT_1}-e^{-sT_2}\right]$$

or
$$I(s)\left[R+Ls+\frac{1}{Cs}\right] = \frac{E}{s}\left[e^{-sT_1}-e^{-sT_2}\right]$$

or
$$I(s)\left[\frac{RCs+LCs^2+1}{Cs}\right] = \frac{E}{s}\left[e^{-sT_1}-e^{-sT_2}\right]$$

or
$$I(s) = \frac{EC(e^{-sT_1}-e^{-sT_2})}{RCs+LCs^2+1}$$

or
$$I(s) = \frac{E}{L}\times\frac{e^{-sT_1}-e^{-sT_2}}{s^2+\frac{R}{L}s+\frac{1}{LC}} \quad \text{...(9.98)}$$

Let the roots of the quadratic equation $s^2 + \frac{R}{L}s+\frac{1}{LC}$ be x_1 and x_2 respectively. Then equation (9.98) becomes,

$$I(s) = \frac{E}{L}\times\frac{e^{-sT_1}-e^{-sT_2}}{(s-x_2)(s-x_2)} \quad \text{...(9.99)}$$

Case I : If x_1 and x_2 are real and equal

Let $x_1 = x_2 = x$, then equation (9.99) reduces to

$$I(s) = \frac{E}{L} \times \frac{e^{-sT_1} - e^{-sT_2}}{(s-x)^2} \quad ....(9.100)$$

Taking the inverse laplace transformation of equation (9.100) we get

$$i(t) = \frac{E}{L} \times L^{-1}\left[e^{-sT_1} - e^{-sT_2}\right]$$

Case II : If x_1 and x_2 are real and not equal.

Then equation (9.99) can be written as

$$I(s) = \frac{E}{L} \times \frac{e^{-sT_1} - e^{-sT_2}}{(s-x_1)(s-x_2)} = \frac{E}{L}\left[\frac{e^{-sT_1}}{(s-x_1)(s-x_2)} - \frac{e^{-sT_2}}{(s-x_1)(s-x_2)}\right] \quad ...(9.101)$$

Using partial fraction methods, $\frac{1}{(s-x_1)(s-x_2)} = \frac{A}{s-x_1} + \frac{B}{s-x_2}$

where A and B are constants. Solving the equation we get

$$A = -\frac{1}{x_2 - x_1}$$

and

$$B = \frac{1}{x_2 - x_1}$$

Then equation (9.101) can be written as

$$I(s) = \frac{E}{L}\left[\frac{e^{-sT_1}}{(s-x_1)(s-x_2)} - \frac{e^{-sT_2}}{(s-x_1)(s-x_2)}\right]$$

$$= \frac{E}{L}\left[-\frac{1}{x_2-x_1}\frac{e^{-sT_1}}{s-x_1} + \frac{1}{x_2-x_1}\frac{e^{-sT_1}}{s-x_2} + \frac{1}{x_2-x_1}\frac{e^{-sT_2}}{s-x_1} - \frac{1}{x_2-x_1}\frac{e^{-sT_2}}{s-x_2}\right]$$

$$= \frac{E}{L} \times \frac{1}{x_2-x_1}\left[-\frac{e^{-sT_1}}{s-x_1} + \frac{e^{-sT_1}}{s-x_2} + \frac{e^{-sT_2}}{s-x_1} - \frac{e^{-sT_2}}{s-x_2}\right]$$

Taking inverse laplace transformation, we get

$$i(t) = \frac{E}{L} \times \frac{1}{x_2-x_1}\left[-e^{\frac{t-T_1}{x_1}} u(t-T_1) + e^{\frac{t-T_1}{x_2}} u(t-T_1) + e^{\frac{t-T_2}{x_1}} u(t-T_2) - e^{\frac{t-T_2}{x_2}} u(t-T_2)\right] \text{ amp}$$

Case III : If x_1 and x_2 are imaginary.

Substituting $x_1 = A + jB$ and $x_2 = A - jB$ in the expression obtained in Case-II we get,

$$i(t) = \frac{E}{L} \times \frac{1}{-2jB}\left[-e^{\frac{t-T_1}{A+jB}} u(t-T_1) + e^{\frac{t-T_1}{A-jB}} u(t-T_1) + e^{\frac{t-T_2}{A+jB}} u(t-T_2) - e^{\frac{t-T_2}{A-jB}} u(t-T_2)\right] \text{ amp}$$

9.4.5 Analysis of RLC Circuit Excited with Sinusoidal Input

Consider the circuit shown in Figure 9.15. Let at time $t = 0$ the switch K is closed, then the loop equation is given by,

$$Ri(t) + \frac{1}{C}\int i(t)\, dt + L\frac{di}{dt} = \sin \omega t \qquad \text{...(9.102)}$$

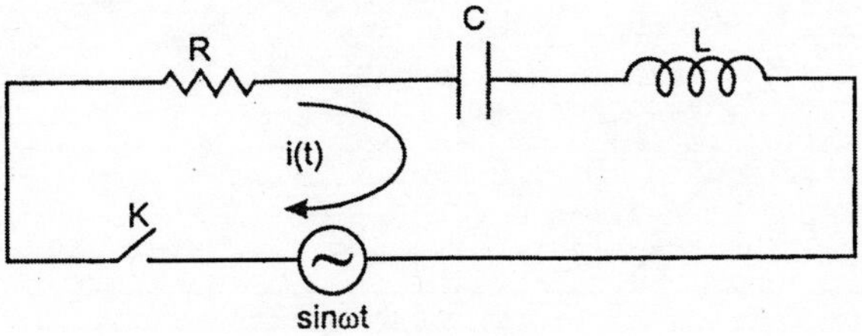

Fig. 9.15

Taking laplace transformation on both sides of equation (9.102) we get

$$RI(s) + L[sI(s) - i(0\_)] + \frac{1}{C}\left[\frac{I(s)}{s} + \frac{e(0\_)}{s}\right] = \frac{\omega^2}{s^2 + \omega^2} \qquad \text{...(9.103)}$$

Assuming all the initial conditions to be zero equation (9.103) reduces to,

$$RI(s) + LsI(s) + \frac{1}{C}\left[\frac{I(s)}{s}\right] = \frac{\omega^2}{s^2 + \omega^2}$$

or $$I(s)\left[R + Ls + \frac{1}{Cs}\right] = \frac{\omega^2}{s^2 + \omega^2}$$

or $$I(s)\left[\frac{RCs + LCs^2 + 1}{Cs}\right] = \frac{\omega^2}{s^2 + \omega^2}$$

or $$I(s) = \frac{\omega^2 Cs}{(s^2 + \omega^2)(RCs + LCs^2 + 1)}$$

or $$I(s) = \frac{\omega^2}{L}\frac{s}{\left(s^2 + \frac{R}{L}s + \frac{1}{LC}\right)} \qquad \text{...(9.104)}$$

Let the roots of the quadratic equation $s^2 + \frac{R}{L}s + \frac{1}{LC}$ be x_1 and x_2 respectively.

Case I : Roots are real and unequal

Let $$\frac{s}{(s+j\omega)(s-j\omega)(s-x_1)(s-x_2)} = \frac{A}{s+j\omega} + \frac{B}{s-j\omega} + \frac{C}{s-x_1} + \frac{D}{s-x_2}$$

where A, B, C and D are constants

Solving by using partial fraction we get

$$A = \frac{(x_1x_2 - \omega^2) - j\omega(x_1 + x_2)}{(x_1x_2 - \omega^2)^2 + (x_1\omega + x_2\omega)^2}$$

$$B = \frac{(x_1x_2-\omega^2)+j\omega(x_1+x_2)}{(x_1x_2-\omega^2)^2+(x_1\omega+x_2\omega)^2}$$

$$C = \frac{x_1}{(x_1^2+\omega^2)\times(x_1-x_2)}$$

$$D = \frac{x_2}{(x_2^2+\omega^2)\times(x_2-x_1)}$$

Then equation becomes

$$I(s) = \frac{s}{(s+j\omega)(s-j\omega)(s-x_1)(s-x_2)} = \frac{A}{s+j\omega}+\frac{B}{s-j\omega}+\frac{C}{s-x_1}+\frac{D}{s-x_2}$$

$$= \frac{(x_1x_2-\omega^2)-j\omega(x_1+x_2)}{(x_1x_2-\omega^2)^2+(x_1\omega+x_2\omega)^2}\frac{1}{s+j\omega}+\frac{(x_1x_2-\omega^2)+j\omega(x_1+x_2)}{(x_1x_2-\omega^2)^2+(x_1\omega+x_2\omega)^2}\frac{1}{s-j\omega}$$

$$+\frac{x_1}{(x_1^2+\omega^2)\times(x_1-x_2)}\frac{1}{s-x_1}+\frac{x_2}{(x_2^2+\omega^2)\times(x_2-x_1)}\frac{1}{s-x_2}$$

Taking inverse laplace transformation, we get

$$I(t) = \frac{(x_1x_2-\omega^2)-j\omega(x_1+x_2)}{(x_1x_2-\omega^2)^2+(x_1\omega+x_2\omega)^2}\times e^{-j\omega t}+\frac{(x_1x_2-\omega^2)+j\omega(x_1+x_2)}{(x_1x_2-\omega^2)^2+(x_1\omega+x_2\omega)^2}\times e^{j\omega t}$$

$$+\frac{x_1}{(x_1^2+\omega^2)\times(x_1-x_2)}\times e^{x_1t}+\frac{x_2}{(x_2^2+\omega^2)\times(x_2-x_1)}\times e^{x_2t}$$

Case II : Roots are imaginary

This case can be solved by substituting $x_1 = A + jB$ and $x_2 = A + jB$ in the expression obtained in **Case I**

SOLVED EXAMPLES

EXAMPLE 9.19. *For the circuit shown in Figure E9.19, find an expression for transient current, when the switch K is closed at time t = 0 sec.*

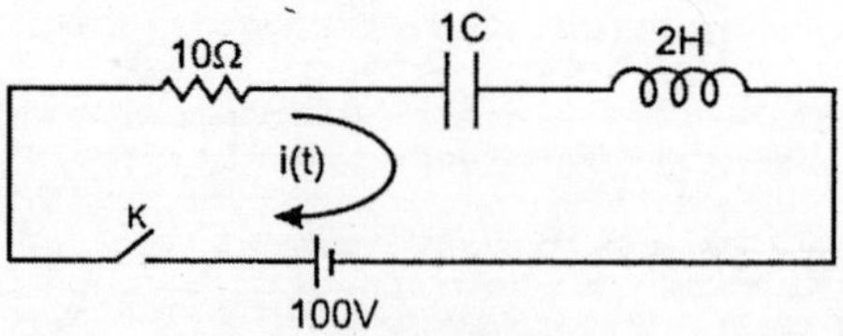

Fig. E9.19

SOLUTION. The direction of the current is shown in Figure E10.19(*a*).

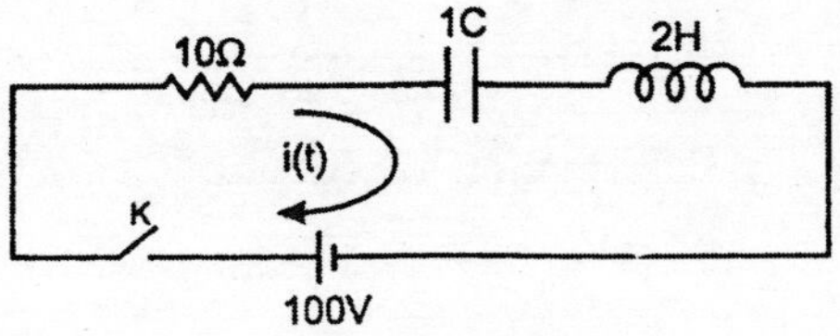

Fig. E9.19(a)

The loop equation is given by,

$$Ri(t) + \frac{1}{C}\int i(t)\ dt + L\frac{di}{dt} = E\ u(t) \quad ...(i)$$

Taking laplace transformation on both sides of equation (*i*), we get

$$RI(s) + L[sI(s) - (0\_)] + \frac{1}{C}\left[\frac{I(s)}{s} + \frac{e(0\_)}{s}\right] = \frac{E}{s} \quad ...(ii)$$

Assuming all the initial conditions to be zero, equation (*ii*) reduces to

$$RI(s) + L[sI(s) + \frac{1}{C}\left[\frac{I(s)}{s}\right] = \frac{E}{s}$$

or $$I(s)\left[R + Ls + \frac{1}{Cs}\right] = \frac{E}{s}$$

or $$I(s)\left[\frac{RCs + LCs^2 + 1}{Cs}\right] = \frac{E}{s}$$

or $$I(s) = \frac{EC}{RCs + LCs^2 + 1}$$

or $$I(s) = \frac{E/L}{s^2 + \frac{R}{L}s + \frac{1}{LC}} \quad ...(iii)$$

Substituting the values, $R = 10\Omega$, $C = 1C$, $L = 2H$ and $E = 100V$ in equation (*iii*), we get

$$I(s) = \frac{50}{s^2 + 5s + 0.5}$$

$$I(s) = \frac{50}{(s + 0.11)(s + 4.89)} \quad ...(iv)$$

Let, $$\frac{50}{(s + 0.11)(s + 4.89)} = \frac{A}{s + 0.11} + \frac{B}{s + 4.89} \quad ...(v)$$

where A and B are constants.

Solving equation (v) we get

$$A = \left(\frac{50}{4.78}\right) = 01.46 \quad ...(vi)$$

and $$B = -\left(\frac{50}{4.78}\right) = -10.46 \quad ...(vii)$$

Then substituting the values of equations (*vi*) and (*vii*) in equation (*v*) we get

$$I(s) = 10.46\left[\frac{1}{(s+0.11)}-\frac{1}{(s+4.89)}\right] \quad \text{...(viii)}$$

Taking the inverse laplace transformation of equation (*viii*) we get

$$i(t) = \frac{E}{L}\left[e^{-0.11t}+e^{-4.89t}\right] \text{ amp}$$

EXAMPLE 9.20. *For the circuit shown in Figure E9.20, find an expression for transient current, when the switch K is closed at time t = 0 sec. Assume the value of* $e_0 = 2V$ *and* $i(0) = 2A$.

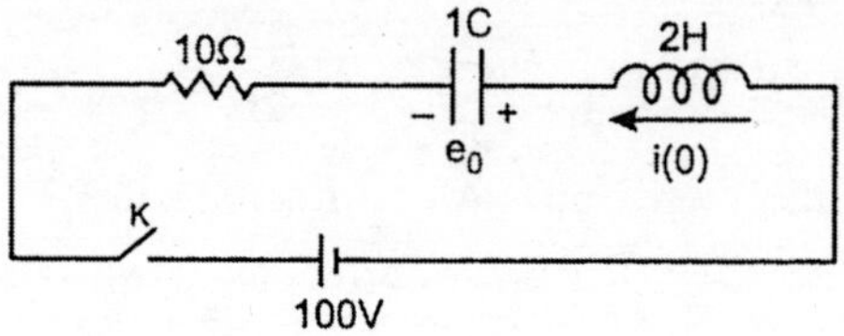

Fig. E9.20

SOLUTION. The direction of the current is shown in Figure E9.20(*a*).

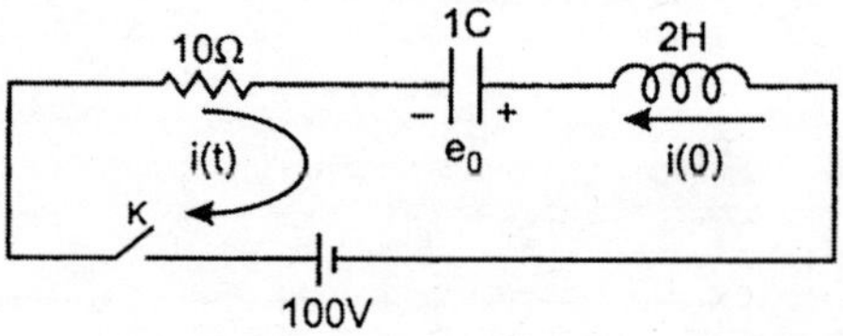

Fig. E9.20(a)

The loop equation is given by,

$$Ri(t) + \frac{1}{C}\int i(t)\, dt + L\frac{di}{dt} = E\ u(t) \quad \text{...(i)}$$

Taking laplace transformation on both sides of equation (*i*) we get,

$$RI(s) + L[sI(s) - (0\_)] + \frac{1}{C}\left[\frac{I(s)}{s}+\frac{e(0\_)}{s}\right] = \frac{E}{s} \quad \text{...(ii)}$$

Substituting the values, $R = 10\Omega$, $C = 1C$, $L = 2H$, $E = 100V$, $e_0 = 2V$ and $i(0) = 2A$ in equation (*ii*), we get

$$10I(s) + 2[sI(s) + 2] + 1\left[\frac{I(s)}{s}-\frac{2}{s}\right] = \frac{100}{s}$$

or $$I(s)\left[10+2s+\frac{1}{s}\right] = \frac{100}{s}+\frac{2}{s}-4$$

or $$I(s)\left[\frac{2s^2+10s+1}{s}\right] = \frac{102-4s}{s}$$

or $$I(s) = \frac{102-4s}{2s^2+10s+1} = \frac{102-4s}{(s+4.89)(s+0.1)}$$

Let $$\frac{102-4s}{(s+4.89)(s+0.1)} = \frac{A}{s+4.89}+\frac{B}{s+0.1} \quad \text{...(iii)}$$

Solving equation (*iii*) we get

$$A = -25.38 \quad ...(iv)$$

and

$$B = -21.38 \quad ...(v)$$

Then substituting the values of equations (*iv*) and (*v*) in equation (*iii*) we get

$$I(s) = \frac{102-4s}{(s+4.89)(s+0.1)} = -\frac{25.38}{s+4.89}+\frac{21.38}{s+0.1} \quad ...(vi)$$

Taking the inverse laplace transformation of equation (*vi*) we get

$$i(t) = [-25.38e^{-4.89t} + 21.38e^{-0.1t}] \text{ amp}$$

EXAMPLE 9.21. *For the circuit shown in Figure E9.21, find the loop currents.*

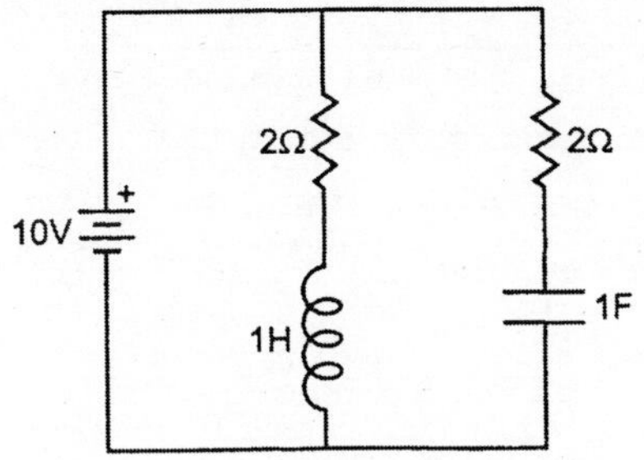

Fig. E9.21

SOLUTION. The loop currents are indicated in Figure E9.21(*a*).

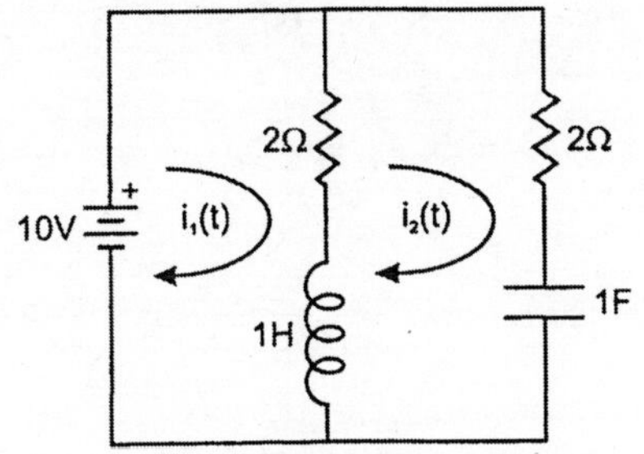

Fig. E9.21(a)

Writing the loop equations we get

Loop 1 :

$$2i_1(t) + 1\frac{di_1(t)}{dt} - 2i_2(t) - 1\frac{di_2(t)}{dt} = 10$$

$$2I_1(s) + sI_1(s) - 2I_2(s) - sI_2(s) = \frac{10}{s}$$

$$I_1(s)\,[2 + s] - I_2(s)\,[2 + s] = \frac{10}{s} \quad ...(i)$$

Loop 2 :

$$4i_2(t) - 2i_1(t) + 1\int_0^t i_2(t)dt + 1\frac{di_2(t)}{dt} - 1\frac{di_1(t)}{dt} = 0$$

$$4I_2(s) - 2I_1(s) + \frac{1}{s}\,I_2(s) + sI_2(s) - sI_1(s) = 0$$

$$I_2(s)\left[4+s+\frac{1}{s}\right] - I_1(s)\,[2 + s] = 0 \quad ...(ii)$$

From equation (*ii*) we get

$$I_2(s)\left[4+s+\frac{1}{s}\right] = I_1(s)\ [2 + s] \qquad ...(iii)$$

Substituting equation (*iii*) in equation (*i*) we get

$$I_2(s)\left[4+s+\frac{1}{s}-2-s\right] = \frac{10}{s}$$

or

$$I_2(s)\left[2+\frac{1}{s}\right] = \frac{10}{s}$$

or

$$I_2(s)\ \left[2+\frac{1}{s}\right] = \frac{10}{2s+1} = \frac{5}{s+0.5} \qquad ...(iv)$$

Taking inverse laplace transformation of equation (*iv*) we get

$$i_2(t) = 5e^{-0.5t} \text{ amp}$$

Substituting equation (*iv*) in equation (*iii*) we get

$$\frac{10}{2s+1}\times\left[4+s+\frac{1}{s}\right] = I_1(s)\ [2 + s]$$

or

$$I_1(s) = \frac{5s^2+20s+5}{s(s+0.5)(2+s)} \qquad ...(v)$$

Let

$$\frac{5s^2+20s+5}{s(s+0.5)(2+s)} = \frac{A}{s}+\frac{B}{s+0.5}+\frac{C}{s+2}$$

Solving we get

$$A = 5,\ B = 5,\ C = -5$$

Then equation (*v*) becomes

$$I_1(s) = \frac{5s^2+20s+5}{s(s+0.5)(2+s)} = \frac{5}{s}+\frac{5}{s+0.5}-\frac{5}{s+2} \qquad ...(vi)$$

Taking inverse laplace transformation of equation (*vi*) we get

$$i_1(t) = 5[u(t) + e^{-0.5t} + e^{-2t}] \text{ amp}$$

EXAMPLE 9.22. *For the circuit shown in Figure E9.22, find the* V_0.

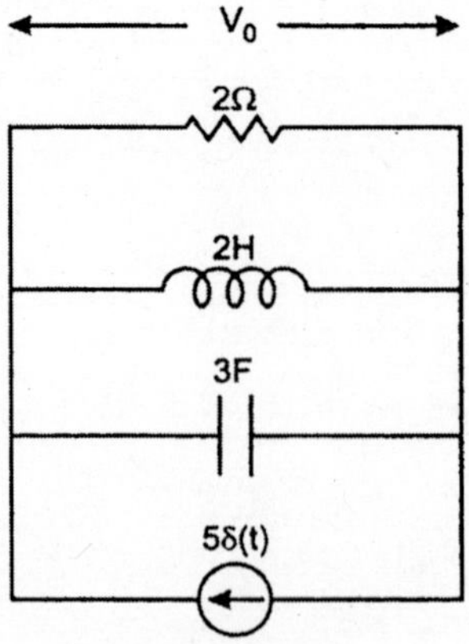

Fig. E9.22

SOLUTION. The loop currents are indicated in Figure E9.22(*a*).

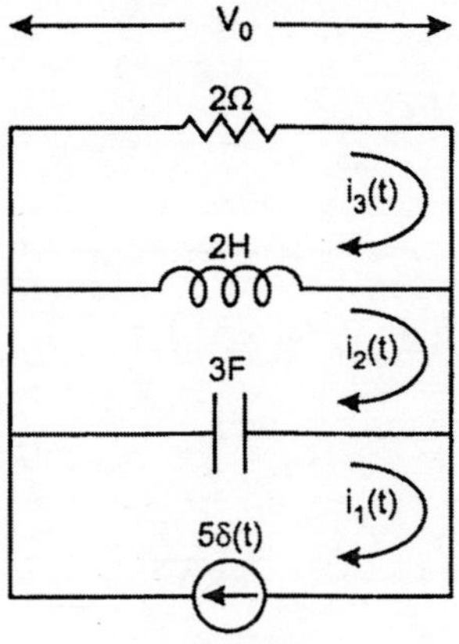

Fig. E9.22(a)

Writing the loop equations we get

Loop 1 : $$i_1(t) = 5\delta(t) \quad \text{....}(i)$$

Taking inverse laplace transformation of equation (*i*) we get

$$I_1(s) = 5 \quad \text{...}(ii)$$

Loop 2 : $$\frac{1}{C}\int_0^t i_2(t)dt - \frac{1}{C}\int_0^t i_1(t)dt + L\frac{di_2(t)}{dt} - L\frac{di_3(t)}{dt} = 0 \quad \text{...}(iii)$$

Taking inverse laplace transformation of equation (*iii*) we get

$$I_2(s)\left[\frac{1}{Cs} + Ls\right] - \frac{1}{Cs}\, I_1(s) - LsI_3(s) = 0 \quad \text{...}(iv)$$

Substituting equation (*ii*) in equation (*iv*) we get

$$I_2(s)\left[\frac{1}{Cs} + Ls\right] - LsI_3(s) = \frac{5}{Cs} \quad \text{...}(v)$$

Loop 3 : $$2i_3(t) + L\frac{di_3(t)}{dt} - L\frac{di_2(t)}{dt} = 0 \quad \text{...}(vi)$$

Taking inverse laplace transformation of equation (*vi*) we get

$$I_3(s)[2 + Ls] - LsI_2(s) = 0$$

$$I_3(s)[2 + Ls] = LsI_2(s)$$

$$I_2(s) = \frac{2 + Ls}{Ls}\, I_3(s) \quad \text{...}(vii)$$

Substituting equation (*vii*) in equation (*v*) we get

$$I_3(s)\frac{2 + Ls}{Ls}\left[\frac{1}{Cs} + Ls\right] - LsI_3(s) = \frac{5}{Cs}$$

$$I_3(s)\left[\frac{(2 + Ls)\times(1 + CLs^2)}{LCs^2} - Ls\right] = \frac{5}{Cs}$$

$$I_3(s)\left[\frac{2+2CLs^2+Ls}{LCs}\right] = \frac{5}{C}$$

$$I_3(s) = \frac{5Ls}{2CLs^2+Ls+2}$$

$$V_0(s) = \frac{1}{Cs}I_3(s) = \frac{5L}{C}\times\frac{1}{2CLs^2+Ls+2}$$

Let the roots of the quadratic equation be x_1 and x_2.

$$V_0(s) = \frac{5L}{C}\times\frac{1}{(s+0.08+j0.398)(s+0.08-j0.398)}$$

$$V_0(s) = \frac{5L}{C}\times\frac{1}{2j0.398}\left[-\frac{1}{s+0.08+j0.398}+\frac{1}{s+0.08-j0.398}\right]$$

$$V_0(t) = \frac{10}{3}\times\frac{1}{2j0.398}\left[-e^{-(0.08+j0.398)t}+e^{-(0.08-j0.398)t}\right] \text{ V}$$

EXAMPLE 9.23. *For the circuit shown in Figure E9.23, the switch was in position 1 for a long period of time, then at time t = 0sec it is moved to position 2. Find an expression for transient current when the switch is in position 2.*

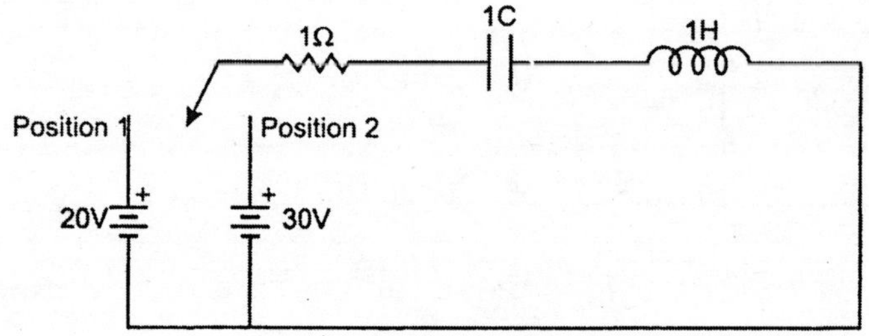

Fig. E9.23

SOLUTION. When the switch was in position-1 :

The current through the inductor $i(0) = 0$

and drop across the capacitor $e_0 = 20$ V

When the switch was in position-2, we get the circuit shown in Figure E10.23(*a*).

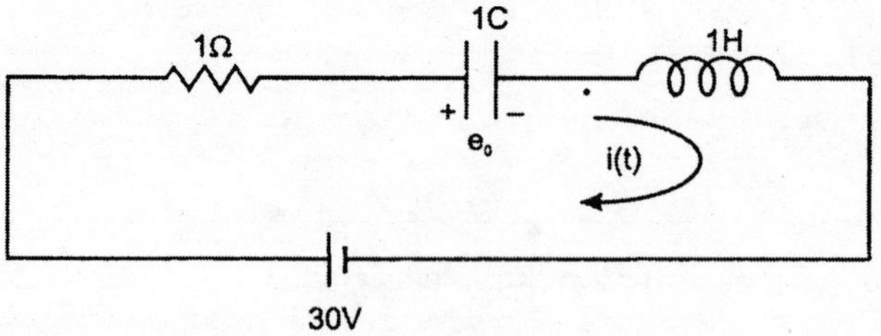

Fig. E9.23(a)

Writing the KVL equation we get

$$Ri(t)+L\frac{di(t)}{dt}+\frac{1}{C}+\int_{-\infty}^{+\infty} i(t)\,dt = 30$$

$$Ri(t) + L\frac{di(t)}{dt} + \frac{1}{C}\int_{-\infty}^{0} i(t)\, dt + \frac{1}{C}\int_{0}^{\infty} i(t)\, dt = 30$$

$$Ri(t) + L\frac{di(t)}{dt} + e_0 + \frac{1}{C}\int_{0}^{\infty} i(t)\, dt = 30$$

Taking the laplace transform we get

$$RI(s) + LsI(s) + \frac{e_0}{s} + \frac{1}{Cs}I(s) = \frac{30}{s}$$

$$I(s)\left[R + Ls + \frac{1}{Cs}\right] = \frac{10}{s}$$

Substituting the values of $R = 1\Omega$, $L = 1\text{H}$ and $C = 1$ F we get

$$I(s) = \frac{10}{s^2+s+1} = \frac{10}{(s+0.5-j0.865)(s+0.5+j0.865)}$$

$$= \frac{10j}{1.73}\left[\frac{1}{s+0.5+j0.865} - \frac{1}{s+0.5-j0.865}\right]$$

Taking inverse laplace transform we get,

$$i(t) = \frac{10j}{1.73}\,[e^{-(0.5+j0.865)t} - e^{-(0.5+j0.865)t}] \text{ amp}$$

EXAMPLE 9.24. *For the circuit shown in Figure E9.24, the switch was closed at time $t = 0$, find the drop across 10 Ω.*

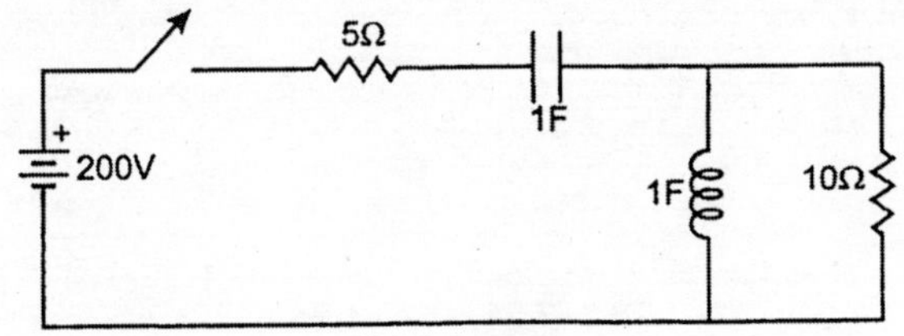

Fig. E9.24

SOLUTION. The loop currents are indicated in Figure E10.24(*a*).

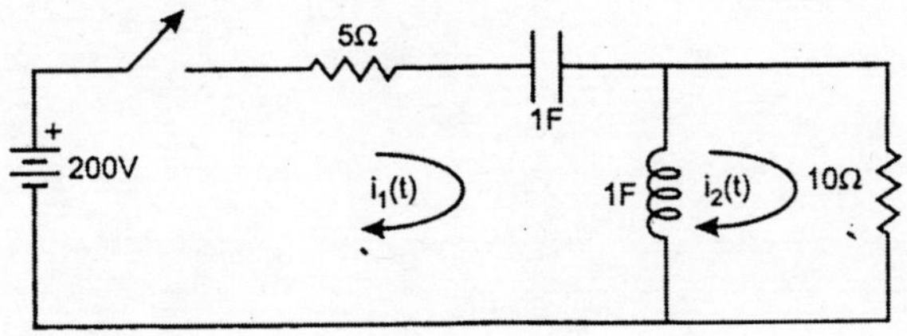

Fig. E9.24(a)

Writing the loop equations

Loop 1 : $$5i_1(t) + \frac{1}{C}\int_0^{\infty} i_1(t)dt + L\frac{di_1(t)}{dt} - L\frac{di_2(t)}{dt} = 200 \quad ...(i)$$

Taking laplace transformation we get

$$I_1(s)\left[5+s+\frac{1}{s}\right] - sI_2(s) = \frac{200}{s} \qquad ...(ii)$$

$$I_1(s)\left[\frac{s^2+5s+1}{s}\right] - sI_2(s) = \frac{200}{s} \qquad ...(iii)$$

Loop 2 :

$$10i_2(t) + L\frac{di_2(t)}{dt} - L\frac{di_1(t)}{dt} = 0 \qquad ...(iv)$$

Taking laplace transformation we get

$$10I_2(s) + LsI_2(s) - LsI_1(s) = 0 \qquad ...(v)$$

$$I_2(s)[10 + s] - sI_1(s)$$

$$I_1(s) = \frac{1}{s}I_2(s)[10 + s] \qquad ...(vi)$$

Substituting equation (*vi*) in equation (*iii*) we get

$$I_2(s)\ \frac{10+s}{s}\left[\frac{s^2+5s+1}{s}\right] - sI_2(s) = \frac{200}{s}$$

$$I_2(s) = \frac{200}{(s-7.093-j1.83)(s-7.093+j1.83)(s+0.186)}$$

$$= \frac{A}{s-7.093-j1.83} + \frac{B}{s-7.093+j1.83} + \frac{C}{s+0.186}$$

Solving the partial fraction

$$A = -1.78 - j7.06$$
$$B = -1.78 - j7.06$$
$$C = 3.55$$

Equation (*vii*) becomes

$$I_2(s) = \frac{(-1.78-j7.06)}{s-7.093-j1.83} + \frac{(-1.78+j7.06)}{s-7.093+j1.83} + \frac{3.55}{s+0.186} \qquad ...(viii)$$

Taking inverse laplace transformation of equation (*viii*) we get

$$i_2(t) = (-1.78 - j7.06)\ e^{(7.093+j1.83)t} + (-1.78 + j7.06)\ e^{(7.093-j1.83)t} + 3.55e^{-0.186t} \text{ amp}$$

$$\text{Drop across } 10\ \Omega = 10 \times i_2(t)$$

$$= 10 \times [(-1.78 - j7.06)\ e^{(7.093+j1.83)t} + (-1.78 + j7.06)\ e^{(7.093-j1.83)t} + 3.55e^{-0.186t}] \text{ Volt}$$

EXERCISES

1. A series RL circuit gets connected to the following voltage source after the switch is closed.
   (*i*) Step input of 10 V (*ii*) Input of $10e^t$ V
   Assume $R = 10\ \Omega$ and $L = 3$ H.
2. A series RC circuit gets connected to the following voltage source after the switch is closed.
   (*i*) $tu(t)$ V (*ii*) $u(t) - u(t - 1)$
   Assume $R = 10\ \Omega$ and $C = 1$ F.

3. For the circuit shown in Figure Q9.3, the switch k was closed at time $t = 0$, connecting the circuit to an impulse source of $10\delta(t)$. Find the complete expression for the current after the switch is closed. Assume the initial current through the inductor is $i(0) = 2A$ (the direction of $i(0)$ is indicated in Figure E9.2).

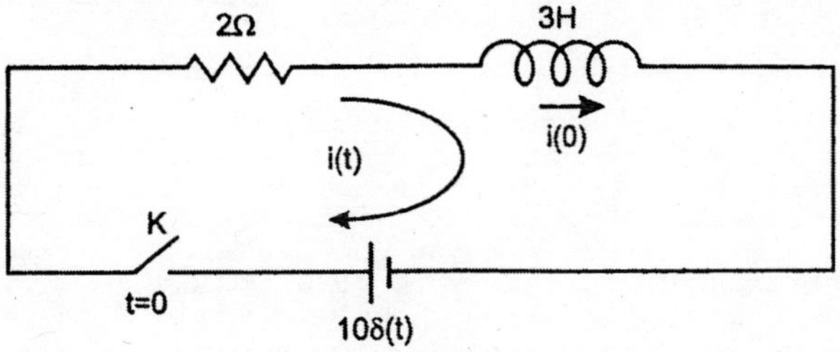

Fig. Q9.3

4. For the circuit shown in Figure Q9.4, find the loop currents assuming all initial conditions to be zero.

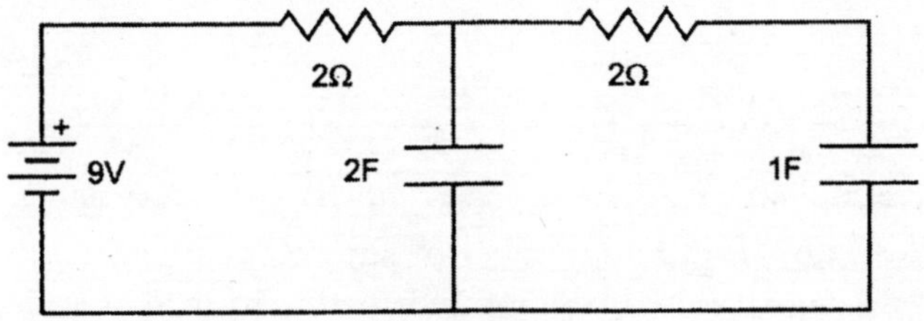

Fig. Q9.4

5. For the circuit shown in Figure Q9.5, find an expression for transient current, when the switch K is closed at time $t = 0$ sec. Assume the value of $e_0 = 2$ V and $i(0) = 2$ A.

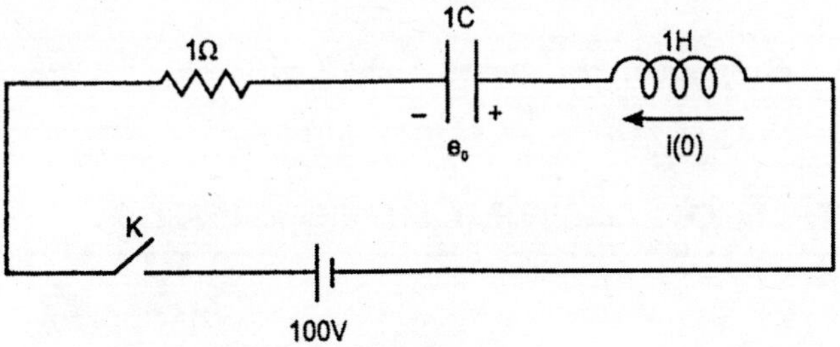

Fig. Q9.5

Chapter 10

FOURIER ANALYSIS

10.1 INTRODUCTION

Fourier analysis is applied mainly to complex periodic signals, not sinusoidal shape. These signals can be decomposed into an infinite number of sinusoidal signals each having a different amplitude and frequency. A signal when represented in this form it is called a Fourier series. A function when represented in Fourier series it takes the form as shown in equation (10.1)

$$f(t) = a_0 + \sum_{n=1}^{\infty} a_n \cos n\,\omega t + \sum_{n=1}^{\infty} b_n \sin n\,\omega t \qquad ...(10.1)$$

where, a_0, a_n and b_n are called Fourier coefficients. The equation (10.1) is called the trigonometric form of Fourier series.

10.2 DETERMINATION OF FOURIER COEFFICIENTS

$$a_0 = \frac{1}{T}\int_0^T f(t)dt \qquad ...(10.2)$$

$$a_n = \frac{2}{T}\int_0^T f(t)\ \sin\ n\omega t\ dt \qquad ...(10.3)$$

$$b_n = \frac{2}{T}\int_0^T f(t)\ \cos\ n\omega t\ dt \qquad ...(10.4)$$

10.3 EVEN AND ODD FUNCTION

For any function $f(x)$, if

(i) $f(x) = f(-x)$ it is called even function or symmetric function.

For an even function $b_n = 0$.

(ii) $f(-x) = -f(x)$ it is called odd function or skew symmetric function.

For an odd function $a_0 = a_n = 0$.

10.4 EXPONENTIAL FORM OF FOURIER SERIES

We know that,

$$\sin n\omega t = \frac{e^{jn\omega t} - e^{-jn\omega t}}{2j} \qquad ...(10.5)$$

$$\cos n\omega t = \frac{e^{jn\omega t} + e^{-jn\omega t}}{2} \qquad ...(10.6)$$

Substituting equations (10.5) and (10.6) in equation (10.1) we get

$$f(t) = a_0 + \sum_{n=1}^{\infty} \frac{a_n}{2}\left(e^{jn\omega t} - e^{-jn\omega t}\right) + \sum_{n=1}^{\infty} \frac{b_n}{2j}\left(e^{jn\omega t} - e^{-jn\omega t}\right)$$

$$f(t) = a_0 + \sum_{n=1}^{\infty} \left[\frac{1}{2}(a_n - jb_n)\, e^{jn\omega t} + \frac{1}{2}(a_n + jb_n)\, e^{-jn\omega t}\right] \qquad ...(10.7)$$

Let
$$A_n = \frac{a_n - jb_n}{2} \qquad ...(10.8)$$

and
$$B_n = \frac{a_n + jb_n}{2} \qquad ...(10.9)$$

Substituting equations (10.8) and (10.9) in equation (10.7) we get

$$f(t) = a_0 + \sum_{n=1}^{\infty} \left[A_n e^{jn\omega t} + B_n e^{-jn\omega t}\right]$$

$$f(t) = a_0 + \sum_{n=1}^{\infty} A_n e^{jn\omega t} + \sum_{n=1}^{\infty} B_n e^{-jn\omega t} \qquad ...(10.10)$$

A_n can be seen as the coefficients of the positive frequency and B_n the coefficient of the negative frequency. We can get rid of the constant term a_0 from equation (10.10) by expanding the range of the second term from 0 to ∞ instead of 1 to ∞. Thus equation (10.10) becomes

$$f(t) = \sum_{n=0}^{\infty} A_n e^{jn\omega t} + \sum_{n=1}^{\infty} B_n e^{-jn\omega t} \qquad ...(10.11)$$

Equation (10.11) can be further simplified and written as

$$f(t) = \sum_{n=-\infty}^{\infty} C_n e^{jn\omega t} \qquad ....(10.12)$$

Equation (10.12) called the exponential or the complex form of the Fourier Series.

where,
$$C_n = \frac{1}{2T}\int_{-T}^{T} f(t)e^{jn\omega t}\,dt \qquad \text{...(10.13)}$$

The magnitude and phase of C_n is given by

$$C_n = \sqrt{A_n^2 + B_n^2} \qquad \text{...(10.14)}$$

$$\phi_n = \tan^{-1}\left(\frac{B_n}{A_n}\right) \qquad \text{...(10.15)}$$

SOLVED EXAMPLES

EXAMPLE 10.1. *Find the Fourier series expansion for the function*

$$f(x) = x \qquad -\pi < x < \pi$$

SOLUTION. The function $f(x) = x$ is an odd function thus $a_0 = a_n = 0$. So,

$$b_n = \frac{2}{\pi}\int_0^{\pi} f(x)\sin nx\,dx$$

$$= \frac{2}{\pi}\int_0^{\pi} x\sin nx\,dx$$

$$= \frac{2}{\pi}\left[\frac{-n\cos nx}{n} + \frac{\sin nx}{n^2}\right]_0^{\pi}$$

$$= -\frac{2\cos nx}{n}$$

EXAMPLE 10.2. *Find the Fourier series expansion for the function*

$$f(x) = x^2 \quad -\pi < x < \pi$$

SOLUTION.

$$a_0 = \frac{1}{\pi}\int_{-\pi}^{\pi} f(x)dx = \frac{1}{\pi}\int_{-\pi}^{\pi} x^2 dx = \frac{2}{\pi}\left[\frac{x^3}{3}\right]_{-\pi}^{\pi} = \frac{2\pi^2}{3}$$

$$a_n = \frac{1}{\pi}\int_{-\pi}^{\pi} x^2\cos nx\,dx = \frac{2}{\pi}\int_0^{\pi} x^2\cos nx\,dx$$

$$= \frac{2}{\pi}\left[\frac{x^2\sin nx}{n} + \frac{2x\cos nx}{n^2} - \frac{2\sin nx}{n^3}\right]_0^{\pi}$$

$$= \frac{2}{\pi}\left[\frac{2\pi\cos nx}{n^2}\right] = \frac{4(-1)^n}{n^2}$$

$$b_n = \frac{1}{\pi}\int_{-\pi}^{\pi} x^2\sin nx\,dx = 0 \qquad \therefore\ x^2 \sin x \text{ is an odd function}$$

EXAMPLE 10.3. *Find the Fourier series expansion for the function*

$$f(x) = e^x \qquad \text{for } -\pi < x < \pi$$

SOLUTION.

$$a_0 = \frac{1}{\pi}\int_{-\pi}^{\pi} f(x)\,dx = \frac{1}{\pi}\int_{-\pi}^{\pi} e^x\,dx$$

$$= \frac{1}{\pi}[e^{\pi} - e^{-\pi}] = \frac{2\sinh\pi}{\pi}$$

$$a_n = \frac{1}{\pi}\int_{-\pi}^{\pi} f(x)\cos nx\,dx = \frac{1}{\pi}\int_{-\pi}^{\pi} e^x \cos x\, nx\,dx$$

$$= \frac{1}{\pi}\left[\frac{e^x}{1^2+n^2}(\cos nx + n\sin nx)\right]_{-\pi}^{\pi}$$

$$= \frac{\cos n\pi(e^{\pi} - e^{-\pi})}{\pi(n^2+1)} = \frac{(-1)^n\, 2\sinh\pi}{\pi(n^2+1)}$$

$$b_n = \frac{1}{\pi}\int_{-\pi}^{\pi} f(x)\sin nx\,dx$$

$$= \frac{1}{\pi}\int_{-\pi}^{\pi} e^x \sin nx\,dx$$

$$= \frac{1}{\pi}\left[\frac{e^x}{1+n^2}(\sin nx - n\cos nx)\right]_{-\pi}^{\pi}$$

$$= \frac{n(-1)^n(e^{-\pi} - e^{\pi})}{\pi(n^2+1)}$$

$$= \frac{2n(-1)^n \sinh\pi}{\pi(n^2+1)}$$

EXAMPLE 10.4. *Find the Fourier series expansion for the function*

$$f(x) = \begin{cases} \pi + 2x & -\pi < x < 0 \\ \pi - 2x & 0 < x < \pi \end{cases}$$

SOLUTION. The above function is an even function, so $b_n = 0$

$$a_0 = \frac{2}{\pi}\int_0^{\pi} (\pi - 2x)dx = 0$$

$$a_n = \frac{2}{\pi}\int_0^{\pi} (\pi - 2x)\cos nx\,dx$$

$$= \frac{4}{n^2\pi}[-\cos nx]_0^{\pi}$$

$$= \frac{4}{n^2\pi}[-(-1)^n + 1]$$

$$= \begin{bmatrix} 0 & \text{if } n \text{ is even} \\ \dfrac{8}{n^2\pi} & \text{if } n \text{ is odd} \end{bmatrix}$$

EXAMPLE 10.5. *Find the Fourier series expansion for the function*

$$f(x) = \begin{bmatrix} -x^2 & -\pi < x < 0 \\ x^2 & 0 < x < \pi \end{bmatrix}$$

SOLUTION. The above function is an odd function, so $a_0 = a_n = 0$

$$b_n = \frac{2}{\pi}\int_0^{\pi} f(x) \sin nx \, dx$$

$$= \frac{2}{\pi}\left[\frac{-x\cos nx}{n} + \frac{2x\sin nx}{n^2} - \frac{2\cos nx}{n^3}\right]_0^{\pi}$$

$$= \frac{2}{\pi}\left[\frac{-\pi^2(-1)^n}{n} + 2\left(\frac{1-(-1)^n}{n^3}\right)\right]$$

EXAMPLE 10.6. *Find a Fourier series expansion for the function*

$$f(x) = \begin{bmatrix} -1 \text{ for } -\pi < x < 0 \\ 1 \text{ for } 0 < x < \pi \end{bmatrix}$$

SOLUTION. The above function is an odd function so $a_0 = a_n = 0$

$$b_n = \frac{2}{\pi}\int_0^{\pi} f(x) \sin nx \, dx$$

$$= -\left.\frac{2\cos nx}{n\pi}\right|_0^{\pi} = \frac{4}{n\pi}$$

EXAMPLE 10.7. *Find the Fourier series expansion for the waveform shown in Figure E10.7.*

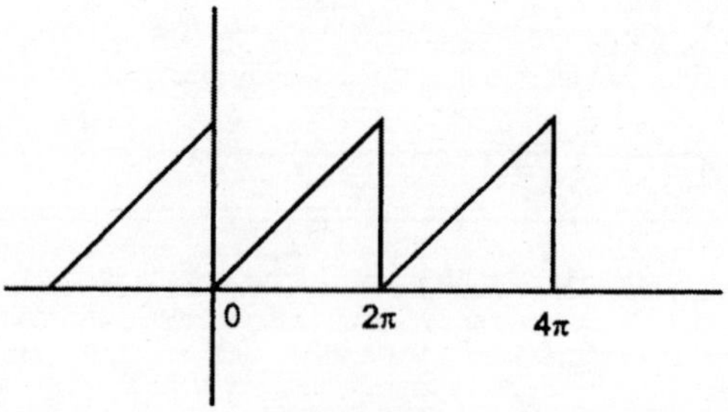

Fig. E10.7

SOLUTION. From Figure E10.7, we can write

$$f(x) = x \qquad 0 < x < 2\pi$$

Then

$$a_0 = \frac{1}{\pi}\int_0^{2\pi} f(x)\,dx = \frac{1}{\pi}\int_0^{2\pi} x\,dx = 2\pi$$

$$a_n = \frac{1}{\pi}\int_0^{2\pi} f(x)\cos nx\,dx = \frac{1}{\pi}\int_0^{2\pi} x\cos nx\,dx$$

$$= \frac{1}{\pi}\left[x\frac{\sin nx}{n} - 1\left(-\frac{\cos nx}{n^2}\right)\right]_0^{2\pi} = 0$$

$$b_n = \frac{1}{\pi}\int_0^{2\pi} f(x)\sin nx\,dx = \frac{1}{\pi}\int_0^{2\pi} x\sin nx\,dx$$

$$= \frac{1}{\pi}\left[\left(-\frac{x\cos nx}{n}\right) - \left(-\frac{\sin nx}{n^2}\right)\right]_0^{2\pi} = -\frac{2}{\pi}$$

EXAMPLE 10.8. *Show that the Fourier series coefficients for the square wave studied earlier are*

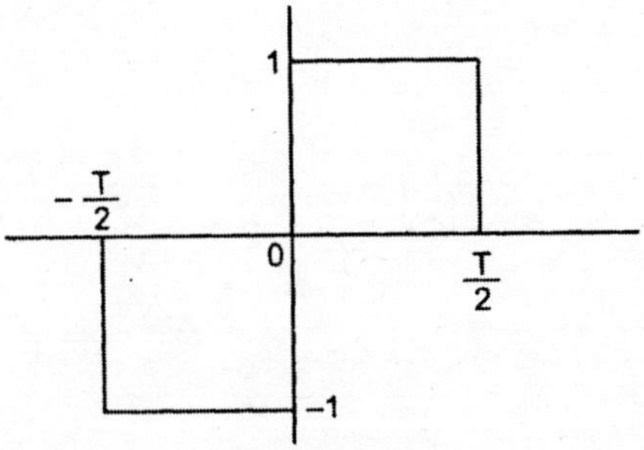

Fig. E10.8

SOLUTION. From Figure E10.8, we can write

$$f(t) = \begin{Bmatrix} 1, & 0 < t < \frac{T}{2} \\ -1, & \frac{-T}{2} < t < 0 \end{Bmatrix}$$

Here the Fourier coefficients $a_0 = a_n = 0$ and

$$b_n = \frac{2}{T}\left[\int_0^{T/2} \sin n\,\omega t\,dt - \int_{-T/2}^{0} \sin n\,\omega t\,dt\right]$$

$$= \frac{2}{n\omega T}\left[-\cos n\,\omega t\,\Big|_0^{T/2} - \left(-\cos n\,\omega t\,\Big|_{-T/2}^{0}\right)\right]$$

$$= \frac{4}{n\omega T}\left(1 - \cos\left(\frac{n\omega T}{2}\right)\right)$$

$$= \frac{2}{n\pi}\,(1 - \cos n\pi)$$

$$= \begin{bmatrix} \frac{4}{n\pi} & n \text{ odd} \\ 0 & n \text{ even} \end{bmatrix}$$

EXAMPLE 10.9. *Find the Fourier series coefficients for the wave shown in Figure E10.9.*

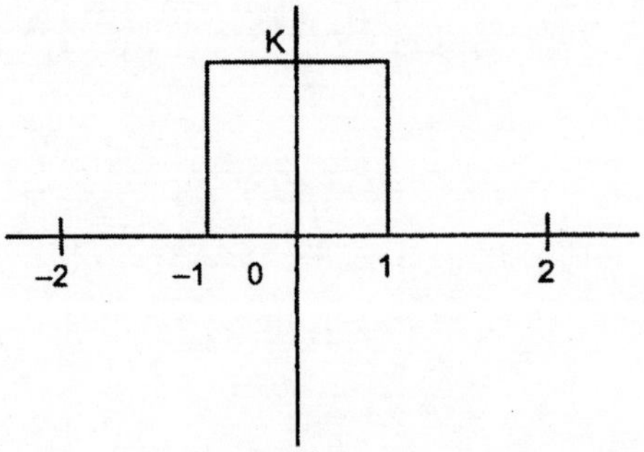

Fig. E10.9

SOLUTION. From Figure E10.9, we can write

$$f(t) = \begin{bmatrix} 0 & -2 < x < -1 \\ K & -1 < x < 1 \\ 0 & 1 < x < 2 \end{bmatrix}$$

Then
$$a_0 = \frac{1}{4}\int_{2}^{2} f(x)\,dx = \frac{1}{4}\int_{-1}^{1} K\,dx = \frac{K}{2}$$

$$a_n = \frac{1}{2}\int_{-2}^{2} f(x)\cos\frac{n\pi x}{2}\,dx = \frac{1}{2}\int_{-1}^{1} K\cos\frac{n\pi x}{2}\,dx$$

$$= \frac{2K}{n\pi}\sin\frac{n\pi}{2}$$

Thus we can say that when n is even the value of $a_n = 0$. Again it is an even function so

$$b_n = 0$$

EXAMPLE 10.10. *Find the Fourier series coefficients for the wave shown in Figure E10.10.*

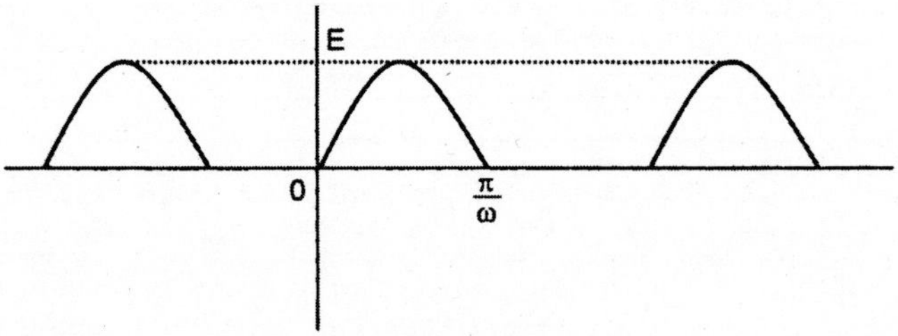

Fig. E10.10

SOLUTION. From Figure E10.10, we can write

$$f(x) = \begin{bmatrix} 0 & \text{for } -\frac{\pi}{\omega} < x < 0 \\ E\sin\omega t & 0 < x < \frac{\pi}{\omega} \end{bmatrix}$$

Then
$$a_0 = \frac{\omega}{2\pi}\int_{0}^{\pi/\omega} E\sin\omega t\,dt = \frac{E}{\pi}$$

and
$$a_n = \frac{\omega}{\pi}\int_0^{\pi/\omega} E \sin \omega t \cos \omega t \, dt$$

$$= \frac{\omega E}{2\pi}\int_0^{\pi/\omega} [\sin (1+n)\omega t + \sin (1-n)\omega t]\, dt \qquad ...(i)$$

If $n = 1$ then from equation (*i*) we get

$$a_n = 0$$

For $n > 1$ equation (*i*) becomes,

$$a_n = \frac{\omega E}{2\pi}\left[-\frac{\cos(1+n)\omega t}{(1+n)\omega} - \frac{\cos(1-n)\omega t}{(1-n)\omega}\right]_0^{\frac{\pi}{\omega}} \qquad ...(ii)$$

Again if n is odd the from equation (*ii*) we get

$$a_n = 0$$

If n is even equation (*ii*) becomes,

$$a_n = \frac{2E}{(1+n)(1-n)\pi}$$

Proceeding in the similar manner with b_n we will get

$$b_1 = \frac{E}{2} \quad \text{and} \quad b_n = 0 \text{ for } n > 1$$

EXAMPLE 10.11. *Find the Fourier series expansion for the waveform shown in Figure E10.11.*

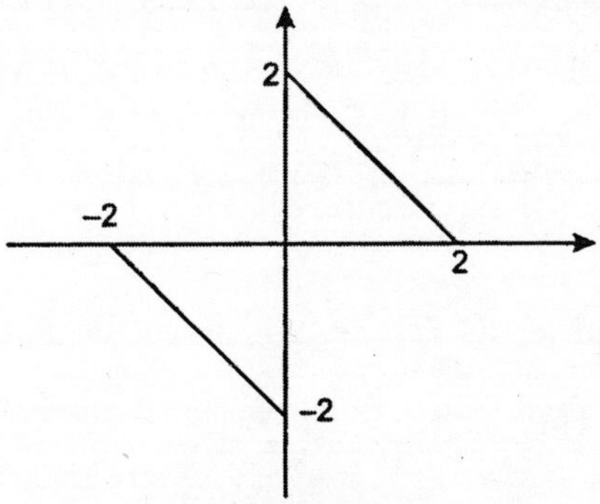

Fig. E10.11

SOLUTION. From Figure E10.11, we can write

$$f(x) = 2 - x \qquad 0 < x < 2$$

Here
$$a_0 = a_n = 0$$

and
$$b_n = \frac{2}{L}\int_0^L f(x) \sin \frac{n\pi x}{L}\, dx \quad \text{here } L = 2$$

$$= \int_0^2 (2-x)\sin\frac{n\pi x}{2}\, dx$$

Using integration by parts we get

$$b_n = \frac{4}{n\pi}$$

Thus the Fourier series is

$$f(x) = \frac{4}{\pi}\sum_{n=1}^{\infty}\frac{1}{n}\sin\frac{n\pi x}{2}$$

10.5 FOURIER TRANSFORMS

Fourier transform is an integral transformation of any function $f(t)$ from time domail to frequency domain. Fourier transforms are used to analyze a non periodic signal. The Fourier transform of any signal $f(t)$ is given by

$$F[\omega] = F[f(t)] = \int_{-\infty}^{\infty} f(t)\, e^{-j\omega t}\, dt$$

and the equation

$$f(t) = F^{-1}[F(\omega)] = \frac{1}{2\pi}\int_{-\infty}^{\infty} F(\omega) e^{j\omega t}\, d\omega$$

gives the inverse Fourier transforms.

10.5.1 Properties of Fourier Transform

(i) Linearity Property

Consider two functions $f_1(t)$ and $f_2(t)$. Let the Fourier transform of the two functions be $F_1(\omega)$ and $F_2(\omega)$ respectively. Then from linearity property

$$F[a_1 f_1(t) + a_2 f_2(t)] = a_1 F_1(\omega) + a_2 F_2(\omega)$$

(ii) Scaling Property

Consider a function $f(t)$ and let its Fourier transform be $F(\omega)$. Then from scaling property,

$$F[f(at)] = \frac{1}{|a|} F\left(\frac{\omega}{a}\right)$$

(iii) Shifting Property

Consider a function $f(t)$ and let its Fourier transform be $F(\omega)$. Then from shifting property,

$$F[f(t - k)] = e^{-j\omega k}\, F(\omega)$$

(iv) Amplitude Modulation Property

Consider a function $f(t)$ and let its Fourier transform be $F(\omega)$. Then from amplitude modulation property,

$$F[f(t)\, e^{j\omega_0 t}] = F(\omega - \omega_0)$$

(v) Differentiation Property

Consider a function $f(t)$ and let its Fourier transform be $F(\omega)$. Then from differentiation property,

$$F\,[f'(t)] = j\omega F(\omega)$$

(vi) Reversal Property

Consider a function $f(t)$ and let its Fourier transform be $F(\omega)$. Then from reversal property,

$$F[f(-t)] = F(-\omega) = F^*(\omega)$$

(vii) Duality Property

Consider a function $f(t)$ and let its Fourier transform be $F(\omega)$. Then from duality property the Fourier transform $F(t)$ is $2\pi f(-\omega)$

(viii) Convolution Theorem or Faltung Theorem

Consider two functions $f_1(t)$ and $f_2(t)$. Let the Fourier transform of the two functions be $F_1(\omega)$ and $F_2(\omega)$ respectively. The convolution of the two functions $f_1(t)$ and $f_2(t)$ is defined as

$$y(t) = f_1 * f_2 = \frac{1}{\sqrt{2\pi}} \int_{-\infty}^{\infty} f_1(t)\, f_2(x - t)\, dt$$

In terms of Fourier transform

$$y(\omega) = F[f_1(t) * f_2(t)] = F_1(\omega)\, F_2(\omega)$$

SOLVED EXAMPLES

EXAMPLE 10.12. *Find the Fourier transform of* $f(t) = te^{-at}\, u(t)$

SOLUTION.

$$F[f(t)] = \int_{-\infty}^{\infty} te^{-at}\, u(t)\, e^{-j\omega t}\, dt$$

$$= \int_{0}^{\infty} te^{(-j\omega+a)t}\, dt$$

$$= \left[\frac{e^{-(j\omega+a)t}(-(j\omega+a)t-1)}{(-(j\omega+a))^2}\right]_0^{\infty}$$

$$= \frac{1}{(-(j\omega+a))^2}$$

EXAMPLE 10.13. *Find the Fourier transform of* $f(t) = e^{-2t}$ *for* $t \geq 0$*. Also find its magnitude.*

SOLUTION.

$$F[f(t)] = \int_{-\infty}^{\infty} f(t)\, e^{-j\omega t}\, dt$$

$$= \int_{-\infty}^{\infty} e^{-2t}\, e^{-j\omega t}\, dt = \int_{-\infty}^{\infty} e^{-(2+j\omega)t}\, dt$$

$$= \frac{1}{2 + j\omega}$$

$$|\, F[f(t)]\, | = \frac{1}{\sqrt{4 + \omega^2}}$$

EXAMPLE 10.14. *Find the Fourier transform of $f(t) = \delta(t) + \delta(t - k)$ where k is constant.*

SOLUTION.

$$F[f(t)] = \int_{-\infty}^{\infty} (\delta(t) + \delta(t-k))\, e^{-j\omega t}\, dt$$

$$= \int_{-\infty}^{\infty} \delta(t)\, e^{-j\omega t}\, dt + \int_{-\infty}^{\infty} \delta(t-k)\, e^{-j\omega t}\, dt$$

$$= 1 + e^{-j\omega k}$$

EXAMPLE 10.15. *Find the Fourier transform of the waveform shown in Figure E10.15.*

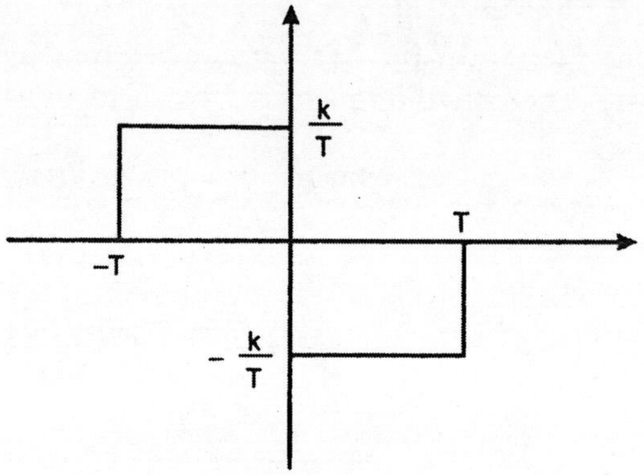

Fig. E10.15

SOLUTION. From Figure E10.15, we can write

$$f(t) = \begin{cases} \dfrac{k}{T} & -T < t < 0 \\ -\dfrac{k}{T} & 0 < t < T \\ 0 & \text{Otherwise} \end{cases}$$

Then

$$F[f(t)] = \int_{-T}^{0} \frac{k}{T} e^{-j\omega T}\, dt + \int_{0}^{T} -\frac{k}{T} e^{-j\omega T}\, dt$$

$$= \frac{k}{T}\left[-\frac{1}{j\omega} + \frac{e^{j\omega T}}{j\omega} + \frac{e^{-j\omega T}}{j\omega} - \frac{1}{j\omega}\right]$$

$$= \frac{k}{j\omega T}\,(e^{j\omega T} + e^{-j\omega T} - 2)$$

EXAMPLE 10.16. *Find the inverse Fourier transform of* $f(x) = \begin{cases} x & |x| \le a \\ 0 & |x| \ge a \end{cases}$

SOLUTION.

$$F[f(x)] = \frac{1}{\sqrt{2\pi}} \int_{-\infty}^{\infty} e^{j\omega x} f(x)\, dx$$

$$= \frac{1}{\sqrt{2\pi}} \int_{-a}^{a} x e^{j\omega t}\, dx$$

$$= \frac{1}{\sqrt{2\pi}}\left[\frac{xe^{-j\omega x}}{j\omega} - \frac{e^{-j\omega x}}{(j\omega)^2}\right]_{-a}^{a}$$

$$= \frac{2j}{\omega^2}\frac{1}{\sqrt{2\pi}}\,[\sin \omega a - a\omega \cos \omega a]$$

EXAMPLE 10.17. *Find the inverse Fourier transform of* $f(x) = \begin{matrix} 1-x^2 & |x|<a \\ 0 & |x|>a \end{matrix}$

SOLUTION.

$$F[f(x)] = \frac{1}{\sqrt{2\pi}}\int_{-\infty}^{\infty} e^{j\omega x} f(x)\,dx$$

$$= \frac{1}{\sqrt{2\pi}}\int_{-a}^{a} (1 - x^2)\,.\,e^{j\omega x}\,dx$$

$$= -\frac{4}{2\pi}\left[\frac{\omega\cos\omega - \sin\omega}{\omega^3}\right]$$

EXAMPLE 10.18. *The signal shown in Figure E10.18, is applied to a series RLC circuit. Find an expression for the current i(t) flowing through the circuit. Assume V = 3.14 Volt and ω = 1 rad/sec.*

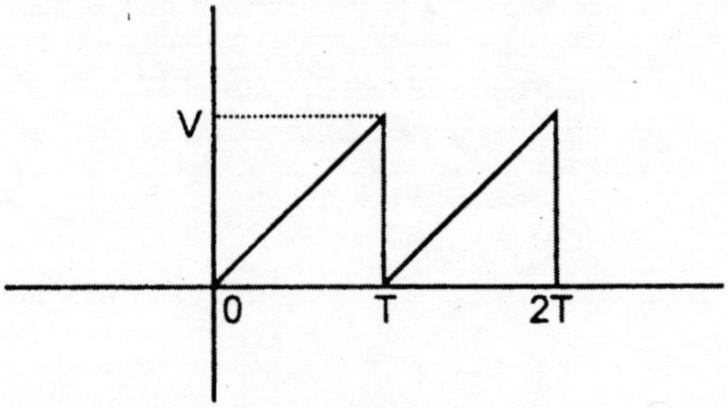

Fig. E10.18

SOLUTION. From Figure E10.18, we can write

$$f(t) = \frac{V}{T}\,t \qquad 0 < t < T$$

The Fourier coefficients are

$$a_0 = \frac{1}{T}\int_0^T f(t)\,dt = \frac{V}{2}$$

$$a_n = \frac{2}{T}\int_0^T f(t)\cos n\omega t\,dt = \frac{2}{T}\int_0^T \frac{Vt}{T}\cos n\omega t\,dt$$

$$= \frac{2V}{T^2}\left[T\sin n\omega T + \frac{\cos n\omega T}{n^2\omega^2} - \frac{1}{n^2\omega^2}\right] = 0$$

Since $\sin n2\pi = 0$ and $n2\pi = 1$

$$b_n = \frac{2}{T}\int_0^T f(t)\sin n\omega t\, dt = \frac{2}{T}\int_0^T \frac{Vt}{T}\sin n\omega t\, dt$$

$$= \frac{2V}{T^2}\left[T - \frac{\cos n\omega t}{n\omega^2}\Big|_0^T + \frac{\sin n\omega t}{n^2\omega^2}\Big|_0^T\right] = -\frac{V}{n\pi}$$

Then $$f(t) = \frac{V}{2} - \frac{V}{\pi}\left[\sin\omega t + \frac{1}{2}\sin 2\omega t + \frac{1}{3}\sin 3\omega t + \ldots\ldots\right] \quad \ldots(i)$$

Substituting $V = 3.14$ Volt and $\omega = 1$ rad/sec in equation (*i*) we get

$$f(t) = 1.57 - \left[\sin t + \frac{1}{2}\sin 2t + \frac{1}{3}\sin 3t + \ldots\ldots\right] \quad \ldots(ii)$$

Again $$y_{j\omega} = \frac{1}{1 + j\omega L - \frac{j}{\omega C}} \text{ then,}$$

$$y_{j1} = \frac{1}{1 + j1 - j} = 1e^{j0}$$

$$y_{j2} = \frac{1}{1 + j2 - \frac{j}{2}} = 0.55e^{-j54.30}$$

$$y_{j3} = \frac{1}{1 + j3 - \frac{j}{3}} = 0.33e^{-j82}$$

and
$$V_1 = 1e^{j0} \text{ volt}$$
$$V_2 = \frac{1}{2}e^{j0} \text{ volt}$$
$$V_3 = \frac{1}{3}e^{j0} \text{ volt}$$

Thus $$i(t) = -\left[\sin t + \frac{0.55}{2}\sin(2t - 54.3) + \frac{0.33}{3}\sin(3t - 82) + \ldots\ldots\right]$$

EXAMPLE 10.19. *The signal shown in Figure 10.18, is applied to a series RC circuit. Find the phase and magnitude of the current i(t), flowing through the circuit. Assume R = 1 Ω and C = 1 F.*

SOLUTION. From Example 10.18, we can write

$$f(t) = \frac{V}{2} - \frac{V}{\pi}\left[\sin\omega t + \frac{1}{2}\sin 2\omega t + \frac{1}{3}\sin 3\omega t + \ldots\ldots\right] \quad \ldots(i)$$

Equation (*i*) can again be written as

$$f(t) = \frac{V}{2} - \frac{V}{\pi}\sum_{n=1}^{\infty}\frac{1}{n}\sin n\omega t \quad \ldots(ii)$$

Since there is a capacitor in the network it will block the d.c. component if $f(t)$, *i.e.* $\frac{V}{2}$

So the voltage that will appear across the circuit is

$$f_1(t) = \frac{V}{2} - \frac{V}{\pi}\sum_{n=1}^{\infty}\frac{1}{n}\sin n\omega t$$

Thus,

$$i(t) = \frac{f_1(t)}{R+\dfrac{1}{jn\omega C}} = \frac{f_1(t)}{1+\dfrac{1}{jn\omega}} = \frac{f_1(t)}{1-\dfrac{j}{n\omega}} = \frac{f_1(t)n\omega}{n\omega - j}$$

$$= \frac{f_1(t)n\omega}{\sqrt{n^2\omega^2+1}} \angle \tan^{-1}\left(\frac{1}{n\omega}\right)$$

EXAMPLE 10.20. *The signal shown in Figure E10.20, is applied to a series RL circuit. Find the phase and magnitude of the current, flowing through the circuit. Assume R = 1 Ω and L = 1 H.*

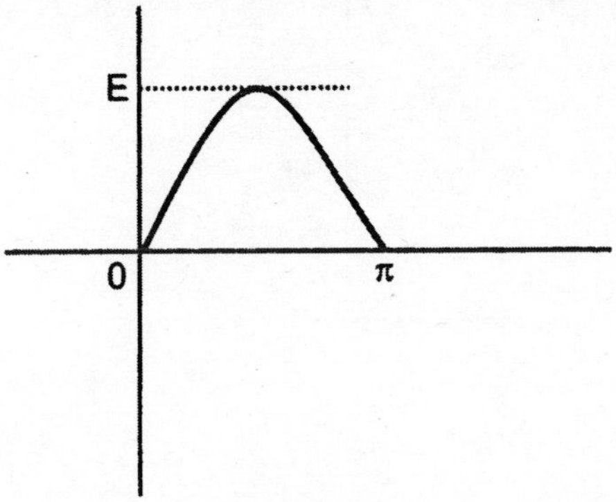

Fig. E10.20

SOLUTION. From

$$v(t) = A \sin t \qquad 0 < t < \pi$$

Then

$$v(j\omega) = \int_{-\infty}^{\infty} v(t)e^{-j\omega t}\, dt = A\int_0^{\pi} e^{-j\omega t} \sin t\, dt$$

$$= A\left[\frac{1+\cos\omega\pi}{1-\omega^2} - j\frac{A\sin\omega\pi}{1-\omega^2}\right]$$

Thus

$$|\, v(j\omega)\, | = 2A\left|\frac{\cos\omega\dfrac{\pi}{2}}{1-\omega^2}\right|$$

and

$$\tan\phi(j\omega) = \tan\left(-\frac{\omega\pi}{2}\right)$$

Thus

$$\phi(j\omega) = -\frac{\omega\pi}{2}$$

Then

$$I(j\omega) = \frac{v(j\omega)}{R+j\omega L} = \frac{|\, v(j\omega)\,|\angle\phi(j\omega)}{\sqrt{R^2+\omega^2L^2}\angle\tan^{-1}\left(\dfrac{\omega L}{R}\right)} \qquad \text{...}(i)$$

Substituting R = 1 Ω and L = 1 H in equation (*i*) we get

$$I(j\omega) = \frac{|\, v(j\omega)\,|\angle\phi(j\omega)}{\sqrt{1+\omega^2}\angle\tan^{-1}\omega}$$

$$= 2A\left|\frac{\cos\omega\dfrac{\pi}{2}}{1-\omega^2}\right| \cdot \frac{1}{\sqrt{1+\omega^2}} \angle -\frac{\omega\pi}{2} - \tan^{-1}\omega$$

EXAMPLE 10.21. *The signal shown in Figure 10.21, is applied to a series RC circuit. Find the phase and magnitude of the current, flowing through the circuit. Assume R = 1Ω and C = 1F.*

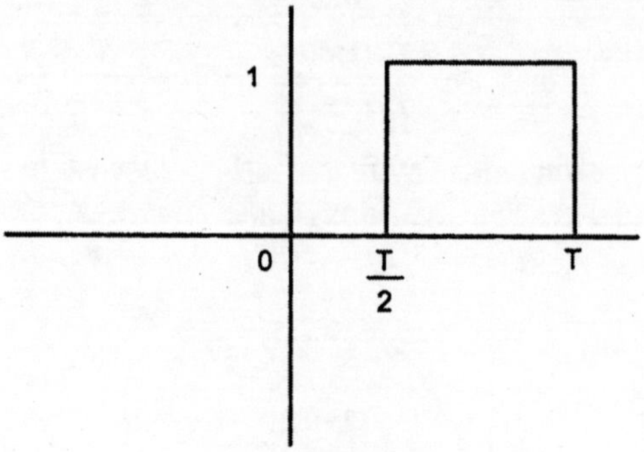

Fig. E10.21

SOLUTION. From Figure E10.21, we can write

$$v(t) = \begin{cases} 0 & 0 < t < \frac{T}{2} \\ 1 & \frac{T}{2} < t < T \end{cases}$$

Then the Fourier coefficients are

$$a_0 = \frac{1}{T}\int_{\frac{T}{2}}^{T} dt = \frac{1}{2}$$

$$a_n = \frac{2}{T}\int_{\frac{T}{2}}^{T} \cos n\omega t \, dt = 0$$

$$b_n = \frac{2}{T}\int_{\frac{T}{2}}^{T} \sin n\omega t \, dt = \frac{1}{n\pi}\,[1 - (-1)^n]$$

$$b_n = \begin{cases} 0 & \text{when } n \text{ is even} \\ -\frac{2}{3\pi} & \text{when } n \text{ is odd} \end{cases}$$

Thus $$v(t) = \frac{1}{2} - \frac{2}{\pi}\left[\sin \omega t + \frac{1}{3}\sin 3\omega t + \frac{1}{5}\sin 5\omega t + \ldots\ldots\right]$$

Since there is a capacitor in the network it will block the d.c. component if $f(t)$, *i.e.*, $\frac{1}{2}$

So the voltage that will appear across the circuit is

$$v_1(t) = -\frac{2}{\pi}\left[\sin \omega t + \frac{1}{3}\sin 3\omega t + \frac{1}{5}\sin 5\omega t + \ldots\ldots\right]$$

Then $$V_1 = 1e^{j0} \text{ volt}$$

$$V_3 = \frac{1}{3}e^{j0} \text{ volt}$$

$$V_5 = \frac{1}{5}e^{j0} \text{ volt}$$

and

$$y_{(j\omega)} = \frac{1}{1+\dfrac{1}{jn\omega C}}$$

Then

$$y_{(j1)} = \frac{1}{1+j} = \frac{1}{\sqrt{2}} \; e^{j45}$$

$$y_{(j3)} = \frac{1}{1-\dfrac{j}{3}} = 0.95 \; e^{-j18.26}$$

$$y_{(j5)} = \frac{1}{1-\dfrac{j}{5}} = 0.99 \; e^{-j11.3}$$

Thus

$$I_1 = V_1 y_{(j1)} = 0.707e^{j45}$$

$$I_3 = V_2 y_{(j3)} = 0.318e^{j18.26}$$

$$I_5 = V_5 y_{(j5)} = 0.2e^{j11.3}$$

Then

$$i(t) = -\frac{2}{\pi}\,[0.707 \sin (t + 45) + 0.381 \sin (3t + 18.26) + 0.2 \sin (5t + 11.30)]$$

EXERCISES

1. The signal shown in Figure Q10.1, is applied to a series RC circuit. Find the phase and magnitude of the current, flowing through the circuit. Assume $R = 1\Omega$ and $L = 1F$.

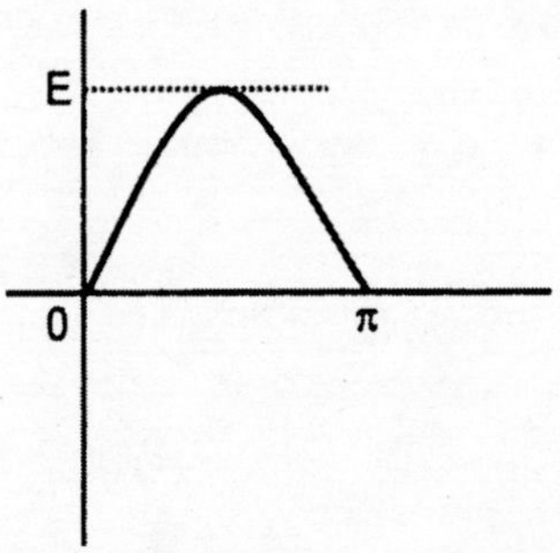

Fig. Q10.1

2. The signal shown in Figure Q10.2, is applied to a series RL circuit. Find an expression for the current $i(t)$ flowing through the circuit. Assume $V = 3.14$ Volt and $\omega = 1$ rad/sec.

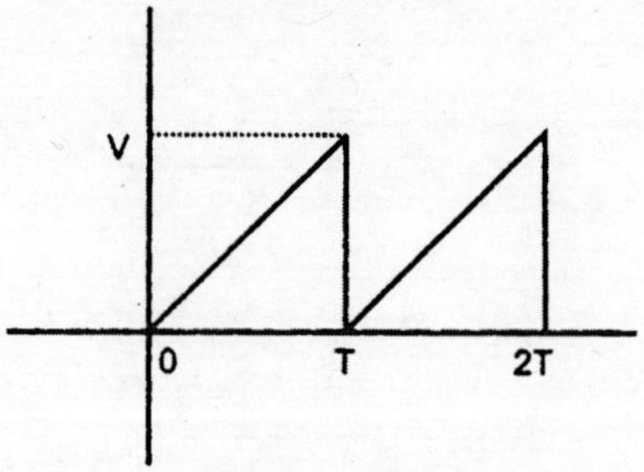

Fig. Q10.2

3. State and explain with example Convolution Theorem.
4. Show that the Fourier Series coefficients for the wave shown in Figure Q.10.4.

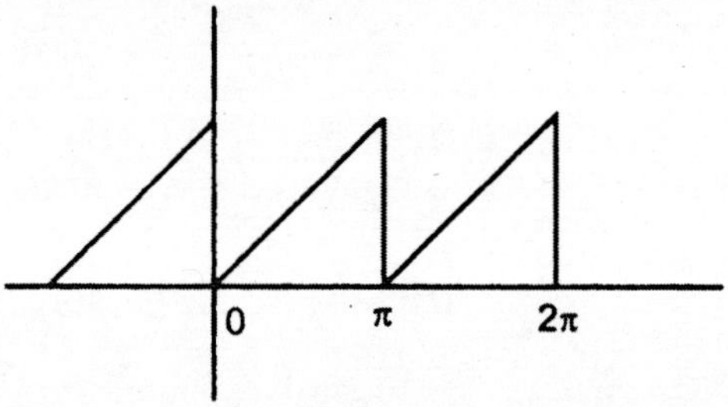

Fig. Q10.4

Chapter **11**

TWO PORT NETWORK

11.1 INTRODUCTION

The network or circuits that we have discussed so far in our previous chapters are one-port networks. A two port that will be discussed in this chapter has four terminals and it is generally represented by a block as shown in Figure 11.1. Here the direction of the current is taken arbitrarily. There are six different parameters of two-port network that connects the four variables of Figure 11.1 in the next section we will discuss these parameters in details.

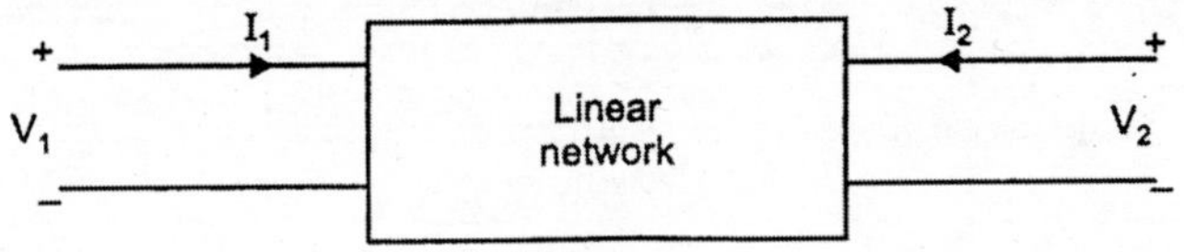

Fig. 11.1

11.2 z PARAMETERS OR IMPEDANCE PARAMETERS

$$V_1 = z_{11}I_1 + z_{12}I_2 \quad ...(11.1)$$

$$V_2 = z_{21}I_1 + z_{22}I_2 \quad ...(11.2)$$

Writing equations (11.1) and (11.2) in matrix form we get

$$\begin{bmatrix} V_1 \\ V_2 \end{bmatrix} = \begin{bmatrix} z_{11} & z_{12} \\ z_{21} & z_{22} \end{bmatrix} \begin{bmatrix} I_1 \\ I_2 \end{bmatrix}$$

Substituting $I_2 = 0$, in equation (11.1), we get $z_{11} = \left.\frac{V_1}{I_1}\right|_{I_2=0}$

Substituting $I_2 = 0$, in equation (11.2), we get $z_{21} = \left.\dfrac{V_2}{I_1}\right|_{I_2=0}$

Substituting $I_1 = 0$, in equation (11.1), we get $z_{12} = \left.\dfrac{V_1}{I_2}\right|_{I_1=0}$

Substituting $I_1 = 0$, in equation (11.2), we get $z_{22} = \left.\dfrac{V_2}{I_2}\right|_{I_1=0}$

where Z_{11} is called the open circuit input impedance and its unit is ohm (Ω).
Z_{12} is called the open circuit transfer impedance and its unit is ohm (Ω).
Z_{21} is called the open circuit transfer impedance and its unit is ohm (Ω).
Z_{22} is called the open circuit output impedance and its unit is ohm (Ω).

11.3 y PARAMETERS OR ADMITTANCE PARAMETERS

$$I_1 = y_{11}V_1 + y_{12}V_2 \qquad ...(11.3)$$

$$I_2 = y_{21}V_1 + y_{22}V_2 \qquad ...(11.4)$$

Writing equations (11.3) and (11.4) in matrix form we get

$$\begin{bmatrix} I_1 \\ I_2 \end{bmatrix} = \begin{bmatrix} y_{11} & y_{12} \\ y_{21} & y_{22} \end{bmatrix} \begin{bmatrix} V_1 \\ V_2 \end{bmatrix}$$

Substituting $V_2 = 0$, in equation (11.3), we get $y_{11} = \left.\dfrac{I_1}{V_1}\right|_{V_2=0}$

Substituting $V_2 = 0$, in equation (11.4), we get $y_{21} = \left.\dfrac{I_2}{V_1}\right|_{V_2=0}$

Substituting $V_1 = 0$, in equation (11.3), we get $y_{12} = \left.\dfrac{I_1}{V_2}\right|_{V_1=0}$

Substituting $V_1=0$, in equation (11.4), we get $y_{22} = \left.\dfrac{I_2}{V_2}\right|_{V_1=0}$

where y_{11} is called the short circuit input admittance and its unit is mho.
y_{12} is called the short circuit transfer admittance and its unit is mho.
y_{21} is called the short circuit transfer admittance and its unit is mho.
y_{22} is called the short circuit output admittance and its unit is mho.

11.4 h PARAMETERS OR HYBRID PARAMETERS

$$V_1 = h_{11}I_1 + h_{12}V_2 \quad \text{....(11.5)}$$

$$I_2 = h_{21}I_1 + h_{22}V_2 \quad \text{....(11.6)}$$

Writing equations (11.5) and (11.6) in matrix form we get

$$\begin{bmatrix} V_1 \\ I_2 \end{bmatrix} = \begin{bmatrix} h_{11} & h_{12} \\ h_{21} & h_{22} \end{bmatrix} \begin{bmatrix} I_1 \\ V_2 \end{bmatrix}$$

Substituting $V_2 = 0$, in equation (11.5), we get $h_{11} = \left.\dfrac{V_1}{I_1}\right|_{V_2=0}$

Substituting $V_2 = 0$, in equation (11.6), we get $h_{21} = \left.\dfrac{I_2}{I_1}\right|_{V_2=0}$

Substituting $I_1 = 0$, in equation (11.5), we get $h_{12} = \left.\dfrac{V_1}{V_2}\right|_{I_1=0}$

Substituting $I_1 = 0$, in equation (11.6), we get $h_{22} = \left.\dfrac{I_2}{V_2}\right|_{I_1=0}$

where h_{11} is called the short circuit input impedance and its unit is ohm (Ω).

h_{12} is called the open circuit reverse voltage gain and its unit less.

h_{21} is called the short circuit forward current gain and its unit less.

h_{22} is called the open circuit output admittance and its unit is mho.

11.5 g PARAMETERS OR INVERSE HYBRID PARAMETERS

$$I_1 = g_{11}V_1 + g_{12}I_2 \quad \text{...(11.7)}$$

$$V_2 = g_{21}V_1 + g_{22}I_2 \quad \text{...(11.8)}$$

Writing equations (11.7) and (11.8) in matrix form we get

$$\begin{bmatrix} I_1 \\ V_2 \end{bmatrix} = \begin{bmatrix} g_{11} & g_{12} \\ g_{21} & g_{22} \end{bmatrix} \begin{bmatrix} V_1 \\ I_2 \end{bmatrix}$$

Substituting $I_2 = 0$, in equation (11.7), we get $g_{11} = \left.\dfrac{I_1}{V_1}\right|_{I_2=0}$

Substituting $I_2 = 0$, in equation (11.8), we get $g_{21} = \left.\dfrac{V_2}{V_1}\right|_{I_2=0}$

Substituting $V_1 = 0$, in equation (11.7), we get $g_{12} = \left.\dfrac{I_1}{I_2}\right|_{V_1=0}$

Substituting $V_1 = 0$, in equation (11.8), we get $g_{22} = \left.\frac{V_2}{I_2}\right|_{V_1=0}$

where g_{11} is called the open circuit input admittance and its unit is mho.

g_{12} is called the short circuit reverse current gain and its unit less.

g_{21} is called the open circuit forward voltage gain and its unit less.

g_{22} is called the short circuit output impedance and its unit is ohm(Ω).

11.6 ABCD PARAMETERS OR TRANSMISSION PARAMETERS

$$V_1 = AV_2 - BI_2 \quad \text{...(11.9)}$$

$$I_1 = CV_2 - DI_2 \quad \text{...(11.10)}$$

Writing equations (11.9) and (11.10) in matrix form we get

$$\begin{bmatrix} V_1 \\ I_1 \end{bmatrix} = \begin{bmatrix} A & B \\ C & D \end{bmatrix} \begin{bmatrix} V_2 \\ -I_2 \end{bmatrix}$$

Substituting $I_2 = 0$, in equation (11.9), we get $A = \left.\frac{V_1}{V_2}\right|_{I_2=0}$

Substituting $I_2 = 0$, in equation (11.10), we get $C = -\left.\frac{I_1}{V_2}\right|_{I_2=0}$

Substituting $V_2 = 0$, in equation (11.9), we get $B = -\left.\frac{V_1}{I_2}\right|_{V_2=0}$

Substituting $V_2 = 0$, in equation (11.10), we get $D = \left.\frac{I_1}{I_2}\right|_{V_2=0}$

where A is called reverse voltage ratio and its unit less.

B is called the transfer impedance and its unit is ohm (Ω).

C is called the transfer admittance and its unit is mho.

D is called the reverse current ratio and its unit less.

11.7 ABCD PARAMETERS OR INVERSE TRANSMISSION PARAMETERS

$$V_2 = aV_1 - bI_1 \quad \text{...(11.11)}$$

$$I_2 = cV_1 - dI_1 \quad \text{...(11.12)}$$

Writing equations (11.11) and (11.12) in matrix form we get

$$\begin{bmatrix} V_2 \\ I_2 \end{bmatrix} = \begin{bmatrix} a & b \\ c & d \end{bmatrix} \begin{bmatrix} V_1 \\ -I_1 \end{bmatrix}$$

Substituting $I_1 = 0$, in equation (11.11), we get $a = \left.\frac{V_2}{V_1}\right|_{I_1=0}$

Substituting $I_1 = 0$, in equation (11.12), we get $c = \left.\frac{I_2}{V_1}\right|_{I_1=0}$

Substituting $V_1 = 0$, in equation (11.11), we get $b = -\left.\frac{V_2}{I_1}\right|_{V_1=0}$

Substituting $V_1 = 0$, in equation (11.12), we get $d = -\left.\frac{I_2}{I_1}\right|_{V_1=0}$

where a is called the open circuit voltage gain and its unit less.

b is called the negative short circuit transfer impedance and its unit is ohm (Ω).

c is called the open circuit transfer admittance and its unit is mho.

d is called the negative short circuit current gain and its unit less.

11.8 RELATIONSHIP BETWEEN PARAMETERS OF TWO PORT NETWORK

11.8.1 z Parameters in Terms of y Parameters

Writing equations (11.3) and (11.4) again,

$$I_1 = y_{11}V_1 + y_{12}V_2 \qquad \text{...(11.13)}$$

$$I_2 = y_{21}V_1 + y_{22}V_2 \qquad \text{...(11.14)}$$

From equation (11.3), we get

$$V_1 = \frac{I_1 - y_{12}V_2}{y_{11}} \qquad \text{...(11.13a)}$$

Substituting equation (11.13a) in equation (11.4) we get

$$I_2 = y_{21} \times \left(\frac{I_1 - y_{12}V_2}{y_{11}}\right) + y_{22}V_2 \qquad \text{...(11.14a)}$$

Solving equation (11.14a) we get

$$V_2 = \frac{y_{21}}{y_{11}y_{22} - y_{21}y_{12}} I_1 + \frac{y_{11}}{y_{11}y_{22} - y_{21}y_{12}} I_2 \qquad \text{...(11.15)}$$

Comparing equation (11.15) with equation (11.2) we get

$$z_{21} = \frac{y_{21}}{y_{11}y_{22} - y_{21}y_{12}} \qquad \text{...(11.16)}$$

and

$$z_{22} = \frac{y_{11}}{y_{11}y_{22} - y_{21}y_{12}} \qquad \text{...(11.17)}$$

Again from equation (11.3) we get

$$V_2 = \frac{I_1 - y_{11}V_1}{y_{12}} \qquad \text{...(11.18)}$$

Substituting equation (11.18) in equation (11.4) we get

$$I_2 = y_{21}V_1 + y_{22}\left(\frac{I_1 - y_{11}V_1}{y_{12}}\right) \qquad ...(11.19)$$

Solving equation (11.19) we get

$$V_1 = \frac{y_{22}}{y_{11}y_{22} - y_{21}y_{12}} I_1 - \frac{y_{12}}{y_{11}y_{22} - y_{21}y_{12}} I_2 \qquad ...(11.20)$$

Comparing equation (11.18) with equation (11.1) we get

$$z_{11} = \frac{y_{22}}{y_{11}y_{22} - y_{21}y_{12}} \qquad ...(11.21)$$

and

$$z_{12} = \frac{y_{12}}{y_{11}y_{22} - y_{21}y_{12}} \qquad ...(11.22)$$

11.8.2 z Parameters in Terms of h Parameters

Writing equations (11.5) and (11.6) again,

$$V_1 = h_{11}I_1 + h_{12}V_2 \qquad ...[\text{Eqn } 11.5]$$

$$I_2 = h_{21}I_1 + h_{22}V_2 \qquad ...[\text{Eqn } 11.6]$$

From equation (11.5), we get

$$V_2 = \frac{V_1 - h_{11}I_1}{h_{12}} \qquad ...(11.23)$$

Substituting equation (11.23) in equation (11.6) we get

$$I_2 = h_{21}I_1 + h_{22} \times \left(\frac{V_1 - h_{11}I_1}{h_{12}}\right) \qquad ...(11.24)$$

Solving equation (11.24), we get

$$V_1 = \frac{h_{11}h_{22} - h_{12}h_{21}}{h_{22}} I_1 + \frac{h_{12}}{h_{22}} I_2 \qquad ...(11.25)$$

Comparing equation (11.25) with equation (11.1) we get

$$z_{11} = \frac{h_{11}h_{22} - h_{12}h_{21}}{h_{22}} \qquad ...(11.26)$$

and

$$z_{12} = \frac{h_{12}}{h_{22}} \qquad ...(11.27)$$

From equation (11.6) we get

$$V_2 = -\frac{h_{21}}{h_{22}} I_1 + \frac{1}{h_{22}} I_2 \qquad ...(11.28)$$

Comparing equation (11.28) with equation (11.2) we get,

$$z_{21} = -\frac{h_{21}}{h_{22}} \qquad ...(11.29)$$

and

$$z_{22} = \frac{1}{h_{22}} \qquad ...(11.30)$$

11.8.3 z Parameters in Terms of g Parameters

Writing equations (11.7) and (11.8) again,

$$I_1 = g_{11}V_1 + g_{12}I_2 \quad \text{...[Eqn 11.7]}$$

$$V_2 = g_{21}V_1 + g_{22}I_2 \quad \text{...[Eqn 11.8]}$$

From equation (11.7), we get

$$V_1 = \frac{1}{g_{11}}I_1 - \frac{g_{12}}{g_{11}}I_2 \quad \text{...(11.31)}$$

Comparing equation (11.31) with equation (11.1) we get

$$z_{11} = \frac{1}{g_{11}} \quad \text{...(11.32)}$$

and

$$z_{12} = -\frac{g_{12}}{g_{11}} \quad \text{...(11.33)}$$

Substituting equation (11.31) in equation (11.8) we get

$$V_2 = \frac{g_{21}}{g_{11}}I_1 + \left(\frac{g_{11}g_{22} - g_{12}g_{22}}{g_{11}}\right)I_2 \quad \text{(11.34)}$$

Comparing equation (11.34) with equation (11.2) we get

$$z_{21} = \frac{g_{21}}{g_{11}} \quad \text{...(11.35)}$$

and

$$z_{22} = \left(\frac{g_{11}g_{22} - g_{12}g_{22}}{g_{11}}\right) \quad \text{...(11.36)}$$

11.8.4 z Parameters in Terms of ABCD Parameters

Writing equations (11.9) and (11.10) again,

$$V_1 = AV_2 - BI_2 \quad \text{...[Eqn 11.9]}$$

$$I_1 = CV_2 - DI_2 \quad \text{...[Eqn 11.10]}$$

From equation (11.10), we get

$$V_2 = I_1\,\frac{1}{C} + \frac{D}{C}I_2 \quad \text{...(11.37)}$$

Substituting equation (11.37) in equation (11.9), we get

$$V_1 = \frac{A}{C}I_1 + \frac{AD - BC}{C}I_2 \quad \text{...(11.38)}$$

Comparing equations (11.38) and (11.1), we get

$$z_{11} = \frac{A}{C} \quad \text{...(11.39)}$$

and

$$z_{12} = \frac{AD - BC}{C} \quad \text{...(11.40)}$$

Comparing equations (11.37) and (11.2), we get

$$z_{21} = \frac{1}{C} \qquad \text{...(11.41)}$$

and

$$z_{22} = \frac{D}{C} \qquad \text{...(11.42)}$$

11.8.5 z Parameters in Terms of abcd Parameters

Writing equations (11.11) and (11.12) again,

$$V_2 = aV_1 - bI_1 \qquad \text{...[Eqn 11.11]}$$

$$I_2 = cV_1 - dI_1 \qquad \text{...[Eqn 11.12]}$$

From equation (11.12), we get

$$V_1 = \frac{d}{c}I_1 + \frac{1}{c}I_2 \qquad \text{...(11.43)}$$

Comparing equations (11.43) and (11.1), we get

$$z_{11} \frac{d}{c} \qquad \text{...(11.44)}$$

and

$$z_{12} = \frac{1}{c} \qquad \text{...(11.45)}$$

Substituting equation (11.43) in equation (11.11), we get

$$V_2 = \frac{ad - bc}{c}I_1 + \frac{a}{c}I_2 \qquad \text{...(11.46)}$$

Comparing equations (11.46) and (11.2), we get

$$z_{21} \frac{ad - bc}{c} \qquad \text{...(11.47)}$$

and

$$z_{22} = \frac{a}{c} \qquad \text{...(11.48)}$$

11.8.6 y Parameters in Terms of z Parameters

Writing equations (11.1) and (11.2) again,

$$V_1 = z_{11}I_1 + z_{12}I_2 \qquad \text{...[Eqn 11.1]}$$

$$V_2 = z_{21}I_1 + z_{22}I_2 \qquad \text{...[Eqn 11.2]}$$

From equation (11.1), we get

$$I_1 = \frac{V_1 - z_{12}I_2}{z_{11}}$$

Substituting equation (11.49) in equation (11.2), we get

$$V_2 = z_{21}\left[\frac{V_1 - z_{12}I_2}{z_{11}}\right] + z_{22}I_2 \qquad \text{(11.49)}$$

or

$$I_2 = -\frac{z_{21}}{z_{11}z_{22} - z_{12}z_{21}}V_1 + \frac{z_{11}}{z_{11}z_{22} - z_{21}z_{12}}V_2 \qquad \text{...(11.50)}$$

Comparing equations (11.50) and (11.4), we get

$$y_{21} = -\frac{z_{21}}{z_{11}z_{22} - z_{12}z_{21}} \quad ...(11.51)$$

and

$$y_{22} = -\frac{z_{11}}{z_{11}z_{22} - z_{21}z_{12}} \quad ...(11.52)$$

Again from equation (11.1), we get

$$I_2 = \frac{V_1 - z_{11}I_1}{z_{12}} \quad ...(11.53)$$

Substituting equation (11.53) in equation (11.2), we get

$$V_2 = z_{21}I_1 + z_{22}\left[\frac{V_1 - z_{11}I_1}{z_{12}}\right]$$

or

$$I_1 = -\frac{z_{22}}{z_{11}z_{22} - z_{12}z_{21}}V_1 - \frac{z_{12}}{z_{11}z_{22} - z_{21}z_{12}}V_2 \quad ...(11.54)$$

Comparing equations (11.54) and (11.3), we get

$$y_{11} = \frac{z_{22}}{z_{11}z_{22} - z_{12}z_{21}} \quad ...(11.55)$$

and

$$y_{12} = \frac{z_{12}}{z_{11}z_{22} - z_{21}z_{12}} \quad ...(11.56)$$

11.8.7 y Parameters in Terms of h Parameters

Writing equations (11.5) and (11.6) again,

$$V_1 = h_{11}I_1 + h_{12}V_2 \quad ...[\text{Eqn } 11.5]$$

$$I_2 = h_{21}I_1 + h_{22}V_2 \quad ...[\text{Eqn } 11.6]$$

From equation (11.5), we get

$$I_1 = \frac{1}{h_{11}}V_1 - \frac{h_{12}}{h_{11}}V_2 \quad ...(11.57)$$

Comparing equations (11.57) and (11.3), we get

$$y_{11} = \frac{1}{h_{11}} \quad ...(11.58)$$

and

$$y_{12} = \frac{h_{12}}{h_{11}} \quad ...(11.59)$$

Substituting equation (11.57) in equation (11.6), we get

$$I_2 = h_{21}\left[\frac{1}{h_{11}}V_1 - \frac{h_{12}}{h_{11}}V_2\right] + h_{22}V_2$$

or

$$I_2 = \frac{h_{21}}{h_{11}}V_1 + \frac{h_{11}h_{22} - h_{12}h_{21}}{h_{11}}V_2 \quad ...(11.60)$$

Comparing equations (11.60) and (11.4), we get

$$y_{21} = \frac{h_{21}}{h_{11}} \qquad ...(11.61)$$

and

$$y_{22} = \frac{h_{11}h_{22} - h_{12}h_{21}}{h_{11}} \qquad ...(11.62)$$

11.8.8 y Parameters in Terms of g Parameters

Writing equations (11.7) and (11.8) again,

$$I_1 = g_{11}V_1 + g_{12}I_2 \qquad ...[Eqn\ 11.7]$$

$$V_2 = g_{21}V_1 + g_{22}I_2 \qquad ...[Eqn\ 11.8]$$

From equation (11.8), we get

$$I_2 = -\frac{g_{21}}{g_{22}}V_1 + \frac{1}{g_{22}}V_2 \qquad ...(11.63)$$

Comparing equations (11.63) and (11.4), we get

$$y_{21} = -\frac{g_{21}}{g_{22}} \qquad ...(11.64)$$

and

$$y_{22} = \frac{1}{g_{22}} \qquad ...(11.65)$$

Substituting equation (11.63) in equation (11.7), we get

$$I_1 = \left[\frac{g_{11}g_{22} - g_{12}g_{21}}{g_{22}}V_1 + \frac{g_{12}}{g_{22}}V_2\right] \qquad ...(11.66)$$

Comparing equations (11.66) and (11.3), we get

$$y_{11} = \frac{g_{11}g_{22} - g_{12}g_{21}}{g_{22}} \qquad ...(11.67)$$

and

$$y_{12} = \frac{g_{12}}{g_{22}} \qquad ...(11.68)$$

11.8.9 y Parameters in Terms of ABCD Parameters

Writing equations (11.9) and (11.10) again,

$$V_1 = AV_2 - BI_2 \qquad ...[Eqn\ 11.9]$$

$$I_1 = CV_2 - DI_2 \qquad ...[Eqn\ 11.10]$$

From equation (11.9), we get

$$I_2 = -\frac{1}{B}V_1 + \frac{A}{B}V_2 \qquad ...(11.69)$$

Comparing equations (11.69) and (11.4), we get

$$y_{21} = -\frac{1}{B} \qquad ...(11.70)$$

and

$$y_{22} = \frac{A}{B} \qquad ...(11.71)$$

Substituting equation (11.69) in equation (11.10), we get

$$I_1 = \left[+\frac{D}{B}V_1 - \frac{AD-BC}{B}V_2\right] \qquad \text{...(11.72)}$$

Comparing equations (11.72) and (11.3), we get

$$y_{11} = \frac{D}{B} \qquad \text{...(11.73)}$$

and

$$y_{12} = -\frac{AD-BC}{B} \qquad \text{...(11.74)}$$

11.8.10 y Parameters in Terms of abcd Parameters

Writing equations (11.11) and (11.12) again,

$$V_2 = aV_1 - bI_1 \qquad \text{...[Eqn 11.11]}$$

$$I_2 = cV_1 - dI_1 \qquad \text{...[Eqn 11.12]}$$

From equation (11.11), we get

$$I_1 = \frac{a}{b}V_1 - \frac{1}{b}V_2 \qquad \text{...(11.75)}$$

Comparing equations (11.75) and (11.3), we get

$$y_{11} = \frac{a}{b} \qquad \text{...(11.76)}$$

and

$$y_{12} = -\frac{1}{b} \qquad \text{...(11.77)}$$

Substituting equation (11.75) in equation (11.12), we get

$$I_2 = \left[-\frac{ad-bc}{b}V_1 + \frac{d}{b}V_2\right] \qquad \text{...(11.78)}$$

Comparing equations (11.78) and (11.4), we get

$$y_{21} = -\frac{ad-bc}{b} \qquad \text{...(11.79)}$$

and

$$y_{22} = -\frac{d}{b} \qquad \text{...(11.80)}$$

11.8.11 h Parameters in Terms of z Parameters

Writing equations (11.1) and (11.2) again,

$$V_1 = z_{11}I_1 + z_{12}I_2 \qquad \text{...[Eqn 11.1]}$$

$$V_2 = z_{21}I_1 + z_{22}I_2 \qquad \text{...[Eqn 11.2]}$$

From equation (11.2), we get

$$I_2 = \frac{1}{z_{22}}V_2 - \frac{z_{21}}{z_{22}}I_1 \qquad \text{...(11.81)}$$

Substituting equation (11.81) in equation (11.1), we get

$$V_1 = Z_{11}I_1 + z_{12}\left[\frac{1}{z_{22}}V_2 - \frac{z_{21}}{z_{22}}I_1\right]$$

$$V_1 = \frac{z_{11}z_{22} - z_{12}z_{21}}{z_{22}} I_1 + \frac{z_{12}}{z_{22}} V_2 \qquad \text{...(11.82)}$$

Comparing equations (11.82) and (11.5), we get

$$h_{11} = \frac{z_{11}z_{22} - z_{12}z_{21}}{z_{22}} \qquad \text{...(11.83)}$$

and

$$h_{12} = \frac{z_{12}}{z_{22}} \qquad \text{...(11.84)}$$

Comparing equations (11.81) and (11.6), we get

$$h_{22} = \frac{1}{z_{22}} \qquad \text{...(11.85)}$$

and

$$h_{21} = -\frac{z_{21}}{z_{22}} \qquad \text{...(11.86)}$$

11.8.12 h Parameters in Terms of y Parameters

Writing equations (11.3) and (11.4) again,

$$I_1 = y_{11}V_1 + y_{12}V_2 \qquad \text{...[Eqn 11.3]}$$

$$I_2 = y_{21}V_1 + y_{22}V_2 \qquad \text{...[Eqn 11.4]}$$

From equation (11.3), we get

$$V_1 = \frac{1}{y_{11}} I_1 - \frac{y_{12}}{y_{11}} V_2 \qquad \text{...(11.87)}$$

Comparing equations (11.87) and (11.5), we get

$$h_{11} = \frac{1}{y_{11}} \qquad \text{...(11.88)}$$

and

$$h_{12} = -\frac{y_{12}}{y_{11}} \qquad \text{...(11.89)}$$

Substituting equation (11.87) in equation (11.4), we get

$$I_2 = y_{21}\left[\frac{1}{y_{11}} I_1 - \frac{y_{12}}{y_{11}} V_2\right] + y_{22}V_2$$

$$I_2 = \frac{y_{21}}{y_{11}} I_1 + \frac{y_{11}y_{22} - y_{12}y_{21}}{y_{11}} V_2 \qquad \text{...(11.90)}$$

Comparing equations (11.90) and (11.6), we get

$$h_{22} = \frac{y_{11}y_{22} - y_{21}y_{12}}{y_{11}} \qquad \text{...(11.91)}$$

and

$$h_{21} = \frac{y_{21}}{y_{11}} \qquad \text{...(11.92)}$$

11.8.13 h Parameters in Terms of g Parameters

Writing equations (11.7) and (11.8) again,

$$I_1 = g_{11}V_1 + g_{12}I_2 \qquad \text{...[Eqn 11.7]}$$

$$V_2 = g_{21}V_1 + g_{22}I_2 \quad \text{...[Eqn 11.8]}$$

From equation (11.7), we get

$$I_2 = \frac{1}{g_{12}}I_1 - \frac{g_{11}}{g_{12}}V_1 \quad \text{...(11.93)}$$

Again from equation (11.8), we get

$$V_1 = \frac{1}{g_{21}}V_2 - \frac{g_{22}}{g_{21}}I_2 \quad \text{...(11.94)}$$

Substituting equation (11.94) in equation (11.93), we get

$$I_2 = \frac{1}{g_{12}}I_1 - \frac{g_{11}}{g_{12}}\left[\frac{1}{g_{21}}V_2 - \frac{g_{22}}{g_{21}}I\right]$$

or

$$I_2 = -\frac{g_{21}}{g_{11}g_{22} - g_{12}g_{21}}I_1 + \frac{g_{11}}{g_{11}g_{22} - g_{12}g_{21}}V_2 \quad \text{...(11.95)}$$

Comparing equations (11.6) and (11.95), we get

$$h_{21} = -\frac{g_{21}}{g_{11}g_{22} - g_{21}g_{12}} \quad \text{...(11.96)}$$

and

$$h_{22} = \frac{g_{11}}{g_{11}g_{22} - g_{21}g_{12}} \quad \text{...(11.97)}$$

Substituting equation (11.93) in equation (11.94), we get

$$V_1 = \frac{1}{g_{21}}V_2 - \frac{g_{22}}{g_{21}}\left[\frac{1}{g_{12}}I_1 - \frac{g_{11}}{g_{12}}V_1\right]$$

or

$$V_1 = -\frac{g_{21}}{g_{11}g_{22} - g_{21}g_{12}}V_2 + \frac{g_{22}}{g_{11}g_{22} - g_{21}g_{12}}I_1 \quad \text{...(11.98)}$$

Comparing equations (11.5) and (11.98), we get

$$h_{11} = \frac{g_{22}}{g_{11}g_{22} - g_{12}g_{21}} \quad \text{...(11.99)}$$

and

$$h_{12} = -\frac{g_{21}}{g_{11}g_{22} - g_{12}g_{21}} \quad \text{...(11.100)}$$

11.8.14 h Parameters in Terms of ABCD Parameters

Writing equations (11.9) and (11.10) again,

$$V_1 = AV_2 - BI_2 \quad \text{...[Eqn 11.9]}$$

$$I_1 = CV_2 - DI_2 \quad \text{...[Eqn 11.10]}$$

From equation (11.10), we get

$$I_2 = -\frac{1}{D}I_1 + \frac{C}{D}V_2 \quad \text{...(11.101)}$$

Comparing equations (11.101) and (11.6), we get

$$h_{21} = -\frac{1}{D} \quad \text{...(11.102)}$$

and

$$h_{22} = \frac{C}{D} \quad \text{...(11.103)}$$

Substituting equation (11.101), in equation (11.9), we get

$$V_1 = AV_2 - B\left[-\frac{1}{D}I_1 + \frac{C}{D}V_2\right]$$

or
$$V_1 = \frac{B}{D}I_1 + \frac{AD-BC}{D}V_2 \qquad \text{...(11.104)}$$

Comparing equations (11.104) and (11.5), we get

$$h_{11} = \frac{B}{D} \qquad \text{...(11.105)}$$

and
$$h_{12} = \frac{AD-BC}{D} \qquad \text{...(11.106)}$$

11.8.15 h Parameters in Terms of abcd Parameters

Writing equations (11.11) and (11.12) again

$$V_2 = aV_1 - bI_1 \qquad \text{...[Eqn 11.11]}$$
$$I_2 = cV_1 - dI_1 \qquad \text{...[Eqn 11.12]}$$

From equation (11.11), we get

$$V_1 = \frac{b}{a}I_1 + \frac{1}{a}V_2 \qquad \text{...(11.107)}$$

Comparing equations (11.107) and (11.5), we get

$$h_{11} = \frac{b}{a} \qquad \text{...(11.108)}$$

and
$$h_{12} = \frac{1}{a} \qquad \text{...(11.109)}$$

Substituting equation (11.107) in equation (11.12), we get

$$I_2 = c\left[\frac{b}{a}I_1 + \frac{1}{a}V_2\right] - dI_1 \qquad \text{...(11.110)}$$

or
$$I_2 = \frac{c}{a}V_2 + \frac{cb-ad}{a}I_1 \qquad \text{....(11.111)}$$

Comparing equations (11.111) and (11.6), we get

$$h_{21} = -\frac{ad-bc}{a} \qquad \text{...(11.112)}$$

and
$$h_{22} = \frac{c}{a} \qquad \text{...(11.113)}$$

11.8.16 g Parameters in Terms of z Parameters

Writing equations (11.1) and (11.2) again,

$$V_1 = z_{11}I_1 + z_{12}I_2 \qquad \text{...[Eqn 11.1]}$$
$$V_2 = z_{21}I_1 + z_{22}I_2 \qquad \text{...[Eqn 11.2]}$$

From equation (11.1), we get

$$I_1 = \frac{1}{z_{11}}V_1 - \frac{z_{12}}{z_{11}}I_2 \qquad \text{...(11.114)}$$

Comparing equations (11.114) and (11.7), we get

$$g_{11} = \frac{1}{z_{11}} \quad ...(11.115)$$

and
$$g_{12} = -\frac{z_{12}}{z_{11}} \quad ...(11.116)$$

Substituting equation (11.114) in equation (11.2), we get

$$V_2 = z_{21}\left[\frac{1}{z_{11}}V_1 - \frac{z_{12}}{z_{11}}I_2\right] + z_{22}I_2$$

or
$$V_2 = \frac{z_{21}}{z_{11}}V_1 + \frac{z_{11}z_{22} - z_{21}z_{12}}{z_{11}}I_2 \quad ...(11.117)$$

Comparing equations (11.117) and (11.8), we get

$$g_{21} = \frac{z_{21}}{z_{11}} \quad ...(11.118)$$

and
$$g_{22} = \frac{z_{11}z_{22} - z_{21}z_{12}}{z_{11}} \quad ...(11.119)$$

11.8.17 g Parameters in Terms of y Parameters

Writing equations (11.3) and (11.4) again,

$$I_1 = y_{11}V_1 + y_{12}V_2 \quad ...[Eqn\ 11.3]$$
$$I_2 = y_{21}V_1 + y_{22}V_2 \quad ...[Eqn\ 11.4]$$

From equation (11.4), we get

$$V_2 = -\frac{y_{21}}{y_{22}}V_1 + \frac{1}{y_{22}}I_2 \quad ...(11.120)$$

Comparing equations (11.120) and (11.8), we get

$$g_{21} = -\frac{y_{12}}{y_{22}} \quad ...(11.121)$$

and
$$g_{22} = \frac{1}{y_{22}} \quad ...(11.122)$$

Substituting equation (11.120) in equation (11.3) we get

$$I_1 = y_{11}V_1 + y_{12}\left[-\frac{y_{21}}{y_{22}}V_1 + \frac{1}{y_{22}}I_2\right]$$

or
$$I_1 = \frac{y_{11}y_{22} - y_{12}y_{21}}{y_{22}}V_1 + \frac{y_{12}}{y_{22}}I_2 \quad ...(11.123)$$

Comparing equations (11.123) and (11.7), we get

$$g_{11} = \frac{y_{11}y_{22} - y_{12}y_{21}}{y_{22}} \quad ...(11.124)$$

and $$g_{12} = \frac{y_{12}}{y_{22}} \qquad ...(11.125)$$

11.8.18 g Parameters in Terms of h Parameters

Writing equations (11.5) and (11.6) again,

$$V_1 = h_{11}I_1 + h_{12}V_2 \qquad ...[\text{Eqn } 11.5]$$

$$I_2 = h_{21}I_1 + h_{22}V_2 \qquad ...[\text{Eqn } 11.6]$$

From equation (11.6), we get

$$I_1 = \frac{1}{h_{21}}I_2 - \frac{h_{22}}{h_{21}}V_2 \qquad ...(11.126)$$

From equation (11.5), we get

$$V_2 = \frac{1}{h_{12}}V_1 - \frac{h_{11}}{h_{12}}I_1 \qquad ...(11.127)$$

Substituting equation (11.127) in equation (11.126), we get

$$I_1 = \frac{1}{h_{21}}I_2 - \frac{h_{22}}{h_{21}}\left[\frac{1}{h_{12}}V_1 - \frac{h_{11}}{h_{12}}I_1\right]$$

or $$I_1 = \frac{h_{22}}{h_{11}h_{22} - h_{12}h_{21}}V_1 - \frac{h_{12}}{h_{11}h_{22} - h_{12}h_{21}}I_2 \qquad ...(11.128)$$

Comparing equations (11.128) and (11.7), we get

$$g_{11} = \frac{h_{22}}{h_{11}h_{22} - h_{12}h_{21}} \qquad ...(11.129)$$

and $$g_{12} = -\frac{h_{12}}{h_{11}h_{22} - h_{12}h_{21}} \qquad ...(11.130)$$

Substituting equation (11.126) in equation (11.127), we get

$$V_2 = \frac{1}{h_{12}}V_1 - \frac{h_{11}}{h_{12}}\left[\frac{1}{h_{21}}I_2 - \frac{h_{22}}{h_{21}}V_2\right]$$

or $$V_2 = -\frac{h_{21}}{h_{11}h_{22} - h_{21}h_{12}}V_1 + \frac{h_{11}}{h_{11}h_{22} - h_{21}h_{12}}I_2 \qquad ...(11.131)$$

Comparing equations (11.131) and (11.8), we get

$$g_{21} = -\frac{h_{21}}{h_{11}h_{22} - h_{12}h_{21}} \qquad ...(11.132)$$

and $$g_{22} = \frac{h_{11}}{h_{11}h_{22} - h_{12}h_{21}} \qquad ...(11.133)$$

11.8.19 g Parameters in Terms of ABCD Parameters

Writing equations (11.9) and (11.10) again,

$$V_1 = AV_2 - BI_2 \qquad ...[\text{Eqn } 11.9]$$

$$I_1 = CV_2 - DI_2 \quad \text{...[Eqn 11.10]}$$

From equation (11.9), we get

$$V_2 = \frac{1}{A}V_1 + \frac{B}{A}I_2 \quad \text{...(11.134)}$$

Comparing equations (11.134) and (11.8), we get

$$g_{21} = \frac{1}{A} \quad \text{...(11.135)}$$

and

$$g_{22} = \frac{B}{A} \quad \text{...(11.136)}$$

Substituting equation (11.134) in equation (11.10) we get

$$I_1 = C\left[\frac{1}{A}V_1 + \frac{B}{A}I_2\right] - DI_2$$

or

$$I_1 = \frac{C}{A}V_1 - \frac{AD-BC}{A}I_2 \quad \text{...(11.137)}$$

Comparing equations (11.137) and (11.7), we get

$$g_{11} = \frac{C}{A} \quad \text{...(11.138)}$$

and

$$g_{12} = -\frac{AD-BC}{A} \quad \text{...(11.139)}$$

11.8.20 g Parameters in Terms of abcd Parameters

Writing equations (11.11) and (11.12) again,

$$V_2 = aV_1 - bI_1 \quad \text{...[Eqn 11.11]}$$

$$I_2 = cV_1 - dI_1 \quad \text{...[Eqn 11.12]}$$

From equation (11.12), we get

$$I_1 = \frac{c}{d}V_1 - \frac{1}{d}I_2 \quad \text{...(11.140)}$$

Comparing equations (11.140) and (11.7), we get

$$g_{11} = \frac{c}{d} \quad \text{...(11.141)}$$

and

$$g_{12} = -\frac{1}{d} \quad \text{...(11.142)}$$

Substituting equation (11.140) in equation (11.11), we get

$$V_2 = aV_1 - b\left[\frac{c}{d}V_1 - \frac{1}{d}I_2\right]$$

or

$$V_2 = \frac{ad-bc}{d}V_1 + \frac{b}{d}I_2 \quad \text{...(11.143)}$$

Comparing equations (11.143) and (11.8), we get

$$g_{21} = \frac{ad-bc}{d} \quad \text{...(11.144)}$$

and
$$g_{22} = \frac{b}{d} \qquad \text{...(11.145)}$$

11.8.21 ABCD Parameters in Terms of z Parameters

Writing equations (11.1) and (11.2) again,

$$V_1 = z_{11}I_1 + z_{12}I_2 \qquad \text{...[Eqn 11.1]}$$
$$V_2 = z_{21}I_1 + z_{22}I_2 \qquad \text{...[Eqn 11.2]}$$

From equation (11.2), we get

$$I_1 = \frac{1}{z_{21}}V_2 - \frac{z_{22}}{z_{21}}I_2 \qquad \text{...(11.146)}$$

Comparing equations (11.146) and (11.10), we get

$$C = \frac{1}{z_{21}} \qquad \text{...(11.147)}$$

and
$$D = \frac{z_{22}}{z_{21}} \qquad \text{...(11.148)}$$

Substituting equation (11.146) in equation (11.1), we get

$$V_1 = z_{11}\left[\frac{1}{z_{21}}V_2 - \frac{z_{22}}{z_{21}}I_2\right] + z_{12}I_2$$

or
$$V_1 = \frac{z_{11}}{z_{21}}V_2 - \frac{z_{11}z_{22} - z_{21}z_{12}}{z_{21}}I_2 \qquad \text{...(11.149)}$$

Comparing equations (11.149) and (11.9), we get

$$A = \frac{z_{11}}{z_{21}} \qquad \text{...(11.150)}$$

and
$$B = \frac{z_{11}z_{22} - z_{21}z_{12}}{z_{21}} \qquad \text{...(11.151)}$$

11.8.22 ABCD Parameters in Terms of y Parameters

Writing equations (11.3) and (11.4) again,

$$I_1 = y_{11}V_1 + y_{12}V_2 \qquad \text{...[Eqn 11.3]}$$
$$I_2 = y_{21}V_1 + y_{22}V_2 \qquad \text{...[Eqn 11.4]}$$

From equation (11.4), we get

$$V_1 = -\frac{y_{22}}{y_{21}}V_2 + \frac{1}{y_{21}}I_2 \qquad \text{...(11.152)}$$

Comparing equations (11.152) and (11.9), we get

$$A = -\frac{y_{22}}{y_{21}} \qquad \text{...(11.153)}$$

and
$$B = -\frac{1}{y_{21}} \qquad \text{...(11.154)}$$

Substituting equation (11.152) in equation (11.3), we get

$$I_1 = y_{11}\left[-\frac{y_{22}}{y_{21}}V_2 + \frac{1}{y_{21}}I_2\right] + y_{12}V_2$$

or

$$I_1 = -\frac{y_{11}y_{22} - y_{12}y_{21}}{y_{21}}V_2 + \frac{y_{11}}{y_{21}}I_2 \qquad \text{...(11.155)}$$

Comparing equations (11.155) and (11.10), we get

$$C = -\frac{y_{11}y_{22} - y_{12}y_{21}}{y_{21}} \qquad \text{...(11.156)}$$

and

$$D = -\frac{y_{11}}{y_{21}} \qquad \text{...(11.157)}$$

11.8.23 ABCD Parameters in Terms of h Parameters

Writing equations (11.5) and (11.6) again,

$$V_1 = y_{11}I_1 + y_{12}V_2 \qquad \text{...[Eqn 11.5]}$$

$$I_2 = y_{21}I_1 + y_{22}V_2 \qquad \text{...[Eqn 11.6]}$$

From equation (11.6), we get

$$I_1 = -\frac{h_{22}}{h_{21}}V_2 + \frac{1}{h_{21}}I_2 \qquad \text{...(11.158)}$$

Comparing equations (11.158) and (11.10), we get

$$C = -\frac{h_{22}}{h_{21}} \qquad \text{...(11.159)}$$

and

$$D = -\frac{1}{h_{21}} \qquad \text{...(11.160)}$$

Substituting equation (11.158) in equation (11.5), we get

$$V_1 = h_{11}\left[-\frac{h_{22}}{h_{21}}V_2 + \frac{1}{h_{21}}I_2\right] + h_{12}V_2$$

or

$$V_1 = -\frac{h_{11}h_{22} - h_{12}h_{21}}{h_{21}}V_2 + \frac{h_{11}}{h_{21}}I_2 \qquad \text{...(11.161)}$$

Comparing equations (11.161) and (11.9), we get

$$A = -\frac{h_{11}h_{22} - h_{12}h_{21}}{h_{21}} \qquad \text{...(11.162)}$$

$$B = -\frac{h_{11}}{h_{21}} \qquad \text{...(11.163)}$$

11.8.24 ABCD Parameters in Terms of g Parameters

Writing equations (11.7) and (11.8) again,

$$I_1 = g_{11}V_1 + g_{12}I_2 \qquad \text{...[Eqn 11.7]}$$

$$V_2 = g_{21}V_1 + g_{22}I_2 \qquad \text{...[Eqn 11.8]}$$

From equation (11.8), we get

$$V_1 = \frac{1}{g_{21}}V_2 - \frac{g_{22}}{g_{21}}I_2 \qquad \text{...(11.164)}$$

Comparing equations (11.164) and (11.9), we get

$$A = \frac{1}{g_{21}} \qquad \text{...(11.165)}$$

and

$$B = \frac{g_{22}}{g_{21}} \qquad \text{...(11.166)}$$

Substituting equation (11.166) in equation (11.7), we get

$$I_1 = g_{11}\left[\frac{1}{g_{21}}V_2 - \frac{g_{22}}{g_{21}}I_2\right] + g_{12}I_2$$

or

$$I_1 = \frac{g_{11}}{g_{21}}V_2 - \frac{g_{11}g_{22} + g_{12}g_{21}}{g_{21}}I_2 \qquad \text{...(11.167)}$$

Comparing equations (11.167) and (11.10), we get

$$C = \frac{g_{11}}{g_{21}} \qquad \text{...(11.168)}$$

and

$$D = \frac{g_{11}g_{22} + g_{12}g_{21}}{g_{21}} \qquad \text{...(11.169)}$$

11.8.25 ABCD Parameters in Terms of abcd Parameters

Writing equations (11.11) and (11.12) again,

$$V_2 = aV_1 - bI_1 \qquad \text{...[Eqn 11.11]}$$

$$I_2 = cV_1 - dI_1 \qquad \text{...[Eqn 11.12]}$$

From equation (11.11), we get

$$V_1 = \frac{1}{a}V_2 + \frac{b}{a}I_1 \qquad \text{...(11.170)}$$

From equation (11.12), we get

$$I_1 = \frac{c}{d}V_1 - \frac{1}{d}I_2 \qquad \text{...(11.171)}$$

Substituting equation (11.171) in equation (11.170), we get

$$V_1 = \frac{1}{a}V_2 + \frac{b}{a}\left[\frac{c}{d}V_1 - \frac{1}{d}I_2\right]$$

or

$$V_1 = \frac{d}{ad - bc}V_2 - \frac{b}{ad - bc}I_2 \qquad \text{...(11.172)}$$

Comparing equations (11.172) and (11.9), we get

$$A = \frac{d}{ad - bc} \qquad \text{...(11.173)}$$

and $$B = \frac{b}{ad-bc} \qquad ...(11.174)$$

Substituting equation (11.170) in equation (11.171), we get

$$I_1 = \frac{c}{d}\left[\frac{1}{a}V_2 + \frac{b}{a}I_1\right] - \frac{1}{d}I_2$$

or $$I_1 = \frac{c}{ad-bc}V_2 - \frac{a}{ad-bc}I_2 \qquad ...(11.175)$$

Comparing equations (11.175) and (11.10), we get

$$C = \frac{c}{ad-bc} \qquad ...(11.176)$$

and $$D = \frac{a}{ad-bc} \qquad ...(11.177)$$

11.8.26 abcd Parameters in Terms of z Parameters

Writing equations (11.1) and (11.2) again,

$$V_1 = z_{11}I_1 + z_{12}I_2 \qquad ...[\text{Eqn } 11.1]$$
$$V_2 = z_{21}I_1 + z_{22}I_2 \qquad ...[\text{Eqn } 11.2]$$

From equation (11.1), we get

$$I_2 = \frac{1}{z_{12}}V_1 - \frac{z_{11}}{z_{12}}I_1 \qquad ...(11.178)$$

Comparing equations (11.178) and (11.12), we get

$$c = \frac{1}{z_{12}} \qquad ...(11.179)$$

and $$d = \frac{z_{11}}{z_{12}} \qquad ...(11.180)$$

Substituting equation (11.178) in equation (11.2), we get

$$V_2 = z_{21}I_1 + z_{22}\left[\frac{1}{z_{12}}V_1 - \frac{z_{11}}{z_{12}}I_1\right]$$

or $$V_2 = \frac{z_{22}}{z_{12}}V_1 - \frac{z_{11}z_{22} - z_{12}z_{21}}{z_{12}}I_1 \qquad ...(11.181)$$

Comparing equations (11.181) and (11.11), we get

$$a = \frac{z_{22}}{z_{12}} \qquad ...(11.182)$$

and $$b = \frac{z_{11}z_{22} - z_{12}z_{21}}{z_{12}} \qquad ...(11.183)$$

11.8.27 abcd Parameters in Terms of y Parameters

Writing equations (11.3) and (11.4) again,

$$I_1 = y_{11}V_1 + y_{12}V_2 \qquad ...[\text{Eqn } 11.3]$$

$$I_2 = y_{21}V_1 + y_{22}V_2 \qquad \text{...[Eqn 11.4]}$$

From equation (11.3), we get

$$V_2 = -\frac{y_{11}}{y_{12}}V_1 + \frac{1}{y_{12}}I_1 \qquad \text{...(11.184)}$$

Comparing equations (11.184) and (11.11), we get

$$a = -\frac{y_{11}}{y_{12}} \qquad \text{...(11.185)}$$

and

$$b = -\frac{1}{y_{12}} \qquad \text{...(11.186)}$$

Substituting equation (11.184) in equation (11.4) we get

$$I_2 = y_{21}V_1 + y_{22}\left[-\frac{y_{11}}{y_{12}}V_1 + \frac{1}{y_{12}}I_1\right]$$

or

$$I_2 = -\frac{y_{11}y_{22} - y_{12}y_{21}}{y_{12}}V_1 + \frac{y_{22}}{y_{12}}I_1 \qquad \text{...(11.187)}$$

Comparing equations (11.187) and (11.12), we get

$$c = -\frac{y_{11}y_{22} - y_{12}y_{21}}{y_{12}} \qquad \text{...(11.188)}$$

and

$$d = -\frac{y_{22}}{y_{12}} \qquad \text{...(11.189)}$$

11.8.28 abcd Parameters in Terms of h Parameters

Writing equations (11.5) and (11.6) again,

$$V_1 = h_{11}I_1 + h_{12}V_2 \qquad \text{...[Eqn 11.5]}$$

$$I_2 = h_{21}I_1 + h_{22}V_2 \qquad \text{...[Eqn 11.6]}$$

From equation (11.5), we get

$$V_2 = \frac{1}{h_{12}}V_1 - \frac{h_{11}}{h_{12}}I_1 \qquad \text{...(11.190)}$$

Comparing equations (11.190) and (11.11), we get

$$a = \frac{1}{h_{12}} \qquad \text{...(11.191)}$$

and

$$b = \frac{h_{11}}{h_{12}} \qquad \text{...(11.192)}$$

Substituting equation (11.190) in equation (11.6), we get

$$I_2 = h_{21}I_1 + h_{22}\left[\frac{1}{h_{12}}V_1 - \frac{h_{11}}{h_{12}}I_1\right]$$

or

$$I_2 = \frac{h_{22}}{h_{12}}V_1 - \frac{h_{11}h_{22} - h_{12}h_{21}}{h_{12}}I_1 \qquad \text{...(11.193)}$$

Comparing equations (11.193) and (11.12), we get

$$c = \frac{h_{22}}{h_{12}} \quad ...(11.194)$$

and
$$d = \frac{h_{11}h_{22} - h_{12}h_{21}}{h_{12}} \quad ...(11.195)$$

11.8.29 abcd Parameters in Terms of g Parameters

Writing equations (11.7) and (11.8) again,

$$I_1 = g_{11}V_1 + g_{12}I_2 \quad ...[\text{Eqn } 11.7]$$

$$V_2 = g_{21}V_1 + g_{22}I_2 \quad ...[\text{Eqn } 11.8]$$

From equation (11.7), we get

$$I_2 = -\frac{g_{11}}{g_{12}}V_1 + \frac{1}{g_{12}}I_1 \quad ...(11.196)$$

Comparing equations (11.196) and (11.12), we get

$$c = -\frac{g_{11}}{g_{12}} \quad ...(11.197)$$

and
$$d = -\frac{1}{g_{12}} \quad ...(11.198)$$

Substituting equation (11.196) in equation (11.8) we get

$$V_2 = g_{21}V_1 + g_{22}\left[-\frac{g_{11}}{g_{12}}V_1 + \frac{1}{g_{12}}I_1\right]$$

or
$$V_2 = -\frac{g_{11}g_{22} - g_{12}g_{21}}{g_{12}}V_1 + \frac{g_{22}}{g_{12}}I_1 \quad ...(11.199)$$

Comparing equations (11.199) and (11.11), we get

$$a = -\frac{g_{11}g_{22} - g_{12}g_{21}}{g_{12}} \quad ...(11.200)$$

and
$$b = -\frac{g_{22}}{g_{12}} \quad ...(11.201)$$

11.8.30 abcd Parameters in Terms of abcd Parameters

Writing equations (11.9) and (11.10) again,

$$V_1 = AV_2 - BI_2 \quad ...[\text{Eqn } 11.9]$$

$$I_1 = CV_2 - DI_2 \quad ...[\text{Eqn } 11.10]$$

From equation (11.9), we get

$$V_2 = \frac{1}{A}V_1 + \frac{B}{A}I_2 \quad ...(11.202)$$

From equation (11.10), we get

$$I_2 = -\frac{1}{D}I_1 + \frac{C}{D}V_2 \qquad ...(11.203)$$

Substituting equation (11.203) in equation (11.202), we get

$$V_2 = \frac{1}{A}V_1 + \frac{B}{A}\left[-\frac{1}{D}I_1 + \frac{C}{D}V_2\right]$$

or

$$V_2 = \frac{D}{AD-BC}V_1 - \frac{B}{AD-BC}I_1 \qquad ...(11.204)$$

Comparing equations (11.204) and (11.11), we get

$$a = \frac{D}{AD-BC} \qquad ...(11.205)$$

and

$$b = \frac{B}{AD-BC} \qquad ...(11.206)$$

Substituting equation (11.202) in equation (11.203), we get

$$I_2 = -\frac{1}{D}I_1 + \frac{C}{D}\left[\frac{1}{A}V_1 + \frac{B}{A}I_2\right]$$

or

$$I_2 = \frac{C}{AD-BC}V_1 - \frac{A}{AD-BC}I_2 \qquad ...(11.207)$$

Comparing equations (11.207) and (11.12), we get

$$c = \frac{C}{AD-BC} \qquad ...(11.208)$$

and

$$d = \frac{A}{AD-BC} \qquad ...(11.209)$$

11.9 INTERCONNECTION BETWEEN THE NETWORKS

Two-port network can be connected either in series, parallel or in cascade combinations. This interconnection of networks can be achieved by any of the six parameter sets, but here in this section we will consider only three parameter sets, z parameter for series connection, y parameter for parallel and transmission parameter for cascade connection.

11.9.1 Series Connection

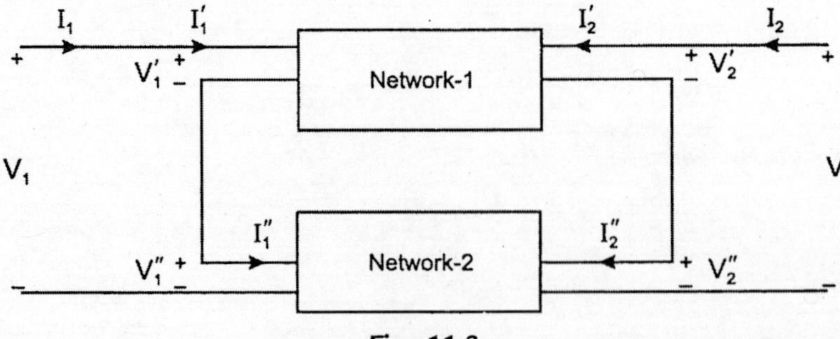

Fig. 11.2

The series connection of two two-port network is shown in Figure 11.2

Then for network-1

$$V_1' = z_{11}'I_1' + z_{12}'I_2' \quad ...(11.210)$$

$$V_2' = z_{21}'I_1' + z_{22}'I_2' \quad ...(11.211)$$

Then for network-2

$$V_1'' = z_{11}''I_1'' + z_{12}''I_2'' \quad ...(11.212)$$

$$V_2'' = z_{21}''I_1'' + z_{22}''I_2'' \quad ...(11.213)$$

For series connection

$$I_1 = I_1' = I_1''$$

and

$$I_2 = I_2' = I_2''$$

Also for series connection

$$V_1 = V_1' + V_1'' \quad ...(11.214)$$

and

$$V_2 = V_2' + V_2'' \quad ...(11.215)$$

Substituting equations (11.210) and (11.212) in equation (11.214) we get

$$V_1 = (z_{11}' + z_{11}'')I_1 + (z_{12}' + z_{12}'')I_2 \quad ...(11.216)$$

and Substituting equations (11.211) and (11.213) in equation (11.215) we get

$$V_2 = (z_{21}' + z_{21}'')I_1 + (z_{22}' + z_{22}'')I_2 \quad ...(11.217)$$

Writing equations (11.216) and (11.217) in matrix form we get

$$\begin{bmatrix} V_1 \\ V_2 \end{bmatrix} = \begin{bmatrix} z_{11} & z_{12} \\ z_{21} & z_{22} \end{bmatrix} \begin{bmatrix} I_1 \\ I_2 \end{bmatrix}$$

where

$$z_{11} = z_{11}' + z_{11}''$$

$$z_{12} = z_{12}' + z_{12}''$$

$$z_{21} = z_{21}' + z_{21}''$$

and

$$z_{22} = z_{22}' + z_{22}''$$

11.9.2 Parallel Connection

The parallel connection of two two-port network is shown in Figure 11.3

Then for network-1

$$I_1' = y_{11}'V_1' + y_{12}'V_2' \quad ...(11.218)$$

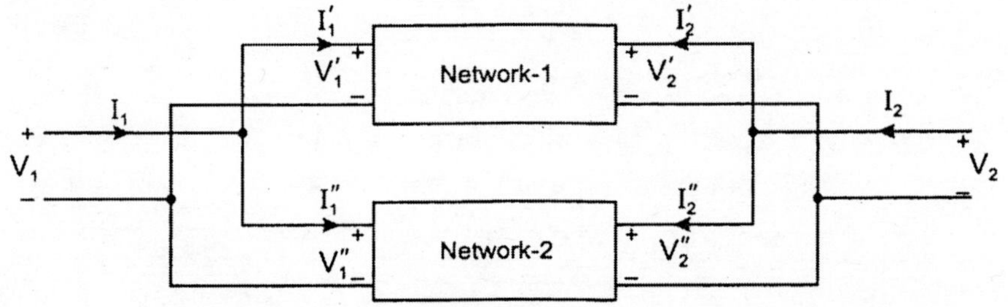

Fig. 11.3

$$I_2' = y_{21}'V_1' + y_{22}'V_2' \quad ...(11.219)$$

Then for network-2

$$I_1'' = y_{11}''V_1'' + y_{12}''V_2'' \quad ...(11.220)$$

$$I_2'' = y_{21}''V_1'' + y_{22}''V_2'' \quad ...(11.221)$$

For series connection

$$V_1 = V_1' = V_1''$$

and

$$V_2 = V_2' = V_2''$$

Also for series connection

$$I_1 = I_1' + I_1'' \quad ...(11.222)$$

and

$$I_2 = I_2' + I_2'' \quad ...(11.223)$$

Substituting equations (11.218) and (11.220) in equation (11.222) we get

$$I_1 = (y_{11}' + y_{11}'')V_1 + (y_{12}' + y_{12}'')V_2 \quad ...(11.224)$$

and Substituting equations (11.219) and (11.221) in equation (11.223) we get

$$I_2 = (y_{21}' + y_{21}'')V_1 + (y_{22}' + y_{22}'')V_2 \quad ...(11.225)$$

Writing equations (11.224) and (11.225) in matrix form we get

$$\begin{bmatrix} I_1 \\ I_2 \end{bmatrix} = \begin{bmatrix} y_{11} & y_{12} \\ y_{21} & y_{22} \end{bmatrix} \begin{bmatrix} V_1 \\ V_2 \end{bmatrix}$$

where,

$$y_{11} = y_{11}' + y_{11}''$$

$$y_{12} = y_{12}' + y_{12}''$$

$$y_{21} = y_{21}' + y_{21}''$$

and

$$y_{22} = y_{22}' + y_{22}''$$

11.9.3 Cascade Connection

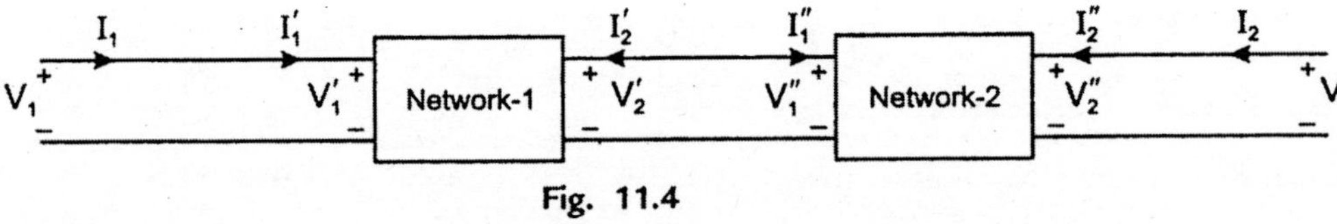

Fig. 11.4

The cascade connection of two two-port network is shown in Figure 11.4

Then for network-1

$$V_1' = A'V_2' - B'I_2' \quad ...(11.226)$$

$$I_1' = C'V_2' - D'I_2' \quad ...(11.227)$$

Writing the above two equation in matrix form we get

$$\begin{bmatrix} V_1' \\ I_1' \end{bmatrix} = \begin{bmatrix} A' & B' \\ C' & D' \end{bmatrix} \begin{bmatrix} V_2' \\ -I_2' \end{bmatrix}$$

Then for network-2

$$V_1'' = A''V_2'' - B''I_2'' \quad ...(11.228)$$

$$I_1'' = C''V_2'' - D''I_2'' \quad ...(11.229)$$

Writing the above two equation in matrix form we get

$$\begin{bmatrix} V_1'' \\ I_1'' \end{bmatrix} = \begin{bmatrix} A'' & B'' \\ C'' & D'' \end{bmatrix} \begin{bmatrix} V_2'' \\ -I_2'' \end{bmatrix}$$

From Figure 11.4 we get

$$V_1 = V_1' \qquad I_1 = I_1'$$
$$V_2' = V_1'' \qquad -I_2' = I_1''$$
$$V_2'' = V_2 \qquad -I_2'' = -I_2$$

Substituting these in equation (11.226) to equation (11.229), we get

$$\begin{bmatrix} V_1 \\ I_1 \end{bmatrix} = \begin{bmatrix} A' & B' \\ C' & D' \end{bmatrix} \begin{bmatrix} V_1'' \\ I_1'' \end{bmatrix} \quad ...(11.230)$$

and

$$\begin{bmatrix} V_1'' \\ I_1'' \end{bmatrix} = \begin{bmatrix} A'' & B'' \\ C'' & D'' \end{bmatrix} \begin{bmatrix} V_2 \\ -I_2 \end{bmatrix} \quad ...(11.231)$$

Again from equations (11.230) and (11.231), we get

$$\begin{bmatrix} V_1 \\ I_1 \end{bmatrix} = \begin{bmatrix} A' & B' \\ C' & D' \end{bmatrix} \times \begin{bmatrix} A'' & B'' \\ C'' & D'' \end{bmatrix} \begin{bmatrix} V_2 \\ -I_2 \end{bmatrix}$$

SOLVED EXAMPLES

Example 11.1. *For the circuit shown in Figure E11.1, find the following parameters*

(i)	*z parameters*	*(iv)*	*g parameters*
(ii)	*y parameters*	*(v)*	*T parameters*
(iii)	*h parameters*	*(vi)*	*t parameters*

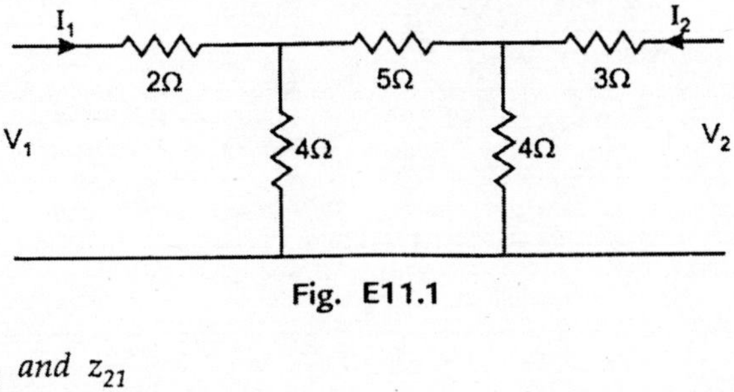

Fig. E11.1

Solution. *To find z_{11} and z_{21}*

To find the above two parameters I_2 should be zero. The loop currents are shown in Figure E11.1(*a*). The loop equations are

Loop 1 : $6I_1 - 4I_3 = V_1$ ...(*i*)

Loop 3 : $13I_3 - 4I_1 = 0$ ...(*ii*)

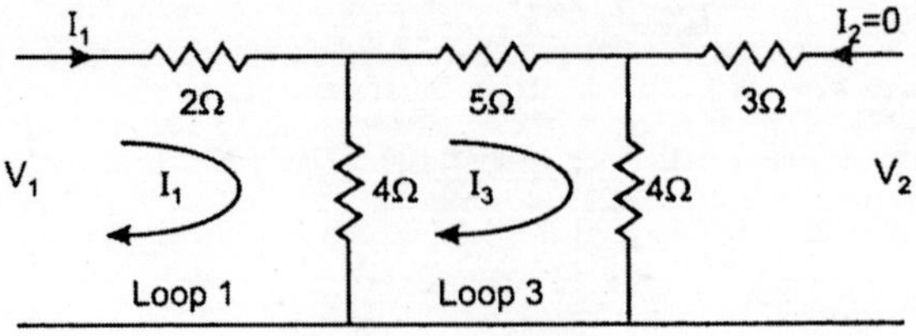

Fig. E11.1(a)

Solving equations (*i*) and (*ii*), we get

$$I_1 = \frac{39V_1}{248} \quad \text{and} \quad I_3 = \frac{2V_1}{31}$$

Then $$V_2 = 4 \times I_3 = \frac{8V_1}{31}$$

Then $$z_{11} = \frac{V_1}{I_1} = \frac{248}{39}\ \Omega \quad \text{and} \quad z_{21} = \frac{V_2}{I_1} = \frac{1984}{1209}\ \Omega$$

To find z_{12} *and* z_{22} :

To find the above two parameters I_1 should be zero.The loop currents are shown in Figure E11.1(*b*). The loop equations are

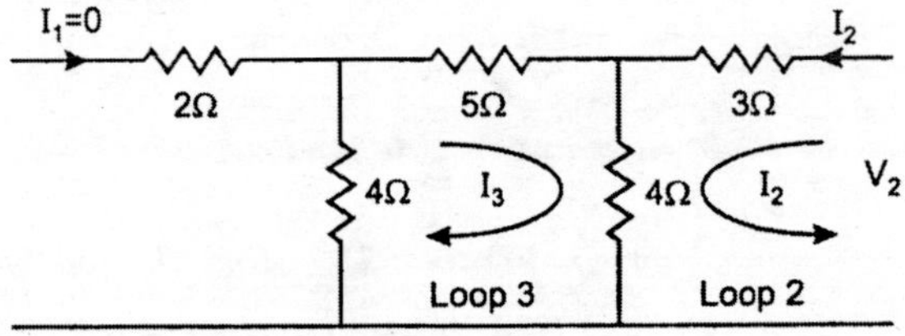

Fig. E11.1(b)

Loop 3 : $13I_3 + 4I_2 = 0$ ...(*iii*)

Loop 2 : $7I_2 + 4I_3 = V_2$ ...(*iv*)

Solving equations (*iii*) and (*iv*), we get

$$I_2 = \frac{52V_2}{300} \quad \text{and} \quad I_3 = -\frac{4V_2}{75}$$

Then $$V_1 = -4 \times I_3 = \frac{16V_2}{75}$$

Then $$z_{12} = \frac{V_1}{I_2} = \frac{4800}{3900}\ \Omega \quad \text{and} \quad z_{22} = \frac{V_2}{I_2} = \frac{300}{52}\ \Omega$$

To find y_{11} *and* y_{21} :

To find the above two parameters V_2 should be zero.The loop currents are shown in Figure E11.1(*c*). The loop equations are

Loop 1 : $6I_1 - 4I_3 = V_1$ ...(*v*)

Loop 3 : $13I_3 - 4I_1 + 4I_2 = 0$ ...(*vi*)

Loop 2 : $7I_2 + 4I_3 = 0$ ...(*vii*)

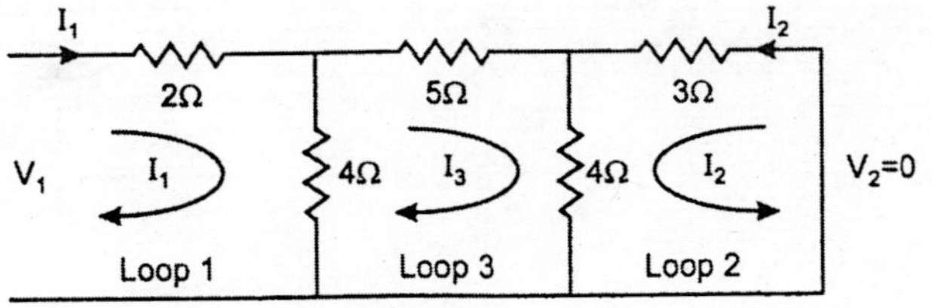

Fig. E11.1(c)

Solving equations (v), (vi) and (vii), we get

$$I_1 = \frac{6975V_1}{31434} \quad I_2 = \frac{8V_1}{169} \quad I_3 = \frac{434V_1}{5239}$$

Then $$y_{11} = \frac{6975}{31434} \text{ and } y_{21} = \frac{8}{169}$$

To find y_{12} and y_{22} :

To find the above two parameters V_1 should be zero.The loop currents are shown in Figure E11.1(d). The loop equations are

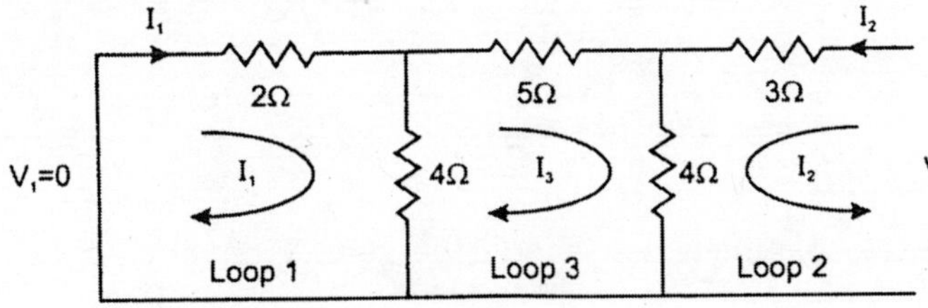

Fig. E11.1(d)

Loop 1 : $6I_1 - 4I_3 = 0$ ...($viii$)

Loop 3 : $13I_3 - 4I_1 + 4I_2 = 0$ ...(ix)

Loop 2 : $7I_2 + 4I_3 = V_2$ ...(x)

Solving equations ($viii$), (ix) and (x), we get

$$I_1 = \frac{48V_2}{1014} \quad I_2 = \frac{31V_2}{169} \quad I_3 = \frac{12V_2}{169}$$

Then $$y_{12} = \frac{58}{1014} \text{ and } y_{22} = \frac{31}{169}$$

To find h_{11} and h_{21} :

To find the above two parameters V_2 should be zero. From the solution of the circuit of Figure E11.1(c), we get

$$I_1 = \frac{6975V_1}{31434} \quad I_2 = \frac{8V_1}{169} \quad I_3 = \frac{434V_1}{5239}$$

Then $$h_{11} = \frac{31434}{6975} \text{ and } y_{21} = \frac{251472}{1178775}$$

To find h_{12} and h_{22} :

To find the above two parameters I_1 should be zero. From the solution of the circuit of Figure E11.1(b), we get

$$I_2 = \frac{52V_2}{300} \qquad I_3 = -\frac{4V_2}{75} \quad \text{and} \quad V_1 = -4 \times I_3 = \frac{16V_2}{75}$$

Then
$$h_{12} = \frac{16}{75} \quad \text{and} \quad y_{22} = \frac{52}{300}$$

To find g_{11} and g_{21} :

To find the above two parameters I_2 should be zero. From the solution of the circuit of Figure E11.1(*a*), we get

$$I_1 = \frac{39V_1}{248} \qquad I_3 = \frac{2V_1}{31} \qquad V_2 = 4 \times I_3 = \frac{8V_1}{31}$$

Then
$$g_{11} = \frac{39}{248} \quad \text{and} \quad g_{21} = \frac{8}{31}$$

To find g_{12} and g_{22} :

To find the above two parameters V_1 should be zero. From the solution of the circuit of Figure E11.1(*d*), we get

$$I_1 = \frac{48V_2}{1014} \qquad I_2 = \frac{31V_2}{169} \qquad I_3 = \frac{12V_2}{169}$$

$$g_{12} = \frac{8112}{31434} \quad \text{and} \quad g_{22} = \frac{169}{31}$$

To find A and C :

To find the above two parameters I_2 should be zero. From the solution of the circuit of Figure E11.1(*a*), we get

$$I_1 = \frac{39V_1}{248} \qquad I_3 = \frac{2V_1}{31} \quad \text{and } V_2 = 4 \times I_3 = \frac{8V_1}{31}$$

$$A = \frac{31}{8} \quad \text{and} \quad C = \frac{1209}{1984}$$

To find B and D :

To find the above two parameters V_2 should be zero. From the solution of the circuit of Figure E11.1(*c*), we get

$$I_1 = \frac{6975V_1}{31434} \qquad I_2 = -\frac{8V_1}{169} \quad \text{and} \quad I_3 = \frac{434V_1}{5239}$$

$$B = \frac{169}{8} \quad \text{and} \quad D = \frac{1178775}{251472}$$

To find a and c :

To find the above two parameters I_1 should be zero. From the solution of the circuit of Figure E11.1(*b*), we get

$$I_2 = \frac{52V_2}{300} \qquad I_3 = -\frac{4V_2}{75} \quad \text{and} \quad V_1 = -4 \times I_3 = \frac{16V_2}{75}$$

$$a = \frac{75}{16} \quad \text{and} \quad c = \frac{52}{300}$$

To find b and d :

To find the above two parameters V_1 should be zero. From the solution of the circuit of Figure E11.1(*d*), we get

$$I_1 = \frac{48V_2}{1014} \qquad I_2 = \frac{31V_2}{169} \qquad I_3 = \frac{12V_2}{169}$$

$$b = \frac{1014}{48} \quad \text{and} \quad d = \frac{8112}{31434}$$

EXAMPLE 11.2. *For the circuit shown in Figure E11.2, find the following parameters*

(*i*) *z parameters* (*ii*) *y parameters*

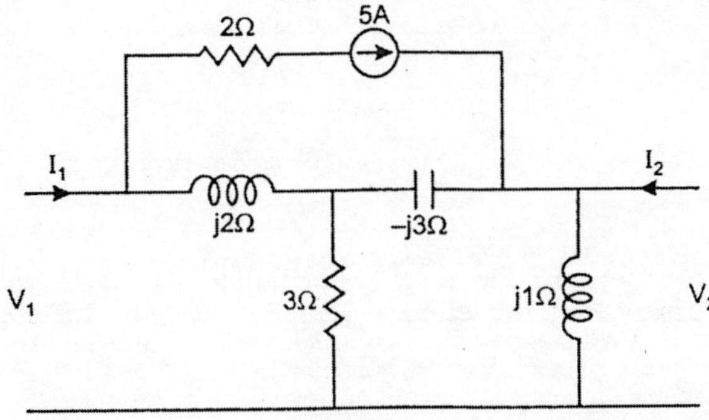

Fig. E11.2

SOLUTION. *To find z_{11} and z_{21}*

To find z_{11} and z_{21}, I_2 should be zero. The circuit with $I_2 = 0$ is shown in Figure E11.2(*a*). The loop currents are indicated in the circuit. The loop equations are.

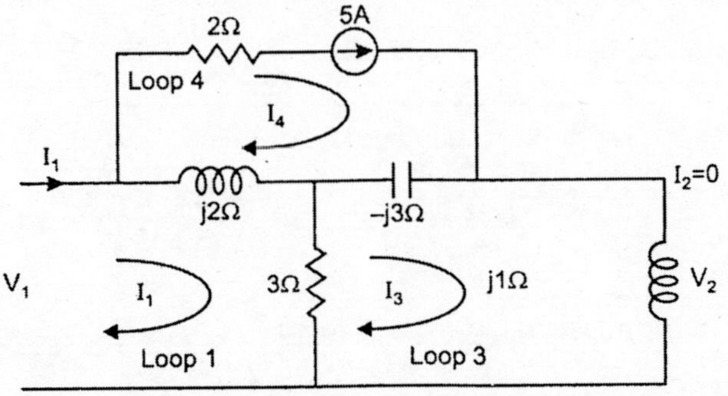

Fig. E11.2(a)

Loop 4 : $I_4 = 5\text{A}$ ...(*i*)

Loop 1 : $(3 + j2)I_1 - 3I_3 - j2I_4 = V_1$ ...(*ii*)

Substituting equation (*i*) in equation (*ii*), we get

$$(3 + j2)I_1 - 3I_3 = V_1 + j10 \qquad ...(iii)$$

Loop 3 : $(3 - j2)I_3 - 3I_1 + j3I_4 = 0$ ...(*iv*)

Substituting equation (*i*) in equation (*iv*), we get

$$(3 - j2)I_3 - 3I_1 = -j15 \qquad \text{...}(v)$$

Solving equations (*iii*) and (*v*) we get

$$I_1 = \frac{(3V_1 + 20) - j(15 + 2V_1)}{4} A$$

and
$$I_3 = \frac{(3V_1 + 30 - j15)}{4} A$$

Thus
$$V_2 = \frac{(j3V_1 + j30 + 15)}{4} V$$

Then
$$z_{11} = \frac{4V_1}{(3V_1 + 20) - j(15 + 2V_1)}$$

and
$$z_{21} = \frac{(j3V_1 + j30 + 15)}{(3V_1 + 20) - j(15 + 2V_1)}$$

To find z_{12} and z_{22} :

To find z_{12} and z_{22}, I_1 should be zero. The circuit with $I_1 = 0$ is shown in Figure E11.2(*b*). The loop currents are indicated in the circuit. The loop equations are

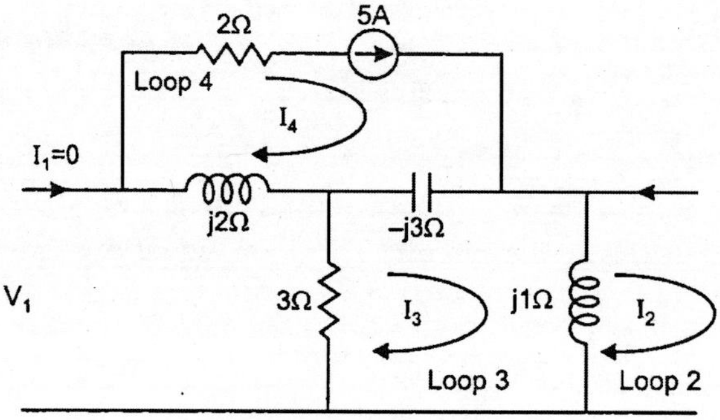

Fig. E11.2(b)

Loop 4 : $I_4 = 5A$ ...(*vi*)
Loop 3 : $(3 - j2)I_3 - j1I_2 = j15$ ...(*vii*)
Loop 2 : $j1I_2 - j1I_3 = -V_2$ ...(*viii*)

Solving equations (*vii*) and (*viii*), we get

$$I_2 = -\frac{15 + 3V_2 - j2V_2}{(3 + j3)} A$$

and
$$I_3 = -\frac{15 - jV_2}{(3 + j3)} A$$

Then from Figure E11.2(*b*), we get

$$V_1 = \frac{25 - j(V_2 + 10)}{1 + j} V$$

Then
$$z_{12} = \frac{25 - j(V_2 + 10)}{1 + j} \times \frac{3 + j3}{15 + 3V_2 - j2V_2}$$

and
$$z_{22} = \frac{V_2(3 + j3)}{15 + 3V_2 - j2V_2}$$

To find y_{11} and y_{21} :

To find y_{11} and y_{21}, V_2 should be zero. The circuit with $V_2 = 0$ is shown in Figure E11.2(*c*). The loop currents are indicated in the circuit. The loop equations are

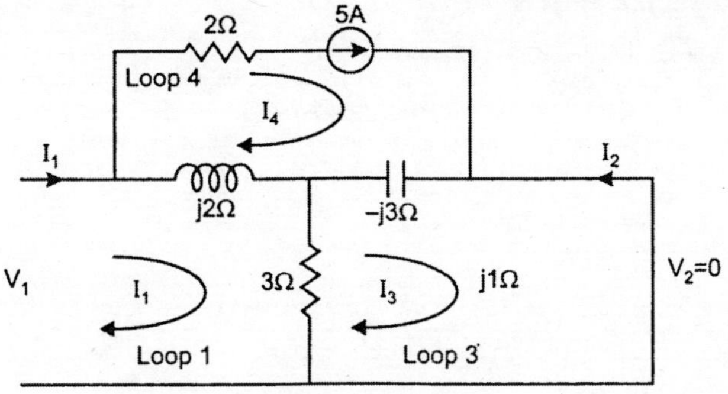

Fig. E11.2(c)

Loop 4 : $I_4 = 5A$ ...(*ix*)

Loop 1 : $(3 + j2)I_1 - 3I_3 = V_1 + j10$ ...(*x*)

Loop 3 : $(3 - j3)I_3 - 3I_1 = -j15$ ...(*xi*)

Solving equations (*x*) and (*xi*) we get

$$I_1 = \frac{(3V_1 + 30) - j(3V_1 + 15)}{6 - j3} A$$

and

$$I_3 = I_2 = \frac{30 + 3V_1 - j5}{6 - j3} A$$

Then

$$y_{11} = \frac{(3V_1 + 30) - j(3V_1 + 15)}{V_1(6 - j3)}$$

and

$$y_{21} = \frac{30 + 3V_1 - j5}{V_1(6 - j3)}$$

To find y_{12} and y_{22} :

To find y_{12} and y_{22}, V_1 should be zero. The circuit with $V_1 = 0$ is shown in Figure E11.2(*d*). The loop currents are indicated in the circuit. The loop equations are

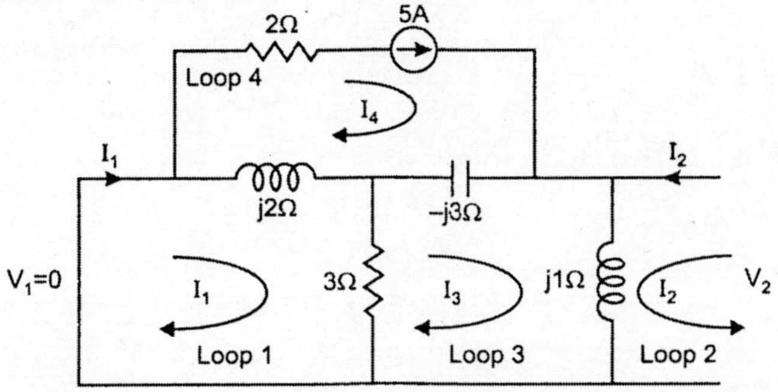

Fig. E11.2(d)

Loop 4 : $I_4 = 5A$ ...(*xii*)

Loop 1 : $(3 + j2)\, I_1 - 3I_3 = j10$ ...(*xiii*)

Loop 3 : $(3 - j2)I_3 - 3I_1 + j1I_2 = -j15$ ...(*xiv*)

Loop 2 : $j1I_2 + j1I_3 = V_2$ ...(*xv*)

Solving equations (*xiii*), (*xiv*) and (*xv*), we get

$$I_1 = \frac{-15 - j30 + j3V_2}{(-3 + j6)} A$$

and $$I_2 = \frac{75 + V_2 + j30 - j12V_2}{(-3 + j6)} A$$

Then $$y_{12} = \frac{-15 - j30 + j3V_2}{V_2(-3 + j6)}$$

and $$y_{22} = \frac{75 + V_2 + j30 - j12V_2}{V_2(-3 + j6)}$$

EXAMPLE 11.3. *Find the T-parameter for the network as shown in Figure E11.3.*

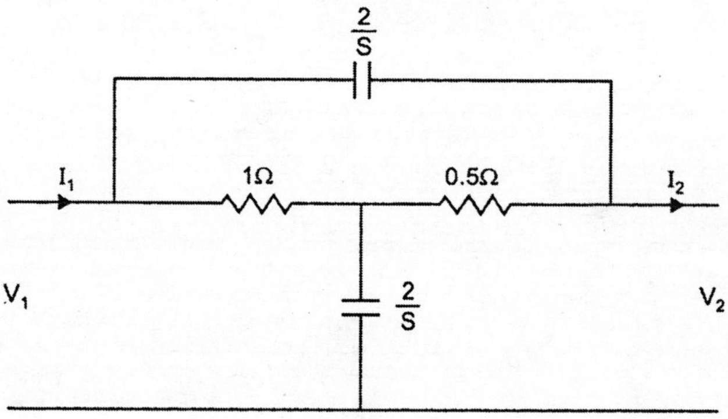

Fig. E11.3

SOLUTION. *When $I_2 = 0$ A*

The circuit is shown in Figure E11.3(*a*), with loop currents. The loop equations are

Loop 1 : $\left(1 + \frac{2}{s}\right) I_1 - I = V_1$ ...(*i*)

Loop 3 : $\left(1.5 + \frac{2}{s}\right) I - I_1 = 0$ ...(*ii*)

Solving equations (*i*) and (*ii*), we get

$$I_1 = \frac{V_1 s(1.5s + 2)}{0.5s^2 + 5s + 4} \quad \text{...(iii)}$$

Then $$V_2 = \frac{V_1(0.5s^2 + 3s + 4)}{0.5s^2 + 5s + 4} \quad \text{...(iv)}$$

From equation (*iv*) we get

$$A = \frac{V_1}{V_2} = \frac{0.5s^2 + 5s + 4}{0.5s^2 + 3s + 4} \quad \text{...(v)}$$

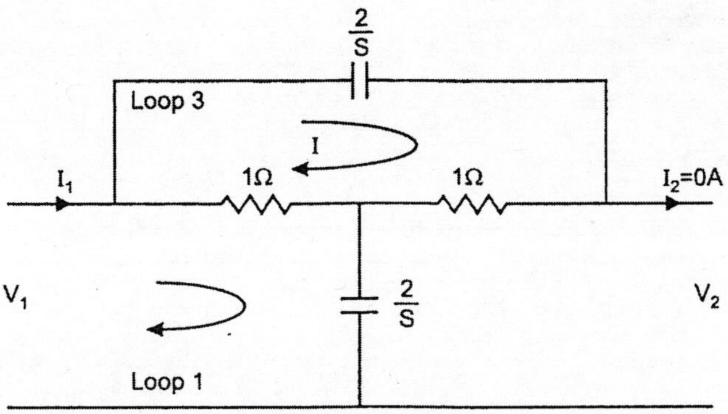

Fig. E11.3(a)

From equation (v), we get

$$V_1 = \frac{0.5s^2+5s+4}{0.5s^2+3s+4} \times V_2 \qquad ...(vi)$$

Substituting equation (vi) in equation (iii) we get

$$I_1 = \frac{s(1.5s+2)}{0.5s^2+5s+4} \times \frac{0.5s^2+5s+4}{0.5s^2+3s+4} \times V_2$$

or

$$C = \frac{I_1}{V_2} = \frac{s(1.5s+2)}{0.5s^2+3s+4}$$

When $V_2 = 0\ V$:

The circuit is shown in Figure E11.3(b), with loop currents. The loop equations are

Loop 1 : $$\left(1+\frac{2}{s}\right)I_1 - \frac{2}{s}I_2 - I = V_1 \qquad ...(vii)$$

Loop 2 : $$\left(0.5+\frac{2}{s}\right)I_2 - \frac{2}{s}I_1 - 0.5I = 0 \qquad ...(viii)$$

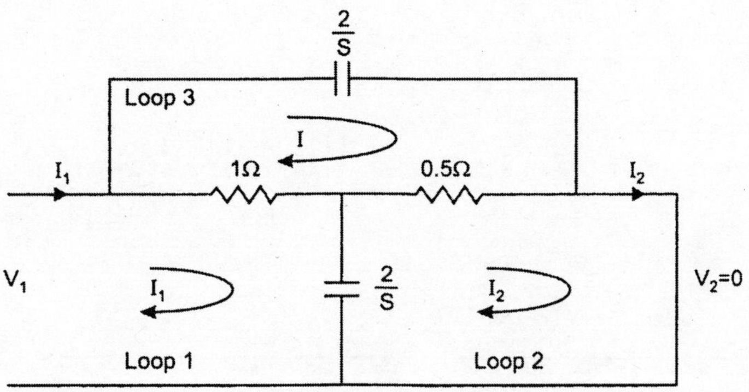

Fig. E11.3(b)

Loop 3 : $$\left(1.5+\frac{2}{s}\right)I - I_1 - 0.5I_2 = 0 \qquad ...(ix)$$

From equation (*ix*) we get

$$I_1 = \left(1.5+\frac{2}{s}\right)I-0.5I_2 \quad ...(x)$$

From equation (*viii*) we get

$$I = \left(1+\frac{4}{s}\right)I_2-\frac{4}{s}I_1 \quad ...(xi)$$

Substituting equation (*xi*) in equation (*x*) we get

$$I_1 = \left(1.5+\frac{2}{s}\right)\times\left(\left(1+\frac{4}{s}\right)I_2-\frac{4}{s}I_1\right)-0.5I_2$$

or

$$\left[\left(1.5+\frac{2}{s}\right)\frac{4}{s}+1\right]I_1 = \left[\left(1.5+\frac{2}{s}\right)\times\left(1+\frac{4}{s}\right)-0.5\right]I_2 \quad ...(xii)$$

Thus

$$D = \frac{I_1}{I_2} = \frac{\left[\left(1.5+\frac{2}{s}\right)\times\left(1+\frac{4}{s}\right)-0.5\right]}{\left[\left(1.5+\frac{2}{s}\right)\frac{4}{s}+1\right]}$$

Substituting equation (*xi*) in equation (*vii*) we get

$$\left(1+\frac{6}{s}\right)I_1-\left(1+\frac{6}{s}\right)I_2 = V_1 \quad ...(xiii)$$

From equation (*xii*) we get

$$I_1 = \frac{\left[\left(1.5+\frac{2}{s}\right)\times\left(1+\frac{4}{s}\right)-0.5\right]}{\left[\left(1.5+\frac{2}{s}\right)\frac{4}{s}+1\right]}I_2 \quad ...(xiv)$$

Substituting equation (*xiv*) in equation (*xiii*) we get

$$\left(1+\frac{6}{s}\right)\times\left[\frac{\left[\left(1.5+\frac{2}{s}\right)\times\left(1+\frac{4}{s}\right)-0.5\right]}{\left[\left(1.5+\frac{2}{s}\right)\frac{4}{s}+1\right]}I_2\right]-\left(1+\frac{6}{s}\right)I_2 = V_1$$

or

$$B = \frac{V_1}{I_2} = \left[\frac{\left[\left(1.5+\frac{2}{s}\right)\times\left(1+\frac{4}{s}\right)-0.5\right]}{\left[\left(1.5+\frac{2}{s}\right)\frac{4}{s}+1\right]}-1\right]$$

EXERCISES

1. Obtain the condition for reciprocity and symmetry of all the six two-port parameters.
2. Obtain the input and output impedance of two-port network in terns of all the six two-port parameters. (Input impedance $= \frac{V_1}{I_1}$ and output impedance $= \frac{V_2}{I_2}$

APPENDIX

MATRIX ALGEBRA

In electrical circuits matrix algebra plays a most important role. The simultaneous equations can easily be solved using matrices, by applying Cramer's Rule. Here in this section some basics of matrix algebra has been discussed.

MATRIX

In a matrix, elements are arranged in the fashion of a rectangular array, as shown in Fig. A1.

$$\begin{bmatrix} a_{11} & a_{12} & \cdots & a_{1n} \\ a_{21} & a_{22} & \cdots & a_{2n} \\ \vdots & \vdots & & \vdots \\ a_{m1} & a_{m2} & \cdots & a_{mn} \end{bmatrix}$$

Fig. A1

Where a_{ij} denotes the elements of the matrix. And the matrix is of order $m \times n$. Where m denotes the number of rows and n denotes the number of columns.

SQUARE MATRIX

If $m = n$ then the matrix is called a square matrix.

SYMMETRIC MATRIX

If the elements of a matrix satisfies the following equality

$$a_{ij} = a_{ji}$$

then the matrix is called a symmetric matrix.

DIAGONAL MATRIX

If for any matrix, except the diagonal elements rest all the elements are zero, then such matrix is called a diagonal matrix. The diagonal elements are $a_{11}, a_{22}, \ldots a_{mn}$.

COLUMN MATRIX

A matrix having only one column is called a column matrix or vector. (Fig. A2)

$$\begin{bmatrix} a_{11} \\ a_{21} \\ \vdots \\ a_{m1} \end{bmatrix}$$

Fig. A2

The matrix shown in Fig. A2, is of order $m \times 1$.

ROW MATRIX

A matrix having only one row is called a row matrix. (Fig. A3)

$$\begin{bmatrix} a_{11} & a_{12} & \cdots & a_{1n} \end{bmatrix}$$

Fig. A3

The matrix shown in Fig. A3, is of order $1 \times n$.

SCALAR

A matrix of order 1×1 is called a scalar.

SCALAR MULTIPLICATION OF A MATRIX

If $A = \begin{bmatrix} a_{11} & a_{12} & \cdots & a_{1n} \\ a_{21} & a_{22} & \cdots & a_{2n} \\ \vdots & \vdots & & \vdots \\ a_{m1} & a_{m2} & \cdots & a_{mn} \end{bmatrix}$, then $k.A = \begin{bmatrix} k.a_{11} & k.a_{12} & \cdots & k.a_{1n} \\ k.a_{21} & k.a_{22} & \cdots & k.a_{2n} \\ \vdots & \vdots & & \vdots \\ k.a_{m1} & k.a_{m2} & \cdots & k.a_{mn} \end{bmatrix}$, where k is a scalar.

PRODUCT OF A MATRIX

If $A = \begin{bmatrix} a_{11} & a_{12} & & a_{1n} \\ a_{21} & a_{22} & & a_{2n} \\ \vdots & \vdots & & \vdots \\ a_{m1} & a_{m2} & \cdots & a_{mn} \end{bmatrix}$ and $B = \begin{bmatrix} b_{11} & b_{12} & & b_{1n} \\ b_{21} & b_{22} & \cdots & b_{2n} \\ \vdots & \vdots & & \vdots \\ b_{m1} & b_{m2} & \cdots & b_{mn} \end{bmatrix}$ then,

$$A \times B = \begin{bmatrix} (a_{11}.b_{11} + a_{12}.b_{21} + \ldots + a_{1n}.b_{m1}) & (a_{11}.b_{12} + a_{12}.b_{22} + \ldots + a_{1n}.b_{m2}) & & (a_{11}.b_{1n} + a_{12}.b_{2n} + \ldots + a_{1n}.b_{mn}) \\ (a_{21}.b_{11} + a_{22}.b_{21} + \ldots + a_{2n}.b_{m1}) & (a_{21}.b_{12} + a_{22}.b_{22} + \ldots + a_{2n}.b_{m2}) & & (a_{21}.b_{1n} + a_{22}.b_{2n} + \ldots + a_{2n}.b_{mn}) \\ \vdots & \vdots & & \vdots \\ (a_{m1}.b_{11} + a_{m2}.b_{21} + \ldots + a_{mn}.b_{m1}) & (a_{m1}.b_{12} + a_{m2}.b_{22} + \ldots + a_{mn}.b_{m2}) & \ldots & (a_{m1}.b_{1n} + a_{m2}.b_{2n} + \ldots + a_{mn}.b_{mn}) \end{bmatrix}$$

TRANSPOSE OF A MATRIX

Transpose of a matrix A can be obtained by interchanging the roes and columns of a given matrix A.

That is if
$$A = \begin{bmatrix} a_{11} & a_{12} & & a_{1n} \\ a_{21} & a_{22} & \cdots & a_{2n} \\ \vdots & \vdots & & \vdots \\ a_{m1} & a_{m2} & \cdots & a_{mn} \end{bmatrix}$$

Then the transpose of A is
$$A' = \begin{bmatrix} a_{11} & a_{21} & \cdots & a_{m1} \\ a_{12} & a_{22} & \cdots & a_{m2} \\ \vdots & \vdots & & \vdots \\ a_{1n} & a_{2n} & \cdots & a_{mn} \end{bmatrix}$$

DETERMINANT OF A MATRIX

Given a matrix
$$A = \begin{bmatrix} a_{11} & a_{12} & a_{13} \\ a_{21} & a_{22} & a_{23} \\ a_{31} & a_{32} & a_{33} \end{bmatrix}$$

Then the determinant of A is given by

$$\det A = a_{11}.(a_{22}.a_{33} - a_{23}.a_{32}) - a_{12}.(a_{21}.a_{33} - a_{23}.a_{31}) + a_{13}.(a_{21}.a_{32} - a_{22}.a_{31})$$

SOLUTION OF A SIMULTANEOUS EQUATION USING CRAMER'S RULE

Let us consider the three simultaneous equations A1, A2 and A3

$$a_1x + b_1y + c_1z = k_1 \quad \ldots(A1)$$
$$a_2x + b_2y + c_2z = k_2 \quad \ldots(A2)$$
$$a_3x + b_3y + c_3z = k_3 \quad \ldots(A3)$$

The three equations can be written in matrix form as shown in A4

$$\begin{bmatrix} a_1 & b_1 & c_1 \\ a_2 & b_2 & c_2 \\ a_3 & b_3 & c_3 \end{bmatrix} \cdot \begin{bmatrix} x \\ y \\ z \end{bmatrix} = \begin{bmatrix} k_1 \\ k_2 \\ k_3 \end{bmatrix} \quad \ldots(A4)$$

Let
$$A = \begin{bmatrix} a_1 & b_1 & c_1 \\ a_2 & b_2 & c_2 \\ a_3 & b_3 & c_3 \end{bmatrix}$$

To find the value of x

Let $$A_x = \begin{bmatrix} k_1 & b_1 & c_1 \\ k_2 & b_2 & c_2 \\ k_3 & b_3 & c_3 \end{bmatrix}$$

Then $$x = \frac{\det A_x}{\det A}$$

To find the value of y

Let $$A_y = \begin{bmatrix} a_1 & k_1 & c_1 \\ a_2 & k_2 & c_2 \\ a_3 & k_3 & c_3 \end{bmatrix}$$

Then $$y = \frac{\det A_y}{\det A}$$

To find the value of z

Let $$A_z = \begin{bmatrix} a_1 & b_1 & k_1 \\ a_2 & b_2 & k_2 \\ a_3 & b_3 & k_3 \end{bmatrix}$$

Then $$z = \frac{\det A_z}{\det A}$$

INDEX

S

T

U

V

Y

Z